Biofuels and Biorefineries

Volume 13

Series Editor

Zhen Fang, Biomass Group, Nanjing Agricultural University, Nanjing, Jiangsu, China

Annual global biomass production is about 220 billion dry tons or 4,500 EJ, equivalent to 8.1 times the world's energy consumption in 2020 (556 EJ). On the other hand, world-proven oil reserves at the end of 2020 reached 1732.4 billion barrels, which can only meet just over 51.5 years of global production. Therefore, alternative resources are needed to both supplement and replace fossil oils as the raw material for transportation fuels, chemicals and materials in petroleum-based industries. Renewable biomass is a likely candidate, because it is prevalent over the Earth and is readily converted to other products. Compared with coal, some of the advantages of biomass are: (i) its carbon-neutral and sustainable nature when properly managed; (ii) its reactivity in biological conversion processes; (iii) its potential to produce bio-oil (ca. yields of 75%) by fast pyrolysis because of its high oxygen content; (iv) its low sulfur and lack of undesirable contaminants (e.g. metals, nitrogen content) (v) its wide geographical distribution and (vi) its potential for creating jobs and industries in energy crop productions and conversion plants. Many researchers, governments, research institutions and industries are developing projects for converting biomass including forest woody and herbaceous biomass into chemicals, biofuels and materials and the race is on for creating new "biorefinery" processes needed for future economies. The development of biorefineries will create remarkable opportunities for the forestry sector, biotechnology, materials, chemical processing industry, and stimulate advances in agriculture. The Biofuels and Biorefineries Series strives to highlight methods and technologies that will help industries to shift to renewable and carbon-neutral resources which will lead to the creation of sustainable society.

Zhen Fang • Richard Lee Smith Jr. • Haixin Guo
Editors

Production of Organic Acids and Alcohols from Agricultural Residues and Food Wastes

Springer

Editors
Zhen Fang
College of Engineering
Nanjing Agricultural University
Nanjing, Jiangsu, China

Richard Lee Smith Jr.
Graduate School of Environmental Studies
Tohoku University
Sendai, Miyagi, Japan

Haixin Guo
Waste management
Agro-Environmental Protection Institute
Ministry of Agriculture and Rural Affairs
Tianjin, Tianjin, China

ISSN 2214-1537 ISSN 2214-1545 (electronic)
Biofuels and Biorefineries
ISBN 978-981-96-1221-5 ISBN 978-981-96-1222-2 (eBook)
https://doi.org/10.1007/978-981-96-1222-2

This Springer imprint is published by the registered company Springer Nature Singapore Pte Ltd.
The registered company address is: 152 Beach Road, #21-01/04 Gateway East, Singapore 189721, Singapore

Editorial Board Members

Aims and Scope of the Series

The Biofuels and Biorefineries Series aims at being a comprehensive and integrated reference for biomass, bioenergy, biofuels, and bioproducts. The Series provides leading global research advances and critical evaluations of methods for converting biomass into biofuels and chemicals. Scientific and engineering challenges in biomass production and conversion are covered that show technological advances and approaches for creating new bio-economies in a format that is suitable for both industrialists and environmental policy decision-makers.

The Biofuels and Biorefineries Series provides readers with clear and concisely written chapters that are peer-reviewed on significant topics in biomass production, biofuels, bioproducts, chemicals, catalysts, energy policy, economics, thermochemical and processing technologies. The text covers major fields of plant science, green chemistry, economics and economy, biotechnology, microbiology, chemical engineering, mechanical engineering and energy.

Preface

Biomass-derived organic acids and alcohols focus on recent technological developments and related challenges surrounding process for industrial applications. This book covers the recent development processes (e.g. catalytic conversion, hydrothermal conversion, fermentation and anaerobic digestion) available for agricultural residues or food wastes conversion into organic acid and alcohols. This book covers the recent development processes (e.g. catalytic conversion, hydrothermal conversion, fermentation and anaerobic digestion) available for agricultural residues or food wastes conversion into organic acid and alcohols. Highlighting the advantages and limitations of different technologies, presenting the challenges, innovative state-of-the-art technologies and assessing future perspectives on the process of biomass-derived organic acid or alcohol production, this book also provides diverse examples of application of biomass-derived organic acid and alcohol, such as production of polymers and amino acids. Overall, the objective of this book is to provide the reader to understanding of the importance of biomass-derived organic acid and alcohols. The scope of text encompasses lignocellulosic biomass, agricultural residues, food wastes, biomass-derived compounds and fundamental carbohydrates with the aim to develop overall production strategies for organic acids and alcohols.

This book is the 13th book of the series entitled, *Biofuels and Biorefineries*, and contains 12 chapters contributed by leading experts in the field. The text is arranged into two key parts:

Part I. Production of Organic Acid from Agricultural Residues and Food Wastes (Chaps. 1–9)

Chapter 1 provides a fundamental understanding of key catalytic synthesis technologies for the production of typical organic acids from agricultural residues and food wastes. Chapter 2 summarizes the catalytic reduction of oxygen-containing functional groups in biomass-derived carboxylic acids with hydrogen gas. This chapter focuses on different types of reaction substrates such as ketocarboxylic acids, hydroxycarboxylic acids, furancarboxylic acids and polyhydroxycarboxylic acids. Chapter 3 focuses on recent advances in the production of organic acids from biomass and its derivatives over heterogeneous metallic catalysts. The content is catalogued by the carbon numbers of organic acids, especially C1–C6 organic acids. Organic acids with more than six carbons constructed by multiple small biomass molecules are outside the scope. Chapter 4 introduces the influence of compositions in feedstocks and operational conditions on organic acids and antimicrobial compounds in aqueous phase. Furthermore, the chapter addressed the potential application of hydrothermal aqueous product in agriculture for phytopathogen control and soil fertilizer. Chapter 5 focuses on bio-oils from different biomass and from different reaction conditions using the same biomass. The acetic acid present in the bio-oil may be separated and utilized and an environmentally friendly scheme for this separation is reported. Chapter 6 lists different catalytic systems for 5-hydroxymethylfurfural (HMF) synthesis, including homogeneous catalytic systems, biphasic catalytic systems, ionic liquids and heterogeneous catalytic systems. This chapter provides a compositional analysis of different agricultural and food residues and compares the recent developments in reaction systems used for the conversion of agricultural and food residues into HMF. A detailed analysis of different catalytic systems involving monophasic, biphasic, ionic liquid (IL) and heterogeneous systems for HMF synthesis from biomass has been provided. Chapter 7 outlines the catalytic conversion process of carbohydrates from biomass into 2,5-furandicarboxylic acid (FDCA) using hydrothermal processing and catalytic oxidation. Chapter 8 explores the versatile potential of FDCA, considering its importance and the challenges associated with its production. It provides a comprehensive overview of the different methods of producing FDCA and concludes with a discussion on the significance of FDCA in sustainable polymer synthesis, emphasizing its role in the production of PEF and other biopolymers. Chapter 9 discusses the current state and prospects of amino acid production from agricultural sidestreams and food wastes. This chapter summarizes hydrolysis options for agricultural sidestreams and food wastes, utilization of monomeric and polymeric compounds, as well as amino acid production using the complex mixtures of substrates present in hydrolysates.

Part II. Production of Biofuels from Agricultural Residues and Food Wastes (Chaps. 10–12)

Chapter 10 introduces the existing research strategies on the catalytic conversion of cellulose to produce aliphatic alcohol, sugar alcohol, ethanol and other organic alcohol value-added chemicals; the catalytic conversion of hemicellulose to produce furfuryl alcohol, sugar alcohol and other organic alcohols; the catalytic conversion of lignin to produce cyclohexanol and other organic alcohols; and the catalytic conversion of glycerol to produce propanol and propylene glycol and other organic alcohol chemicals. Chapter 11 reviews and evaluates a variety of lignocellulose pretreatment methods developed in recent years, including their advantages and drawbacks, pretreatment mechanisms, and offers a summary of the current status of cellulosic ethanol technologies. This chapter also discusses processes of enzymatic hydrolysis and fermentation for fuel ethanol production, as well as a detailed analysis of many factors affecting the process performance. Chapter 12 provides a comprehensive overview of methane production from biowastes via solid-state anaerobic digestion (SS-AD), focusing on its technological principles, influencing factors, optimization strategies, applications and prospects.

The text should be of interest to professionals in academia and industry who are working in the fields of natural renewable materials, platform chemicals, polymer and materials. It can also be used as comprehensive references for university students with backgrounds in chemical engineering, material science, microbial chemistry and environmental engineering.

Nanjing, Jiangsu, China — Zhen Fang
Sendai, Miyagi, Japan — Richard L. Smith Jr.
Tianjin, Tianjin, China — Haixin Guo

Acknowledgements

First and foremost, we would like to cordially thank all the contributing authors for their great efforts in writing and revising the chapters and insuring the reliability of the information given in their chapters. Their contributions have really made this project realizable.

Apart from the efforts of authors, we would also like to acknowledge the individuals listed below for carefully reading the book chapters and giving many constructive comments that significantly improved the quality of the chapters:

Dr. Fernando Pérez García, Norwegian Univ. of Science and Technology, Norway
Dr. Cheng Chen, Wuhan Univ. of Technology, China
Prof. Guanyi Chen, Tianjin Univ., China
Dr. Xi Chen, Shanghai Jiao Tong Univ., China
Dr. Leilei Dai, Univ. of Minnesota, USA
Dr. Jin Deng, Univ. of Science and Technology of China, China
Dr. Jianbin Guo, China Agricultural Univ., China
Prof. Yucai He, Changzhou Univ., China
Dr. Qidong Hou, Nankai Univ., China
Dr. Abid Hussain, Carleton Univ., Canada
Prof. Mingjie Jin, Nanjing Univ. of Science and Technology, China
Prof. Lijian Leng, Central South Univ., China
Prof. Ning Li, Dalian Institute of Chemical Physics, CAS, China
Prof. Qiang Li, Huazhong Agricultural Univ., China
Prof. Zhenglong Li, Zhejiang Univ., China
Prof. Long Liu, Jiangnan Univ., China
Dr. Xianxiang Liu, Hunan Normal Univ., China
Dr. Zhihua Liu, Tianjin Univ., China
Dr. Feng Long, Institute of Forestry Chemical Industry, Chinese Academy of Forestry Sciences, China
Dr. Rafael Mariscal, Institute of Catalysis and Petrochemistry (CSIC), Spain
Dr. Bartoli Mattia, Polytechnic of Turin, Italy
Prof. Dmitry Yu. Murzin, Abo Akademi Univ., Finland

Dr. Jifeng Pang, Dalian Institute of Chemical Physics, CAS, China
Dr. Lili Qian, Jiangsu Univ., China
Dr. Yaojing Qiu, North Carolina State Univ., USA
Prof. Phillip E. Savage, The Pennsylvania State Univ., USA
Dr. Yuchao Shao, Tsinghua Univ., China
Dr. Xing Tang, Xiamen Univ., China
Prof. Congming Tang, Chongqing Univ. of Technology, China
Dr. Hongliang Wang, China Agricultural Univ., China
Dr. Xiaoqi Wang, Hebei Univ. of Technology, China
Prof. Zhongjiang Wang, Northeast Agricultural Univ., China
Dr. Yuming Wen, National Univ. of Singapore, Singapore
Prof. Chunbao Charles Xu, City Univ. of Hong Kong, Hong Kong, China
Prof. Shihui Yang, Hubei Univ., China
Dr. Yang Yang, Shanghai Jiaotong Univ., China
Dr. Yunlei Zhang, Jiangsu Univ., China
Prof. Liang-Fang Zhu, Sichuan Univ., China

We are also grateful to Dr. Mei Hann Lee (Senior Editor, SpringerNature) for her encouragement, assistance and guidance during preparation of the book.

We would like to express our deepest gratitude towards our families for their love, understanding and encouragement, which help us in completion of this project. Especially, ZF would like to dedicate this book to his brother (Mr. Jing Fang), who passed away on November 28, 2023 in Sanming, Fujian, for all of his love and support even during his final illness.

Zhen Fang, August 24, 2024 in Nanjing
Richard L. Smith Jr., April 19, 2023 in Sendai
Haixin Guo, April 19, 2023 in Tianjin

Contents

Editors and Contributors

About the Editors

Zhen Fang is professor and leader of biomass group, College of Engineering, Nanjing Agricultural University. He is the inventor of the "fast hydrolysis" process, the elected fellow of the Canadian Academy of Engineering. Professor Fang specializes in thermal/biochemical conversion of biomass, nanocatalysts synthesis and their applications, pretreatment of biomass for biorefineries and supercritical fluid processes. He obtained his PhDs from China Agricultural University (Beijing) and McGill University (Montreal). Professor Fang is Associate Editor of the international journals, *Biotechnology for Biofuels* and *Journal of Supercritical Fluids (2018–2020)*. He has more than 20-year international research experience at top universities and institutes around the world, including 1 year in Spain (University of Zaragoza), 3 years in Japan (Biomass Technology Research Center, AIST; Tohoku University), and more than 8 years in Canada (McGill) in renewable energy and green technologies.

Richard Lee Smith Jr is professor of Chemical Engineering, Graduate School of Environmental Studies, Research Center of Supercritical Fluid Technology, Tohoku University, Japan. Professor Smith has a strong background in physical properties and separations and obtained his PhD in chemical engineering from the Georgia Institute of Technology (USA). His research focuses on developing green chemical processes especially those that use water and carbon dioxide as the solvents in their supercritical state. He has expertise in physical property measurements, separation techniques with ionic liquids, hydrothermal and solvothermal carbonization, catalysis in reaction systems and has published more than 290 scientific papers, patents, and reports in the field of chemical engineering. Professor Smith is the Asia Regional Editor for the *Journal of Supercritical Fluids* and has served on editorial boards of major international journals associated with properties and energy.

Haixin Guo is Professor at the Agro-Environmental Protection Institute, Ministry of Agriculture and Rural Affairs (Tianjin, China). She obtained her MS degree in Environmental Engineering from Nankai University in China and her Doctor of Philosophy in Environmental Science at Tohoku University. She has published more than 60 scientific papers and has research interests in sustainable catalyst development, hydrothermal reactions including carbonization, solvothermal and mechanochemical pretreatment of biomass and conversion of biomass into chemicals and biofuels.

Contributors

A. Bueno School of Engineering (UPV/EHU), Bilbao, Spain

Mingyang Che School of Chemistry and Chemical Engineering, Hefei University of Technology, Hefei, China

Guanyi Chen Interdisciplinary Innovation Lab for Environment & Energy/School of Mechanical Engineering, Tianjin University of Commerce, Tianjin, China

School of Environmental Science and Engineering, Tianjin University, Tianjin, China

School of Ecology and Environment, Tibet University, Lhasa, China

P. Díaz-Maizkurrena School of Engineering (UPV/EHU), Bilbao, Spain

Marie-Josée Dumont Department of Chemical Engineering, Université Laval, Pavillon Adrien-Pouliot, Québec, QC, Canada

Marco Frediani Department of Chemistry "Ugo Schiff", University of Florence, Florence, Italy

Piero Frediani Department of Chemistry "Ugo Schiff", University of Florence, Florence, Italy

Yao-Bing Huang New Energy School, North China Electric Power University, Beijing, China

A. Iriondo School of Engineering (UPV/EHU), Bilbao, Spain

Mingjie Jin School of Environmental and Biological Engineering, Nanjing University of Science and Technology, Nanjing, China

Nora Junker Genetics of Prokaryotes, Faculty of Biology & Center for Biotechnology (CeBiTec), Bielefeld University, Bielefeld, Germany

Mengying Liu State Key Laboratory of Efficient Production of Forest Resources, Beijing Forestry University, Beijing, China

Beijing Key Laboratory of Lignocellulosic Chemistry, Beijing Forestry University, Beijing, China

Zhidan Liu Laboratory of Environment-Enhancing Energy (E2E), Key Laboratory of Agricultural Engineering in Structure and Environment of Ministry of Agriculture and Rural Affairs, College of Water Resources and Civil Engineering, China Agricultural University, Beijing, China

State Key Laboratory of Efficient Utilization of Agricultural Water Resources, Beijing, China

Taotao Lu School of Chemistry and Chemical Engineering, Hefei University of Technology, Hefei, China

Yilun Luo School of Chemistry and Chemical Engineering, Hefei University of Technology, Hefei, China

Yoshinao Nakagawa Department of Applied Chemistry, School of Engineering, Tohoku University, Sendai, Japan

Valérie Orsat Department of Bioresource Engineering, McGill University, Sainte-Anne-de-Bellevue, QC, Canada

Surabhi Pandey Department of Bioresource Engineering, McGill University, Sainte-Anne-de-Bellevue, QC, Canada

J. Requies School of Engineering (UPV/EHU), Bilbao, Spain

Berna Sariyar Akbulut Department of Bioengineering, Marmara University, Istanbul, Turkey

Zhuohua Sun State Key Laboratory of Efficient Production of Forest Resources, Beijing Forestry University, Beijing, China

Beijing Key Laboratory of Lignocellulosic Chemistry, Beijing Forestry University, Beijing, China

Keiichi Tomishige Department of Applied Chemistry, School of Engineering, Tohoku University, Sendai, Japan

N. Viar School of Engineering (UPV/EHU), Bilbao, Spain

Kui Wang School of Chemistry and Chemical Engineering, Hefei University of Technology, Hefei, China

Smith School of Chemical and Biomolecular Engineering, Cornell University, Ithaca, NY, USA

Zhi Wang School of Environmental Science and Engineering, Tianjin University, Tianjin, China

Volker F. Wendisch Genetics of Prokaryotes, Faculty of Biology & Center for Biotechnology (CeBiTec), Bielefeld University, Bielefeld, Germany

Wenzhu Wu Interdisciplinary Innovation Lab for Environment & Energy/School of Mechanical Engineering, Tianjin University of Commerce, Tianjin, China

Yongdong Xu Laboratory of Environment-Enhancing Energy (E2E), Key Laboratory of Agricultural Engineering in Structure and Environment of Ministry of Agriculture and Rural Affairs, College of Water Resources and Civil Engineering, China Agricultural University, Beijing, China

State Key Laboratory of Efficient Utilization of Agricultural Water Resources, Beijing, China

Mizuho Yabushita Department of Applied Chemistry, School of Engineering, Tohoku University, Sendai, Japan

Beibei Yan School of Environmental Science and Engineering, Tianjin University, Tianjin, China

Xiao Yan College of Mechanical and Electronic Engineering, Northwest A&F University, Yangling, China

Northwest Research Center of Rural Renewable Energy, Exploitation and Utilization of Ministry of Agriculture, Northwest A&F University, Yangling, China

Yiqing Yao College of Mechanical and Electronic Engineering, Northwest A&F University, Yangling, China

Northwest Research Center of Rural Renewable Energy, Exploitation and Utilization of Ministry of Agriculture, Northwest A&F University, Yangling, China

Tong-Qi Yuan State Key Laboratory of Efficient Production of Forest Resources, Beijing Forestry University, Beijing, China

Beijing Key Laboratory of Lignocellulosic Chemistry, Beijing Forestry University, Beijing, China

Xinchuan Yuan School of Environmental and Biological Engineering, Nanjing University of Science and Technology, Nanjing, China

Jiaojiao Zhan School of Chemistry and Chemical Engineering, Hefei University of Technology, Hefei, China

Zhe-Hui Zhang State Key Laboratory of Efficient Production of Forest Resources, Beijing Forestry University, Beijing, China

Beijing Key Laboratory of Lignocellulosic Chemistry, Beijing Forestry University, Beijing, China

Yuxiang Zhao School of Chemistry and Chemical Engineering, Hefei University of Technology, Hefei, China

Ziyan Zhou College of Mechanical and Electronic Engineering, Northwest A&F University, Yangling, China

Northwest Research Center of Rural Renewable Energy, Exploitation and Utilization of Ministry of Agriculture, Northwest A&F University, Yangling, China

Part I
Production of Organic Acid from Agricultural Residues and Food Wastes

Chapter 1
Production of Organic Acids via Catalytic Processes of Agricultural Residues

Wenzhu Wu, Zhi Wang, Beibei Yan, and Guanyi Chen

Abstract Sustainable production of high-value chemicals by utilizing lignocellulosic biomass waste has garnered significant attention over the past decades. This chapter provides an overview of catalytic processes for producing organic acids derived from agricultural residues, as well as their derivatives. Organic acids are classified into monocarboxylic and dicarboxylic acids according to typical carboxylic acid groups. Amino acids are also introduced as an indispensable category. The chemical catalytic process for direct synthesis of sustainable compounds is an effective existing technology, avoiding the bottlenecks of low selectivity towards specific products and long duration that occur in other conventional technologies. The synthesis typically comprises multiple sequential reactions, making multifunctional catalysts essential for the selective activation of particular C-O or C-C bonds. The high oxygen-carbon ratio of lignocellulosic biomass offers the natural advantages for synthesizing these acids; meanwhile, catalyst design, pathway selection, and parameter control are major concerns in research. Progress in traditional catalytic oxidation technologies and novel catalytic systems for various organic acids is summarized. This chapter aims to provide a fundamental understanding of key catalytic synthesis technologies for the production of typical organic acids.

W. Wu
Interdisciplinary Innovation Lab for Environment & Energy/School of Mechanical Engineering, Tianjin University of Commerce, Tianjin, China

Z. Wang · B. Yan
School of Environmental Science and Engineering, Tianjin University, Tianjin, China

G. Chen (✉)
Interdisciplinary Innovation Lab for Environment & Energy/School of Mechanical Engineering, Tianjin University of Commerce, Tianjin, China

School of Environmental Science and Engineering, Tianjin University, Tianjin, China

School of Ecology and Environment, Tibet University, Lhasa, China
e-mail: chengy@tjcu.edu.cn

Z. Fang et al. (eds.), *Production of Organic Acids and Alcohols from Agricultural Residues and Food Wastes*, Biofuels and Biorefineries 13,
https://doi.org/10.1007/978-981-96-1222-2_1

Keywords Organic acids · Catalytic oxidation · Monocarboxylic acid · Dicarboxylic acid · Amino acid · Lignocellulosic biomass

1.1 Introduction

Organic acids, a subset of acidic compounds, serve as vital raw materials with high demand in industries such as leather, chemical and pharmaceutical synthesis, agriculture, and food. Traditionally, they were synthesized through the selective oxidation of fossil-based feedstocks. The swift depletion of fossil resources coupled with growing environmental issues has driven significant research into harnessing the plentiful and renewable lignocellulosic biomass for generating biofuels and chemicals [1]. Biomass, as a replacement for petroleum depletion feedstocks, offers significant potential for producing these chemical compounds and biofuels. Biomass typically contains a higher oxygen content, ranging from 40 to 45 wt% [2]. In contrast to the current petroleum routes that form C−O bonds by inserting oxygen atoms, breaking down oxygen-enriched lignocellulosic biomass into organic acids appears more atom-economical, as it avoids deoxygenation and some defunctionalization. Lignocellulosic biomass is typically composed of three major components: cellulose (40–60 wt%), hemicellulose (15–30 wt%), and lignin (10–25 wt%) [3]. The three main components are distributed unevenly in the biomass cell wall, where cellulose macromolecules clustering regularly to form strong microfibers, which act as the cell wall's framework, while amorphous hemicellulose and lignin fill the interior as bonding materials [3]. Cellulose primarily bonds with hemicellulose or lignin through hydrogen bonds. In contrast, hemicellulose and lignin are connected through hydrogen bonds and covalent bonds. Cellulose containing aliphatic poly-hydroxyl structures could serves as an ideal raw material for producing organic acids and other oxygenates such as polyols and furanic compounds due to the relatively high O/C ratio [1]. Agricultural wastes, as a typical kind of lignocellulosic biomass, are the most widely used feedstocks for biorefineries. Estimates suggest that over 4.8 billion tons of cellulose could be obtained annually from common agricultural and forest residues worldwide, offering abundant feedstocks for the sustainable production of organic acids [4].

Traditionally, several mature technologies for converting biomass into fuels and chemicals rely on the non-selective breakdown of complex biomass into smaller molecular mixtures, such as gasification for syngas, pyrolysis for crude bio-oil, fermentation for biogas or organic compounds, liquefaction or hydrolysis, and so on [5, 6]. These processes require high energy consumption, and often exhibit poor selectivity for specific products or long durations. Therefore, direct catalytic synthesis processes to transform biomass and derived platform compounds into various organic acids have attracted much attention in the last 10 years. The chemical bonds in Initial materials, such as C-C or C-O bonds, are selectively broken then used to directly synthesize building-block chemicals in these processes. Various strategies of direct catalytic synthesis involve not only traditional catalytic oxidation and wet oxidation, but also emerging electro- and photocatalytic technologies (Fig. 1.1) [6].

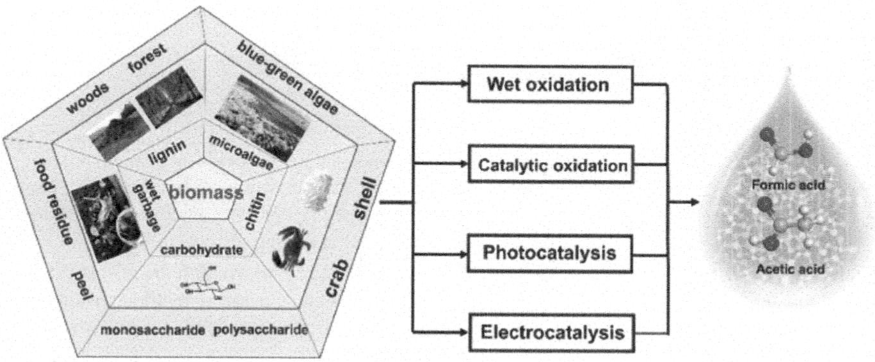

Fig. 1.1 Different methods for the production of organic acids through the oxidation of diverse biomass [6]. (Reprinted with permission from [6], Copyright © 2023 Elsevier)

Numerous acids, such as formic, oxalic, malonic, glycolic, gluconic, glucaric acids, as well as 2,5-furandicarboxylic acid (FDCA) and 5-hydroxymethyl-2-furancarboxylic acid (HFCA), can be derived from the catalytic oxidation of biomass [2]. Formic acid, the simplest organic acid, has prominent advantages as a commonly utilized reductant, environmentally friendly solvent, and a building block in various chemical syntheses. It is a typal monocarboxylic acid. Lactic acid is a typical hydroxy carboxylic acid with abundant annual production worldwide. Gluconic acid and glucaric acid, are both essential chemicals derived from biomass, which widely used in the food industry and pharmaceuticals. Specifically, gluconic acid is usually produced by the oxidation of glucose with O_2, and glucaric acid is an oxidation product of gluconic acid, thus classifying them as sugar acids. Moreover, 2,5-furandicarboxylic acid (FDCA) is increasingly attracting attention due to its potential applications in the production of bio-based polyesters and other polymers. These renewable organic acids exhibit notable biological activity and are easily converted into various chemicals and fuels in subsequent step for high-energy-density chemical products, industrial and agricultural production, and the pharmaceutical industry.

Chemical catalytic routes provide an efficient pathway for producing organic acids derived both from raw biomass and carbohydrate-derived molecules. For instance, gluconic acid can be produced via direct glucose oxidation or through a one-pot conversion of polysaccharides and conversion of lignocellulose using acid–metal bifunctional catalysts. The one-pot conversion of carbohydrates using acid–metal bifunctional catalysts generate furan chemicals. Catalytic oxidative cleavage of C–C bonds in raw materials can generate small molecular organic acids.

The synthesis of these organic acids from lignocellulose or its platform compounds generally involves several tandem reactions, and thus multifunctional catalysts capable of selectively activating specific C−O or C−C bonds are crucial [1]. The design of different catalysts and the optimization of catalytic systems facilitating the selective cleavage of C-C and/or C-O bonds are imperative for the targeted production of organic acids. In terms of efficiently converting cellulose into

Table 1.1 Analysis of the market and applications for commodity chemicals produced by catalytic oxidation of carbohydrates in 2018 [2]

Chemicals	Market price (US $ per metric ton)	Volume (metric tons per year)	Main applications
Formic acid	400	720,000	Pharmaceutical, pesticide, leather tanning, textiles, formate salts, rubber coagulant, food and silage preservative, catalysts and plasticizers, fiber dyeing agents
Acetic acid	450–690	6,500,000	Production of pure terephthalic acid (PTA), ethylene acetate, acetic anhydride, acetate for photographic film production, polyvinyl acetate, and synthetic fibers and fabrics
Glycolic acid	100,000–300,000	15000[a]	Food additives, production for glycine, synthetic fibers, plastics in industry. For plant growth regulator and pesticide. For bacteriostat in pharmaceutical industry
Gluconic acid	760–830	60,000	Food additives, detergent, medicines, raw materials for concrete, and for biodegradable polymers
FDCA	690,000–720,000	40	Primarily serves as a polymer building block, replacing terephthalic acid for producing modern polymers and polyesters with an aromatic moiety

[a] Note: The volume of glycolic acid belongs to the data of USA

organic acids, low reactivity under mild conditions is another challenge due to strong intra- and intermolecular hydrogen bond networks among hydroxyl groups. Developing new catalytic strategies and systems for effective conversions with high reaction selectivity and yield is key to the production of organic acids via catalytic processes.

In this chapter, we focus on the catalytic transformation of agricultural wastes to produce representative high-value organic acids. According to a basic classification criterion, i.e., number of carboxyl groups and ionized H^+, these organic acids are categorized into (1) mono-carboxylic acids, like formic, acetic, lactic, gluconic acid, etc.; (2) dicarboxylic acids, like oxalic, maleic, malonic, FDCA; and (3) amino acids. Generally, lactic acid is also classified as hydroxycarboxylic acids, while FDCA and levulinic acid are classified as 5-hydroxymethylfurfural (HMF)-derived acids. Lactic acid and hydroxycarboxylic acids are the most extensively studied with mature commercial production and technologies. Dicarboxylic acids and amino acids possess a restricted annual production capacity. The cost and annual production capability are shown as follows (Table 1.1) [2]. Developing effective chemical approaches for converting renewable and plentiful feedstocks into amino acids is still in its early stages. Chemical-catalytic routes provide an efficient pathway for producing organic acids, derived both from raw biomass and carbohydrate-derived molecules. Research should focus on enhancing the efficiency of catalytic strategies, promoting the utilization ratio of raw materials, and collaboratively distributing carbon feedstock into various products [7].

1.2 Monocarboxylic Acids

In this section, according to basic molecule structure, simplex hydroxy monocarboxylic acids such as formic acid and acetic acid, hydroxycarboxylic acids like lactic acid, and HMF-derived acids like levulinic acid were introduced. These acids are widely used in agricultural, pharmaceutical, food, adsorbents, and polymer industries.

1.2.1 Formic and Acetic Acid

Formic acid and acetic acid represent the most typical monocarboxylic acids with simplest molecular structures. Formic acid serves as a widely used platform chemical in the agriculture, rubber, pharmaceutical, and leather industries, while also holding promise as a hydrogen carrier. Formic acid is industrially produced by hydrolyzing methyl formate, which is derived from methanol carbonylation. When it comes to carbohydrate biomass, formic acid forms through the oxidation-driven breaking of C-C bonds that connect aldehyde and hydroxyl/aldehyde groups. In the standard process of converting glucose to formic acid, the terminal carbon undergoes oxidation, resulting in the formation of formic acid. This degradation process then repeats with C5 intermediates, progressively cleaving all C-C bonds. Theoretically, this process can yield six molecules of formic acid. Nevertheless, it is challenging to control the selectivity when breaking C-C bonds during the process, as it also produces by-products. Carbon dioxide is the main by-product generated during the decarboxylation process of acidic intermediates. Cellulose can be converted to formic acid through two steps: first, cellulose is hydrolyzed into glucose, which is then oxidized to formic acid [8]. Some typical studies on catalytic oxidation process of biomass to formic acid include alkaline hydrothermal oxidation, polyoxometalate (POMs) catalyzed oxidation, and vanadium salt catalysts [9–12]. Regardless of the catalytic system employed, it is essential to balance oxidation and over-oxidation during formic acid production. Typically, concentrated alkali is typically needed to prevent formic acid decomposition, and H_2O_2 is introduced to boost oxidation. Vanadium salts and POMs are frequently employed as catalysts for the oxidative cleavage of C-C bonds in carbohydrate biomass.

Acetic acid, ranks as second simplest monocarboxylic acid. Generally, compared to formic acid production, achieving selective oxidation of biomass to acetic acid in a single-step reaction strategy is more challenging. To increase the yield of acetic acid, a useful strategy involves adding an acid or base prior to oxidation to produce more transition products that favor acetic acid formation. This approach is known as the two-step strategy [13] (Fig. 1.2).

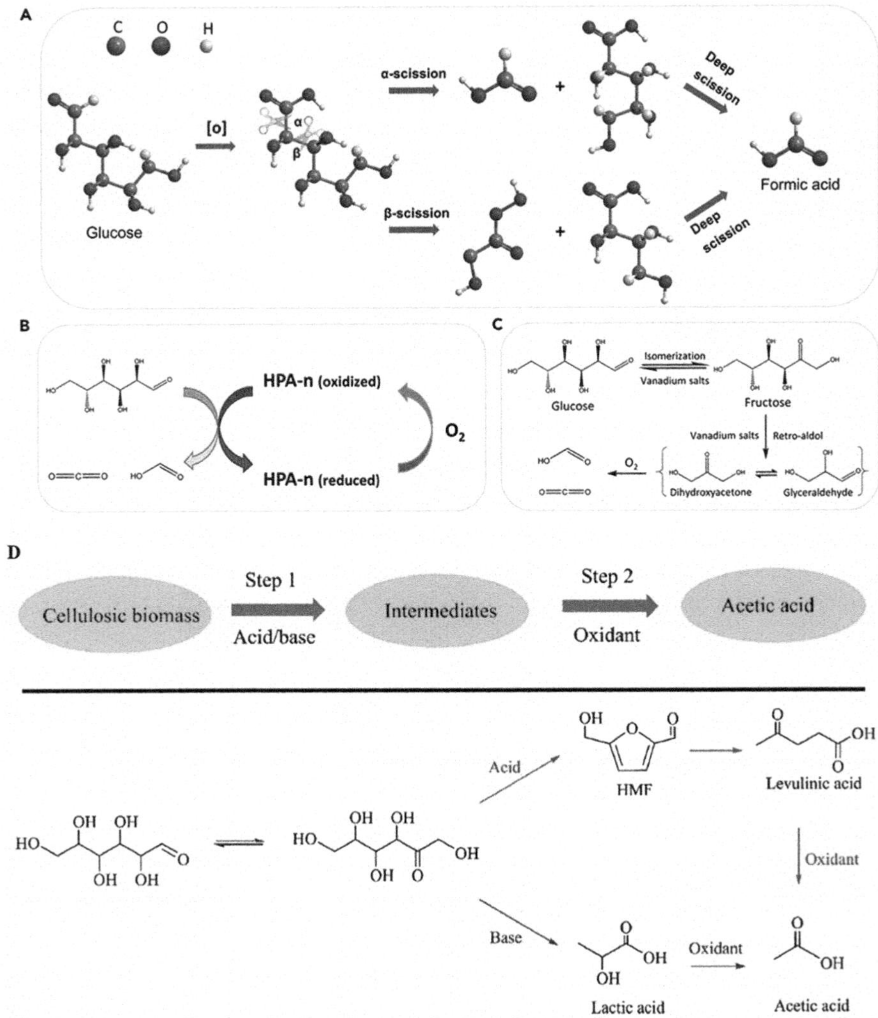

Fig. 1.2 Conversion mechanism depicting the transformation of glucose into formic acid catalyzed under alkali hydrothermal conditions (**a**), polyoxometalates ($H_{3+n}PV_nMo_{12-n}O_{40}$ (HPA-n, n = 0~5)) (**b**), vanadium salts (**c**) [4]. Reprinted with permission from [4], Copyright © 2023 Elsevier, and two-step approach of converting cellulosic biomass into acetic acid (**d**) [6]. (Reprinted with permission from [6], Copyright © 2023 Elsevier)

1.2.2 Lactic Acid

Lactic acid stands out as the key member of hydroxycarboxylic acids in industry. It is considered a crucial raw material for the production of polylactic acid polymers, pharmaceuticals, and cosmetics. Lactic acid contains one carboxylic (-COOH) group and one hydroxyl (-OH) group, with only one H^+ that can be ionized from the molecule, thus categorizing it as a monocarboxylic acid. As the most prominent member of hydroxycarboxylic acids in industry, lactic acid is traditionally produced through microbial fermentation of biomass, a process that is time-consuming and space-intensive [14]. Recently, anaerobic hydrothermal conditions have been employed for lactic acid production to prevent further oxidation of lactic acid. Consequently, various catalytic systems have received attention [15]. For instance, dihydroxyacetone (DHA), glycerol, and glyceraldehyde (GLY) could be served as feedstocks for lactic acid synthesis. Typically, alkali and acids are served as catalysts [16]. For example, Holm et al. have found that Lewis acidic zeolite derivatives, like Sn-Beta in methanol, selectively catalyze the tansformation of glucose, fructose, and sucrose sugars into methyl lactate, a key intermediate for lactic acid production [16, 17]. It should be emphasized that the type of solvent (e.g., methanol and water) and the type of acid (e.g., Brønsted acids and Lewis acids) affect the lactic acid output. Alkaline hydrothermal conversion of lignocellulosic biomass was also drawn much attention. Yan et al. demonstrated that NaOH and $Ca(OH)_2$ are effective in enhancing lactic acid production during the hydrothermal conversion of starch and cellulose [18]. A 27% yield of lactic acid was achieved at 300 °C in 1.5 min with NaOH addition.

1.2.3 Levulinic Acid

Levulinic acid (LA) serves as a versatile platform chemical, capable of synthesizing various valuable compounds, including γ-valerolactone and polymer-related chemicals, etc. [19]. To some extent, levulinic acid can be considered a monocarboxylic acid due to its molecular structure, however, it is more commonly recognized as a biomass-based platform compound. LA demonstrates significant versatility as a platform chemical, thus providing an opportunity to replace petroleum-based chemicals with renewable alternatives. This trait positions it as a viable option for advancing a sustainable pathway in the future [20]. Many cellulosic biomass such as agricultural and forest wastes have been utilized to produce LA [8]. Typically, the most used catalysts for synthesizing LA from biomass include mineral acids (such as H_2SO_4 and HCl), ionic liquids, metal salts, and solid acids (like sulfonated metal oxides, zeolites, and resin materials with SO_3H group (Fig. 1.3). The most widely used approach to synthesize LA from biomass involves controlled dehydrative treatment using acid catalysts [8]. Among them, those with SO_3H group demonstrate particularly unique and effective performance. The reaction rate is increased due to the easy accessibility of catalytic sites (-SO_3H) [21]. Some studies illustrated the production of LA from biomass and macroalgae by using ionic resin catalysts. In addition, Meinita et al. used homogeneous strong acid catalysts, such as HCl, H_2SO_4, and methanesulfonic acid (MSA) for LA synthesis from macroalgae [22]. Generally, LA is produced through a sequence of tandem reactions, involved in the isomerization of glucose to fructose, dehydration of fructose to 5-hydroxymethylfurfural (HMF), subsequent rehydration and

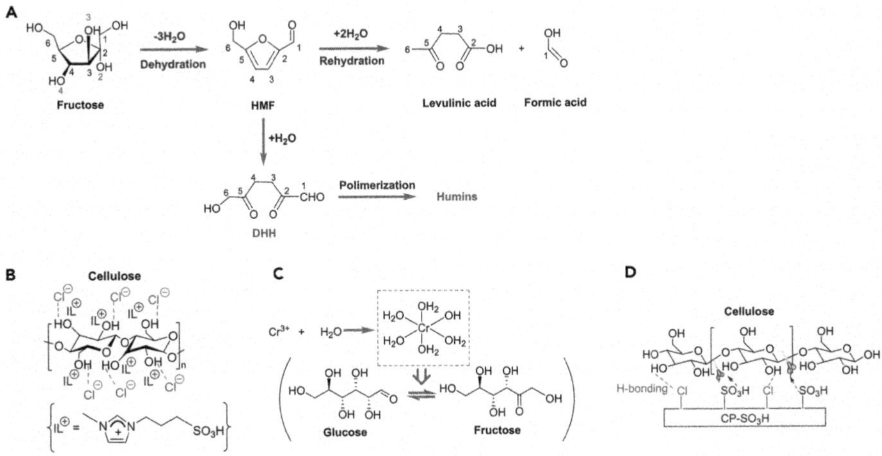

Fig. 1.3 Levulinic acid production from biomass [4]. Reprinted with permission from [4], Copyright © 2023 Elsevier: (**a**) Conversion to levulinic acid under acidic condition, (**b**) Hydrogen bonding between cellulose biomass and ionic liquids (ILs) catalyst, (**c**) Cr(III)-catalyzed isomerization of glucose into fructose, (**d**) Interactions between cellulose and CP-SO_3H catalyst

rearrangement of HMF to LA [19, 23–25]. Overall, the yield of LA is primarily influenced by four processing conditions: acid and solvent concentrations, temperature, and residence time [26].

1.3 Dicarboxylic Acids

Dicarboxylic acids, characterized by having two carboxyl groups, find extensive use across sectors including food processing, agriculture, pharmaceuticals, adsorbent manufacturing, and polymer production [27]. Biomass is increasingly attractive as a feedstock compared to conventional petroleum or fossil-based feedstocks, as removing oxygen from molecules is generally more cost-effective than adding oxygen through selective oxidation reactions [28]. As shown in Fig. 1.4, a variety of dicarboxylic acids can be prepared by agricultural residues and food waste, including oxalic acid, malonic acid, maleic acid, succinic acid, 2,5-furandicarboxylic acid, glucaric acid, and fumaric acid. Several widely used dicarboxylic acids are summarized below:

1.3.1 Oxalic Acid

Oxalic acid is a significant chemical employed as a mordant in the dyeing process, as well as a bleach for pulping wood and rust removal. In general, it has been reported

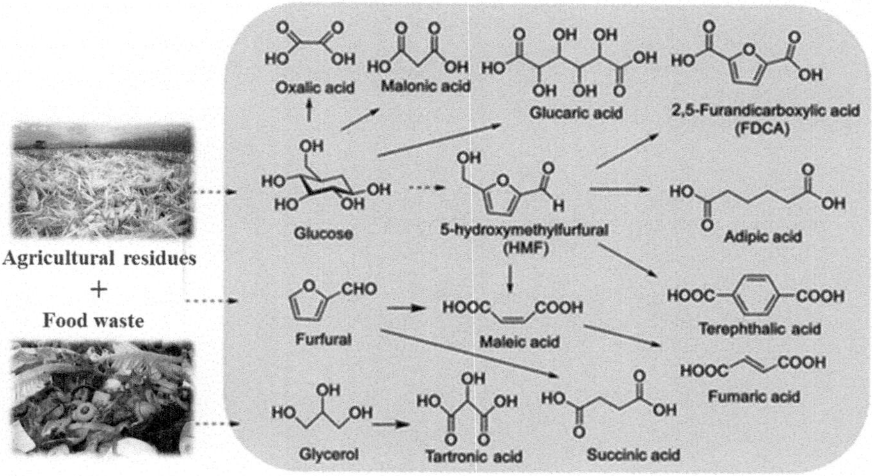

Fig. 1.4 Dicarboxylic acids produced from agricultural residues and food waste [29]. (Reprinted with permission from [29], Copyright © 2021 American Chemical Society)

that the synthesis of oxalic acid from biomass cellulose includes two methods: one involves using nitric acid oxidation technology to catalyze the conversion of cellulose to oxalic acid, but the corrosive nature of nitric acid requires strict equipment standards; the other method catalyzes the cellulose substrate directly to oxalic acid in a high-temperature alkali solution [30]. A yield of 41.5% oxalic acid yield and an oxalic acid productivity of 3.86 $mol_{oxalic\ acid}\ g_{cat}^{-1}\ h^{-1}$ could be obtained in an aqueous NaOH solution (2 M) at 200 °C with CuO catalyst [31].

It should be emphasized that oxalates are a direct product of two processes for producing oxalic acid, which requires the release of free acid through acidification. Incorrect discharge of wastewater may result in secondary pollution. Therefore, developing and perfecting highly selective, green processes for generating oxalic acid from biomass continues to be a formidable task. In addition, glycerol and some C6 derived cellulose substrates [32], including glucose, fructose, and 5-hydroxymethylfurfural (HMF), can also serve as substrates for oxalic acid synthesis (Fig. 1.5).

1.3.2 Maleic Acid

Maleic acid (MAc) and maleic anhydride (MA) are crucial raw materials in the fields of food, pharmaceuticals, pesticides, etc., and as vital intermediates within the chemical manufacturing, with an annual output of millions tons [34]. Currently, maleic acid cannot be produced by biochemical route, industrially it is produced from fossil materials such as butane, benzene, etc., with vanadium oxide or vanadium-phosphorus oxide as catalysts at a reaction temperature of 450–500 °C

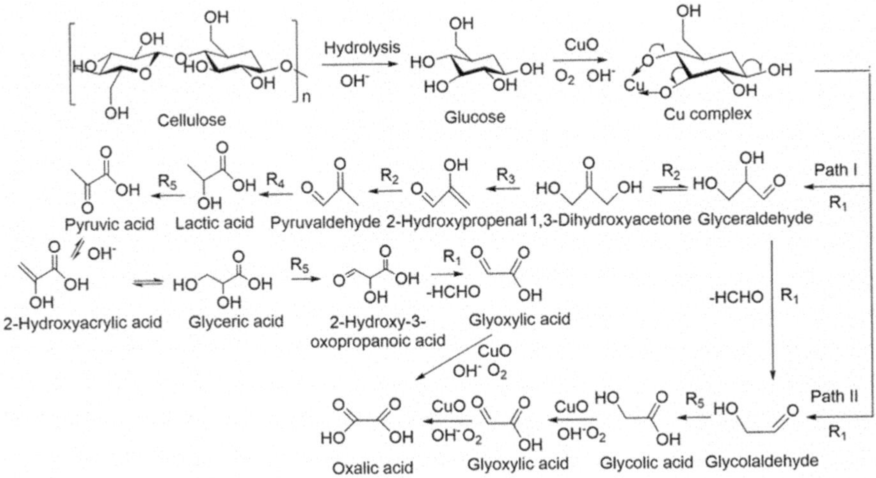

Fig. 1.5 Potential reaction of cellulose to oxalic acid [33]. Reprinted with permission from [33], Copyright © 2016 American Chemical Society. Note: R_1 retro-aldol condensation; R_2 keto-enol tautomerization; R_3 elimination of water; R_4 benzilic acid rearrangement; R_5 base-catalyzed oxidation by water

[35]. The synthesis of maleic acid derived from sustainable biomass materials, rather than petroleum-based feedstocks, is greatly preferred for the advancement of sustainable chemical production. There are two routes to synthesize organic acids from sugar, namely the non-furan pathway and the furan pathway. The maleic acid production process is based on HMF or furfural with glucose/xylose as starting material, a typical furan-related route.

Maleic acid is usually produced by catalytic oxidation of furfural in both the gas and liquid phases, with the liquid-phase process occurring in the aqueous phase, where MAc is preferentially formed at temperatures more moderate than those for gas-phase oxidation. MA and MAc could be interconverted via dehydration and hydration processes, meaning that producing one inherently grants access to the other [36]. The reaction system is essential in the oxidation process of furfural. Furfural oxidation to maleic acid yield in $Cu(NO_3)_2/H_3PMo_{12}O_{40}$-dioxygen catalytic system was 49% [37], FeT(p-Cl)PPCl-dioxygen was 44% [38]. Another approach is the conversion of HMF and furfural obtained from biomass to MA by selective oxidation. HMF offers multiple benefits compared to furfural. HMF can be synthesized from abundant natural resources such as cellulose, glucose, and fructose [39]. HMF has more functional structures than furfural. Presently, catalysts based on vanadium are primarily utilized for converting HMF into MA [40]. With the addition of vanadium, the efficiency of electron transfer between catalyst and reactant molecules is higher, while aiding in the oxidative severance of the C-C bonds in HMF and its intermediates throughout the conversion to MA [41] (Fig. 1.6).

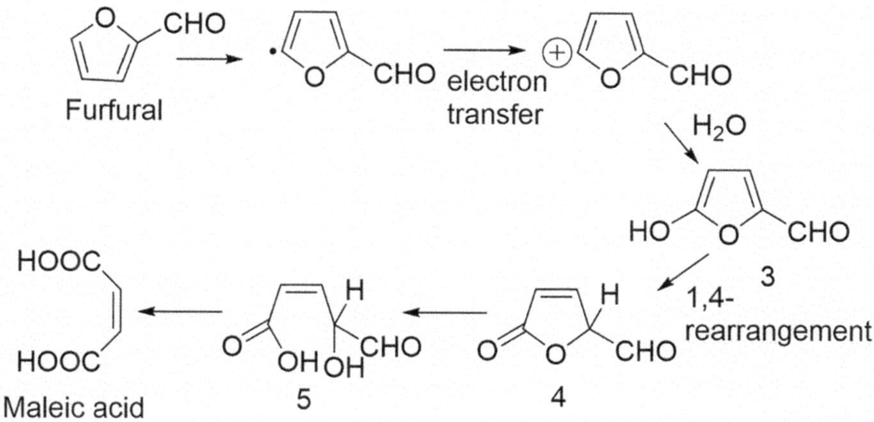

Fig. 1.6 Formation of maleic acid from furfural using mesoporous KIT-6-supported vanadyl pyrophosphate and zirconium pyrophosphate catalysts [29]. (Reprinted with permission from [29], Copyright © 2021 American Chemical Society)

1.3.3 Malonic Acid

Malonic acid, a dicarboxylic acid containing three carbons, is a pivotal chemical employed across a wide range of industrial chemistry applications [42]. Conventionally, malonic acid is synthesized from fossil resources through a petrochemical process. The byproduct, sodium cyanide, is dangerous and environmentally unfriendly. With the growing concern over energy and environmental impacts intensifying, the production of malonic acid from renewable feedstocks has garnered significant interest on a global scale [43].

Malonic acid can be prepared from cellulose in a homogeneous solution at temperatures between 160 and 300 °C without a catalyst. The yield of malonic acid can reach about 40%, along with the formation of three other valuable chemicals (formic acid, acetic acid, and lactic acid) [44]. Figure 1.7 depicts the process where cellulose is hydrolyzed into glucose. Following this, glucose is converted into malonic acid, glyceraldehyde, 1,3-dihydroxyacetone through retro-aldol condensation (C3–C4 scission) and isomerization. Meanwhile, 1,3-dihydroxyacetone can also be converted to malonic acid through a sequence of isomerization and oxidation reactions.

1.3.4 Succinic Acid

Succinic acid is a key ingredient in the chemical industry, acting as a precursor for various high-value compounds including N-methyl-2-pyrrolidone, 1,4-butanediol, γ-butyrolactone, 2-pyrrolidone, and tetrahydrofuran. For more than a decade,

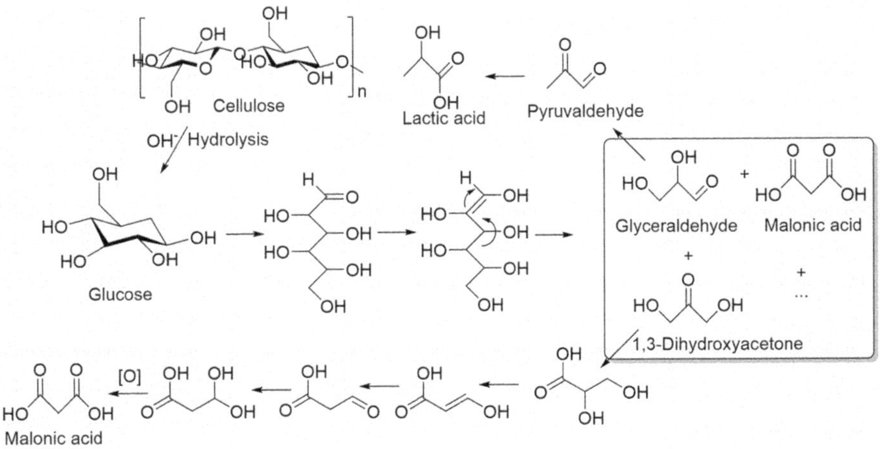

Fig. 1.7 Formation mechanism of cellulose to malonic acid [44]. (Reprinted with permission from [44], Copyright © 2014 American Chemical Society)

succinic acid has consistently been included in the top 12 building block chemicals by the U.S. Department of Energy [45]. Succinic acid can be synthesized from biomass-derived derivatives such as glucose and furfural.

Typically, glucose can be fermented to produce tartaric acid, which can be further hydrogenated into succinic acid. Vlachos et al. utilized a carbon black-supported molybdenum oxide (MoOx/C) catalyst with the assistance of HBr in acetic acid solvent to transform gluconic acid into succinic acid with a yield of 87% [46]. Usually, gluconic acid undergoes C-O cleavage to form malic acid via the synergistic action of MoOx/C and HBr, which is then hydrogenated to succinic acid. Furfural is another precursor for the manufacture of succinic acid, which can be produced by the removal of water from xylose found in corn cobs. Brønsted acid catalysts, such as the Amberlyst-15 resin, have superior results for the conversion of furfural to succinic acid [47]. Ebitani et al. employed the Amberlyst-15 solid acid catalyst with a 30% mass fraction of H_2O_2 as the oxidant to achieve the reaction of furfural oxidation to succinic acid [48]. Succinic acid was synthesized with an isolated yield of 68%, achieving over 99% conversion of furfural following a 36-h reaction using Amberlyst-15 (1.0 g) and 30% H_2O_2 (80 mmol) in an aqueous medium (30 mL) at a temperature of 80 °C. It should be noted that the problem of inactivation of solid acids is a hindrance to this process, which is attributed to the constant leaching of acid sites during the reaction.

1.3.5 2,5-Furandicarboxylic Acid (FDCA)

2,5-furandicarboxylic acid (FDCA) is considered a potential environmentally friendly raw material that can substitute petroleum-based monomers or the synthesis

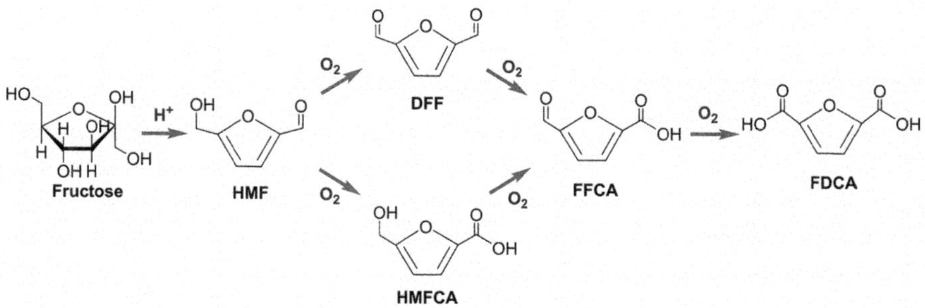

Fig. 1.8 Production of 2,5-furandicarboxylic acid (FDCA) from biomass [4]. (Reprinted with permission from [4], Copyright © 2023 Elsevier)

of polymers including polyamides, polyesters, and polyurethanes. In the typical process of converting fructose to FDCA, two primary steps are involved: first, the dehydration of fructose to produce HMF, followed by the oxidation of HMF to yield FDCA (Fig. 1.8). Catalysts based on Pd, Pt, and Au have been documented to achieve the aerobic oxidation of HMF to FDCA under moderate conditions [49]. Nonetheless, there are ongoing challenges that necessitate a reduction in cost and an enhancement of the reaction rate. For instance, the conditions of harsh production for O_2/air ratio, temperature, time are 1–5.5 MPa, 100–180 °C, and 5–48 h respectively [50]. Transition metal oxides represent a viable substitute for precious metals in the aerobic oxidation of 5-HMF to FDCA due to characteristics such as affordability, lower metal-O bond energies, varied oxidation states, and enhanced activation of molecular oxygen [51]. Notably, to rival noble metal-based catalysts, improvements in the activity, stability, and selectivity of these transition metal oxides are necessary. These enhancements can be achieved through various strategies such as modulating the valence states of metal ions, forming bimetallic constructs, and modifying surface electronic structures [52]. The choice of catalyst support also plays a pivotal role, influencing both the reaction kinetics and pathways. For instance, graphene is favored as a support material due to its superior electrical conductivity, mechanical strength, stability, and extensive surface area [53]. According to the study of Rao et al. [50], dense CuOx-CoOy heterostructure nanoparticles were successfully synthesized on graphene surfaces using a one-pot method under mild conditions. The resulting catalyst was effective in selectively oxidizing 5-HMF to FDCA in an aqueous solution using NaClO as an oxidant at 30 ° C, showing performance comparable to or even surpassing that of reported noble metal catalysts. Hence, an optimal combination (e.g. optimal surface electronic structure and efficient catalytic kinetics) could lead to highly efficient and selective oxidization of HMF to FDCA.

1.3.6 Gluconic and Glucaric Acid

Sugar acids rank among the top 30 globally in terms of high-value chemicals and find wide applications in the food, pharmaceutical, agriculture, and construction industries [54]. A typical method for converting cellulose to sugar acids like gluconic acid and glucaric acid involves hydrolysis to break the glycosidic bonds of cellulose, releasing glucose units, followed by selective oxidation of glucose to form these sugar acids (Fig. 1.9a). The selective transformation of monosaccharides into sugar acids is involved in two major pathways, including the oxidation of the aldehyde group and the alcohol group [4]. However, the challenge persists in developing environmentally friendly and efficient catalytic systems remains challenging, primarily due to issues with the oxidative cleavage of C-C bonds in monosaccharides and the risk of over-oxidation of sugar acids, as illustrated in Fig. 1.9b and c.

Gluconic acid, derived from the oxidation of glucose, is valued for its eco-friendly properties and biodegradable characteristics. It has a wide range of applications, including serving as a water-soluble cleaning agent and chelating agent. Gluconic acid is the most prominent member of sugar acids as well as a kind of polyhydric-hydroxycarboxylic acids. Numerous methods for producing gluconic acid from glucose have been reported, encompassing chemical and electrochemical catalysis, as well as enzymatic biocatalysis. Conventional practices for producing gluconic acid and glucaric acid using fungi, bacteria, or enzymes face challenges in scaling up due to issues such as low productivity, susceptibility to contamination, wastewater production, lack of recycling, and the need for neutralization, etc. [57]. The targeted oxidation of biomass to produce sugar acids has generated heightened interest among both the academic and industrial communities. While, in the chemical catalytic processes of sugar acids production from biomass, selectively converting cellulose biomass into the desired short-chain organic acids poses challenges. Monosaccharides, the basic units of cellulose and hemicellulose found in inedible lignocellulosic biomass, hold promise as feedstocks for producing sugar acids [58, 59].

The primary reaction pathway converting glucose precursor into gluconic acid entails oxidizing the aldehyde group (-CHO) at the C1 position of glucose to a carboxyl group (-COOH), resulting in the formation of gluconic acid [57]. Because of high activity [60], noble metals such as Pt, Pd, Au, and Ru, are extensively employed as catalysts for selective oxidation of monosaccharide to gluconic acid, targeting the activation and oxidation of only the terminal carbonyl group [61]. Compared to Pd and Pt, Au exhibits higher efficiency in gluconic acid production because of the low sensitivity to oxygen poisoning and catalytic activity across a broad pH range [60]. In comparison, the direct conversion of cellulose or cellobiose requires bifunctional catalysts that could combine hydrolytic and oxidative functions together [62, 63]. For instance, Rafaïdeen et al. utilized carbon-supported Pd/Au bimetallic nanoparticles (Pd0-3/Au0-7/C) as electrocatalysts for glucose and xylose oxidation

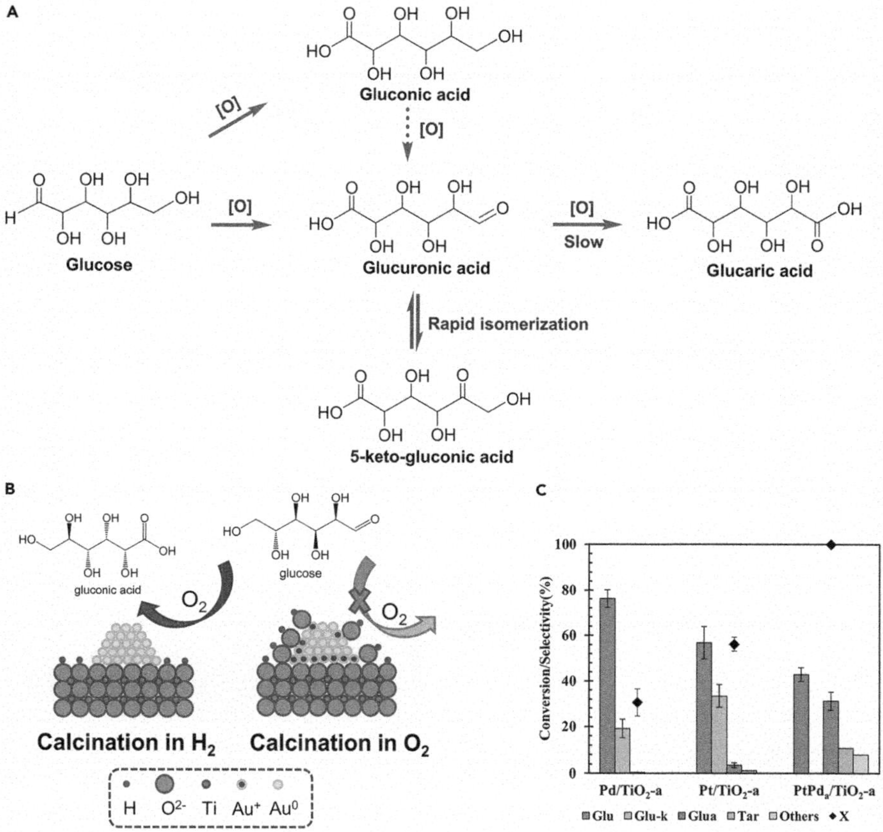

Fig. 1.9 Sugar acids production from biomass: (**a**) Oxidation of glucose to gluconic acid and glucaric acid [55]. Reprinted with permission from [55], Copyright © 2016 American Chemical Society, (**b**) glucose oxidation to gluconic acid utilizing Au/TiO_2 catalyst reproduced from [56], and (**c**) selective products in glucose oxidation to glucaric acid using various catalysts [55]. Reprinted with permission from [55], Copyright © 2016 American Chemical Society. Note: *Glu* gluconic acid; *Glu-k* 5-keto-gluconic acid; *Glua* glucaric acid; *Tar* tartronic acid. Other products include lactic, glyceric, glycolic, oxalic, and formic acids

in alkaline aqueous media [64]. Additionally, carbon materials and metal oxides are frequently utilized as supports for synthesizing supported Au catalysts [56]. TiO_2, Al_2O_3, ZrO_2, CeO_2, SiO_2 and active carbon are typical supports [56]. Compared to these traditional supports, the functional carbon nanotube (CNT) is more suitable to act as the support, showing a significantly higher yield of acids. The sp^2 carbon network on the surface of CNT may facilitate interactions with the aromatic furan ring in HMF or its intermediates [65]. Furthermore, the oxygen-rich functional groups on the CNT surface, particularly carbonyl/quinone and phenolic groups, can serve as redox mediators, promoting hydrogen transfer of platinum nanoparticles, accelerating metal recovery, and thus facilitating the progression of the reaction [66]. Recently, solid acids severed as supports exhibit outstanding performance. For instance, metals (e.g. Au) coupled with solid acids, e.g., polyoxometalates (POMs) exhibit outstanding performance. The acidic support facilitated the hydrolysis of cellulose into glucose, while metals played a role in the subsequent oxidation of glucose to gluconic acid [67]. To tackle the cost concern associated with noble metals, two potential avenues have been proposed: utilizing secondary metals to inform bimetallic catalysts and exploring alternative utilization of non-noble metal-based catalysts. The catalytic activity of more affordable transition metal (e.g. Fe, Ni, and Co) catalysts in oxidation reactions has generated considerable interest. Ni is extensively utilized in the oxidation of acetic acid [68]. For instance, Li et al. proposed a nickel-embedded carbon nanotube (Ni-NCNT) catalyst that efficiently converted the monosaccharides into gluconic acid under mild conditions [69]. To reduce energy requirements and avoid harsh reaction condition, methods such as ultrasound, microwave, and solar energy have been employed to enhance glucose oxidation through radical and/or thermal activation processes [70–72]. In addition, other new catalysts such as ionic liquids (ILs), POMs, bimetal, and co-catalysts show promising selectivity in breaking specific bonds [4].

Glucaric acid is a further oxidized product of gluconic acid. It is a flexible, sustainable platform-chemical for food additives, plastics, and nylons [73]. Many studies reported gluconic acid is the predominant product of glucose oxidation, owing to the oxidation of gluconic acid to glucaric acid occurs slowly [74]. In the generation of glucaric acid, the hydroxyl group (-OH) at the C6 position of gluconic acid undergoes oxidation first to an aldehyde (-CHO) and subsequently to a carboxyl group, resulting in formation of glucuronic acid and then glucaric acid (shown in Fig. 1.9) [55]. The yield of glucaric acid tends to be lower than gluconic acid as the oxidation of −OH is less thermodynamically favorable than that of −CHO [74]. Reports indicated that glucose can be oxidized to glucaric acid using electrochemical methods or stoichiometric oxidation with NaOCl, NaBr, bleaches, and mineral acids [75, 76]. The controllable oxidation of glucose i.e., the ratio of gluconic acid and glucaric acid primarily results from electrochemical oxidation and improved mass transfer [75]. Heterogeneous catalytic oxidation of glucose presents an alternative route from glucose to glucaric acid [77]. The solution pH, metal species metallic promoter, physicochemical properties of support, etc., are key factors in achieving good catalytic performance and affecting glucaric acid production [57, 78]. Lee et al. proposed a commercial Pt/C catalyst for aerobic oxidation of glucose in water, reaching up to 74% yield of glucaric acid [74]. In acidic solutions, gluconic acid is the main product, while in strongly basic conditions, the selectivity for glucaric acid diminishes due to C–C bond cleavage, resulting in shorter-chain carboxylic acids. Moreover, elevated temperatures and higher Pt loadings further reduce the selectivity for glucaric acid.

1.4 Amino Acids

Amino acids are a series of organic substances having an acidic carboxylic group (-COOH) and a basic amino group ($-NH_2$), with significant differences in molecular structure compared to the aforementioned organic acids. Amino acids serve as the fundamental components of proteins, play a crucial role in the food and medicine industries. The global amino acids market has been steadily expanding in recent years, reflecting the necessity for sustainable manufacturing processes and a significant opportunity for innovation in amino acid synthesis technologies. The traditional methods for amino acid production include chemical and biological synthesis. The biological processes, including enzymatic and fermentative synthesis, were the dominant technologies for amino acid production. However, it is noted that the biological methods require the super-clean or sterile conditions and suffer from the complex separation and extraction steps, resulting in high energy consumption and low production efficiency.

Adolph Strecker firstly raised chemical method of Strecker synthesis for α-amino acid synthesis in 1850, which used non-renewable aldehydes as substrates, thus facing challenges in the manufacturing of amino acids from biomass-related feedstocks. In recent years, a new approach to chemocatalysis method was pioneered for α-amino acids synthesis utilizing lignocellulose derivatives as substrates [79]. Chemical processes could effectively yield a range of α-amino acids, such as valine, phenylalanine, leucine, aspartic acid, and alanine. In this process, the lignocellulose derivatives, such as α-hydroxyl acids, could be converted into α-amino acids prepared by heterogeneous catalysts based on Ru and dehydrogenation-reductive amination. α-hydroxy acids, serving as a crucial platform compound, are easily accessible from a variety of lignocellulose ingredients. Besides, the glycerol and furfural, two oxygen-containing platform molecules derived from biomass lignocellulose, were also shown to be convertible into alanine and proline respectively, via chemocatalytic methods. The α-amino acids obtained by chemocatalytic synthesis are shown in Table 1.2. In comparison to Strecker synthesis and biological synthesis, while the chemocatalytic method is still in its infancy, it enables direct utilization of renewable biomass feedstock and displays the huge possibility to raise the efficiency of synthesis, thus attracting increasing attention from academic and industrial communities.

The intensive studies on the chemocatalytic synthesis of amino acids using lignocellulose derivatives, especially α-hydroxyl acids as starting substrate develop rapidly. This development is based on catalytic system design and catalyst selection. Catalytic transformation technologies have evolved from conventional thermocatalytic to photocatalytic and electrocatalytic pathways [84]. The reaction processes and applications of these conversions from α-hydroxy acids to α-amino acids are summarized and discussed as follows (Fig. 1.10).

Thermocatalytic conversion

Deng et al. developed the first thermocatalytic conversion system in which the α-hydroxyl acids from lignocellulosic biomass were directly transformed into

Table 1.2 The summarizations of chemocatalytic methods for converting α-hydroxyl acids to α-amino acids

Amino acids production	Yield (%)	Substrate	Catalyst	Reaction conditions (temperature, K; pressure, MPa; duration, h)	Ref.
Thermocatalytic conversion					
Alanine	70	Lactic acid	Ru/N-CNTs	493 K, 1 MPa H_2	[80]
Alanine	64	Lactic acid	Ru/Ni@C	473 K, 1.5 MPa H_2	[81]
Alanine	62	Lactic acid	Ru/CNT modified with Ni	1 MPa H_2, 493 K, 2 h	[79]
Alanine	43	Glucose	Ru/CNT	1 MPa H_2, 493 K, 2 h	[79]
Valine	48	α-hydroxyl-3 methylbutyric acid	Ru/CNT	1 MPa H_2, 493 K, 2 h	[79]
Valine	60	α-hydroxyl-3-methylbutyric acid	Ru/CNT modified with Ni	1 MPa H_2, 493 K, 2 h	[79]
Leucine	49	α-hydroxyisocaproic acid	Ru/CNT	1 MPa H_2, 493 K, 2 h	[79]
Leucine	69	α-hydroxyisocaproic acid	Ru/CNT	1 MPa H_2, 493 K, 2 h, NaOH, KOH	[79]
Aspartic acid	27	2-hydroxysuccinic acid	Ru/CNT	1 MPa H_2, 493 K, 2 h, NaOH, KOH	[79]
Phenylalanine	30	3-phenyllactic acid	Ru/CNT modified with Ni	1 MPa H_2, 493 K, 2 h	[79]
α-aminobutyric acid	66	α-hydroxyl butyric acid	Ru/CNT	1 MPa H_2, 493 K, 2 h	[79]
α-aminobutyric acid	73	α-hydroxyl butyric acid	Ru/CNT modified with Ni	1 MPa H_2, 493 K, 2 h	[79]
Photocatalytic conversion					
Leucine	31.6	α-hydroxyisocaproic acid	CdS nanosheet	323 K, 0.1 MPa N_2, visible light, 8 h	[82]
Valine	21.6	α-hydroxy-3-methylbutyric acid	CdS nanosheet	323 K, 0.1 MPa N_2, visible light, 8 h	[82]
Phenylalanine	22.4	Phenyllactic acid	CdS nanosheet	323 K, 0.1 MPa N_2, visible light, 8 h	[82]
α-aminobutyric acid	23.2	α-hydroxybutyric acid	CdS nanosheet	323 K, 0.1 MPa N_2, visible light, 8 h	[82]
Electrocatalytic conversion					
Alanine	54.8	Pyruvic acid and NO_3^-	PdCu nanobead wire	2 M KNO_3, 100 mM pyruvic acid, −1.7 V (vs. ag/AgCl), flow rate 30 mL·min^{-1}	[83]

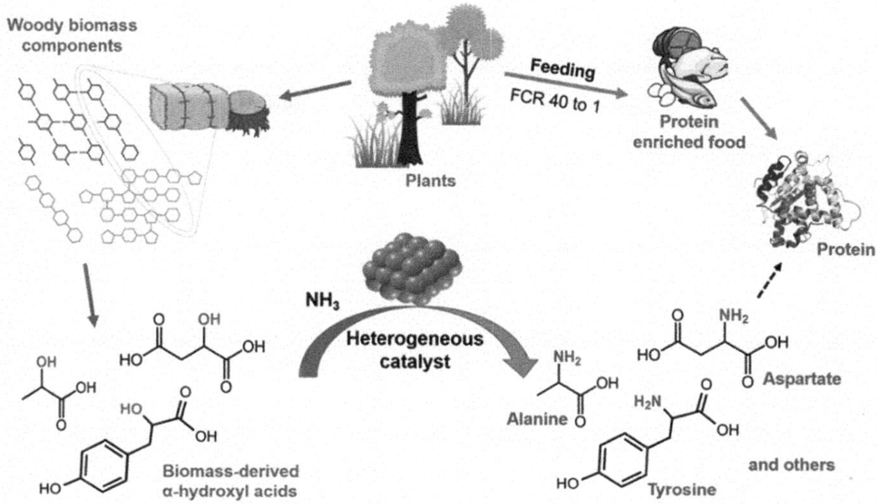

Fig. 1.10 Catalytic conversion of α-hydroxy acids from biomass into amino acids [79]. (Reprinted with permission from [79], Copyright © 2018 National Academy of Sciences)

α-amino acids via the heterogeneous catalyst following the dehydrogenation-reductive amination [79]. There are two possible routes for dehydrogenation-reductive amination. In route I, the occurrence of dehydrogenation in α-hydroxyl acids firstly transforms hydroxyl acids into α-ketonic acid, which creates an imine when it interacts with ammonia. The imine is finally converted into α-amino acids via hydrogenation. In another route, S_N2 substitution of the ^-OH group by a $^-NH_2$ group occurs in the α-hydroxy acids, which subsequently transforms into α-amino acids through dehydrogenation. For both of the reaction routes, it has been demonstrated that the dehydrogenation is the rate-determining step, thus placing the significant demands on dehydrogenation catalysts. Due to their well-established capacity for catalyzing dehydrogenation and hydrogenation reactions, noble metals, such as Pt, Pd as well as Ru, are set as the vital candidates. However, among the several noble metal nanoparticles including Ru, Rh, Pt, Ir, and Pd, only Ru-based catalysts exhibit a higher efficiency in conversion of lactic acid to aqueous ammonia. Further analysis reveals the presence of NH_3 results in that Ru displays a distinctive enhancement effect on dehydrogenation activity, leading to superior performance in the reaction of dehydrogenation-reductive amination. On the contrary, the capacity of additional noble metal-based catalysts to dehydrate is suppressed by NH_3.

Based on the above findings, multiple Ru-based catalysts have been developed for the conversion of α-hydroxyl acids to α-amino acids in a thermocatalytic system (Table 1.2). For example, Deng et al. loaded Ru nanoparticles onto carbon nanotube (CNTs) to construct a Ru/CNT catalyst, which efficiently catalyzed the lactic acid to alanine in NH_3 aqueous solution at 1 MPa H_2 and 493 K. After a 2-h reaction, the maximum yield of alanine reached 62% [79]. Following this, substantial research endeavors were directed toward enhancing α-amino acid generation efficiency by optimizing Ru-based catalysts. Xie et al. embellished Ru nanoparticles on N-doped carbon nanotubes and obtained a Ru/N-CNTs catalyst, the maximum alanine yield of which is improved to 70% at 180 °C and 10 bar H_2 for 2 h [80].

Photocatalytic conversion

Notably, the dehydrogenation in dehydrogenation-reductive amination of α-hydroxyl acids needs to overcome an energy barrier via high temperature in thermocatalytic conversion system, which not only increases the energy consumption but also is prone to generating the side products. To this end, the photocatalytic conversion system was developed in 2020 for transform α-hydroxy acids into α-amino acids under mild conditions [82]. Using a CdS nanosheet semiconductor, Song et al. performed α-hydroxyl acids conversion via photocatalytic for α-amino acid production, in which the alanine productivity converting from lactic acid reached 10.5 mmol h^{-1} $g_{cat}{}^{-1}$ in the condition of 50 °C and 1 bar N_2 [82]. In comparison to thermocatalytic conversion, only inert gases of N_2 at low pressure is required to flush into photocatalytic reaction system, rather than high-pressure hydrogen. Additionally, the reaction temperature is lower, thus significantly decreasing energy consumption and representing a green method. The reaction pathway of α-hydroxyl acids to α-amino acid by photocatalysis was similar to that of thermocatalytic conversion. Specifically, the process of α-hydroxyl acids dehydrogenation is activated by photogenerated holes, which transform α-hydroxyl acids into α-ketonic acid. The occurrence of self-addition of α-ketonic acids with NH_3 results in the formation of α-imines. The α-imines are finally reduced to α-amino acids by photogenerated electrons.

A group of α-amino acids, such as α-aminobutyric acid, alanine, valine, phenylalanine, as well as leucine, were successfully synthesized via photocatalysis from biomass-derived hydroxyl acids (e.g. phenyllactic acid, α-hydroxyisocaproic acid, α-hydroxybutyric acid, and α-hydroxy-3-methylbutyric acid) over CdS nanosheets (Table 1.2). Note that the productivity of α-amino acids in photocatalytic conversion is higher than that of the thermocatalytic conversion, while maximal the α-amino acids yield of photocatalytic conversion (<30%) was lower. This might result from the existence of hydrogen evolution reaction in photocatalytic conversion system, which serves as the major side reaction consuming photogenerated electrons.

Electrocatalytic conversion

The electrocatalytic conversion for α-amino acids synthesis generally employs α-ketonic acid (e.g. Pyruvic acid) as a substrate. For example, Wu et al. conducted the alanine electrosynthesis utilizing pyruvic acid and $NO_3{}^-$ as feedstock, which was catalyzed by a PdCu nanobead wire in cathode with a posing potential of −1.7 V (vs. Ag/AgCl) [83]. This reaction is a tandem catalytic process. In the PdCu catalyst, Cu demonstrates remarkable proficiency in reducing $NO_3{}^-$, yet it exhibits limited hydrogenation activity. And the Pd displays the strong hydrogenation capabilities. When loading the catalyst of PdCu on cathode, the $NO_3{}^-$ in cathode electrolyte is first electro-reduced to the $^*NH_2OH$ intermediate catalyzed by Cu. The intermediate of *NH_2OH subsequently reacts with another substrate, pyruvic acid, leading to the formation of pyruvic oxime. Finally, the hydrogenation of pyruvic oxime is conducted via electroreduction catalyzed by Pd, ultimately producing the alanine. Based on these findings, this work further regulated the metal components in PdCu catalysts, and reflected that the ideal proportion of Pd to Cu was 1:1, which performed alanine electrocatalytic synthesis with the maximum yield of 54.8% [83].

1.5 Conclusions and Future Outlook

Biomass catalytic transformations represent eco-friendly and sustainable synthetic routes for the production of organic acids, which have developed remarkably with significant outcomes. Multifunctional catalysts are essential for creating new, environmentally friendly processes that transform biomass into various kinds of organic acids. Reaction mechanisms provide a comprehensive understanding of the roles played by catalysts and enrich chemistry to cleave C-C and/or C-O bonds in a regioselective manner. The connections between performance and structure, particularly in activity and selectivity, are also examined. It provides ideas for future catalyst design. In short, numerous new catalysts and catalytic systems have been introduced developed to convert renewable biomass feedstocks, especially those from agricultural residues and carbohydrate. The conversion of raw agricultural wastes faces more challenges in terms of technical stability and simplicity compared to platform compounds. The challenges are from three aspects: (1) The complex compositions, involving varying content ratios and properties, result in inconsistent catalytic conversion efficiency, making it difficult to unify reaction conditions and necessitating multi-functional catalysts; (2) The extremely stable lignocellulose often makes pretreatment necessary to improve its reactivity before catalytic conversion; (3) A small amount of ash and protein in agricultural wastes may cause side reactions and reduce selectivity of desired products. For the production of some monocarboxylic acids, like formic acid, the efficiency of the transformation is no longer an important challenge. The main challenge is to come up with an economically viable process that has low costs, simple handling, easy product separation, and catalyst recycling [85]. For most organic acids, the development of efficient, cost-effective, and homo-/heterogeneous catalytic systems will remain a priority for a long time in the future. Suitable design is also an important proposition to be able to ensure the activity and selectivity of the catalyst under mild conditions. Additionally, the valuable pretreatment technologies for biomass before conversion could be warranted to investigate. In terms of environmental sustainability, by-products and waste streams generated during the synthesis of organic acids should also be effectively managed.

Acknowledgments Apart from the efforts of authors, we would also like to acknowledge the individuals listed below for providing plenty of information and giving many constructive comments that significantly improved the quality of the chapter.

Dr. Haixiao Guo, Tianjin University of Commerce, Tianjin, People's Republic of China.

Dr. Baocai Zhang, Tianjin University, Tianjin, People's Republic of China.

Dr. Chen Chen, Tianjin University of Commerce, Tianjin, People's Republic of China.

References

1. Li S, Deng W, Wang S, Wang P, An D, Li Y, Zhang Q, Wang Y. Catalytic transformation of cellulose and its derivatives into functionalized organic acids. ChemSusChem. 2018;11(13): 1995–2028. https://doi.org/10.1002/cssc.201800440.
2. Zhang Z, Huber GW. Catalytic oxidation of carbohydrates into organic acids and furan chemicals. Chem Soc Rev. 2018;47(4):1351–90. https://doi.org/10.1039/c7cs00213k.
3. Wang S, Dai G, Yang H, Luo Z. Lignocellulosic biomass pyrolysis mechanism: a state-of-the-art review. Prog Energy Combust Sci. 2017;62:33–86. https://doi.org/10.1016/j.pecs.2017.05.004.
4. Yan W, Guan Q, Jin F. Catalytic conversion of cellulosic biomass to harvest high-valued organic acids. iScience. 2023;26(10):107933. https://doi.org/10.1016/j.isci.2023.107933.
5. Zhou C-H, Xia X, Lin C-X, Tong D-S, Beltramini J. Catalytic conversion of lignocellulosic biomass to fine chemicals and fuels. Chem Soc Rev. 2011;40(11):5606–30. https://doi.org/10.1039/c1cs15124j.
6. Chen Y, Yang Y, Liu X, Shi X, Wang C, Zhong H, Jin F. Sustainable production of formic acid and acetic acid from biomass. Mol. Catal. 2023;545:113199. https://doi.org/10.1016/j.mcat.2023.113199.
7. Huo J, Shanks BH. Bioprivileged molecules: integrating biological and chemical catalysis for biomass conversion. Annu Rev Chem Biomol Eng. 2020;11(1):63–85. https://doi.org/10.1146/annurev-chembioeng-101519-121127.
8. Lin H, Su J, Liu Y, Yang L. Catalytic conversion of lignocellulosic biomass to value-added organic acids in aqueous media. In: Fan J, editor. Application of hydrothermal reactions to biomass conversion. Berlin, Heidelberg, New York: Springer; 2014. p. 109–38. https://doi.org/10.1007/978-3-642-54458-3_5.
9. Jin F, Yun J, Li G, Kishita A, Tohji K, Enomoto H. Hydrothermal conversion of carbohydrate biomass into formic acid at mild temperatures. Green Chem. 2008;10:612–5. https://doi.org/10.1039/B802076K.
10. Xu J, Zhang H, Zhao Y, Yang Z, Yu B, Xu H, Liu Z. Heteropolyanion-based ionic liquids catalysed conversion of cellulose into formic acid without any additives. Green Chem. 2014;16: 4931–5. https://doi.org/10.1039/C4GC01252F.
11. Wei G-H, Lu T, Liu H-Y, Bai J-X, Wang Q, Li G-Y, Liang Y-H. Exploring the continuous cleavage-oxidation mechanism of the catalytic oxidation of cellulose to formic acid: a combined experimental and theoretical study. Fuel. 2023;341:127667. https://doi.org/10.1016/j.fuel.2023.127667.
12. Wang W, Wang YP, Wu Y, Yang X, Wu Y, Liu Q, Ren S, Marsh KN. Catalytic conversion of biomass-derived carbohydrates to formic acid using molecular oxygen. Green Chem. 2014;16: 719–26. https://doi.org/10.1039/C4GC00145A.
13. Jin F, Kishita A, Moriya T, Enomoto H, Sato N. A new process for producing Ca/Mg acetate deicer with Ca/Mg waste and acetic acid produced by wet oxidation of organic waste. Chem Lett. 2002;31:88–9. https://doi.org/10.1246/cl.2002.88.

14. Coman SM, Verziu M, Tirsoaga A, Jurca B, Teodorescu C, Kuncser V, Parvulescu VI, Scholz G, Kemnitz E. NbF_5–AlF_3 catalysts: design, synthesis, and application in lactic acid synthesis from cellulose. ACS Catal. 2015;5:3013–26. https://doi.org/10.1021/acscatal.5b00282.
15. Dusselier M, Van Wouwe P, Dewaele A, Makshina E, Sels BF. Lactic acid as a platform chemical in the biobased economy: the role of chemocatalysis. Energy Environ Sci. 2013;6(5):1415–42. https://doi.org/10.1039/c3ee00069a.
16. Li L, Shen F, Smith RL, Qi X. Quantitative chemocatalytic production of lactic acid from glucose under anaerobic conditions at room temperature. Green Chem. 2017;19:76–81. https://doi.org/10.1039/C6GC02443B.
17. Holm M, Taarning E. Conversion of sugars to lactic acid derivatives using heterogeneous zeotype catalysts. Science. 2010;328(5978):602–5. https://doi.org/10.1126/science.1183990.
18. Yan X, Jin F, Tohji K, Kishita A, Enomoto H. Hydrothermal conversion of carbohydrate biomass to lactic acid. AICHE J. 2010;56(10):2727–33. https://doi.org/10.1002/aic.12193.
19. Liu L, Li Z, Hou W, Shen H. Direct conversion of lignocellulose to levulinic acid catalyzed by ionic liquid. Carbohydr Polym. 2018;181:778–84. https://doi.org/10.1016/j.carbpol.2017.11.078.
20. Park Y, Jeong G-T. Production of levulinic acid from macroalgae by hydrothermal conversion with ionic resin catalyst. Bioresour Technol. 2024;402:130778. https://doi.org/10.1016/j.biortech.2024.130778.
21. Yu F, Thomas J, Smet M, Dehaen W, Sels BF. Molecular design of sulfonated hyperbranched poly(arylene oxindole)s for efficient cellulose conversion to levulinic acid. Green Chem. 2016;18(6):1694–705. https://doi.org/10.1039/c5gc01971k.
22. Meinita MDN, Amron A, Trianto A, Harwanto D, Caesarendra W, Jeong G-T, Choi J-S. Levulinic acid production from macroalgae: production and promising potential in industry. Sustain For. 2021;13(24):13919. https://doi.org/10.3390/su132413919.
23. Velasco Calderon JC, Jiang S, Mushrif SH. Understanding the effect of solvent environment on the interaction of hydronium ion with biomass derived species: a molecular dynamics and metadynamics investigation. ChemPhysChem. 2021;22(21):2222–30. https://doi.org/10.1002/cphc.202100485.
24. Choudhary V, Mushrif SH, Ho C, Anderko A, Nikolakis V, Marinkovic NS, Frenkel AI, Sandler SI, Vlachos DG. Insights into the interplay of Lewis and Bronsted acid catalysts in glucose and fructose conversion to 5-(hydroxymethyl)furfural and levulinic acid in aqueous media. J Am Chem Soc. 2013;135(10):3997–4006. https://doi.org/10.1021/ja3122763.
25. Zuo Y, Zhang Y, Fu Y. Catalytic conversion of cellulose into levulinic acid by a sulfonated chloromethyl polystyrene solid acid catalyst. ChemCatChem. 2014;6(3):753–7. https://doi.org/10.1002/cctc.201300956.
26. Jin F. Application of hydrothermal reactions to biomass conversion. Berlin, Heidelberg, New York: Springer; 2014. https://doi.org/10.1007/978-3-642-54458-3.
27. Chae TU, Ahn JH, Ko Y-S, Kim JW, Lee JA, Lee EH, Lee SY. Metabolic engineering for the production of dicarboxylic acids and diamines. Metab Eng. 2020;58:2–16. https://doi.org/10.1016/j.ymben.2019.03.005.
28. Huber GW, Iborra S, Corma A. Synthesis of transportation fuels from biomass: chemistry, catalysts, and engineering. Chem Rev. 2006;106(9):4044–98. https://doi.org/10.1021/cr068360d.
29. Wan Y, Lee J-M. Toward value-added dicarboxylic acids from biomass derivatives via thermocatalytic conversion. ACS Catal. 2021;11(5):2524–60. https://doi.org/10.1021/acscatal.0c05419.
30. Chadwick SS. Ullmann's encyclopedia of industrial chemistry. Ref Serv Rev. 1988;16(4):31–4. https://doi.org/10.1108/eb049039.
31. Beverskog B, Puigdomenech I. Revised pourbaix diagrams for copper at 25 to 300°C. J Electrochem Soc. 1997;144(10):3476–83. https://doi.org/10.1149/1.1838036.

32. Ventura M, Williamson D, Lobefaro F, Jones MD, Mattia D, Nocito F, Aresta M, Dibenedetto A. Sustainable synthesis of oxalic and succinic acid through aerobic oxidation of C6 polyols under mild conditions. ChemSusChem. 2018;11(6):1073–81. https://doi.org/10.1002/cssc.201702347.
33. Jiang Z, Zhang Z, Song J, Meng Q, Zhou H, He Z, Han B. Metal-oxide-catalyzed efficient conversion of cellulose to oxalic acid in alkaline solution under low oxygen pressure. ACS Sustain Chem Eng. 2016;4(1):305–11. https://doi.org/10.1021/acssuschemeng.5b01212.
34. Li X, Ho B, Lim DSW, Zhang Y. Highly efficient formic acid-mediated oxidation of renewable furfural to maleic acid with H_2O_2. Green Chem. 2017;19(4):914–8. https://doi.org/10.1039/c6gc03020c.
35. Dente M, Pierucci S, Tronconi E, Cecchini M, Ghelfi F. Selective oxidation of n-butane to maleic anhydride in fluid bed reactors: detailed kinetic investigation and reactor modelling. Chem Eng Sci. 2003;58(3–6):643–8. https://doi.org/10.1016/s0009-2509(02)00590-0.
36. Rodenas Y, Mariscal R, Fierro JLG, Martín Alonso D, Dumesic JA, López Granados M. Improving the production of maleic acid from biomass: TS-1 catalysed aqueous phase oxidation of furfural in the presence of γ-valerolactone. Green Chem. 2018;20(12):2845–56. https://doi.org/10.1039/c8gc00857d.
37. Shi S, Guo H, Yin G. Synthesis of maleic acid from renewable resources: catalytic oxidation of furfural in liquid media with dioxygen. Catal Commun. 2011;12(8):731–3. https://doi.org/10.1016/j.catcom.2010.12.033.
38. Huang Y, Wu C, Yuan W, Xia Y, Liu X, Yang H, Wang H. Catalytic aerobic oxidation of biomass-based furfural into maleic acid in aqueous phase with metalloporphyrin catalysts. J Chin Chem Soc. 2017;64(7):786–94. https://doi.org/10.1002/jccs.201700004.
39. Shi N, Liu Q, Zhang Q, Wang T, Ma L. High yield production of 5-hydroxymethylfurfural from cellulose by high concentration of sulfates in biphasic system. Green Chem. 2013;15(7):1960–71. https://doi.org/10.1039/c3gc40667a.
40. Wang M, Ma J, Liu H, Luo N, Zhao Z, Wang F. Sustainable productions of organic acids and their derivatives from biomass via selective oxidative cleavage of C–C bond. ACS Catal. 2018;8(3):2129–65. https://doi.org/10.1021/acscatal.7b03790.
41. Lan J, Lin J, Chen Z, Yin G. Transformation of 5-Hydroxymethylfurfural (HMF) to maleic anhydride by aerobic oxidation with heteropolyacid catalysts. ACS Catal. 2015;5(4):2035–41. https://doi.org/10.1021/cs501776n.
42. Paone E, Mauriello F. New catalytic routes for the sustainable production of food preservatives from biomass. Chem Catal. 2022;2(9):2122–4. https://doi.org/10.1016/j.checat.2022.08.002.
43. Gu S, Zhao Z, Yao Y, Li J, Tian C. Designing and constructing a novel artificial pathway for malonic acid production biologically. Front Bioeng Biotechnol. 2021;9:820507. https://doi.org/10.3389/fbioe.2021.820507.
44. Yan L, Qi X. Degradation of cellulose to organic acids in its homogeneous alkaline aqueous solution. ACS Sustain Chem Eng. 2014;2(4):897–901. https://doi.org/10.1021/sc400507s.
45. Becker J, Lange A, Fabarius J, Wittmann C. Top value platform chemicals: bio-based production of organic acids. Curr Opin Biotechnol. 2015;36:168–75. https://doi.org/10.1016/j.copbio.2015.08.022.
46. Fu J, Vasiliadou ES, Goulas KA, Saha B, Vlachos DG. Selective hydrodeoxygenation of tartaric acid to succinic acid. Cat Sci Technol. 2017;7(21):4944–54. https://doi.org/10.1039/c7cy01374d.
47. Harmer MA, Sun Q. Solid acid catalysis using ion-exchange resins. Appl Catal A Gen. 2001;221(1):45–62. https://doi.org/10.1016/s0926-860x(01)00794-3.
48. Choudhary H, Nishimura S, Ebitani K. Metal-free oxidative synthesis of succinic acid from biomass-derived furan compounds using a solid acid catalyst with hydrogen peroxide. Appl Catal A Gen. 2013;458:55–62. https://doi.org/10.1016/j.apcata.2013.03.033.
49. Davis SE, Zope BN, Davis RJ. On the mechanism of selective oxidation of 5-hydroxymethylfurfural to 2,5-furandicarboxylic acid over supported Pt and au catalysts. Green Chem. 2012;14(1):143–7. https://doi.org/10.1039/c1gc16074e.

50. Rao M, Wang M, Zheng X, Pan D, Hong C, Chen K, Lin L. One-pot preparation of CuOx-CoOy heterostructure decorated on graphene for highly efficient oxidation of 5-hydroxymethylfurfural (5-HMF) to 2,5-furandicarboxylic acid (FDCA). Chem Eng Sci. 2022;262:118004. https://doi.org/10.1016/j.ces.2022.118004.
51. Hansen TS, Sádaba I, García-Suárez EJ, Riisager A. Cu catalyzed oxidation of 5-hydroxymethylfurfural to 2,5-diformylfuran and 2,5-furandicarboxylic acid under benign reaction conditions. Appl Catal A Gen. 2013;456:44–50. https://doi.org/10.1016/j.apcata.2013.01.042.
52. Gu D, Jia C-J, Weidenthaler C, Bongard H-J, Spliethoff B, Schmidt W, Schüth F. Highly ordered mesoporous cobalt-containing oxides: structure, catalytic properties, and active sites in oxidation of carbon monoxide. J Am Chem Soc. 2015;137(35):11407–18. https://doi.org/10.1021/jacs.5b06336.
53. Szabo T, Berkesi O, Forgo P, Josepovits K, Sanakis Y, Petridis D, Dekany I. Evolution of surface functional groups in a series of progressively oxidized graphite oxides. Chem Mater. 2006;18:2740–9. https://doi.org/10.1021/cm060290w.
54. Ma J, Zhong L, Peng X, Sun R. d-Xylonic acid: a solvent and an effective biocatalyst for a three-component reaction. Green Chem. 2016;18(6):1738–50. https://doi.org/10.1039/c5gc01727k.
55. Jin X, Zhao M, Vora M, Shen J, Zeng C, Yan W, Thapa PS, Subramaniam B, Chaudhari RV. Synergistic effects of bimetallic PtPd/TiO_2 nanocatalysts in oxidation of glucose to glucaric acid: structure dependent activity and selectivity. Ind Eng Chem Res. 2016;55:2932–45. https://doi.org/10.1021/acs.iecr.5b04195.
56. Guo S, Fang Q, Li Z, Zhang J, Zhang J, Li G. Efficient base-free direct oxidation of glucose to gluconic acid over TiO_2-supported gold clusters. Nanoscale. 2019;11:1326–34. https://doi.org/10.1039/c8nr08282f.
57. Zhang Q, Wan Z, Yu IKM, Tsang DCW. Sustainable production of high-value gluconic acid and glucaric acid through oxidation of biomass-derived glucose: a critical review. J Clean Prod. 2021;312:127745. https://doi.org/10.1016/j.jclepro.2021.127745.
58. Guo X, Guo F, Li Y, Zheng Z, Xing Z, Zhu Z, Liu T, Zhang X, Jin Y. Dehydration of D-xylose into furfural over bimetallic salts of heteropolyacid in DMSO/H_2O mixture. Appl Catal A Gen. 2018;558:18–25. https://doi.org/10.1016/j.apcata.2018.03.027.
59. Naidu DS, Hlangothi SP, John MJ. Bio-based products from xylan: a review. Carbohydr Polym. 2018;179:28–41. https://doi.org/10.1016/j.carbpol.2017.09.064.
60. Donoeva B, Masoud N, Jongh PE. Carbon support surface effects in the gold-catalyzed oxidation of 5-Hydroxymethylfurfural. ACS Catal. 2017;7(7):4581–91. https://doi.org/10.1021/acscatal.7b01456.
61. Wang YR, Van de Vyver S, Sharma KK, Roman-Leshkov Y. Insights into the stability of gold nanoparticles supported on metal oxides for the base-free oxidation of glucose to gluconic acid. Green Chem. 2014;16:719–26. https://doi.org/10.1039/C3GC42074F.
62. Zhang J, Liu X, Hedhili MN, Zhu Y, Han Y. Highly selective and complete conversion of cellobiose to gluconic acid over Au/Cs2HPW12O40 nanocomposite catalyst. ChemCatChem. 2011;3:1294–8. https://doi.org/10.1002/cctc.201100137.
63. An D, Ye A, Deng W, Zhang Q, Wang Y. Selective conversion of cellobiose and cellulose into gluconic acid in water in the presence of oxygen, catalyzed by polyoxometalate-supported gold nanoparticles. Chem Eur J. 2012;18:2938–47. https://doi.org/10.1002/chem.201102228.
64. Rafaïdeen T, Baranton S, Coutanceau C. Highly efficient and selective electrooxidation of glucose and xylose in alkaline medium at carbon supported alloyed PdAu nanocatalysts. Appl Catal B Environ. 2019;243:641–56. https://doi.org/10.1016/j.apcatb.2018.10.014.
65. Zhou C, Deng W, Wan X, Zhang Q, Yang Y, Wang Y. Functionalized carbon nanotubes for biomass conversion: the base-free aerobic oxidation of 5-hydroxymethyl-furfural to 2,5-furandicarboxylic acid over platinum supported on a carbon nanotube catalyst. ChemCatChem. 2015;7:2853–63. https://doi.org/10.1002/cctc.201500516.

66. Deng W, Wang Y, Yan N. Production of organic acids from biomass resources. Curr Opin Green Sustain Chem. 2016;2:54–8. https://doi.org/10.1016/j.cogsc.2016.10.002.
67. Eblagon KM, Pereira MFR, Figueiredo JL. One-pot oxidation of cellobiose to gluconic acid. Unprecedented high selectivity on bifunctional gold catalysts over mesoporous carbon by integrated texture and surface chemistry optimization. Appl Catal B Environ. 2016;184:381–96. https://doi.org/10.1016/j.apcatb.2015.12.023.
68. Jin C, Ma L, Sun W, Han P, Tan X, Wu H, Liu M, Jin H, Wu Z, Wei H, Sun C. Single-atom nickel confined nanotube superstructure as support for catalytic wet air oxidation of acetic acid. Commun Chem. 2019;2:1–7. https://doi.org/10.1038/s42004-019-0190-9.
69. Li Z, Li D, Ma J, Zou R, Li X, Liu C, Peng X. Highly selective oxidation of monosaccharides to sugar acids by nickel-embedded carbon nanotubes under mild conditions. Renew Energy. 2021;175:650–9. https://doi.org/10.1016/j.renene.2021.05.003.
70. Amaniampong PN, Trinh QT, De Oliveira Vigier DQ, Dao NH, Tran Y, Wang MP, Sherburne FJ. Synergistic effect of high-frequency ultrasound with cupric oxide catalyst resulting in a selectivity switch in glucose oxidation under argon. J Am Chem Soc. 2019;141:14772–9. https://doi.org/10.1021/jacs.9b07626.
71. Rautiainen S, Lehtinen P, Vehkamäki M, Niemelä K, Kemell M, Heikkilä TR. Microwave-assisted base-free oxidation of glucose on gold nanoparticle catalysts. Catal Commun. 2016;74: 115–8. https://doi.org/10.1016/j.catcom.2016.01.003.
72. Zhuge Y, Fan G, Lin Y, Yang L, Li F. A hybrid composite of hydroxyapatite and Ca-Al layered double hydroxide supported Au nanoparticles for highly efficient base-free aerobic oxidation of glucose. Dalton Trans. 2019;48:9161–72. https://doi.org/10.1039/c9dt01367h.
73. Lee J, Jung S, Kim YT, Kim HJ, Kim K-H. Catalytic and electrocatalytic conversion of glucose into value-added chemicals. Renew Sust Energ Rev. 2023;181:113337. https://doi.org/10.1016/j.rser.2023.113337.
74. Lee J, Saha B, Vlachos DG. Pt catalysts for efficient aerobic oxidation of glucose to glucaric acid in water. Green Chem. 2016;18(13):3815–22. https://doi.org/10.1039/c6gc00460a.
75. Bin D, Wang H, Li J, Wang H, Yin Z, Kang J, He B, Li Z. Controllable oxidation of glucose to gluconic acid and glucaric acid using an electrocatalytic reactor. Electrochim Acta. 2014;130: 170–8. https://doi.org/10.1016/j.electacta.2014.02.128.
76. Ibert MI, Merbouh N, Bruckner C. Determination of the side-products formed during the nitroxide-mediated bleach oxidation of glucose to glucaric acid. Carbohydr Res. 2002;337: 1059–63. https://doi.org/10.1016/s0008-6215(02)00043-7.
77. Okatsu H, Kinoshita N, Akita T, Ishida T, Haruta M. Deposition of gold nanoparticles on carbons for aerobic glucose oxidation. Appl Catal A Gen. 2009;369(1–2):8–14. https://doi.org/10.1016/j.apcata.2009.08.013.
78. Comotti M, Pina CD, Rossi M. Mono- and bimetallic catalysts for glucose oxidation. J Mol Catal A Chem. 2006;251(1–2):89–92. https://doi.org/10.1016/j.molcata.2006.02.014.
79. Deng W, Wang Y, Zhang S, Gupta KM, Hülsey MJ, Asakura H, Liu L, Han Y, Karp EM, Beckham GT, Dyson PJ, Jiang J, Tanaka T, Wang Y, Yan N. Catalytic amino acid production from biomass-derived intermediates. Proc Natl Acad Sci USA. 2018;115(20):5093–8. https://doi.org/10.1073/pnas.1800272115.
80. Xie Z, Chen B, Peng F, Liu M, Liu H, Yang G, Han B. Highly efficient synthesis of amino acids by amination of bio-derived hydroxy acids with ammonia over Ru supported on N-doped carbon nanotubes. ChemSusChem. 2020;13(21):5683–9. https://doi.org/10.1002/cssc.202001561.
81. Xin H, Xiu Z, Liu S, Wang H, Wang C, Ma L, Liu Q. Efficient conversion of lactic acid to alanine over noble metal supported on Ni@C catalysts. RSC Adv. 2022;12(26):16847–59. https://doi.org/10.1039/d2ra02514k.
82. Song S, Qu J, Han P, Hulsey MJ, Zhang G, Wang Y, Wang S, Chen D, Lu J, Yan N. Visible-light-driven amino acids production from biomass-based feedstocks over ultrathin CdS nanosheets. Nat Commun. 2020;11(1):4681. https://doi.org/10.1038/s41467-020-18532-3.

83. Wu J, Xu L, Kong Z, Gu K, Lu Y, Wu X, Zou Y, Wang S. Integrated tandem electrochemical-chemical-electrochemical coupling of biomass and nitrate to sustainable alanine. Angew Chem Int Ed Engl. 2023;62(45):e202311196. https://doi.org/10.1002/anie.202311196.
84. Ding M, Song S, Li X. A perspective on renewable production of amino acids from biomass through the chemocatalytic method. Green Chem. 2024;26:4468–76. https://doi.org/10.1039/d4gc00846d.
85. Chen X, Liu Y, Wu J. Sustainable production of formic acid from biomass and carbon dioxide. Mol Catal. 2020;483:110716. https://doi.org/10.1016/j.mcat.2019.110716.

Chapter 2
Reductive Conversion of Extrinsic Functional Groups in Biomass-Derived Carboxylic Acids

Yoshinao Nakagawa, Mizuho Yabushita, and Keiichi Tomishige

Abstract Biomass-derived carboxylic acids frequently contain functional groups such as C=O, OH and furan ring, and the selective removal with retention of carboxyl groups is a useful reaction in biomass valorization. This chapter summarizes the catalytic reduction of such oxygen-containing functional groups in biomass-derived carboxylic acids with H_2. Reduction of C=O in levulinic acid is relatively easy with simple metal catalysts to γ-valerolactone by hydrogenation and cyclization and with metal + acid catalysts to valeric acid by hydrodeoxygenation. However, these approaches cannot be applied to wide range of substrates. The use of hydrogen halides as catalysts or co-catalysts with metal is a more general approach, applicable to hydrodeoxygenation of lactic acid, furancarboxylic acids and sugar acids to the corresponding simple acids and diacids, although the corrosive nature of hydrogen halides is a problem. Specific catalysts have been reported for each type of substrates, such as Mo catalysts for lactic acid, Pt-MoO_x/TiO_2 catalyst for furancarboxylic acids, and noble metal + Re catalysts for sugar acids.

Keywords Heterogeneous catalysis · Hydrogenolysis · Hydrogenation · Bifunctional catalyst

2.1 Introduction

Biomass resources except lipids contain large amount of oxygen atoms. Many biomass-derived carboxylic acids, especially those produced from sugars or lignocellulose including agricultural residues by fermentation or chemical processes, also contain large number of oxygen atoms in addition to the carboxyl groups [1]. Such oxygen atoms are sometimes unnecessary, especially for the utilization as fuel additives and monomers, and should be removed by reduction reactions. Among the reductants, molecular hydrogen (H_2) is the best one in practical view, because of

Y. Nakagawa (✉) · M. Yabushita · K. Tomishige
Department of Applied Chemistry, School of Engineering, Tohoku University, Sendai, Japan
e-mail: yoshinao@erec.che.tohoku.ac.jp

Z. Fang et al. (eds.), *Production of Organic Acids and Alcohols from Agricultural Residues and Food Wastes*, Biofuels and Biorefineries 13,
https://doi.org/10.1007/978-981-96-1222-2_2

the low cost, high atom efficiency and potential production from renewable sources. Here, we summarize the systems for the reduction of biomass-derived carboxylic acids at the extrinsic oxygen-containing functional groups with H_2. Chemical reduction with H_2 requires catalysts, and control of the chemo- and regio-selectivity strongly relies on the catalyst development. Each following section explains effective catalysts for each type of substrates: ketocarboxylic acids in Sect. 2.2, hydroxycarboxylic acids in Sect. 2.3, furancarboxylic acids in Sect. 2.4 and polyhydroxycarboxylic acids in Sect. 2.5. Sections 2.2 and 2.3 mainly explain the reaction of very representative substrate in each category, namely levulinic acid and lactic acid, respectively. The production of each substrate from biomass resources and the main utilization of the reduction products are described in the begining of each section or subsection. Reactions in later sections tend to be more difficult and require more sophisticated catalysts. Except for the reduction of ketocarboxylic acids, most effective catalysts have been developed recently, in 2010s and 2020s.

2.2 Reduction of Ketone Group: Levulinic Acid Hydrogenation/Hydrodeoxygenation

The most typical molecule that contains both carboxyl and ketone groups is levulinic acid (IUPAC name: 4-oxopentanoic acid). Levulinic acid can be synthesized from hexoses by dehydration and hydrolysis in water solvent with acid catalysts (Fig. 2.1) [2–4]. 5-Hydroxymethylfurfural, which is frequently abbreviated as HMF, is an intermediate of levulinic acid formation and is also a very important platform chemical. Synthesis of levulinic acid from pentose is also possible via dehydration to furfural, hydrogenation to furfuryl alcohol and then hydration-rearrangement. Based on the huge potential of supply, levulinic acid has been regarded as one of the top candidates of platform chemicals in biomass upgrading [1, 5].

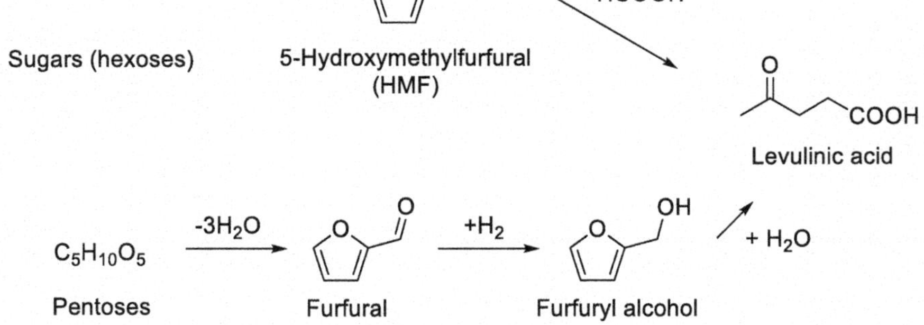

Fig. 2.1 Production of levulinic acid from sugars

2.2.1 Reduction of Levulinic Acid to γ-Valerolactone

The reduction of ketone first produces alcohol. In the case of levulinic acid, the hydrogenation product is 4-hydroxyvaleric acid (IUPAC name: 4-hydroxypentanoic acid). This hydroxyacid is easily dehydrated to γ-valerolactone (GVL) (Fig. 2.2). GVL is a useful chemical that can be utilized as a food additive, a green solvent, a fuel additive, and so on [6, 7]. The reduction of levulinic acid or its esters is the only practical method to synthesize GVL. Generally, ketone groups are more reactive than carboxyl groups toward hydrogenation. Therefore, the catalysts for hydrogenation of levulinic acid to GVL are not special. Wide range of active metals in groups 8–11 in periodic table can be used for the hydrogenation catalysts. In view of the catalyst cost, Ni and Cu catalysts have been mainly tested [8]. The selection of supports is also not critical: conventional supports such as Al_2O_3, SiO_2, TiO_2 and ZrO_2 can be used. Acidic supports are also acceptable, although alcohol solvent should not be combined to suppress ether formation. Strongly basic supports are generally not good because the substrate is a carboxylic acid. The typical reaction conditions are 373–473 K temperature and 1–5 MPa H_2. Noble metal catalysts such as Ru can decrease the reaction temperature below 373 K. The hydrogenation to GVL can proceed almost quantitatively. Catalyst stability is a major issue in practical view. Water solvent, which is usually used in the synthesis of levulinic acid, can rather promote the leaching of catalytically active species, especially non-noble metals [9]. Keeping the metallic state of active element is necessary to prevent the leaching.

GVL molecule has one asymmetric carbon atom (γ- (4-) position) while levulinic acid is a symmetric molecule. Optically pure GVL is present in nature and is important in fragrance and pharmaceutical uses [10]. Enantioselective hydrogenation of levulinic acid to optically pure GVL has been investigated mainly with homogeneous Ru catalysts [10, 11]. SEGPHOS (2*H*,2'*H*-[4,4'- -Bi-1,3-benzodioxole]-5,5′-diyl)bis(diphenylphosphane)) [10] and BINAP (bis (diphenylphosphino)-1,1′-binaphthyl) [12] ligands have been reported to be effective to obtain excellent enantioselectivity (99% ee) (Fig. 2.3a). The systems using homogeneous Ni catalysts have been recently reported (up to 95% ee; Fig. 2.3b) [13, 14]. The use of heterogeneous catalysts for enantioselective hydrogenation of levulinic acid has been tested, with modification of the catalyst surface with chiral ligands; however, the enantioselectivity was still moderate (up to 60% ee) [15].

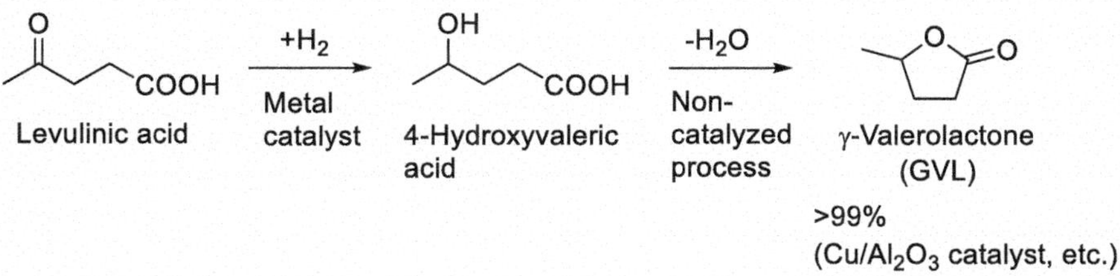

Fig. 2.2 Hydrogenation of levulinic acid to γ-valerolactone (GVL)

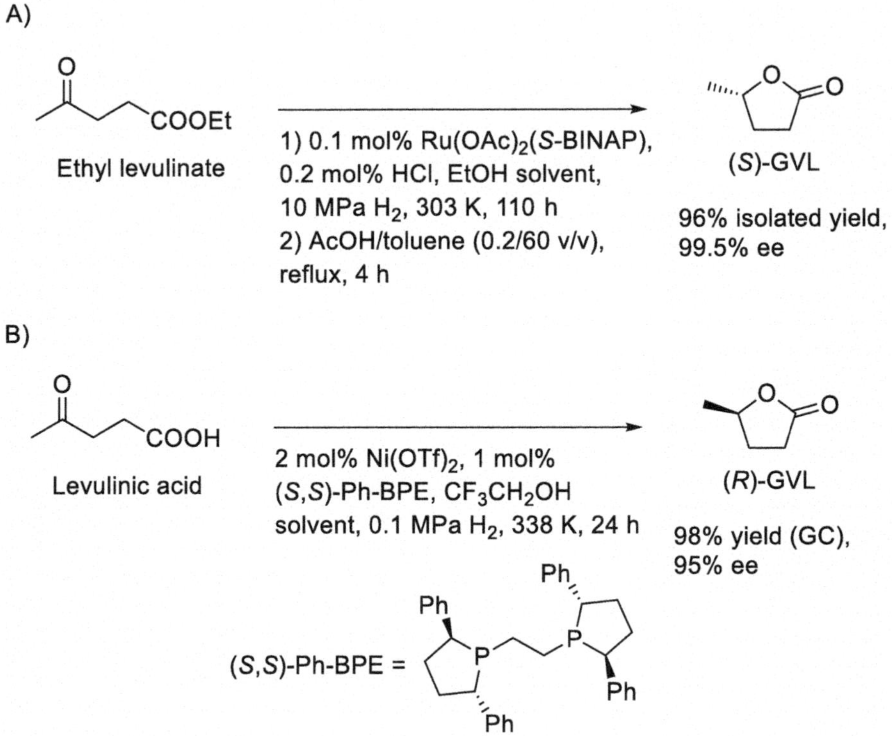

Fig. 2.3 Synthesis of optically pure (*S*)-GVL and (*R*)-GVL [12, 13]

2.2.2 Reduction of Levulinic Acid or γ-Valerolactone to Valeric Acid

Total reduction of the ketone group in levulinic acid produces valeric acid (IUPAC name: pentanoic acid). Valeric acid is not a main component of triglycerides, and hydrodeoxygenation of levulinic acid is the easiest method to synthesize renewable valeric acid. Esters of valeric acid have been used as perfumes, and in addition they can be components of transportation fuels [16]. The reaction route to valeric acid from levulinic acid is summarized in Fig. 2.4 [17]. γ-Valerolactone (GVL) is first produced in levulinic acid reduction. Next, the ring-opening of GVL to pentenoic acids occurs, and then the pentenoic acids are hydrogenated to valeric acid. The ring-opening reaction is catalyzed by Brønsted acid, and the hydrogenation proceeds over metal catalysts. The combination of strong acid such as H-ZSM-5 zeolite with metal is a common approach to develop effective catalysts (Table 2.1) [18–24]. The strong acid also catalyzes the dehydration of levulinic acid to angelica lactones; however, they are anyway converted to GVL by hydrogenation. Because of the formation of olefinic intermediates (pentenoic acids and angelica lactones), coking is a severe problem in the synthesis of valeric acid. Noble metal catalysts

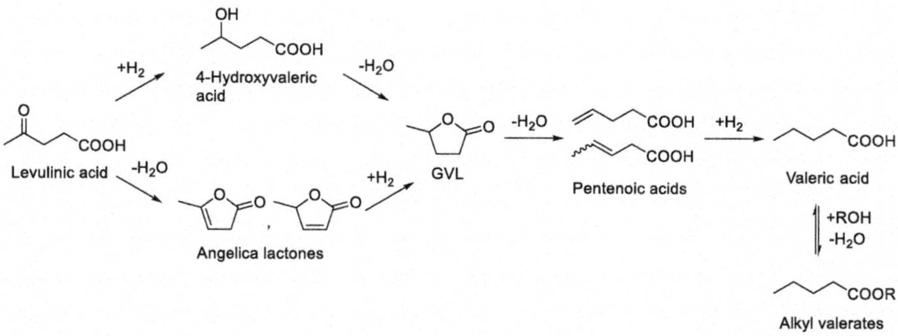

Fig. 2.4 Reaction route of valeric acid production from levulinic acid over metal + acid catalysts

Table 2.1 Selected reported systems of levulinic acid synthesis from levulinic acid

Catalyst	Solvent	$P(H_2)$/MPa	Temp./K	Product (selectivity/%)	Ref.
Ru/SBA-SO_3H	EtOH	4	523	VA + EV (90)	[18]
Ru/H-ZSM-5	1,4-Dioxane	4	473	VA (66), EV (25)	[19]
Pt/H-ZSM-5	Neat	0.8	473	VA (99)	[20]
Pt/H-ZSM-5	Water	1	473	VA (91)	[21]
Pd/H-ZSM-5	EtOH	4	513	EV (91)	[22]
Co/H-ZSM-5	EtOH	3	483	VA (23), EV (74)	[23]
Ni/H-ZSM-5	EtOH	3	513	VA + EV (90)	[24]
Nb-Cu/ZrO_2	Water	3	423	VA (97)	[25]
Ni-P	Toluene	5	623	VA (94)	[27]
Pd/C + $Hf(OTf)_4$	*n*-octane	5	423	VA (92)	[30]

VA valeric acid; *EV* ethyl valerate

are typically used to rapidly hydrogenate the pentenoic acid intermediate. The acidity control such as that by neutralization of too strong acid sites has been also investigated to increase the catalyst life [24]. Typical reaction conditions are about 473–523 K temperature and 1–5 MPa H_2. The solvents used in reported systems include water, 1,4-dioxane, alkanes and alcohols. Solventless systems have been also reported. Alcohols have been more used in the literature; however, in this case the esters are the main products. Systems using non-noble metal catalysts more frequently used alcohol solvents. A recent report enabled the synthesis of valeric acid from aqueous levulinic acid by using Nb-Cu catalysts [25]. Niobium oxide is known as water-tolerant acid catalyst and is used as unique support for many hydrodeoxygenation reactions [26]. Nickel or cobalt phosphide catalysts have been also investigated for the synthesis of valeric acid, based on their good resistance to acidic conditions [27, 28].

Historically, the process for the valeric acid production from levulinic acid was first proposed as cascade ones composed of GVL production from levulinic acid and transformation of GVL to valeric acid as separated steps [29]. One merit of GVL as substrate is the easy application of a gas-phase flow system due to the lower boiling point of GVL than that of levulinic acid. The direct conversion systems of levulinic acid usually used a batch reactor system. The valeric acid yield reached over 90% from both levulinic acid or GVL [17]. Another merit of GVL as substrate is that the amount of acidic compounds in the reactor can be decreased, especially when an alcohol solvent is used to obtain the ester instead of free valeric acid.

The approach using metal catalysts supported on strong acids has not been applied to wide scope of substrates. The severe reaction conditions prevent attracting attention by researchers of organic synthesis. Yao Fu and co-workers reported that the combination of Pd/C and metal triflate as a heterogenous hydrogenation catalyst and homogeneous Lewis acid catalysts, respectively, can effectively catalyze the ring-opening and hydrogenation of lactones to straight-chain carboxylic acids at relatively lower temperature (around 408 K) [30]. This type of catalysts had been developed for the hydrodeoxygenation of ketones or ethers to alkanes [31]. The research group of Yao Fu further generalized the reaction system of Pd/C and metal triflate catalysts to hydrodeoxygenation of ketones or ethers at γ- or more distant positions of carboxylic acids to give alkyl carboxylic acids [32].

2.3 Reduction of α-Hydroxy Group of Carboxylic Acids: Hydrodeoxygenation of Lactic Acid

The approach of the combination of metal and strong acid catalysts in Sect. 2.2.2 does not work well for the hydrodeoxygenation of C-O bond at the α-position of carboxyl group. This is because the carbonyl in carboxyl group increases the acidity of α-hydrogen (-C<u>H</u>(OH)-COOH) and then strengthens the C-O bond. The most typical molecule with such structure is lactic acid. The hydrodeoxygenation of lactic acid to propanoic acid, which is currently industrially manufactured from petroleum and used in many fields [33], is a difficult reaction. Most studies for the lactic acid reduction rather targeted the reduction of the carboxyl group to 1,2-propanediol which can proceed over various metal catalysts [34], and for the reaction of the hydroxy group of lactic acid the oxidation to pyruvic acid has been a main topic [34, 35].

2.3.1 Dehydration of Lactic Acid to Acrylic Acid

For simple alcohol substrates, the hydrodeoxygenation by the metal + acid catalysts is composed of acid-catalyzed dehydration and metal-catalyzed hydrogenation. The problem here is the difficulty in the dehydration of lactic acid to acrylic acid (Fig. 2.5) with simple acid catalysts. Nevertheless, acrylic acid is an important chemical in petrochemistry, such as the manufacture of sodium polyacrylate as superabsorbent polymer, and there has been large interest in the synthesis of acrylic

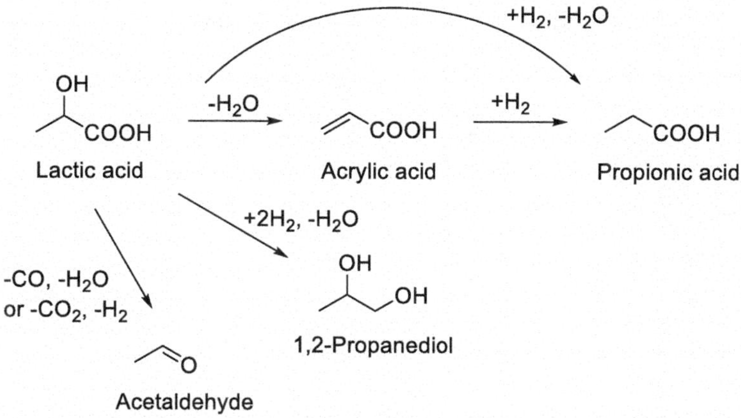

Fig. 2.5 Reaction of conversion of lactic acid over acid and/or metal catalysts under H_2

acid from biomass [36–38]. Lactic acid is produced in large scale by fermentation and can be also synthesized chemically [39]. The transformation of lactic acid to acrylic acid is an attractive synthesis method of acrylic acid because it does not consume costly reagents and only co-produces water. While acrylic acid itself is a valuable compound, further hydrogenation of acrylic acid to propionic acid is an easy hydrogenation reaction, and the overall process of the dehydration and the hydrogenation can be regarded as the hydrodeoxygenation of lactic acid to propionic acid. Other methods to synthesize acrylic acid or propionic acid from biomass include dehydration and oxidation of glycerol to acrylic acid [40, 41].

Simple acids cannot dehydrate lactic acid to acrylic acid. Too strong acid sites promote decarbonylation or decarboxylation reaction to give acetaldehyde [41]. Acid-base bifunctional catalysts are necessary to remove hydroxy group and hydrogen atom at 2- and 3-positions, respectively, at the same time. However, the base site strongly binds the carboxyl group of lactic acid molecule. The typical catalysts contain alkali or alkali-earth metal cations as mobile basic centers. The lactic acid molecule is adsorbed as metal lactate, which is then dehydrated to acrylate (Fig. 2.6). On the other hand, metal lactate is proposed as a key species for another major byproduct, 2,3-pentanedione, which is formed by decarboxylative condensation of two lactic acid molecules [42]. The reported catalysts for lactic acid dehydration to acrylic acid include alkali earth phosphates [43–45], barium sulfate [46, 47], and alkali salts of zeolites [48–51] (Table 2.2). The reaction has been carried out at high temperature (>573 K) under flow conditions. Recent studies have tended to use alkali metal-zeolite catalysts such as Na-Y and Na-ZSM-5. The performance of zeolite catalysts much depends on the structure type, silica-alumina ratio, and composition of counter-cations. The best catalysts showed over 70% selectivity to acrylic acid, and the selectivity was kept high at total conversion of lactic acid. Generally, the zeolite catalysts showed deactivation during hours of uses, but they can be regenerated by calcination [48–51].

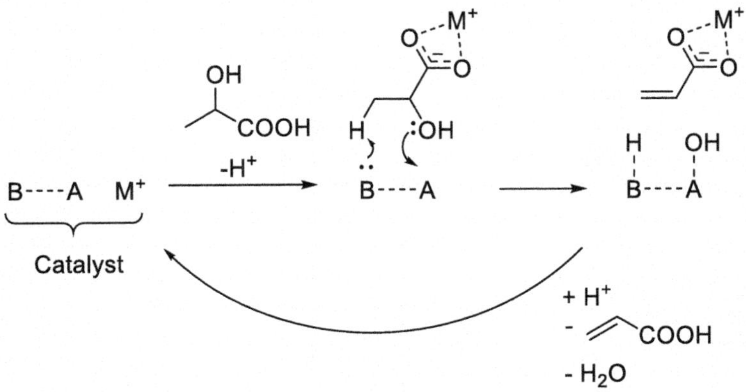

Fig. 2.6 Dehydration of lactic acid to acrylic acid over acid-base catalysts containing alkali metal (M^+). A: acid site; B: base site

Table 2.2 Selected reported systems of lactic acid dehydration

Catalyst	Conc. of lactic acid	Flow conditions	Temp./ K	Conv./ %	Acrylic acid selectivity/ %	Ref.
$Ca_5(PO_4)_3OH$	38% aq.	*W*/*F* 167 g h mol^{-1} (WHSV$_{\text{lactic acid}}$ 0.54 h^{-1})	623	90	78	[43]
$Ca_2P_2O_7$	50% aq.	WHSV 3 h^{-1}	648	100	74	[44]
$Ba_2P_2O_7$	20% aq.	WHSV 1.8 h^{-1}	673	99.7	76	[45]
$BaSO_4$	20% aq.	WHSV 1.8 h^{-1}	673	99.8	74	[46]
$BaSO_4$	20% aq.	WHSV 2.8 h^{-1}	673	100	81	[47]
Na-Y	10% aq.	WHSV 6 h^{-1}	623	100	75	[48]
KOH-Ca-NaY	20% aq.	WHSV 0.48 h^{-1}	623	100	84	[49]
$K_{0.92}Na_{0.03}$ZSM-5	36% aq.	WHSV$_{\text{lactic acid}}$ 2.1 h^{-1}	633	96	81	[50]
Na-ZSM-5	10% aq.	LHSV 0.67 h^{-1}	598	100	79	[51]

W/*F* ratio of catalyst weight (*W*) to lactic acid flow rate (*F*); *WHSV* weight hourly space velocity; *LHSV* liquid hourly space velocity

2.3.2 One-Step Production of Propionic Acid from Lactic Acid

As described above, metal-catalyzed hydrodeoxygenation usually reduces the lactic acid molecule at the carboxyl group. In order to selectively and reductively dissociate the single C-O(H) bond at the 2-position of lactic acid, there has been two approaches. One approach is the reverse Mars van Krevelen mechanism. Reverse Mars van Krevelen mechanism is composed of two steps: the reduced catalyst reacts with substrate to give product and oxidized catalyst, and then the oxidized catalyst is reduced by the reductant (Fig. 2.7) [52]. In the case of lactic acid deoxygenation, the

Fig. 2.7 Reverse Mars van Krevelen mechanism [52]

$$\begin{cases} R\text{-}OH + MO_{x-1} \longrightarrow R\text{-}H + MO_x \\ MO_x + H_2 \longrightarrow MO_{x-1} + H_2O \end{cases}$$

Table 2.3 Selected reported heterogeneous catalytic systems of lactic acid deoxygenation to propionic acid

Catalyst	Feed	Temp./K	Conv./%	Propionic acid selectivity/%	Ref.
Fe_3O_4	20% aq., flow, N_2	663	94.3	57.5	[53]
MoO_3	20% aq., flow, N_2	663	96.8	62.8	[54]
MoO_3	10% aq., batch, H_2 (3 MPa)	488	86.2	55.1	[55]
MoS_2	10% aq., batch, H_2 (3 MPa)	488	59.1	81.4	[55]
MoS_2/MoO_3	10% aq., batch, H_2 (3 MPa)	488	90.4	88.4	[55]
Pd/MoO_3	10% aq., batch, H_2 (3 MPa)	488	81.3	92.4	[56]

reduced catalyst contains anion vacancy sites, which are filled with oxygen atom originated from the hydroxy group. The C-OH bond at the 2-position is weaker than the other carbon-oxygen bonds in lactic acid, and therefore this bond is preferably dissociated. An important point is that the reduction step of lactic acid does not involve activated hydrogen species. If an activated hydrogen species has a central role of the reduction reaction, the hydrogen species preferably reacts with the C=O double bond to reduce the carboxyl group. Typical active catalysts for the lactic acid deoxygenation are iron oxide, molybdenum oxide and molybdenum sulfide (Table 2.3) [53–56]. Molybdenum-based catalysts are more selective to propionic acid. Molybdenum sulfide catalysts are more selective than molybdenum oxide catalysts; however, the use of sulfide catalysts has a risk of contamination of product stream with H_2S. In the case of molybdenum oxide catalysts, the reduction of Mo with H_2 is not fast; rather, the lactic acid molecule itself works as the reducing agent for MoO_3. The consumption of lactic acid as the reductant for MoO_3 decreases the selectivity of deoxygenation. To increase the activity and selectivity of molybdenum oxide catalysts, addition of Pd was reported [56]. Pd catalysts have high activation ability of H_2, but the hydrogenation activity of carboxyl group is very low. The addition of Pd can accelerate the reduction step of catalyst with H_2 to suppress the consumption of substrate as reducing agent. However, the Pd-added catalyst was less stable: the activity was gradually decreased in reuses.

Another approach of hydrodeoxygenation of lactic acid to propionic acid is the use of iodide anion (I^-) [57]. The detailed reaction route is shown in Fig. 2.8. The reaction was carried out in acetic acid solvent containing NaI catalyst. The lactic acid molecule is first acetylated at the hydroxy group to 2-acetoxypropionic acid. Next, nucleophilic substitution with I^- produces 2-iodopropionic acid, and then propionic

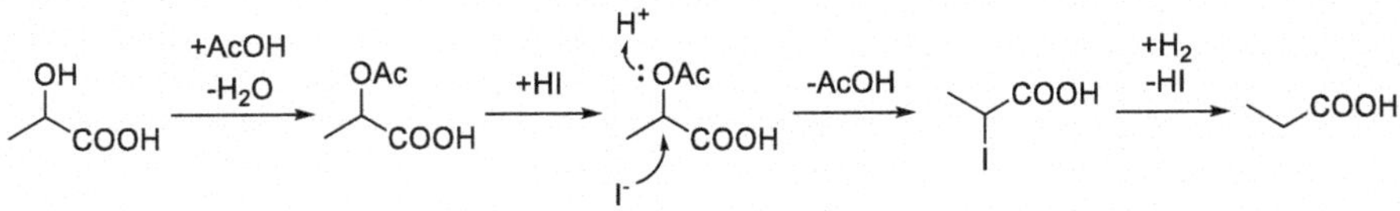

Fig. 2.8 Lactic acid hydrodeoxygenation mediated by iodide ion in acetic acid

acid is formed by C-I hydrogenolysis. The hydrogenolysis of 2-iodopropionic acid is a fast non-catalytic radical-chain reaction (confirmed by radical-trapping agent), and therefore only 2-acetoxypropionic acid was observed as an intermediate during the reaction. Very high yield of propionic acid was obtained by this reaction (even >99%; conditions: 0.25 M lactic acid in acetic acid, 1 equiv. NaI, 2.8 MPa H_2, 493 K, 8 h). Glycolic acid (IUPAC name: 2-hydroxyacetic acid) could be converted to acetic acid in >99% yield under similar conditions (0.25 M glycolic acid in propionic acid, 1 equiv. NaI, 2.8 MPa H_2, 493 K, 1 h). Similar systems have been used for the hydrodeoxygenation of more complex molecules [58], which will be explained in each section of this chapter. The problem of this system is the handling of corrosive homogeneous HI leading to high capital cost.

2.4 Reduction of Furan Ring or Tetrahydrofuran Ring with Carboxy Group

Compounds with a furan ring are important class of biomass-derived chemicals, because furfural and HMF (5-hydroxymethylfurfural) can be synthesized by dehydration of sugars (Fig. 2.1). The furan ring has aromatic nature, and the hydroxymethy group in HMF is easily oxidized like that in benzyl alcohols. The aldehyde groups in furfural and HMF are also easily oxidized, and therefore 2-furancarboxylic acid (abbreviated as FCA; "furoic acid" is another name) and 2,5-furandicarboxylic acid (abbreviated as FDCA) are relatively easily obtained from biomass (Fig. 2.9) [59, 60]. In particular, the polymer of FDCA and ethylene glycol, called as PEF (poly(ethylene 2,5-furandicarboxylate)), has attracted much attention as an alternative to poly(ethylene terephthalate) (PET) resin [61]. Considering the large demand of PET, it is likely that FDCA will be produced in large scale in near future.

While there are numerous reports on the reduction of furfural, HMF and their condensates with other molecules [62–64], reduction of FCA and FDCA has been less investigated. Nevertheless, the reduction of furan ring in FCA and FDCA without reduction of the carboxyl group is meaningful because the products of such reactions cannot be produced by reduction of furfural or HMF [65].

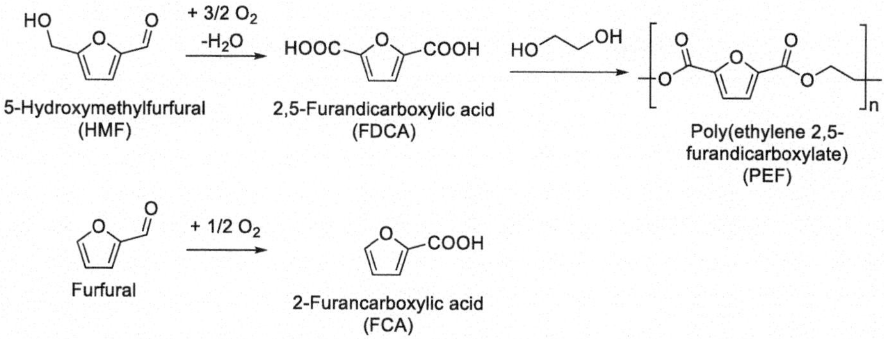

Fig. 2.9 Biomass-derived furancarboxylic acids

Fig. 2.10 Hydrogenation of furancarboxylic acids

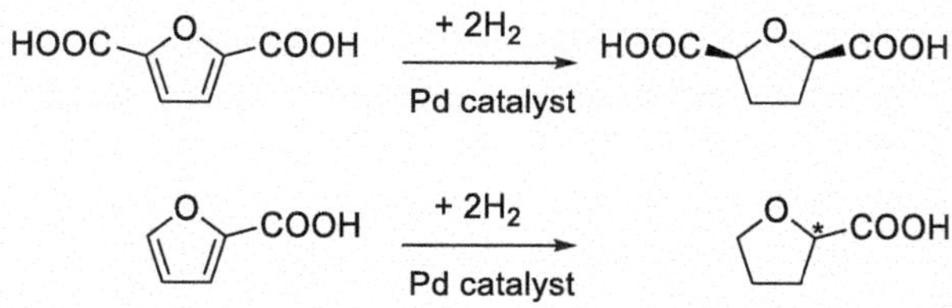

2.4.1 Hydrogenation of the Furan Ring in FCA and FDCA

Generally, hydrogenation of C=C bonds is easier than that of carboxyl group, even if the C=C bonds have aromatic nature. The hydrogenation of the furan ring in FCA and FDCA produces tetrahydrofuran-2-carboxylic acid and tetrahydrofuran-2,5-dicarboxylic acid, respectively (Fig. 2.10). These saturated carboxylic acids have stereoisomers, because the carbon atoms bonding the carboxyl group are asymmetric, while FCA and FDCA do not have asymmetric carbon atoms.

Typical catalysts for hydrogenation of furan ring are supported Pd such as Pd/C. Palladium is known to have high activity in C=C hydrogenation and very low activity in dissociation of C-O and C-C bonds; these properties are suitable to the selective hydrogenation at the C=C bonds. The hydrogenation proceeds almost quantitatively [66]. For the FDCA hydrogenation, the product is only *cis*-isomer of tetrahydrofuran-2,5-dicarboxylic acid, based on the general *syn* hydrogenation mechanism over metal surfaces. For the hydrogenation of FCA, the product is racemic mixture. There have been only a few reports testing asymmetric hydrogenation of FCA to tetrahydrofuran-2-carboxylic acid, and the reported optical purity was not high (up to 42% ee) [67]. The synthesis of optically pure tetrahydrofuran-2-carboxylic acid is easier by kinetic resolution of the racemic mixture of its ester with enzymatic hydrolysis [68, 69].

2.4.2 Hydrogenolysis of Tetrahydrofuran-2-Carboxylic Acid and Tetrahydrofuran-2,5-Dicarboxylic Acid

Further reduction of tetrahydrofuran-2-carboxylic acid and tetrahydrofuran-2,5-dcarboxylic acid can give useful compounds (Fig. 2.11). Among the potential products, 5-hydroxyvaleric acid (or its lactone form, δ-valerolactone (DVL)) and adipic acid are useful because they have functional groups at both ends of the carbon chain. Adipic acid has been industrially manufactured from petroleum in large scale as a monomer of polyamides, and the development of synthesis route from biomass is desired [70, 71]. Hydrodeoxygenation of FDCA explained here is one promising route, and other possible routes include production of K-A oil (a precursor of adipic acid in petrochemistry) from lignin [72], hydrodeoxygenation of sugar acids which will be explained later, oxidative C-C cleavage reactions of 1,2-difunctionalized cyclohexanes [73], and fermentative synthesis [74].

The hydrogenolysis of tetrahydrofuran-2-carboxylic acid and tetrahydrofuran-2,5-dicarboxylic acid to 5-hydroxyvaleric acid and adipic acid, respectively, require dissociation of C-O bonds at the α-position of the carboxyl group. Such reactions are difficult like the case of lactic acid. Furthermore, in the synthesis of 5-hydroxyvaleric acid, the reactions at the δ-position and those at the carboxyl group itself must be suppressed. No effective catalysts have been reported with such selectivity. On the other hand, the selective C-O hydrogenolysis at the α-position of the terminal functional group is possible in the case of tetrahydrofurfuryl alcohol (THFA) where the carboxyl group of tetrahydrofuran-2-carboxylic acid is replaced with –

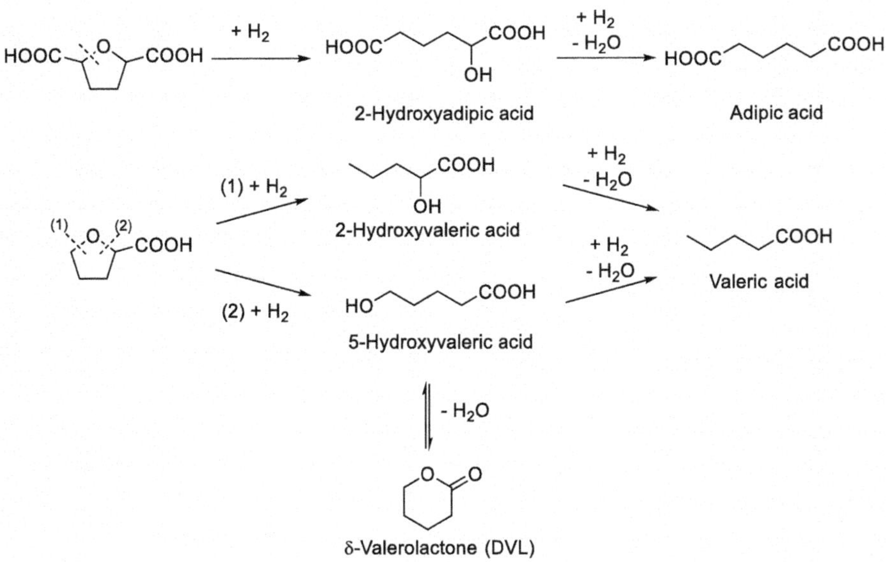

Fig. 2.11 Hydrogenolysis products of tetrahydrofuran-2,5-dicarboxylic acid and tetrahydrofuran-2-carboxylic acid

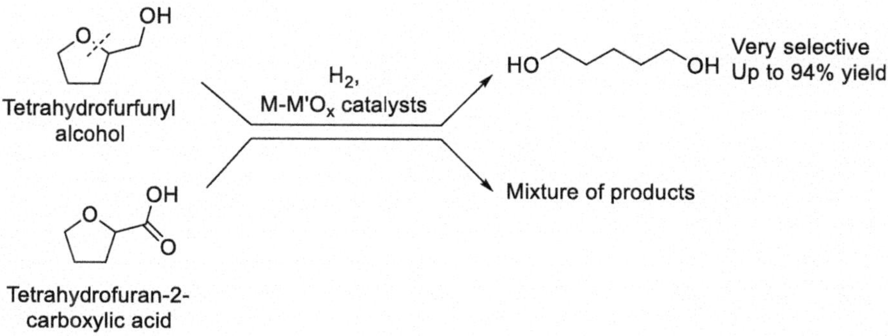

Fig. 2.12 Comparison between hydrogenolysis reactions of tetrahydrofurfuryl alcohol and tetrahydrofuran-2-carboxylic acid. *M* noble metal (Rh, Ir, Pt); *M'* oxophilic transition metal (Re, Mo, W)

CH_2OH group (Fig. 2.12) [75–77]. The selective hydrogenolysis of THFA to 1,5-pentanediol uses metal-oxide-modified noble metal catalysts (M-M'O_x; M = Rh, Ir, Pt; M' = Re, Mo, W) or Ni-LnO_x (Ln: rare-earth metal) catalysts, and more than 90% yields of 1,5-pentanediol were obtained with both types of catalysts. The M-M'O_x type catalysts were tested for the reduction of tetrahydrofuran-2-carboxylic acid [78]. Surely the ratio of 1,5-difunctionalized products to 1,2-difunctionalized products was high; however, the selectivity to 5-hydroxyvaleric acid + DVL was low because of the hydrogenation of the carboxyl group. Both the C-O hydrogenolysis and the carboxyl group hydrogenation involves hydride-like activated hydrogen species, and the reactivity of carboxyl group is higher. The synthesis of DVL is rather easier by oxidation of 1,5-pentanediol [79], Baeyer-Villiger oxidation of cyclopentanone [80], and direct reduction of FCA which will be explained later in this chapter. Cyclopentanone as well as 1,5-pentanediol can be produced by reduction of furfural [62, 81].

The synthesis of adipic acid from tetrahydrofuran-2,5-dicarboxylic acid requires the dissociation of two C-O bonds. This reaction is easier than that of 5-hydroxyvaleric acid (+ DVL) formation from tetrahydrofuran-2-carboxylic acid because the regioselectivity is not necessary. The use of iodide anion is an effective reaction route, which has been recently intensively investigated by Gilkey et al. [82–85]. Earlier reports used HI as iodide source. In order to avoid the use of highly corrosive HI, the combinations of alkali iodide and solid acids have been tested in recent works. The combination of Nafion and LiI gave adipic acid yield of 87% (conditions: 1 wt% substrate concentration in propionic acid, effective proton 0.3 M of Nafion, 0.3 M LiI, 3.4 MPa H_2, 433 K, 2 h) [84]. Propionic acid is a good solvent in this iodide-mediated hydrodeoxygenation. Like the case of lactic acid hydrodeoxygenation in acetic acid, propionic acid reversibly reacted with 2-hydroxyadipic acid, the partially-reduced intermediate, to form the ester. However, the substrate does not directly react with the organic acid solvent. Nevertheless, the organic acid solvent is essential in the conversion of tetrahydrofuran-2,5-dicarboxylic acid. Organic acid solvents likely play a role in tuning the acid/base

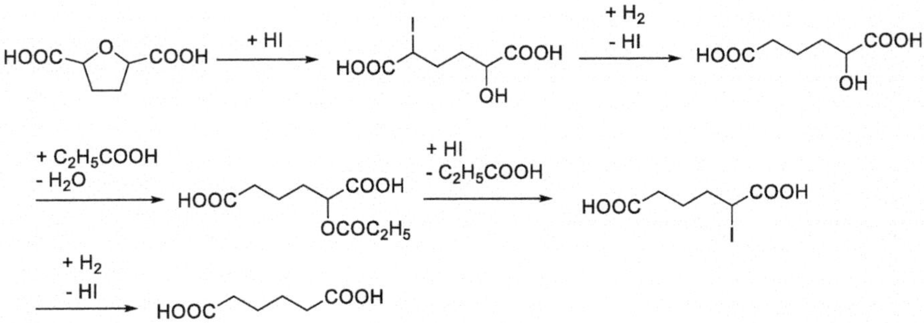

Fig. 2.13 Hydrodeoxygenation of tetrahydrofuran-2,5-dicarboxylic acid with acid + I^- catalyst system

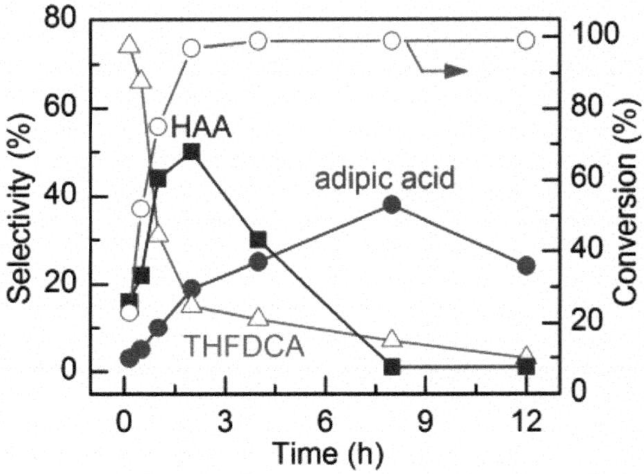

Fig. 2.14 Time course of 2,5-furandicarboxylic acid (FDCA) reduction over $Pt/Nb_2O_5{\cdot}xH_2O$ catalyst. Reaction conditions: FDCA 0.10 mmol, catalyst 20 mg, H_2O 5 mL, H_2 3 MPa, 473 K. *THFDCA* tetrahydrofuran-2,5-dicarboxylic acid; *HAA* 2-hydroxyadipic acid. (Reprinted with permission from ref. [86], Copyright © 2019 The Royal Society of Chemistry)

properties of the system, to promote the protonation at the oxygen atom in the tetrahydrofuran ring of the substrate molecule (Fig. 2.13).

Another approach to adipic acid from tetrahydrofuran-2,5-dicarboxylic acid is the use of acid-metal bifunctional $Pt/Nb_2O_5{\cdot}xH_2O$ catalyst reported by Wei et al. [86]. As mentioned in Sect. 2.2.2, niobium oxide is a good solid acid catalyst that can work in water solvent. The time course of the reaction of FDCA over the $Pt/Nb_2O_5{\cdot}xH_2O$ catalyst is shown in Fig. 2.14. The main product changed: first tetrahydrofuran-2,5-dicarboxylic acid, 2-hydroxyadipic acid, and then adipic acid. The result means that the hydrogenation of the furan ring occurred first, and then the hydrogenolysis of C-O bonds proceeded step by step. However, the highest adipic acid yield (38%) was not high, because of the overreaction of adipic acid. Another problem of this reaction system was the too low concentration of FDCA (about 0.3 wt% aq.) because of the low solubility of FDCA in water.

2.4.3 Hydrogenolysis of Furan Ring Before Hydrogenation to Tetrahydrofuran Ring

Some metal catalysts, especially Pt, have activity in direct ring-opening hydrogenolysis of furanic compounds [77, 87]. The catalyst does not need acidic property, in contrast to the ring-opening reaction over Pt/$Nb_2O_5{\cdot}xH_2O$ catalyst mentioned above. The ring-opening occurs only for the substrates with a furan ring, and compounds with a tetrahydrofuran ring instead of the furan ring are not reactive. Ring-opening hydrogenolysis of FCA to 5-hydroxyvaleric acid and its lactone form (DVL) was first reported in 1949 by using Adams catalyst (hydrated PtO_2) and acetic acid solvent [88], but details of the conditions and numerical results were not provided. This reaction had been forgotten until recent days when the conversion of biomass-derived compounds attracts much attention. In 2019, two research groups almost at the same time published detailed study of ring-opening hydrogenolysis of FCA to 5-hydroxyvaleric acid or its derivatives [89, 90]. The effect of active metal was tested in both reports, and Pt was much superior in view of ring opening hydrogenolysis than other noble metals which mainly catalyzed ring hydrogenation to tetrahydrofuran-2-carboxylic acid. The group of Tomishige and co-workers investigated the effects of supports for Pt catalysts, solvents and reaction conditions [89]. For the support, Al_2O_3, ZrO_2, CeO_2, TiO_2, SiO_2 and C were tested, and Pt/Al_2O_3 showed the best activity and selectivity to 5-hydroxyvaleric acid and its derivatives (DVL and ester of 5-hydroxyvaleric acid). The solvent effect was investigated with Pt/Al_2O_3 catalyst. Acetic acid and methanol were good solvent in view of selectivity. Acetic acid, which was also used in the old paper, showed the highest selectivity to DVL (66%); however, the activity of Pt/Al_2O_3 was lower in acetic acid than those in other solvents such as alcohols and water. Competitive adsorption of acetic acid and FCA molecule, both of which are carboxylic acids, on the catalyst surface may decrease the reaction rate. In alcohol solvent, the target product became the mixture of DVL, 5-hydroxyvaleric acid and ester of 5-hydroxyvavleric acid. The sum of selectivity of Pt/Al_2O_3 catalyst to these target products in various solvents decreased in the following order: methanol (62%) > ethanol (57%) > 1-propanol (54%) > 2-propanol (51%) > *t*-butyl alcohol (44%) > 1,4-dioxane (39%) ~ THF (38%) ~ water (37%). Except water, the solvents with high polarity seem to be better. The γ-Al_2O_3 support is well known to be transformed to boehmite (AlOOH) phase in hot water solvent, and the phase transformation may influence the performance. On the other hand, the group of Haichao Liu and co-workers reported the reaction in water solvent [90]. The tested supports for the Pt catalysts were SiO_2, TiO_2 and ZrO_2. The Pt/SiO_2 catalyst showed the highest selectivity to the target product, 5-hydroxyvaleric acid. The highest yield of 5-hydroxyvaleric acid reached 78%. The best results obtained by the two studies are summarized in Fig. 2.15. The synthesis of 5-hydroxyvaleric acid or DVL by FCA reduction has advantage over other biomass-based methods such as oxidation of 1,5-pentanediol and Baeyer-Villiger oxidation of cyclopentanone in view of reagent cost: Furfural to DVL via FCA requires 3 equiv. of H_2 and 1/2 equiv. of

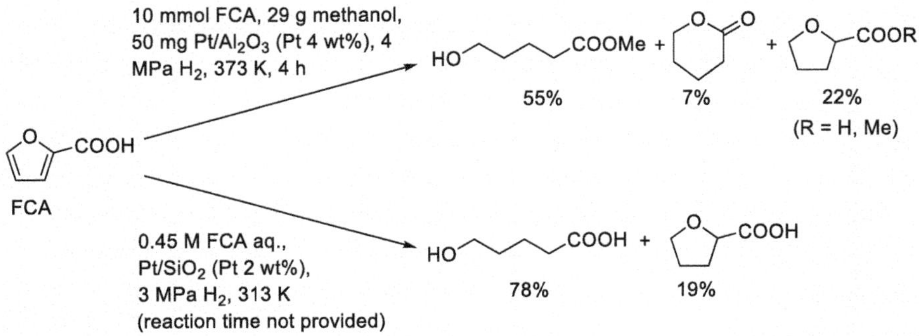

Fig. 2.15 Ring-opening hydrogenolysis of 2-furancarboxylic acid (FCA) [89, 90]

O_2; furfural to DVL via 1,5-pentanediol requires 4 equiv. of H_2 and 1 equiv. of O_2; furfural to DVL via cyclopentanone requires 3 equiv. of H_2 and 1 equiv. of peroxide.

In both reports, the main by-product over the best catalysts (Pt/Al_2O_3 and Pt/SiO_2) was tetrahydrofuran-2-carboxylic acid (or its ester, in alcohol solvents) in all solvents and under all conditions. Ring-opening hydrogenolysis of tetrahydrofuran-2-carboxylic acid did not occur at all over both catalysts. The product of dissociation of the other C-O bond in the furan ring (2-hydroxyvaleric acid) was not formed at all in any runs.

Ring-opening hydrogenolysis has been recently intensively investigated for the reduction of furfuryl alcohol to pentanediol [87, 91]. There are several similarities between ring-opening hydrogenolysis reactions of FCA and furfuryl alcohol: the effectiveness of Pt catalysts; ring-hydrogenation as a side reaction; no reactivity of tetrahydrofuran derivatives; and water or alcohol solvent. There are also different points for these ring-opening hydrogenolysis reactions: furfuryl alcohol hydrogenolysis has effectiveness of Cu- and Co-based catalysts in addition to Pt-based ones, co-formation of 1,2-pentanediol and 1,5-pentanediol in almost all systems, and positive role of basic supports. Although not realized, development of inexpensive Cu or Co-based catalysts is promising for FCA hydrogenolysis.

The similarity between FCA hydrogenolysis and furfuryl alcohol hydrogenolysis suggests that the reaction mechanisms are similar. However, there are little consensus for the reaction mechanisms for the ring-opening hydrogenolysis of furan compounds. Many papers proposed each own mechanism for the same reaction. Roughly speaking, the proposed reaction mechanisms can be divided to two categories (Fig. 2.16) [77]: (i) direct dissociation of C-O bond in the furan ring by the attack of active hydrogen species on metal surface, and (ii) partial hydrogenation of furan ring to dihydrofuran compounds followed by C-O dissociation by E2 (in basic conditions) or E1 (over acidic conditions). In the mechanism (i), the substrate molecule strongly interacts with the metal surface. Activated hydrogen species is bonded to oxygen atom of the furan ring, and one C-O bond is dissociated. The carbon atom derived from the dissociated C-O is bonded to the metal surface and stabilized (R-C-O-R' + H^+ + * → R-C-* + HO-R'). In the mechanism (ii), the C (sp^3)-O bond formed by the partial hydrogenation becomes weaker than the C

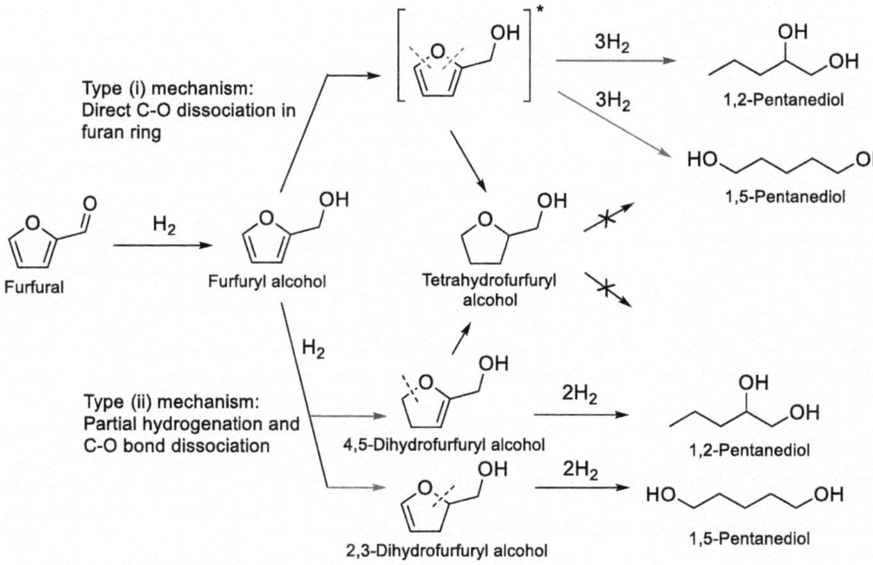

Fig. 2.16 Proposed mechanisms of ring-opening hydrogenolysis of furfuryl alcohol. For 2-furancarboxylic acid (FCA), the $-CH_2OH$ group is replaced with $-COOH$. (Reprinted from [77] with Creative Commons Attribution 4.0 International License https://creativecommons.org/licenses/by/4.0/)

(sp^2)-O in furan ring and then dissociated. Both mechanisms have several weaknesses to explain the reaction results. In particular, how the regioselectivity of C-O dissociation is determined is not explained well; in the case of FCA, the reason why 2-hydroxyvaleric acid is not formed at all is unclear. Further studies are necessary to draw the whole picture of mechanism of the furan ring hydrogenolysis.

When the ring hydrogenolysis systems of FCA to 5-hydroxyvaleric acid is applied to FDCA, the product becomes 2-hydroxyadipic acid. The report by Wei et al. for $Pt/Nb_2O_5{\cdot}xH_2O$ catalyst mentioned above [86] also described the FDCA hydrogenolysis over various supported Pt catalysts showing good 2-hydroxyadipic acid yield, such as Pt/Al_2O_3 (65% yield) and Pt/SiO_2 (69% yield). The highest reported yield was 73% with Pt/H-ZSM-5 catalyst. The same ring-opening hydrogenolysis as that of FCA to 5-hydroxyvaleric acid probably occurred in this reaction system. 2-Hydroxyadipic acid was not the target product in this paper, and the catalysts and reaction conditions were not deeply investigated. As mentioned for $Pt/Nb_2O_5{\cdot}xH_2O$ catalyst, the system used too low concentration of substrate in practical view.

There is one variant of ring opening hydrogenolysis of FCA to 5-hydroxyvaleric acid. Asano et al. reported that $Pt\text{-}MoO_x/TiO_2$ catalyst reduces aqueous FCA to valeric acid in about 60% yield (Fig. 2.17) [92, 93]. Similar to many other supported Pt catalysts, the $Pt\text{-}MoO_x/TiO_2$ catalyst did not convert tetrahydrofuran-2-carboxylic acid. Detailed characterization and the relationship between composition and performance suggest that the MoO_x species are dispersed on TiO_2 support, and the

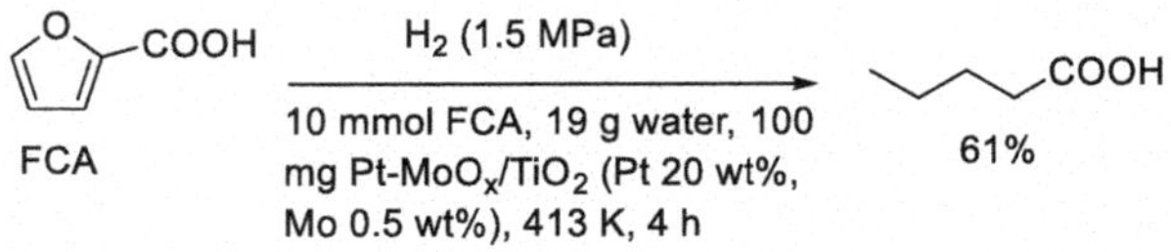

Fig. 2.17 Hydrodeoxygenation of 2-furancarboxylic acid (FCA) over Pt-MoO_x/TiO_2 catalyst [93]

interface between Pt metal particles and MoO_x/TiO_2 is the active site for the hydrodeoxygenation of FCA [93]. A mechanism consisting of ring-opening hydrogenolysis over Pt and reverse Mars van Krevelen mechanism over Mo^{IV} was proposed. The Pt-MoO_x/TiO_2 catalyst was applied to FDCA hydrogenolysis, but the adipic acid yield was not high (21%) [92].

2.5 Reduction of Multiple Hydroxy Groups: Hydrodeoxygenation of Sugar Acids

Some biomass-derived molecules such as sugars have many OH groups. The oxidation of one or both terminal groups of sugars gives sugar acids which still have many OH groups. The hydrodeoxygenation of sugar acids at the OH groups can produce useful less-functionalized carboxylic acids. Synthesis of adipic acid from glucaric acid (saccharic acid of glucose) by total removal of the secondary OH groups is the most desirable reaction in this type of reactions.

Sugar acids are not stable at high temperature especially in the presence of strong acids. The dehydration of sugar acids first produces lactones, and then further dehydration occurs at the α-hydrogen and β-hydroxy groups. The products can be further converted by various reactions such as condensation and decarboxylation. Teng Li et al. reported that the decomposition of glucaric acid without catalyst in water + toluene mixed solvent at 473 K gives furoic acid and furan with 10% and 29% yield, respectively [94]. The yields of these compounds can be increased by addition of homogeneous acid catalysts such as HCl and $SnCl_4$. The proposed reaction route is shown in Fig. 2.18.

The dehydration is difficult to control, and the dehydration + hydrogenation approach with acid + metal bifunctional catalysts is not effective in the selective reductive conversion of sugar acids.

2.5.1 *Hydrodeoxygenation of Sugar Acids Mediated by Hydrogen Halides*

The hydrodeoxygenation using HI is an effective approach for sugar acids, like the cases of furancarboxylic acids and lactic acid. Hexian Shi et al. reported the hydrodeoxygenation of glucaric acid in concentrated HI aq (57 wt%) [95]. While

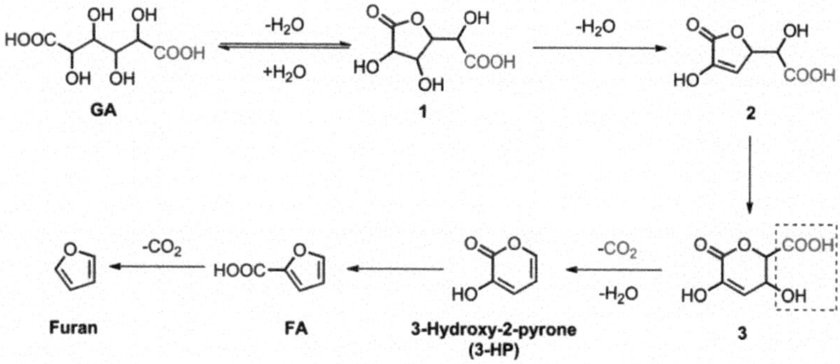

Fig. 2.18 Proposed reaction route from glucaric acid (GA) to 2-furancarboxylic acid ("FA" in this figure; "FCA" in the main text) and furan. (Reprinted with permission from ref. [94], Copyright © 2021 Elsevier)

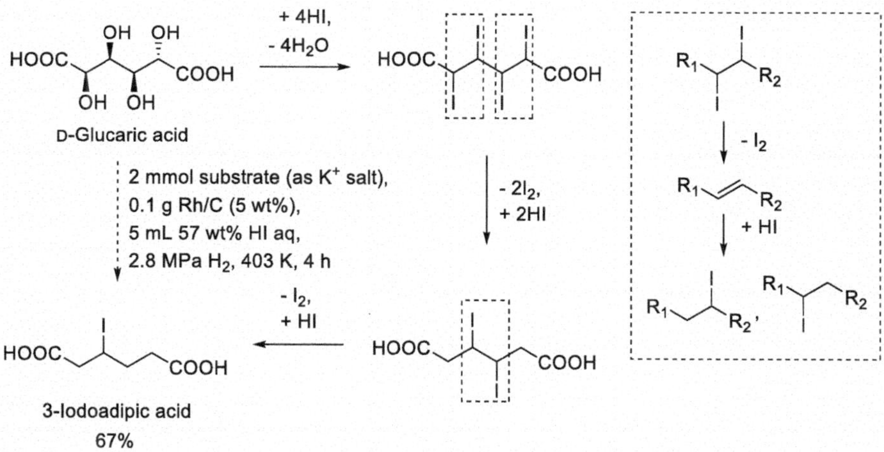

Fig. 2.19 Hydrodeoxygenation/iodization of glucaric acid with HI and Rh catalyst [95]

hydrodeoxygenation reaction proceeded without catalyst, metallic Rh/C catalyst was combined to suppress the formation of I_2. The addition of Rh/C slightly improved the reaction rate. Interestingly, the main product was 3-iodoadipic acid, and the yield reached 67%. The authors proposed that the contribution of hydrodeoxygenation with H_2 was very small, and the formation of 3-iodoadipic acid proceeded via total exchange of hydroxy groups with iodide to 2,3,4,5-tetraiodoadipic acid followed by repeating I_2 elimination and HI addition (Fig. 2.19). The preservation of C-I bond is probably due to the low reaction temperature (403 K) in comparison with the systems of total hydrodeoxygenation (lactic acid to propionic acid with NaI + acetic acid solvent: 493 K [57]; tetrahydrofuran-2,5-dicarboxylic acid to adipic acid with HI: 433 K [83]). Increase of reaction temperature surely increased the adipic acid yield (2.2% at 403 K to 10.0% at 423 K). 3-Iodoadipic acid can be readily converted to 3-hydroxyadipic acid and adipic acid by hydrolysis with NaOH aq (yield >99%)

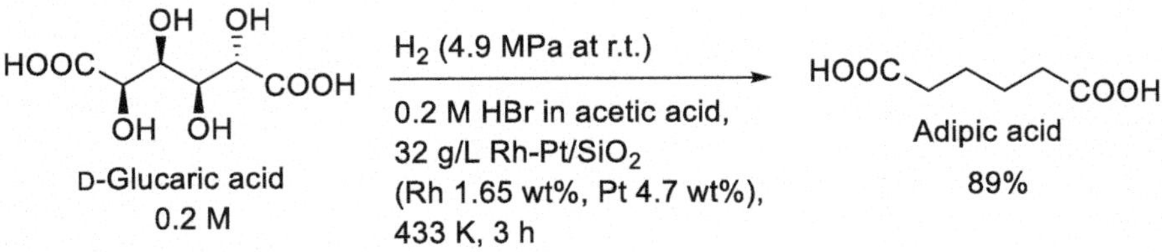

Fig. 2.20 Hydrodeoxygenation of glucaric acid with HBr and Rh-Pt/SiO_2 catalyst [96]

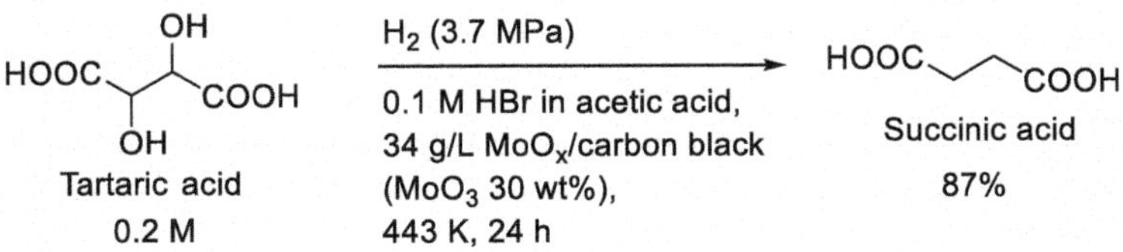

Fig. 2.21 Hydrodeoxygenation of tartaric acid with HBr and MoO_x/C catalyst [97]

and hydrogenolysis with Pd/C (yield 80%), respectively. This is the only report for the chemical synthesis of 3-hydroxyadipic acid from biomass platforms.

Direct synthesis of adipic acid from glucaric acid was reported in the patent invented by Boussie et al. of Rennovia, Inc. (Fig. 2.20) [96]. Hydrobromic acid (HBr) was used instead of HI, and the solvent was acetic acid. 89% yield of adipic acid was obtained in combination with Rh-Pt/SiO_2 catalyst. The consumption amount of reducing agent in the overall process from sugars is 4 equiv. of H_2, which is the same as that consumed in the overall process of adipic acid production via 2,5-furancarboxylic acid (FDCA) explained in Sect. 2.4.

A similar system was reported by the group of Vlachos and co-workers for tartaric acid hydrodeoxygenation to succinic acid with HBr + MoO_x/C catalysts in acetic acid solvent (Fig. 2.21) [97]. The MoO_x/C catalyst should be pre-reduced at high temperature (773 K) to form catalytically active Mo^0 phase. The yield of succinic acid reached 87%.

2.5.2 *Deoxydehydration of Sugar Acids*

Another approach of hydrodeoxygenation of sugar acids is the use of deoxydehydration (frequently abbreviated as DODH). DODH reaction is the reverse reaction of *cis*-dihydroxylation of alkenes: *cis*-vicinal diols are converted to C=C double bond with two-electron reducing agent (Fig. 2.22) [98]. Many sugar-derived compounds have vicinal diol moiety, and the DODH reaction can decrease the oxygen content selectively. Typical DODH systems use homogeneous high-valent Re catalysts such as CH_3ReO_3. Typical reducing agents are Ph_3P, metal powder, sodium sulfite, and secondary alcohols such as 3-octanol. These reducing agents and homogeneous expensive Re catalysts are not allowed for the practical use as

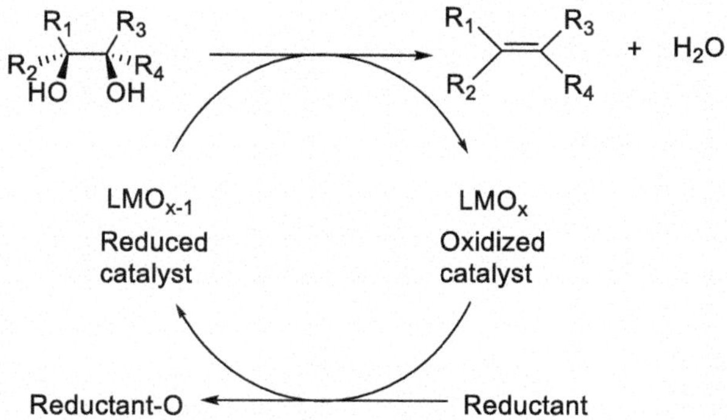

Fig. 2.22 Deoxydehydration (DODH) reaction

substitute for petrochemical processes. Recent development in the DODH reaction has enabled the use of heterogeneous catalysts, cheaper Mo-based catalysts, and atom-efficient reducing agent such as H_2 [99–101].

Sugar acids are one important class of substrates of DODH. The most important target product is adipic acid, which is produced by double DODH of saccharic acids such as glucaric acid followed by hydrogenation. The reported systems of adipic acid synthesis from sugar acids by DODH + hydrogenation are summarized in Table 2.4 [102–108]. While glucaric acid is the most abundant sugar acid, mucic acid which is the saccharic acid of galactose has been also frequently used as the substrate [101, 104, 105, 108]. The DODH product of mucic acid is stable *trans*, *trans*-isomer of muconic acid (2,4-hexadiendioic acid), and mucic acid is reactive toward DODH among saccharic acids. The co-loading of Re and platinum-group metal on supports is effective to obtain adipic acid from sugar acids in one-pot system [105, 107, 108]. Typical solvents are alcohols which also work as a reducing agent and a capping agent of carboxylic acids. Problems in the reported systems are low turnover number (TON) and low catalyst stability. Most supported Re catalysts for DODH reaction are plagued by leaching of Re species into the reaction media [109]. The DODH reaction is catalyzed by high-valent Re species, which are generally soluble in polar solvents. Recently, CeO_2-supported Re catalysts combined with a metallic promoter have been shown to have high DODH activity with H_2 as the reducing agent [110–113]. The basic CeO_2 strongly binds Re species to suppress leaching, and the catalysts are reusable after re-oxidation as regeneration [110–112, 114]. However, the CeO_2-supported Re catalysts do not work well for sugar acids or their esters. More active and stable catalysts that can use H_2 as the reducing agent are necessary for the industrial use of DODH. The use of cheaper Mo catalysts is also desirable. The use of solid Mo catalysts for DODH + hydrogenation of tartaric acid to succinic acid has been already reported [115, 116]. In particular, noble-metal-modified MoO_x/TiO_2 are effective DODH catalysts using H_2 as the reducing agent [116, 117]. Excellent yield (92%) of succinic acid was reported with MoO_x-Pd/TiO_2 catalyst and H_2 [116], although the exact reaction condition was not

Table 2.4 Examples of deoxydehydration (DODH) + hydrogenation systems for adipic acid synthesis

Substrate	Catalyst	Reducing agent	Temp./ %	Product (yield/%)	TON	Catalyst stability	Ref.
MCA	CH_3ReO_3 + TsOH + Pt/C	3-pentanol	473	Adipate (75)	15	No data	[102]
Glucarate-6,3-lactone	$KReO_4$ + Pd/C + H_3PO_4 + C	Methanol + H_2 (0.5 MPa)	423	Adipate (88)	88	No data	[103]
MCA-1,4-lactone	ReO_x/ZrO_2 (DODH), Pd/C (hydrogenation)	1-butanol + H_2 (0.1 MPa)	393, 393	Adipate (82) after two cycles of DODH and hydrogenation	8.3	No data	[104]
MCA	Pt-ReO_x/C	2-propanol	443	Adipate (85)	24	Leaching observed for Re	[105]
GAA	Re/C + Pd/C	Methanol + H_2 (0.5 MPa)	393	Adipate (25)	2.1	Leaching observed for Re	[106]
GAA	Pd-ReO_x/C + Amberlyst	Methanol + H_2 (2 MPa)	383	Adipate (95)	18	No data	[107]
MCA	Ir-ReO_x/C	2-propanol	473	Adipate (63)	18	Reusable after regeneration	[108]

MCA mucic acid; *GAA* glucaric acid; *TON* turnover number defined as (adipate amount/mol)/(Re amount in catalyst/mol)

provided. The data at lower conversion level suggested the involvement of successive hydrogenolysis by dehydration + hydrogenation in addition to DODH + hydrogenation. Dimeric Mo species as well as monomeric ones on anatase TiO_2 support have been proposed to be catalytically active species in DODH [118, 119]. Further development of well-defined inexpensive Mo-based catalysts is desired.

2.6 Conclusions and Future Outlook

There are several approaches for the reduction of oxygen-containing functional groups in carboxylic acids with retention of carboxyl groups. Hydrogenation of keto group such as that in levulinic acid is easy because ketones are more reactive toward hydrogenation than carboxyl groups. Hydrodeoxygenation of C-O single bond located distant from the carboxyl group is possible with the combination of acid and metal catalysts, if the thermal stability of the substrate is enough. Hydrodeoxygenation of γ-valerolactone or levulinic acid can be converted to valeric acid with this approach. The C-O bond at the α-position of the carboxyl group is more difficult to be dissociated such as that in lactic acid. The acid-base bifunctional catalysts, especially those containing alkali or alkali-earth metals as mobile base sites, are effective in the dehydration of lactic acid to acrylic acid. There are catalysts that can directly remove the OH group in lactic acid via reverse Mars van Krevelen mechanism. However, these two approaches have not been successfully utilized for more complex substrates such as tetrahydrofurancarboxylic acids and sugar acids. One approach that can be applied to wide range of substrates is the utilization of halide ions such as I^- in organic acid solvent. The C-O bonds are replaced with C-X (X: halogen atom) bonds by nucleophilic substitution reaction, and then the hydrogenolysis of C-X bonds occurs without catalysts or supported noble metal catalysts. Total removal of extrinsic functional groups of carboxylic acids is possible with this halide-mediated approach; however, the handling of corrosive hydrogen halide with high concentration is a big problem in practical view.

There are some approaches that are applicable to a specific structure. Ring-opening hydrogenolysis of 2-furancarboxylic acid over Pt catalysts produces 5-hydroxyvaleric acid. The deoxydehydration (DODH) reaction is effective to remove vicinal hydroxy groups in sugar acids. The examples are limited, and there is still plenty of room for development.

The development of catalysts is necessary to improve the performance to the practical level for most approaches. The state-of-the-art systems can give high yield of target product; however, catalyst cost and stability are essential issues for industrial use. Lowering the H_2 pressure and increase of the substrate concentration have much impact on the operation cost. Among potential catalysts, metal-niobium oxide catalysts and metal-molybdenum oxide catalysts with precisely controlled structure of niobium/molybdenum oxide species are promising, because niobium oxide works heterogeneous acid catalyst that can work in water, and molybdenum oxide species can work as active sites for several important reactions such as reverse Mars van Krevelen reactions and DODH reactions. The use of non-noble metal as the site for H_2 activation is also desirable in view of the catalyst cost. Another expansion of the catalytic systems is the defunctionalization of amino acids at amino group or other extrinsic groups with retention of carboxyl groups. This reaction has not been realized yet because the amino group can deactivate the hydrogenolysis catalyst, but the variety of molecule structure in natural amino acids is attractive as the building blocks in organic synthesis.

References

1. Bozell JJ, Petersen GR. Technology development for the production of biobased products from biorefinery carbohydrates—the US Department of Energy's "top 10" revisited. Green Chem. 2010;12:539–54. https://doi.org/10.1039/b922014c.
2. Dutta S, Bhat NS. Recent advances in the value addition of biomass-derived levulinic acid: a review focusing on its chemical reactivity patterns. ChemCatChem. 2021;13:3202–22. https://doi.org/10.1002/cctc.202100032.
3. Sajid M, Farooq U, Bary G, Azim MM, Zhao X. Sustainable production of levulinic acid and its derivatives for fuel additives and chemicals: progress, challenges, and prospects. Green Chem. 2021;23:9198–238. https://doi.org/10.1039/D1GC02919C.
4. Bucchianico DDMD, Wang Y, Buvat JC, Pan Y, Moreno VC, Leveneur S. Production of levulinic acid and alkyl levulinates: a process insight. Green Chem. 2022;24:614–46. https://doi.org/10.1039/D1GC02457D.
5. Millán GG, Hellsten S, Llorca J, Luque R, Sixta H, Balu AM. Recent advances in the catalytic production of platform chemicals from Holocellulosic biomass. ChemCatChem. 2019;11:3033–2042. https://doi.org/10.1002/cctc.201801843.
6. Wright WRH, Palkovits R. Development of heterogeneous catalysts for the conversion of Levulinic acid to γ-valerolactone. ChemSusChem. 2012;5:1657–67. https://doi.org/10.1002/cssc.201200111.
7. Raj T, Chandrasekhar K, Banu R, Yoon JJ, Kumar G, Kim SH. Synthesis of γ-valerolactone (GVL) and their applications for lignocellulosic deconstruction for sustainable green biorefineries. Fuel. 2021;303:121333. https://doi.org/10.1016/j.fuel.2021.121333.
8. Dutta S, Yu IKM, Tsang DCW, Nh YH, Ok YS, Sherwood J, Clark JH. Green synthesis of gamma-valerolactone (GVL) through hydrogenation of biomass-derived levulinic acid using non-noble metal catalysts: a critical review. Chem Eng J. 2019;372:992–1006. https://doi.org/10.1016/j.cej.2019.04.199.
9. Weng R, Yu Z, Xiong J, Lu X. Effects of water in the heterogeneous catalytic valorization of levulinic acid into γ-valerolactone and its derivatives. Green Chem. 2020;22:3013–27. https://doi.org/10.1039/D0GC01082K.
10. Tukacs JM, Fridrich B, Dibó G, Székely E, Mika LT. Direct asymmetric reduction of levulinic acid to gamma-valerolactone: synthesis of a chiral platform molecule. Green Chem. 2015;17:5189–95. https://doi.org/10.1039/C5GC01099C.
11. Omoruyi U, Page S, Hallett J, Miller PW. Homogeneous catalyzed reactions of Levulinic acid: to γ-valerolactone and beyond. ChemSusChem. 2016;9:2037–47. https://doi.org/10.1002/cssc.201600517.
12. Ohkuma T, Kitamura M, Noyori R. Enantioselective synthesis of 4-substituted γ-lactones. Tetrahedron Lett. 1990;31:5509–12. https://doi.org/10.1016/S0040-4039(00)97885-7.

13. Deng CQ, Liu J, Luo JH, Gan LJ, Deng J, Fu Y. Proton-promoted nickel-catalyzed asymmetric hydrogenation of aliphatic ketoacids. Angew Chem Int Ed. 2022;61:e202115983. https://doi.org/10.1002/anie.202115983.
14. Lautens M, Torelli. Asymmetric hydrogenation/cyclization cascade of aliphatic ketoacids: access to chiral lactones. Synfacts. 2022;18:0512. https://doi.org/10.1055/s-0041-1737433.
15. Osawa T, Tanabe Y. Facile synthesis of optically-active γ-valerolactone from levulinic acid and its esters using a heterogeneous enantio-selective catalyst. Catal Lett. 2018;148:824–30. https://doi.org/10.1007/s10562-017-2291-2.
16. Pothu R, Gundeboyina R, Boddula R, Perugopu V, Ma J. Recent advances in biomass-derived platform chemicals to valeric acid synthesis. New J Chem. 2022;46:5907–21. https://doi.org/10.1039/D1NJ05777D.
17. Yu Z, Lu X, Xiong J, Ji N. Transformation of levulinic acid to valeric biofuels: a review on heterogeneous bifunctional catalytic systems. ChemSusChem. 2019;12:3915–30. https://doi.org/10.1002/cssc.201901522.
18. Pan T, Deng J, Xu Q, Xu Y, Guo QX, Fu Y. Catalytic conversion of biomass-derived levulinic acid to valerate esters as oxygenated fuels using supported ruthenium catalysts. Green Chem. 2013;15:2967–74. https://doi.org/10.1039/c3gc40927a.
19. Luo W, Bruijnincx PCA, Weckhuysen BM. Selective, one-pot catalytic conversion of levulinic acid to pentanoic acid over Ru/H-ZSM5. J Catal. 2014;320:33–41. https://doi.org/10.1016/j.jcat.2014.09.014.
20. Kon K, Onodera W, Shimizu K. Selective hydrogenation of levulinic acid to valeric acid and valeric biofuels by a Pt/HMFI catalyst. Cat Sci Technol. 2014;4:3227–34. https://doi.org/10.1039/C4CY00504J.
21. Gu X, Zhang B, Liang H, Ge H, Yang H, Qin Y. Pt/HZSM-5 catalyst synthesized by atomic layer deposition for aqueous-phase hydrogenation of levulinic acid to valeric acid. J Fuel Chem Technol. 2017;45:714–22. https://doi.org/10.1016/S1872-5813(17)30035-X.
22. Muñoz-Olasagasti M, Sañudo-Mena A, Cecilia JA, Granados ML, Maireles-Torres P, Mariscal R. Direct conversion of levulinic acid into valeric biofuels using Pd supported over zeolites as catalysts. Top Catal. 2019;62:579–88. https://doi.org/10.1007/s11244-019-01147-4.
23. Sun P, Gao G, Zhao Z, Xia C, Li F. Stabilization of cobalt catalysts by embedment for efficient production of valeric biofuel. ACS Catal. 2014;4:4136–42. https://doi.org/10.1021/cs501409s.
24. Sun P, Gao G, Zhao Z, Xia C, Li F. Acidity-regulation for enhancing the stability of Ni/HZSM-5 catalyst for valeric biofuel production. Appl Catal B Environ. 2016;189:19–25. https://doi.org/10.1016/j.apcatb.2016.02.026.
25. Karanwal N, Verma D, Butolia P, Kim SM, Kim J. One-pot direct conversion of levulinic acid into high-yield valeric acid over a highly stable bimetallic Nb-Cu/Zr-doped porous silica catalyst. Green Chem. 2020;22:766–87. https://doi.org/10.1039/C9GC03516H.
26. Ji N, Yin J, Rong Y, Li H, Yu Z, Lei Y, Wang S, Diao X. More than a support: the unique role of Nb_2O_5 in supported metal catalysts for lignin hydrodeoxygenation. Cat Sci Technol. 2022;12:3751–66. https://doi.org/10.1039/D2CY00245K.
27. Golubeva MA. In situ generated nickel phosphide based catalysts for hydroprocessing of levulinic acid. Petrol Chem. 2021;61:670–5. https://doi.org/10.1134/S0965544121040022.
28. Yun GN, Takagaki A, Kikuchi R, Oyama ST. Hydrodeoxygenation of gamma-valerolactone on transition metal phosphide catalysts. Cat Sci Technol. 2017;7:281–92. https://doi.org/10.1039/C6CY02252A.
29. Lange JP, Price R, Ayoub PM, Louis J, Petrus L, Clarke L, Gosselink H. Valeric biofuels: a platform of cellulosic transportation fuels. Angew Chem Int Ed. 2010;49:4479–83. https://doi.org/10.1002/anie.201000655.
30. Zhou J, Zhu R, Deng J, Fu Y. Preparation of valeric acid and valerate esters from biomass-derived levulinic acid using metal triflates + Pd/C. Green Chem. 2018;20:3974–80. https://doi.org/10.1039/C8GC01606B.

31. Sutton AD, Waldie FD, Wu R, Schlaf M, Silks LAS III, Gordon JC. The hydrodeoxygenation of bioderived furans into alkanes. Nat Chem. 2013;5:428–32. https://doi.org/10.1038/nchem.1609.
32. Liu X, Li Y, Deng J, Fu Y. Selective hydrodeoxygenation of biomass-derived furfural-acetone to prepare 1-octyl acetate. Green Chem. 2019;21:4532–40. https://doi.org/10.1039/C9GC01767D.
33. Ahmadi K, Khosravi-Darani K, Mortazavian AM. An overview of biotechnological production of propionic acid: from upstream to downstream processes. Electron J Biotechnol. 2017;28:67–75. https://doi.org/10.1016/j.ejbt.2017.04.004.
34. Mäki-Arvela P, Simakova IL, Salmi T, Murzin DY. Production of lactic acid/lactates from biomass and their catalytic transformations to commodities. Chem Rev. 2014;114:1909–71. https://doi.org/10.1021/cr400203v.
35. Wang Z, Zhang Y, Wang Y, Li J, Jia X, Wu Z. Recent progress in glycerol oxidation to lactic acid and pyruvic acid with heterogeneous metal catalysts. Carb Resour Conv. 2024:100250. https://doi.org/10.1016/j.crcon.2024.100250.
36. Nakagawa Y, Yabushita M, Tomishige K. A perspective on catalytic production of olefinic compounds from biomass. RSC Sust. 2023;1:814–37. https://doi.org/10.1039/D3SU00033H.
37. Liu L, Ye P, Bozell JJ. A comparative review of petroleum-based and bio-based acrolein production. ChemSusChem. 2012;5:1162–80. https://doi.org/10.1002/cssc.201100447.
38. Makshina EV, Canadell J, van Krieken J, Peeters E, Dusselier M, Sels BF. Bio-acrylates production: recent catalytic advances and perspectives of the use of lactic acid and their derivates. ChemCatChem. 2019;11:180–201. https://doi.org/10.1002/cctc.201801494.
39. Sun D, Yamada Y, Sato S, Ueda W. Glycerol as a potential renewable raw material for acrylic acid production. Green Chem. 2017;19:3186–213. https://doi.org/10.1039/C7GC00358G.
40. Jin X, Meng K, Zhang G, Liu M, Song Y, Song Z, Yang C. Interfacial catalysts for sustainable chemistry: advances on atom and energy efficient glycerol conversion to acrylic acid. Green Chem. 2021;23:51–76. https://doi.org/10.1039/D0GC02913K.
41. Zhai Z, Li X, Tang C, Peng J, Jiang N, Bai W, Gao H, Liao Y. Decarbonylation of lactic acid to acetaldehyde over aluminum sulfate catalyst. Ind Eng Chem Res. 2014;53:10318–27. https://doi.org/10.1021/ie500988q.
42. Li X, Zhang Y, Chen Z, Cao P, Zou W, Tang C, Dong L, Wang Y. Sustainable production of 2,3-pentanedione: catalytic performance of $Ba_2P_2O_7$ doped with Cs for vapor-phase condensation of lactic acid. Ind Eng Chem Res. 2017;56:14437–46. https://doi.org/10.1021/acs.iecr.7b03595.
43. Matsuura Y, Onda A, Yanagisawa K. Selective conversion of lactic acid into acrylic acid over hydroxyapatite catalysts. Catal Commun. 2014;48:5–10. https://doi.org/10.1016/j.catcom.2014.01.009.
44. Ghantani VC, Dongare MK, Umbarkar SB. Nonstoichiometric calcium pyrophosphate: a highly efficient and selective catalyst for dehydration of lactic acid to acrylic acid. RSC Adv. 2014;4:33319–26. https://doi.org/10.1039/C4RA06429A.
45. Tang C, Peng J, Fan G, Li X, Pu X, Bai W. Catalytic dehydration of lactic acid to acrylic acid over dibarium pyrophosphate. Catal Commun. 2014;43:231–4. https://doi.org/10.1016/j.catcom.2013.10.009.
46. Peng J, Li X, Tang C, Bai W. Barium sulphate catalyzed dehydration of lactic acid to acrylic acid. Green Chem. 2014;16:108–11. https://doi.org/10.1039/C3GC42028K.
47. Li X, Chen Z, Cao P, Pu W, Zou W, Tang C, Dong L. Ammonia promoted barium sulfate catalyst for dehydration of lactic acid to acrylic acid. RSC Adv. 2017;7:54696–705. https://doi.org/10.1039/C7RA10736F.
48. Lari GM, Puértolas B, Frei MS, Mondelli C, Pérez-Ramírez J. Hierarchical NaY zeolites for lactic acid dehydration to acrylic acid. ChemCatChem. 2016;8:1507–14. https://doi.org/10.1002/cctc.201600102.

49. Zhang L, Theng DS, Du Y, Xi S, Huang L, Gao F, Wang C, Chen L, Borgna A. Selective conversion of lactic acid to acrylic acid over alkali and alkaline-earth metal co-modified NaY zeolites. Cat Sci Technol. 2017;7:6101–11. https://doi.org/10.1039/C7CY02142A.
50. Yan B, Tao LZ, Mahmood A, Liang Y, Yu BQ. Potassium-ion-exchanged zeolites for sustainable production of acrylic acid by gas-phase dehydration of lactic acid. ACS Catal. 2017;7:538–50. https://doi.org/10.1021/acscatal.6b01979.
51. Kim J, Bang J, Choi JS, Lim DH, Guk DG, Jae J. Selective conversion of lactic acid to renewable acrylic acid over SDA-free Na-ZSM-5: the critical role of basic sites of sodium oxide. J Catal. 2023;421:271–84. https://doi.org/10.1016/j.jcat.2023.03.024.
52. Xiao X, Bergstrom H, Saenger R, Johnson B, Sun R, Peterson A. The role of oxygen vacancies in biomass deoxygenation by reducible zinc/zinc oxide catalysts. Cat Sci Technol. 2018;8: 1819–27. https://doi.org/10.1039/C7CY02535A.
53. Zhang Y, Li X, Sun L, Tang C. Facile preparation of Fe_3O_4 from Fe_2O_3 via reduction of mixed vapors containing lactic acid and water: catalytic deoxygenation of lactic acid into propionic acid. ChemistrySelect. 2016;1:5002–7. https://doi.org/10.1002/slct.201601195.
54. Li X, Pang J, Zhang J, Yin C, Zou W, Tang C, Dong L. Vapor-phase deoxygenation of lactic acid to biopropionic acid over dispersant-enhanced molybdenum oxide catalyst. Ind Eng Chem Res. 2019;58:101–9. https://doi.org/10.1021/acs.iecr.8b04713.
55. Liu R, Li X, Ma K, Chen Z, Tang C. Sustainable production of bio-propionic acid: synergy between vacancy and thermoelectron in MoS_2/MoO_3 composite-enhanced hydrodeoxygenation of lactic acid. Cat Sci Technol. 2023;13:1221–32. https://doi.org/10.1039/D2CY01913B.
56. Zhao M, Tang C, Ma K, Li X. Synergy of highly dispersed Pd and the oxygen-vacancy-promoted hydrodeoxygenation of lactic acid to propionic acid. New J Chem. 2023;47:13918–28. https://doi.org/10.1039/D3NJ02457A.
57. Liu S, Feng H, Li T, Wang Y, Rong N, Yang W. Highly selective production of propionic acid from lactic acid catalyzed by NaI. Green Chem. 2020;22:7468–75. https://doi.org/10.1039/D0GC02676J.
58. Luján-Montelongo JA, Mateus-Ruiz JB, Valdez-García RM. Iodine and iodide in reductive transformations. Eur J Org Chem. 2023;26:e202201156. https://doi.org/10.1002/ejoc.202201156.
59. Arias PL, Cecilia JA, Gandarias I, Iglesias J, Granados ML, Mariscal R, Morales G, Moreno-Test R, Maireles-Torres P. Oxidation of lignocellulosic platform molecules to value-added chemicals using heterogeneous catalytic technologies. Cat Sci Technol. 2020;10:2721–57. https://doi.org/10.1039/D0CY00240B.
60. Xu C, Paone E, Rodriguez-Padrón D, Luque R, Mauriello F. Recent catalytic routes for the preparation and the upgrading of biomass derived furfural and 5-hydroxymethylfurfural. Chem Soc Rev. 2020;49:4273–306. https://doi.org/10.1039/D0CS00041H.
61. Loos K, Zhang R, Pereira I, Agostinho B, Hu H, Maniar D, Sbirrazzuoli N, Silvestre AJD, Guigo N, Sousa AF. A perspective on PEF synthesis, properties, and end-life. Front Chem. 2020;8:585. https://doi.org/10.3389/fchem.2020.00585.
62. Nakagawa Y, Tamura M, Tomishige K. Catalytic reduction of biomass-derived furanic compounds with hydrogen. ACS Catal. 2013;3:2655–68. https://doi.org/10.1021/cs400616p.
63. Chen S, Wojcieszak R, Dumeignil F, Marceau E, Royer S. How catalysts and experimental conditions determine the selective hydroconversion of furfural and 5-hydroxymethylfurfural. Chem Rev. 2018;118:11023–117. https://doi.org/10.1021/acs.chemrev.8b00134.
64. Nakagawa Y, Liu S, Tamura M, Tomishige K. Catalytic total hydrodeoxygenation of biomass-derived polyfunctionalized substrates to alkanes. ChemSusChem. 2015;8:1114–32. https://doi.org/10.1002/cssc.201403330.
65. Nakagawa Y, Yabushita M, Tomishige K. Reductive conversion of biomass-derived furancarboxylic acids with retention of carboxylic acid moiety. Trans Tianjin Univ. 2021;27:165–79. https://doi.org/10.1007/s12209-021-00284-w.

66. Zaytsev AV, Pickles JE, Harnor SJ, Henderson AP, Alyasiri M, Waddell PG, Cano C, Griffin RJ, Golding BT. Concise syntheses of bridged morpholines. RSC Adv. 2016;6:53955–7. https://doi.org/10.1039/C6RA08737J.
67. Maris M, Huck WR, Mallat T, Baiker A. Palladium-catalyzed asymmetric hydrogenation of furan carboxylic acids. J Catal. 2003;219:52–8. https://doi.org/10.1016/S0021-9517(03)00184-2.
68. Fujima Y, Hirayama Y, Ikunaka M, Nishimoto Y. A scalable chemoenzymatic preparation of (*R*)-tetrahydrofuran-2-carboxylic acid. Tetrahedron Asym. 2003;14:1385–91. https://doi.org/10.1016/S0957-4166(03)00249-0.
69. Uhm K, Lim S, Lim J. Method for optically resolving a racemic alpha-substituted heterocyclic carboxylic acid using enzyme. WO Patent 2001092553. 2001.
70. Rios J, Lebeau J, Yang T, Li S, Lynch MD. A critical review on the progress and challenges to a more sustainable, cost competitive synthesis of adipic acid. Green Chem. 2021;23:3172–90. https://doi.org/10.1039/D1GC00638J.
71. Lang M, Li H. Sustainable routes for the synthesis of renewable Adipic acid from biomass derivatives. ChemSusChem. 2022;15:e202101531. https://doi.org/10.1002/cssc.202101531.
72. Wang X, Arai M, Wu Q, Zhang C, Zhao F. Hydrodeoxygenation of lignin-derived phenolics—a review on the active sites of supported metal catalysts. Green Chem. 2020;22:8140–68. https://doi.org/10.1039/D0GC02610G.
73. Nakagawa Y, Yabushita M, Tomishige K. Oxidative cleavage of C-C bonds in non-aromatic oxygenates with molecular oxygen for synthesis of carboxylic acids. Asian J Org Chem. 2023;12:e202300409. https://doi.org/10.1002/ajoc.202300409.
74. Skoog E, Shin JH, Saez-Jimenez V, Mapelli V, Olsson L. Biobased adipic acid—the challenge of developing the production host. Biotechnol Adv. 2018;36:2248–63. https://doi.org/10.1016/j.biotechadv.2018.10.012.
75. Tomishige K, Nakagawa Y, Tamura M. Selective hydrogenolysis and hydrogenation using metal catalysts directly modified with metal oxide species. Green Chem. 2017;19:2876–924. https://doi.org/10.1039/C7GC00620A.
76. Tomishige K, Nakagawa Y, Tamura M. Design of supported metal catalysts modified with metal oxides for hydrodeoxygenation of biomass-related molecules. Curr Opin Green Sus Chem. 2020;22:13–21. https://doi.org/10.1016/j.cogsc.2019.11.003.
77. Tomishige K, Honda M, Sugimoto H, Liu L, Yabushita M, Nakagawa Y. Recent progress on catalyst development for ring-opening C-O hydrogenolysis of cyclic ethers in the production of biomass-derived chemicals. Carbon Neutrality. 2024;3:17. https://doi.org/10.1007/s43979-024-00090-y.
78. Asano T, Nakagawa Y, Tamura M, Tomishige K. Hydrogenolysis of tetrahydrofuran-2-carboxylic acid over tungsten-modified rhodium catalyst. Appl Catal A Gen. 2020;602:117723. https://doi.org/10.1016/j.apcata.2020.117723.
79. Pinkos R, Bauduin C, Paul A, Fritz G, Wagner H. Process for preparing delta-valerolactone in the gas phase. U.S. Patent 0237806. 2011.
80. ten Brink GJ, Arends IWCE, Sheldon RA. The Baeyer-Villiger reaction: new developments toward greener procedures. Chem Rev. 2004;104:4105–24. https://doi.org/10.1021/cr030011l.
81. Dutta S, Bhat NS. Catalytic transformation of biomass-derived furfurals to cyclopentanones and their derivatives: a review. ACS Omega. 2021;6:35145–72. https://doi.org/10.1021/acsomega.1c05861.
82. Boussie TR, Dias EL, Fresco ZM, Murphy VJ.Production of Adipic acid and derivatives from Carbohydrate-containing materials. U.S. Patent 0317822. 2010.
83. Gilkey MJ, Mironenko AV, Vlachos DG, Xu B. Adipic acid production via metal-free selective hydrogenolysis of biomass-derived tetrahydrofuran-2,5-dicarboxylic acid. ACS Catal. 2017;7:6619–34. https://doi.org/10.1021/acscatal.7b01753.

84. Gilkey MJ, Balakumar R, Vlachos DG, Xu B. Adipic acid production catalyzed by a combination of a solid acid and an iodide salt from biomass-derived tetrahydrofuran-2,5-dicarboxylic acid. Cat Sci Technol. 2018;8:2661–71. https://doi.org/10.1039/C8CY00379C.
85. Gilkey MJ, Cho HJ, Murphy BM, Wu J, Vlachos DG. Catalytic adipic acid production on zeolites from biomass-derived tetrahydrofuran-2,5-dicarboxylic acid. ACS Appl Energy Mater. 2020;3:99–105. https://doi.org/10.1021/acsaem.9b01954.
86. Wei L, Zhang J, Deng W, Xie S, Zhang Q, Wang Y. Catalytic transformation of 2,5-furandicarboxylic acid to adipic acid over niobic acid-supported Pt nanoparticles. Chem Commun. 2019;55:8013–6. https://doi.org/10.1039/C9CC02877C.
87. Sun X, Wen B, Wang F, Zhang W, Zhao K, Liu X. Research advances on the catalytic conversion of biomass-derived furfural into pentanediols. Catal Commun. 2024;187:106864. https://doi.org/10.1016/j.catcom.2024.106864.
88. Smith HA, Fuzek JF. Catalytic hydrogenation of furan and substituted furans on platinum. J Am Chem Soc. 1949;71:415–9. https://doi.org/10.1021/ja01170a013.
89. Asano T, Takagi H, Nakagawa Y, Tamura M, Tomishige K. Selective hydrogenolysis of 2-furancarboxylic acid to 5-hydroxyvaleric acid derivatives over supported platinum catalysts. Green Chem. 2019;21:6133–45. https://doi.org/10.1039/C9GC03315G.
90. Sun Q, Wang S, Liu H. Selective hydrogenolysis of α-C–O bond in biomass-derived 2-furancarboxylic acid to 5-hydroxyvaleric acid on supported Pt catalysts at near-ambient temperature. ACS Catal. 2019;9:11413–25. https://doi.org/10.1021/acscatal.9b04074.
91. Tan J, Su Y, Gao K, Cui J, Wang Y, Zhao Y. Recent advances in the selective hydrogenation of furfural and its derivatives to pentanediol. J Fuel Chem Technol. 2021;49:780–90. https://doi.org/10.1016/S1872-5813(21)60036-1.
92. Asano T, Tamura M, Nakagawa Y, Tomishige K. Selective hydrodeoxygenation of 2-furancarboxylic acid to valeric acid over molybdenum-oxide-modified platinum catalyst. ACS Sust Chem Eng. 2016;4:6253–7. https://doi.org/10.1021/acssuschemeng.6b01640.
93. Asano T, Nakagawa Y, Tamura M, Tomishige K. Structure and mechanism of titania-supported platinum-molybdenum catalyst for hydrodeoxygenation of 2-furancarboxylic acid to valeric acid. ACS Sust Chem Eng. 2019;7:9601–12. https://doi.org/10.1021/acssuschemeng.9b01104.
94. Li T, Sun G, Xiong L, Zheng B, Duan Y, Yu R, Jiang J, Wang Y, Yang W. Transition-metal-free decarboxylation of D-glucaric acid to furan catalyzed by $SnCl_4$ in a biphasic system. Mol Catal. 2021;516:111958. https://doi.org/10.1016/j.mcat.2021.111958.
95. Shi H, Zhang L, Wu Y, Yu R, Peng Y, Wang Y, Li T, Yang W. Production of adipic acid derivatives from D-glucaric acid by hydrodeoxygenation mediated with hydroiodic acid. Catal Lett. 2021;151:338–43. https://doi.org/10.1007/s10562-020-03312-x.
96. Boussie TR, Dias EL, Presco ZM, Murphy VJ, Shoemaker J, Archer R, Jiang H. PRODUCTION OF ADIPIC ACID AND DERVATIVES FROM CARBOHYDRATE-CONTAINING MATERLALS. U.S. Patent 0317823. 2010.
97. Fu J, Vasiliadou ES, Goulas KA, Saha B, Vlachos DG. Selective hydrodeoxygenation of tartaric acid to succinic acid. Cat Sci Technol. 2017;7:4944–54. https://doi.org/10.1039/C7CY01374D.
98. Tshibalonza NN, Monbaliu JCM. The deoxydehydration (DODH) reaction: a versatile technology for accessing olefins from bio-based polyols. Green Chem. 2020;22:48001–4848. https://doi.org/10.1039/D0GC00689K.
99. Petersen AR, Fristrup P. New motifs in deoxydehydration: beyond the realms of rhenium. Chem Eur J. 2017;23:10235–43. https://doi.org/10.1002/chem.201701153.
100. Tomishige K, Nakagawa Y, Tamura M. Taming heterogeneous rhenium catalysis for the production of biomass-derived chemicals. Chin Chem Lett. 2020;31:1071–7. https://doi.org/10.1016/j.cclet.2019.07.014.
101. Jentoft FC. Transition metal-catalyzed deoxydehydration: missing pieces of the puzzle. Cat Sci Technol. 2022;12:6308–58. https://doi.org/10.1039/D1CY02083H.

102. Li X, Wu D, Lu T, Yi G, Su H, Zhang Y. Highly efficient chemical process to convert mucic acid into adipic acid and DFT studies of the mechanism of the rhenium-catalyzed deoxydehydration. Angew Chem Int Ed. 2014;53:4200–4. https://doi.org/10.1002/anie.201310991.
103. Larson RT, Samant A, Chen J, Lee W, Bohn MA, Ohlmann DM, Zuend SJ, Toste FD. Hydrogen gas-mediated deoxydehydration/hydrogenation of sugar acids: catalytic conversion of glucarates to adipates. J Am Chem Soc. 2017;139:14001–4. https://doi.org/10.1021/jacs.7b07801.
104. Lin J, Song H, Shen X, Wang B, Xie S, Deng W, Wu D, Zhang Q, Wang Y. Zirconia-supported rhenium oxide as an efficient catalyst for the synthesis of biomass-based adipic acid ester. Chem Commun. 2019;55:11017–20. https://doi.org/10.1039/C9CC05413H.
105. Jang JH, Ro I, Christopher P, Abu-Omar MM. A heterogeneous Pt-ReO_x/C catalyst for making renewable adipates in one step from sugar acids. ACS Catal. 2021;11:95–109. https://doi.org/10.1021/acscatal.0c04158.
106. Hočevar B, Prašnikar A, Huš M, Grilc M, Likozar B. H_2-free re-based catalytic dehydroxylation of aldaric acid to muconic and adipic acid esters. Angew Chem Int Ed. 2021;60:1244–53. https://doi.org/10.1002/anie.202010035.
107. Deng W, Yan L, Wang B, Zhang Q, Song H, Wang S, Zhang Q, Wang Y. Efficient catalysts for the green synthesis of adipic acid from biomass. Angew Chem Int Ed. 2021;60:4712–9. https://doi.org/10.1002/anie.202013843.
108. Jang JH, Hopper JT, Ro I, Christopher P, Abu-Omar MM. One-step production of renewable adipic acid esters from mucic acid over an Ir-ReO_x/C catalyst with low Ir loading. Cat Sci Technol. 2023;13:714–25. https://doi.org/10.1039/D2CY01144A.
109. Sharkey BE, Jentoft FC. Fundamental insights into deactivation by leaching during rhenium-catalyzed deoxydehydration. ACS Catal. 2019;9:11317–28. https://doi.org/10.1021/acscatal.9b02806.
110. Ota N, Tamura M, Nakagawa Y, Okumura K, Tomishige K. Hydrodeoxygenation of vicinal OH groups over heterogeneous rhenium catalyst promoted by palladium and ceria support. Angew Chem Int Ed. 2015;54:1897–900. https://doi.org/10.1002/anie.201410352.
111. Tazawa S, Ota N, Tamura M, Nakagawa Y, Okumura K, Tomishige K. Deoxydehydration with molecular hydrogen over ceria-supported rhenium catalyst with gold promoter. ACS Catal. 2016;6:6393–7. https://doi.org/10.1021/acscatal.6b01864.
112. Yamaguchi K, Cao J, Betchaku M, Nakagawa Y, Tamura M, Nakayama A, Yabushita M, Tomishige K. Deoxydehydration of biomass-derived polyols over silver-modified ceria-supported rhenium catalyst with molecular hydrogen. ChemSusChem. 2022;15:e202102663. https://doi.org/10.1002/cssc.202102663.
113. Yamaguchi K, Nakagawa Y, Li C, Yabushita M, Tomishige K. Utilization of Ni as a non-Noble-metal co-catalyst for ceria-supported rhenium oxide in combination of deoxydehydration and hydrogenation of vicinal diols. ACS Catal. 2022;12:12582–95. https://doi.org/10.1021/acscatal.2c03042.
114. Cao J, Larasati S, Yabushita M, Nakagawa Y, Wärmå J, Murzin DY, Asada D, Nakayama A, Tomishige K. Kinetic modeling of erythritol deoxydehydration and consecutive hydrogenation over the ReO_x-Pd/CeO_2 catalyst. ACS Catal. 2024;14:1663–77. https://doi.org/10.1021/acscatal.3c04785.
115. Jiang H, Lu R, Si X, Luo X, Xu J, Lu F. Single-site molybdenum catalyst for the synthesis of fumarate. ChemCatChem. 2019;11:4291–6. https://doi.org/10.1002/cctc.201900332.
116. Nacy A, Freitas LFL, Albarracín-Suazo S, Ruiz-Valentín G, Roberts CA, Nikolla E, Pagán-Torres YJ. Selective C-O bond cleavage of bio-based organic acids over palladium promoted MoO_x/TiO_2. ChemCatChem. 2021;13:1294–8. https://doi.org/10.1002/cctc.202001799.
117. Hacatrjan S, Liu L, Gan J, Nakagawa Y, Cao J, Yabushita M, Tamura M, Tomishige K. Titania-supported molybdenum oxide combined with Au nanoparticles as a hydrogen-driven deoxydehydration catalyst of diol compounds. Cat Sci Technol. 2022;12:2146–61. https://doi.org/10.1039/D1CY02144C.

118. Xi Y, Lauterbach J, Pagan-Torres Y, Heyden A. Deoxydehydration of 1,4-anhydroerythritol over anatase TiO_2(101)-supported ReO_x and MoO_x. Cat Sci Technol. 2020;10:3731–8. https://doi.org/10.1039/D0CY00434K.
119. Asada D, Ikeda T, Muraoka K, Nakagawa Y, Tomishige K, Nakayama A. Density functional theory study of deoxydehydration reaction by TiO_2-supported monomeric and dimeric molybdenum oxide catalysts. J Phys Chem C. 2022;126:20375–87. https://doi.org/10.1021/acs.jpcc.2c06018.

Chapter 3
Chemical Conversions of Biomass into Organic Acids over Heterogeneous Metallic Catalysts

Yao-Bing Huang

Abstract Chemical conversions of biomass has provided different types of chemicals, including alcohols, aldehydes, acids, furanics, esters and etc. Among these molecules, organic acids are the most appealing ones with numerous applications in synthetic building blocks, polymer monomers, cosmetic ingredients and food additives. By taking advantage of the native structures of biomass resource, a series of acids with unique structures, e.g. formic acid (AA), acetic acid, succinic acid (SA), levulinic acid (LA), 2,5-furandicarboxylic acid (FDCA), gluconic acid and amino acids, can be selectively obtained. A great deal of achievements have been made in the production of these organic acids from biomass feedstock with either heterogeneous catalysis or homogeneous catalysis processes. Heterogeneous catalysts own several inspiring features and played the central role in establishing the green and economic transformation systems. Thus, a systematic review on the important advances in this area is urgently important.

In this chapter, we aim to systematically review the recent advances on the production of organic acids from biomass and its derivatives over heterogeneous metallic catalysts. The content will be cataloged by the carbon numbers of organic acids, especially C1–C6 organic acids. Organic acids with more than six carbons constructed by multiple small biomass molecules are out of the scope. The feedstock for organic acid production includes polysaccharide (cellulose, chitin, monosaccharides, etc.), platform molecules (5-hydroxylfurfural (HMF), LA, furfural, etc.) and biomass derived molecules (glycerol, guaiacol, furfuryl alcohol, etc.). The catalysts developed for the conversions are supported metal catalysts, metal oxides, heteropoly acids, zeolites, and etc. Detailed information regarding catalyst synthesis, reaction selectivity, and catalytic performance will be revealed. Special attention will be paid to the structure-reactivity relationship between catalytic sites and production distribution. Finally, challenges and prospects of heterogeneous catalysis in organic acid production from biomass are highlighted.

Y.-B. Huang (✉)
New Energy School, North China Electric Power University, Beijing, China
e-mail: hyb@ncepu.edu.cn; hyb123@mail.ustc.edu.cn

Z. Fang et al. (eds.), *Production of Organic Acids and Alcohols from Agricultural Residues and Food Wastes*, Biofuels and Biorefineries 13,
https://doi.org/10.1007/978-981-96-1222-2_3

Keywords Biomass conversion · Heterogeneous catalysts · Organic acids · Reaction mechanism

3.1 Introduction

The conversion of biomass into various chemicals has been a hot research topic over the past few decades, since the as-prepared products can serve as good substitutes to traditional fossil based chemicals and fuels. As for the conversions, two main category strategies are frequently applied to biomass conversions, one is the removal of excess oxygen atoms to increase their energies densities to meet the demands of fuels, while the other is selective modification of the molecule structures to yield various oxygenated compounds. Biomass materials on earth are different in compositions, but typically contains 40–45 wt% of oxygen [1]. This provides a good opportunity for the preparation of oxygen-rich chemicals, eliminating the introduction of extra oxygenated resources frequently occurred in the petroleum refinery process. By taking advantage of different biomass molecule structures, these molecules could be prepared more economically, as the removal of oxygen from a molecule may be more competitive than the introduction of oxygen to the present petroleum based feedstocks.

Organic acids and their derivatives, such as esters and anhydrides, are value-added fine chemicals, and widely used as solvents, cosmetics, polymer monomers, and food additives in all sectors of society. The annual consumption of commodity acids in china is shown in Fig. 3.1 [2], most of which are now readily available from

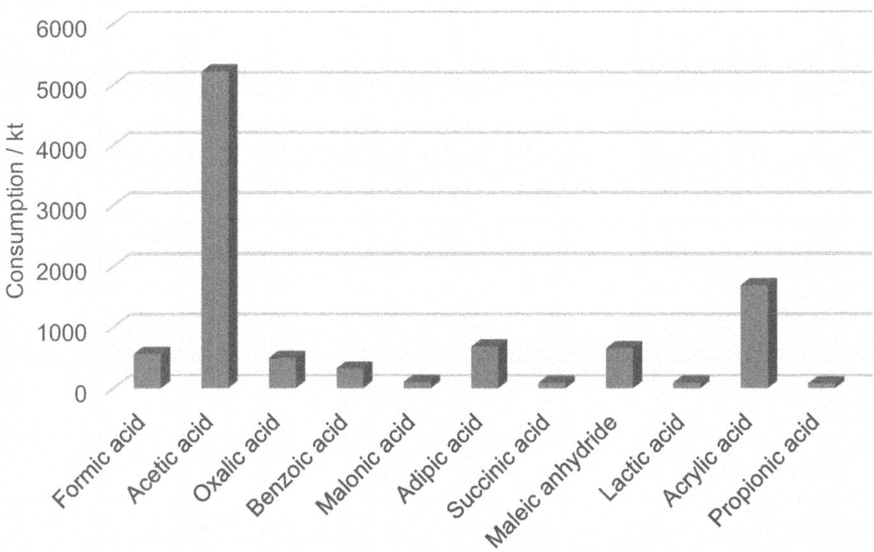

Fig. 3.1 Consumption of different organic acids in China. (Reprinted with permission from [2], Copyright © 2018, ACS)

biomass resource. Among the recent developed systems for biomass based acids production, heterogeneous metal catalysts play central roles with several fascinating advantages, such as high efficiency, good selectivity, convenient catalyst separation, and minimal waste generation. For example, gluconic acid can be obtained by the direct oxidation of glucose, or the one pot transformation of polysaccharides over acid-metal bifunctional catalysts [3]. Catalytic oxidation of HMF, readily available from C6 sugars, can selectively generates 5-hydroxymethyl-2-furancarboxylic acid (HFCA), 5-formyl-2-furancarboxylic acid (FFCA), and FDCA [4]. Due to the high oxygen content of biomass resource, the production of organic acids is economically feasible. According to the data from market, 38 kt of bio-based succinic acid was globally produced annually and the market value is about $108 million, whereas about 40 kt of fossil-based succinic acid are produced with a $100 million market value [5]. The current market price of bio-based succinic acid is about $2860/t, which is relatively higher than that of fossil-based equivalent (around $2500/t).

This chapter mainly focuses on the production of acid chemicals from biomass over heterogeneous catalysts. Recent progress in the development of efficient heterogeneous catalytic systems for the production of various acid are highlighted, including but not limited to formic acid, acetic acid, glycolic acid, malonic acid, oxalic acid, gluconic acid, glucaric acid, HFCA, FFCA, DFF and FDCA.

3.2 C1 Organic Acid

FA is the simplest carboxylic acid with only one carbon atom in the skeleton, which plays an indispensable position in the chemical, textile, leather, pharmaceutical, rubber, road deicing agent and other industries [6]. FA is also used as a hydrogen storage molecule, which can release H_2 after its decomposition over metal catalysts [7]. Traditional FA production mainly relies on methane oxidation, which is not sustainable enough. Therefore, efforts are devoted to the use of renewable biomass for a sustainable production. The oxidation reaction is the key step for FA production with diverse biomass resources.

3.2.1 Formic Acid Production

3.2.1.1 Glucose Oxidation

He et al. used Ag_2O as catalyst to selectively convert glucose to FA and glycolcolic acid under mild conditions [8]. Glucose was selectively converted to FA with an optimal yield of 40.7% and glycolic acid yield of 6.1% at 135 °C in 30 min. The aldehyde group on glucose is firstly oxidized to carboxylic acid group. Subsequently, the breakage of C1–C2 bond in gluconic acid forms FA and xylose. Wu et al. prepared Mg(OH)(OOH) catalyst from H_2O_2 and MgO and subjected it to

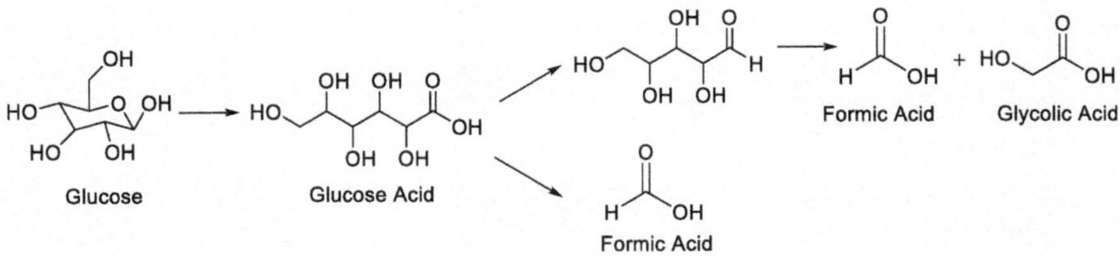

Scheme 3.1 Reaction pathway of glucose oxidation to FA

glucose oxidation [9]. A high FA yield of 78.6% could be obtained at 323 K in 4 h, where FA was produced by the oxidization of –CHO groups of glucose and the cleavage of C-C bond.

Li et al. proposed an effective strategy for the conversion of glucose to FA by using zeolite catalyst in biphasic solvent γ-butyrolactone (GBL) and H_2O [10]. A 54.3% yield of pentose and a 54.7% yield of FA were obtained from glucose at 453 K for 1.5 h. The yields of pentose and FA remarkably increased to 95.6% and 96.8% when fructose was used as the substrate. The synergistic effect of Lewis acids and Brønsted acids was proposed to facilitate the C-C bond of glucose to form FA.

Takagaki et al. used CaO as catalyst for the production of FA from sugars in H_2O_2 solution [11]. The formic acid yields were 52 and 66% from glucose and xylose at 343 K for 0.5 h, respectively. The proposed reaction pathway was illustrated in Scheme 3.1.

3.2.1.2 Glycerol Oxidation

Glycerol is a side product from biodiesel production, but receiving considerable attentions as a source for high-value chemicals. The oxidation of glycerol is one of the appealing routes to generate FA. Chagas et al. reported the use of mesoporous silica (SiO_2) catalyst to catalyze glycerol to FA conversion [12]. The reaction achieved a 90% conversion and an 83% FA selectivity in the presence of H_2O_2 at 150 °C in a continuous flow reactor. The oxygen vacancies on catalyst surface reacted with H_2O_2 to produce highly reactive Si-O oxygen species, which effectively oxidized glycerol to yield FA. This work demonstrated for the first time a metal-free catalytic system for glycerol oxidation to FA.

Liu et al. reported the combination of $Ru(OH)_4$/r-GO and $FeCl_3$ catalysts for glycerol oxidation. A 60% yield of FA was achieved in aqueous solution at 160 °C, 0.5 MPa O_2 for 1 h [13]. The reaction system was also successfully applied to the conversion of glucose, fructose and cellulose, and 50%, 45.7% and 20.3% yields were obtained, respectively. The catalytic role of Ru(OH)4/r-GO was proposed to facilitated the oxidation of dihydroxyacetone (DHA) to acetic acid and formaldehyde, while $FeCl_3$ promoted the conversion of DHA to glycolic acid and formaldehyde, and further to FA production (Scheme 3.2).

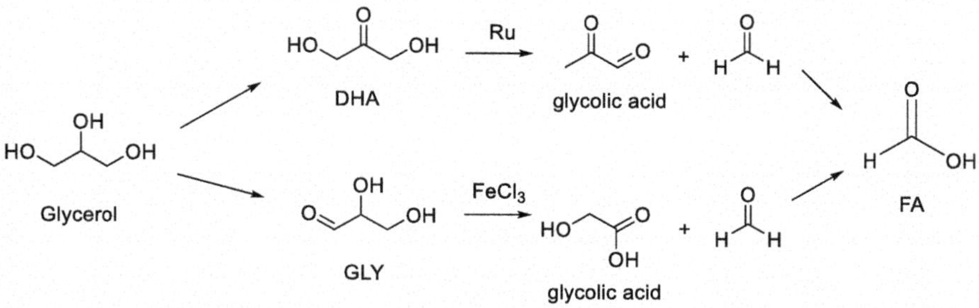

Scheme 3.2 Reaction pathway of oxidation of glycerol into FA

3.3 C2 Organic Acids

Acetic acid and glycolic acid are two representative C2 organic acids from biomass. They are readily prepared from carbohydrates by chemical or biological methods, and can be used in different areas in chemistry industry. Acetic acid is well-known as the main component of vinegar, and can be produced through the fermentation of various agricultural grain such as sticky rice, sorghum and corn [14]. Glycolic acid is an alpha hydroxyl acid derived from sugar cane, and can also be found in plants. It is also an important gradient in the skincare products, which would own a high value if produced from renewable biomass resource.

Over the past years, chemical productions of acetic acid and glycolic acid are more and more concerned due to the easy availability and high efficiency of the catalytic systems. Several biomass resources are reported for chemical production of acetic acid including starch, chitin. For glycolic acid preparation, monosaccharide, glycerol and 1,3-dihydroxyacetone can be used as the materials in the presence of heterogeneous catalysts.

3.3.1 Acetic Acid Production

3.3.1.1 Starch Conversion

Starch is the main component of agricultural grain and can be fermented to ethanol or acetic acid. Recent advance had demonstrated its conversion to acetic acid through chemical conversion methods, providing a new route in this area. Jin et al. used a two-step hydrothermal method to convert carbohydrates, including starch, cellulose, into acetic acid [15]. The first step was a base (i.e. $Ca(OH)_2$) catalyzed decomposition of starch to lactic acid, after which CuO oxidant and NaOH additive was added to the reaction conducted at 300 °C, 1 M NaOH. A maximum acetic acid of 23% was obtained. Cu^{2+} coordinated to the two adjacent oxygen atoms of lactic acid and promoted the cracking of C1-C2 bond to yield acetaldehyde, which was then reacted with OH^- to form the target product. XRD characterization of Cu species after reaction showed that CuO was converted to Cu_2O in the reaction. The acetic acid yields were 26% and 22% from glucose and cellulose, respectively.

3.3.1.2 Chitin Conversion

Chitin is the second largest biomass from discarded crustacean shell. It has a similar sugar structure of cellulose but contains an acetylamino groups in molecule structure. Jin and Yan et al. used CuO and NaOH to catalyze the conversion of chitin to acetic acid, and a maximum yield of 38.1% was obtained at 573 K in 35 min of reaction time [16]. The acetic acid was produced from the hydrolysis of acetyl amide chain of chitin, and the decomposition of C3-C6 intermediates from glucose hydrolysate (Scheme 3.3).

3.3.2 *Glycolic Acid Production*

3.3.2.1 Monosaccharide Conversion

Barakat et al. used mixed zinc vanadium oxides $Zn_3V_2O_8$ as catalyst for the conversion of xylose to glycolic acid [17], which achieved a 60% glycolic acid yield in water at 150 °C for 1 h. A selective C2–C3 bond retro-aldol condensation reaction occurred on xylose molecule in the presence of metal cations (Zn^{2+} or V^{5+}), resulting in the formation of glycolic aldehyde and glycolaldehyde. Glycolaldehyde was then oxidized to the target glycolic acid (Scheme 3.4).

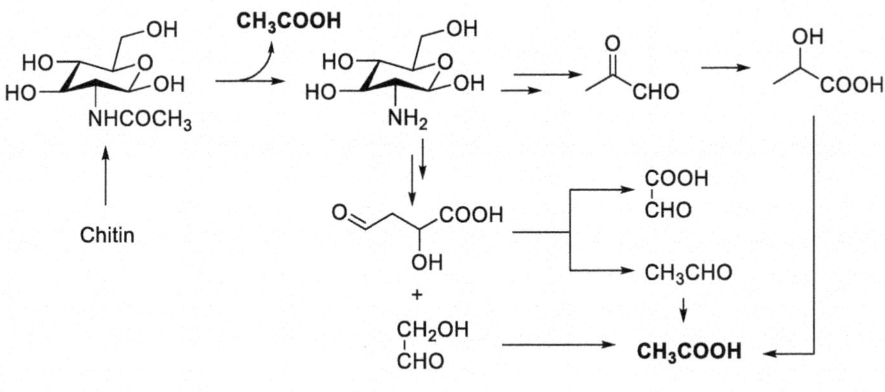

Scheme 3.3 Reaction pathway of oxidation of chitin into acetic acid.

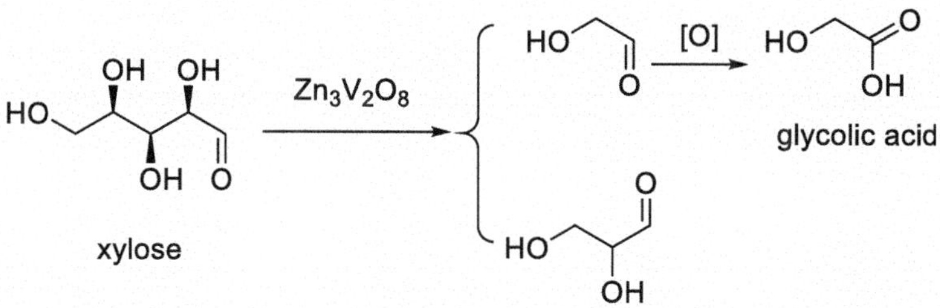

Scheme 3.4 Selective oxidation of xylose to glycolic acid using $Zn_3V_2O_8$ catalyst

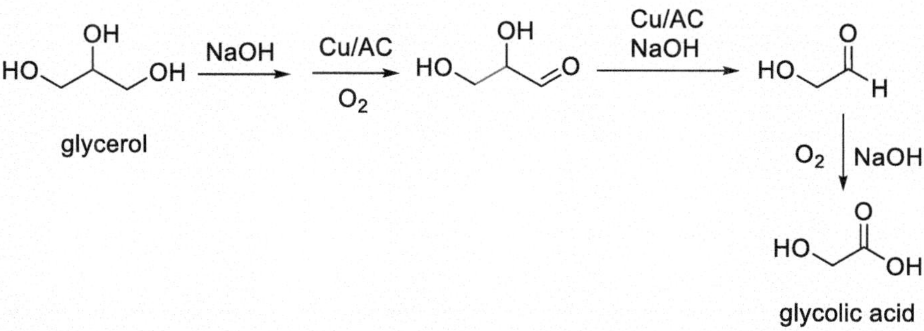

Scheme 3.5 Proposed reaction mechanism for glycerol conversion into glycolic acid

3.3.2.2 Glycerol Conversion

Li and Hu et al. reported the activated carbon (AC) supported Cu catalyst for the conversion of glycerol to glycolic acid [18]. The reaction achieved a 98.6% glycerol conversion and a 68.3% glycolic acid yield in water at 190 °C, 1 MPa O_2. Glycerol underwent the dehydrogenation oxidation to yield glyceraldehyde under O_2 atmosphere, and Cu/AC catalyzed the oxidation of glyceraldehyde to glycolic acid. The activation of O_2 by Cu atom of Cu(111) led to the formation of O radical species (e.g. $^{\bullet}O_2^-$, $^{\bullet}OH$), which acted as the oxidant for the reaction. Apart from that, it was proposed that the high content of surface oxygen vacancy on catalyst contributed to the activation of glycerol and the oxidation of glycolaldehyde to glycolic acid (Scheme 3.5).

Later, the same research group developed Cu/γ-Al_2O_3 for the production of glycolic acid from glycerol [19]. A 71.3% reaction conversion and a 68.6% product selectivity were obtained at 483 K, 1 MPa air in $NaHCO_3$ solution. The base significantly affect the product selectivity. Both $NaHCO_3$ and CuO enhanced the electrophilicity of C2–C3 in glycerol aldehyde and facilitated its cleavage to glycolaldehyde, which was then oxidized to glycolic acid.

1,3-Dihydroxyacetone (DHA), a derivative from glycerol, is also used for glycolic acid production. Shi research group used Cu/Al_2O_3 catalyst for 1,3-dihydroxyacetone conversion to glycolic acid in the presence of H_2O_2, and achieved an 85% glycolic acid yield at room temperature [20]. Cu/Al_2O_3 was proposed to activate H_2O_2 to form $^{\bullet}OH$ radicals, which oxidized DHA to form the $^{\bullet}C(O)CH_2OH$ by the C2–C3 bond cleavage. This specific active species then reacted with $^{\bullet}OH$ to produce glycolic acid. The reaction proceeded without bases or acids, and produced minimal waste water.

3.3.2.3 Cellulose Conversion

Zhang et al. employed MoO_3 catalyst for the conversion of corn stalk cellulose into glycolic acid, which offered a 23.4% glycolic acid yield at 210 °C in 60 min [21]. The reaction pathway is illustrated in Scheme 3.6. The epimerization of glucose to mannose was identified as the key step, whose energy barriers were significantly reduced by using MoO_3 catalyst, and the whole reaction efficiency was thereby enhanced.

3.3.3 Oxalic Acid Production

Oxalic acid is the simplest dicarboxylic acid that can be used in the production of celluloid and rayon, leather manufacture and dressing, extraction of metal cations from rosin, and etc. Normally, it can be selectively produced by the oxidation of cellulose with nitric acid, but associating with the environmental pollution issues. Thus, green oxidant and inexpensive catalytic system are more appealing options in currently developed systems.

3.3.3.1 Cellulose Conversion

Han et al. reported the use of CuO catalyst for the oxidative decomposition of cellulose to oxalic acid in the presence of O_2 [22]. The reaction achieved a 41.5% yield of oxalic acid in NaOH reaction solution at 200 °C, 0.3 MPa O_2 for 2 h. In this reaction, cellulose was firstly depolymerized to glucose, which then coordinated on Cu surface and transformed into glyceraldehyde via C3-C4 cleavage.

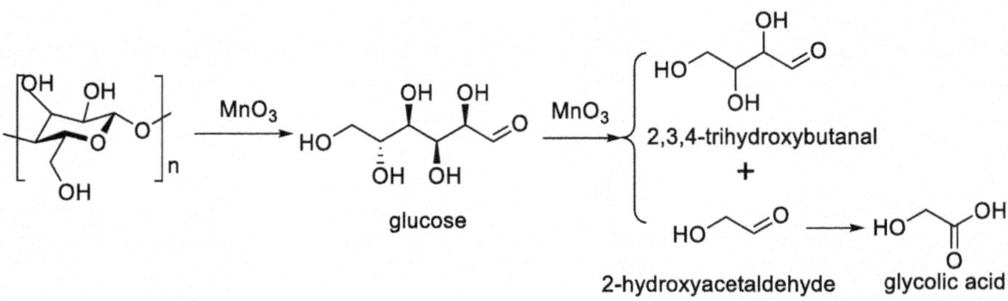

Scheme 3.6 Reaction path for cellulose to glycolic acid conversion

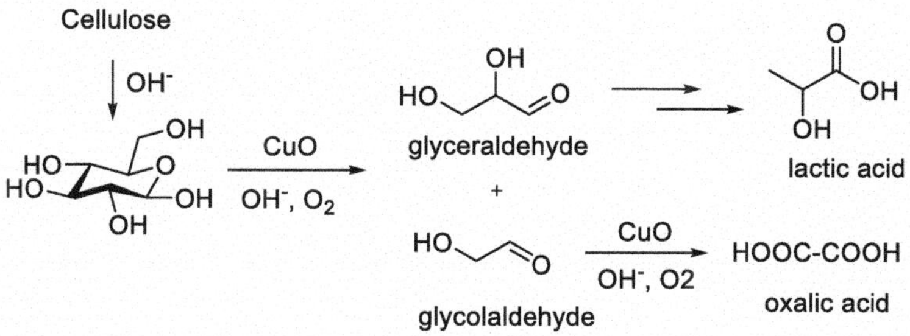

Scheme 3.7 Conversion of cellulose to oxalic acid over Cu catalysts

Glyceraldehyde was further converted to formaldehyde and glycol aldehyde through retro-aldol condensation, and further oxidized to glycolic acid and oxalic acid, respectively (Scheme 3.7).

3.3.3.2 Glycerol Conversion

In addition to cellulose, glycerol is also the feedstock for oxalic acid synthesis. Xu et al. employed Au/CeO_2 catalyst for the oxidation of glycerol in NaOH solution. The main product was oxalic acid and the product yield was 55% with a productivity of 0.76 mmol g^{-1} h^{-1} [23]. The unique property of CeO_2 provided active sites for the cleavage of C-C bond in glycerol, resulting in the production of formic acid and oxalic acid. This catalytic performance was different from that with other Au catalysts (HY, AC, REY supports), which offered tartronic acid as the main product.

3.3.3.3 HMF Conversion

HMF is a platform molecule derived from C6 carbohydrates. Mattia and Dibenedetto et al. reported the oxidation of HMF to oxalic acid with carbon nanotubes supported Fe catalyst (Fe@CNT) in the absence of alkali additive [24]. The reaction achieved a 38.4% yield of OA at 403 K, 10 atm of O_2 in 1.5 h. Reaction pathway study indicated that OA was mainly produced from the C4 intermediated succinic acid, which originated from the cleavage of HMF (Scheme 3.8).

3.4 C3 Organic Acids

3.4.1 Propionic Acid Production

Propionic acid (PA) is a C3 bulk platform chemicals commonly used in food, perfumes, agriculture and chemical synthesis industry. Traditional PA systhesis

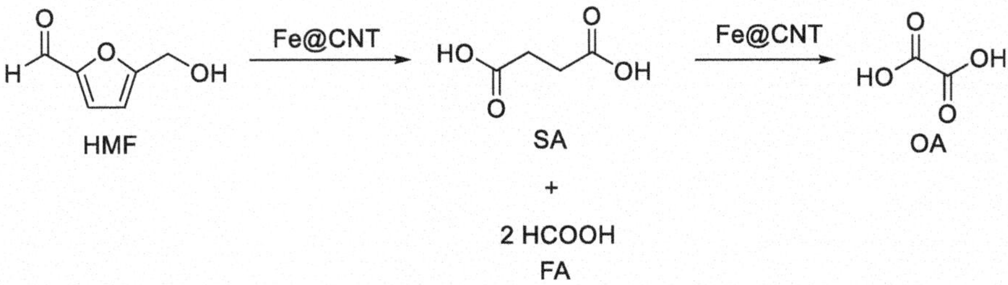

Scheme 3.8 Reaction pathway for 5-HMF to oxalic acid conversion

methods include propionaldehyde oxidation, ethylene carbonylation, ethanol carbonylation and etc. [25]. The oxidation of biomass molecules is more appealing over the past few years.

Huo et al. firstly used metallic Co catalyst in the presence of Zn reductant for the chemical conversion of lactic acid to propionic acid in water [26]. The reaction achieved a highest product yield of 58.8% at 250 °C for 2 h. The in-situ formed ZnO was proposed as the key species to promote the catalytic activity of Co catalyst. Other biomass carbohydrates such as glucose and starch were successfully converted to PA with 16.5%, 15.4%, respectively, under the optimized condition.

Dai and Tang et al. prepared a series of MoO_3/Fe catalysts to convert lactic acid to produce PA. The reaction achieved a 56.7% yield of PA at 390 °C [27]. The reduction temperature of MoO_3 decreased significantly when Fe was introduced and resulting the formation of active MoO_2 species. Both the metal types and metal loading were crucial factors to regulate the reduction degree of MoO_3, which in turn affected the catalytic performance. The catalyst with a Fe/Mo molar ratio of 0.908 offered an excellent stability in a 120 h reaction under a high LHSV of 37 h^{-1} at 390 °C.

3.4.2 Tartronic Acid Production

Tartronic acid is a versatile chemical that can be widely used for the production of biopolymer, anticorrosive protective agents, and pharmaceuticals. Glycerol is the primary biomass feedstock for tartronic acid production through metal catalysis process.

Xu et al. prepared a series of Au nanoclusters immoblized on different supports (CeO_2, HY, NaY, activated carbon) for glycerol oxidation under mild conditions [23]. The reaction attained a 98% conversion and an 80% TA yield over Au/HY catalyst at 60 °C, 0.3 MPa O_2 in 9 h. Further characterization indicated that the small Au nanoparticles was the key to the high oxidation ability of the catalyst (Scheme 3.9).

Evans et al. prepared $LaMnO_3$ perovskite supported Au catalysts for glycerol oxidaiton [28]. The reaction achieved a 44% yield of tartronic acid in a 4:1 NaOH:

Scheme 3.9 Glycerol oxidation to tartronic acid over Au catalyst

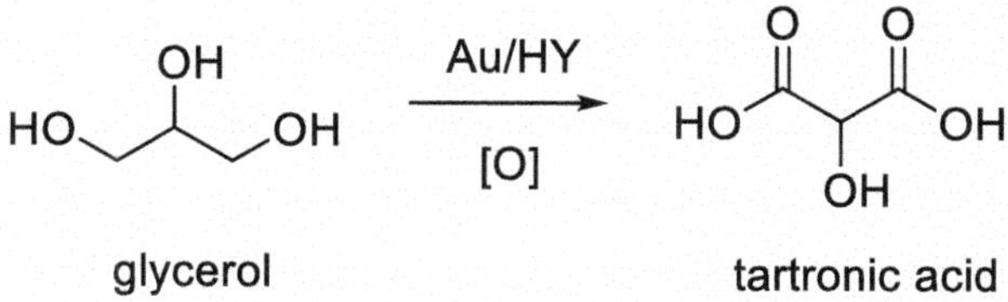

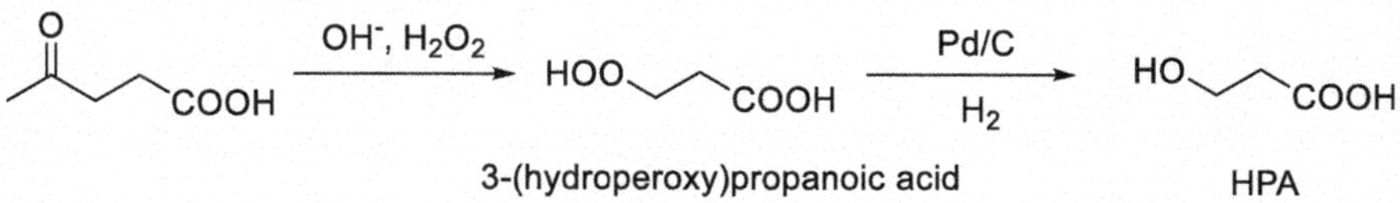

Scheme 3.10 Conversion of levulinic acid to HPA via a two-step process

glycerol solution at 373 K, 3 bar O_2 condition. The use of $LaMnO_3$ support could suppress the dehydration and oxidation pathway to generate lactic acid, regardless of metal used (Au or Pt, or bimetallic AuPt). The direct oxidation of glyceraldehyde intermediate to glyceric acid and further to tartronic acid was proposed as the main reaction route.

3.4.3 3-Hydroxypropanoic Acid Production

3-Hydroxypropanoic acid (HPA) is a C3 organic molecule that can be used for synthesizing acrylic acid, poly-3-HPA, propiolactone, polyesters and other high value compounds [29]. It is mainly prepared from fossil fuel, such as hydration of acrylic acid and the oxidation of allylic alcohol.

LA can serve as the starting material for HPA production. Mascal et al. reported a two-step chemical conversion of LA to HPA [30]. The first step was the oxidation of LA to 3-(hydroperoxy)propanoic acid with KOH and H_2O_2, which was followed by the reduction of 3-(hydroperoxy)propanoic acid to HPA with Pd/C catalyst at room temperature, 3.8 bar H_2 (Scheme 3.10). The total product yield reached 80%, which provided an appealing route to HPA synthesis avoiding the fermentation pathway.

Apart from LA, 1,3-propanediol can also be used for the production of HPA. Ebitani et al. prepared $Au_{0.8}Pd_{0.2}$-PVP/HT catalysts by a polyol reduction method, and used it for the aerobic oxidation of 1,3-propanediol to 3-HPA [31]. The reaction achieved a 42% yield of 3-HPA at 343 K under an O_2 gas flow. Bimetallic catalyst exhibited a higher activity than the monometallic Au/HT catalyst for the aerobic oxidation, with a lower activation energy of 26.1 kJ/mol.

3.4.4 *Acrylic Acid Production*

Acrylic acid is widely used in industry for the synthesis of water absorbent polymers and plastics. It is commonly produced by propylene oxidation [32]. Recently, several studies used biomass resource, such as lactic acid and glycerol, as the raw materials to synthesize acrylic acid.

3.4.4.1 Glycerol Conversion

The oxidative dehydration of glycerol has emerged as a promising strategy for the sustainable production of acrylic acid [33]. Most literatures focused on the use of mixed oxide catalysts, particularly those containing tungsten, vanadium, or molybdenum. Ueda et al. prepared a series of W-V-Nb-O mixed oxides with different V contents by hydrothermal method. Under the optimal conditions of 558 K, the yield of acrylic acid was 54.6% over $W_{2.2}V_{0.4}Nb_{2.4}O_{14}$ catalyst [34]. The introduction of W increased the Brønsted acidity of catalysts, and enhanced the activity of glycerol dehydration to acrolein, while V exhibited high selectivity in the oxidation of acrolein to acrylic acid.

Alternatively, Mascarenhas et al. prepared Fe-MCM-22 catalyst for the oxidative dehydration of glycerol to acrylic acid (Scheme 3.11) [35]. The reaction attained a 57% product yield after 10 h of reaction using 1.2 wt% Fe-MCM-22. The increase of Fe content resulted in a decrease of acid sites density and ET surface area of the catalysts. Fe^{3+} oligomers and iron oxide clusters were identified as the main iron species. The catalyst containing the lowest Fe content showed better activities in acrylic acid production. A balance between acid density and BET surface was revealed as the key to higher glycerol conversion.

3.4.4.2 Lactic Acid Conversion

Tang et al. investigated the dehydration of lactic acid to acrylic acid over various metal sulphates, and found $BaSO_4$ had excellent activity for the reaction [36]. Under the optimal reaction condition of 400 °C, a 74% acrylic acid yield was achieved. The catalyst with a moderate acidity strength ($+3.3 < H_0 < +4.8$) was found to exhibit the highest reactivity. Besides, low reaction temperature favored the dehydration of

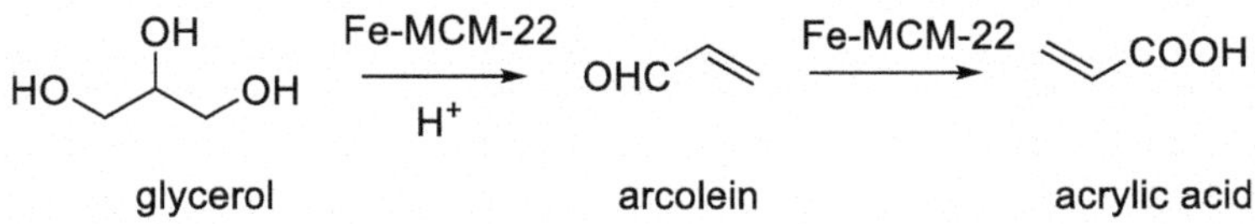

Scheme 3.11 Glycerol conversion to acrylic acid

Scheme 3.12 Conversion of lactic acid to acrylic acid over Ca-HAP catalyst

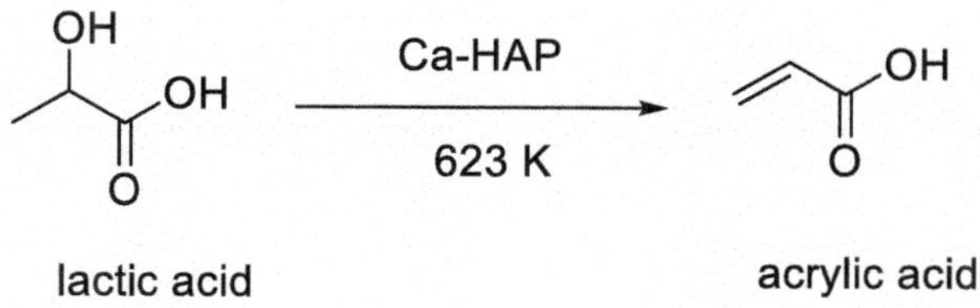

lactic acid to form acrylic acid, while high temperature favored decarbonylation/decarboxylation of lactic acid to form acetaldehyde.

Matsuura et al. reported the conversion of lactic acid to acrylic acid with Ca-HAP catalysts in gas phase (Scheme 3.12) [37]. The reaction provided a 60% yield of acrylic acid at 623 K. The surface characters in acid–base balance of the catalyst could be controlled by changing Ca/P ratios and Na contents in Ca-HAP catalysts.

3.4.5 *Lactic Acid Production*

Lactic acid is one of important chemicals derived from carbohydrates, and has been listed among the top ten sugar-based chemicals. It can be used for the synthesis a range of chemicals such as pyruvic acid, 2,3-pentanedione and acrylic acid [38]. Currently, the raw biomass materials for lactic acid synthesis are glucose and cellulose.

3.4.5.1 Glucose Conversion

Kupila et al. reported the conversion of glucose to lactic acid by biomass-based activated carbon supported metal oxide catalysts ($Sn/Al_{5/2.5}@AC_{SN}$) [39]. The utilization of nitric acid treatment in combination with the doping of Sn and Al oxides provided an active catalyst for lactic acid production, offering a 42% yield of lactic acid at 180 °C and 20 min. An isomerization and retro-aldol condensation reaction occurred in glucose structure and yielded glyceraldehyde and DHA on Lewis acid sites, which underwent a dehydration to pyruvaldehyde and finally to lactic acid.

Jin et al. employed ZnO as a simple and facile catalyst to convert glucose to lactic acid. Owning to the strong Lewis acid and base sites of catalysts, the reaction provided a 41.5% yield of lactic acid yield at 200 °C in 2 h (Scheme 3.13) [40]. The acid sites of catalysts promoted the polarization of carbonyl group and the hydride shift of glucose, while the base sites promotes the intermediate pyruvaldehyde transformation to lactic acid.

3.4.5.2 Cellulose Conversion

Lin et al. developed the $LaCoO_3$ perovskite metal oxide with strong redox properties for the conversion of cellulose to lactic acid [41]. Under a reaction temperature of

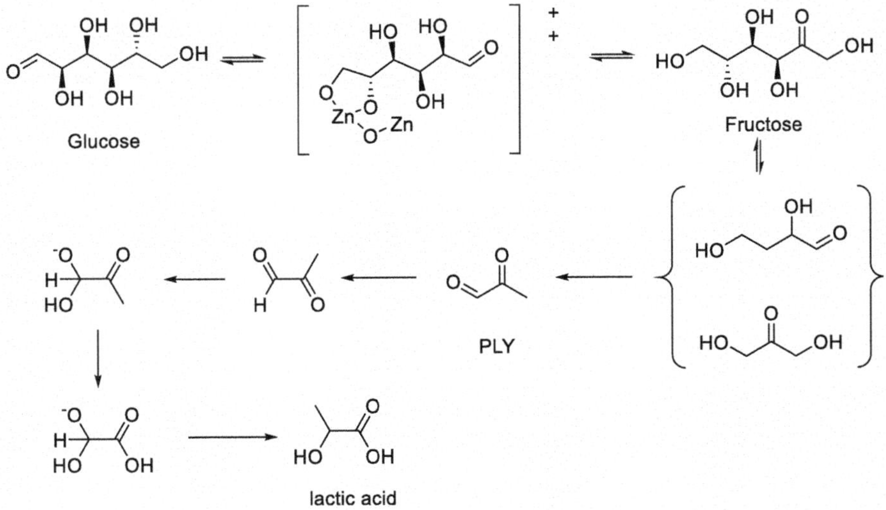

Scheme 3.13 Proposed reaction mechanism for lactic acid production from glucose

240 °C and water solvent, the reaction achieved a 24% yield of lactic acid. A possible redox reaction mechanism was proposed to involve the oxidative decarboxylation of aldose sugars and the reduction of pyruvic acid with the aid of the Co^{3+} and Co^{2+} redox cycles in the $LaCoO_3$ perovskite structure. The lattice oxygen atoms in the $LaCoO_3$ perovskite metal oxides acted as the weak oxidants in the selective conversions of aldoses to lactic acid under inert or even mild reductive atmosphere.

3.5 C4 Organic Acids

3.5.1 Succinic Acid Production

Succinic acid is a C4 chemical that can be used to produce fuel additives, pharmaceuticals, biopolymers, and other chemicals [42, 43]. Succinic acid is traditionally produced through the hydration or hydrogenation of petroleum derived maleic anhydride or maleic acid [43, 44]. Recently, the synthesis of succinic acid from biomass derivatives, e. g. furfural and levulinic acid, has attracted widespread attentions, and heterogeneous metal catalysts have been developed for such purpose.

3.5.1.1 Furfural Conversion

The key step for furfural conversion to succinic acid is a Bayer-Villager oxidation [44], where the C_5 furfural skeleton is converted into a C_4 based furfuran-2-(3H)-oneintermediate [45]. In 2010, Kunioka et al. reported a two-step method for the

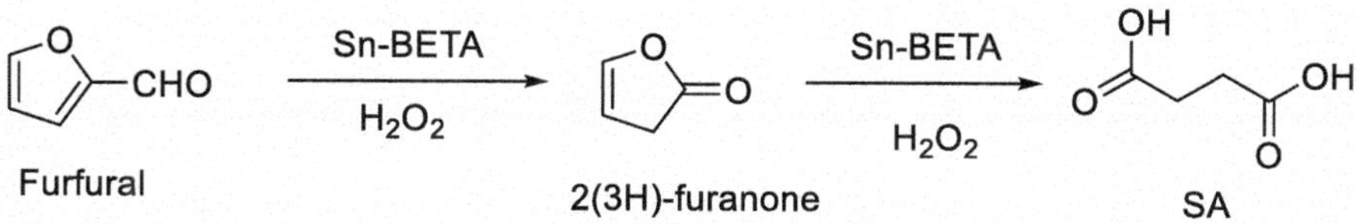

Scheme 3.14 Catalytic oxidation of furfural to bio-based succinic acid

Scheme 3.15 LA catalyzed oxidation to succinic acid over Ru-MNP catalyst

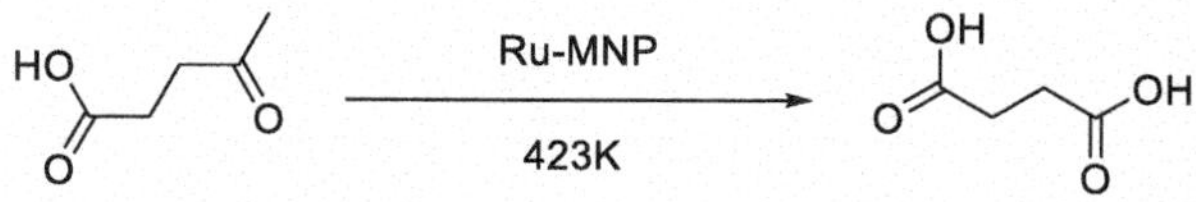

conversion of furfural to succinic acid [46]. By using V_2O_5 catalyst, furfural was firstly oxidized in aqueous solution by $NaClO_3$ to form fumaric acid in 47% yield. Then, fumaric acid was hydrogenated to succinic acid with a final yield of 98% over Pd/C catalyst under 13 MPa H_2 condition.

In 2022, Fukuoka et al. reported Sn-Beta catalyst for SA production from furfural with H_2O_2 as oxidant [47]. The reaction achieved a 53% yield at 50 °C in 6 h. Kinetic model showed that pure acidic Sn-Beta could promote the Baeyer-Villiger oxidation of furfural to form the intermediate 2(3H)-furanone (Scheme 3.14). DRIFTS analysis showed that reaction rate was directly related to the density of Lewis acid sites. In addition, Sn-Beta was incapable of activating H_2O_2, thus inhibiting some of the side reaction and promoting the product yield.

3.5.1.2 LA Conversion

The earliest succinic acid production from LA can be traced back to 1954 [48]. It was reported that V_2O_5 was able to convert LA into succinic acid and succinic anhydride at high temperature (473–673 K) with oxygen atmosphere. The selectivity of succinic acid was up to 81%. In 2012, V_2O_5 was used together with nitric acid as the oxidant for LA oxidation [49]. Under the condition of 313–333 K, the selectivity of succinic acid reached 38% to 52%.

In 2013, Parvulescu and Coman et al. synthesized the Ru/functional silica coated nanoparticle [Ru(III)-MNP] catalyst for the oxidation of LA to succinic acid (Scheme 3.15) [50]. The catalyst was prepared by depositing silica shells on the surface of ferromagnetic core through Fe-O-Si bond. The reaction achieved a 79.3% LA conversion and a 98.6% succinic acid selectivity in water at 423 K, 14 bar O_2 in 6 h. The catalyst could be recycled by a permanent magnet with a high recovery degree of 99%, and the leaching of metal sites was not observed.

3.5.1.3 Others

Succinic acid can also be produced from other bio-based molecules such as glucose and maleic anhydride. In 2016, Coman and Garcia et al. reported a one-pot conversion of glucose to succinic acid over Ru (III)@MNP catalyst [51]. The glucose conversion was 99.9% with an 87.5% selectivity of succinic acid. The addition of n-butylamine efficiently promoted the reaction selectivity, which was proposed to change the oxidation potential of Ru species by an electron-donating effect.

For the hydrogenation of maleic anhydride, Li et al. reported a Ni-Pt catalyst for the hydrogenation of maleic anhydride to produce SA [52]. By using Pt-Ni/Al_2O_3-HY catalyst, the catalyst provided a maximum 80% yield of succinic acid by hydrogenation of maleic anhydride at a mild temperature of 40 °C. The formation of bimetallic Ni-Pt ensembles led to enhanced activity and selectivity for partial hydrogenation.

3.5.2 *Maleic Acid Production*

Maleic acid (MAc) is a key chemical intermediate in the manufacturing of ethylene based copolymers, surface coatings, plasticizers, and drugs. MAc is normally produced by gas-phase oxidation of benzene or butene/butane [43]. The production of MA from renewable cellulose such as furfural and HMF is appealing to scientific community over the past few years.

3.5.2.1 Furfural Conversion

The preparation of MAc from furfural began in the early twentieth century. In 1928, Company Sessions [53] successfully prepared MAc from furfural oxidation using V_2O_5 catalyst at 473–573 K reaction temperatures, and the product yield was only 12.2%.

In 2014, Granados et al. successfully applied titanosilozeolite-1 (TS-1) catalyst for the oxidation of furfural to MAc in H_2O_2 solution [54], and a MAc yield of 78% was obtained at 323 K in 24 h. To reduce the use of H_2O_2, a two-step sequence of reactions was then developed by using TS-1 and Amberlyst-70 consecutively in the reaction, and the product yield reached 80% at a H_2O_2/furfural mol ratio = 4.4 [55]. Later, they developed a biphasic solvent containing γ-valerolactone (GVL) and water forfurfural oxidation [56]. By using TS-1 catalyst, a 70% yield of MAc was obtained. The catalyst was recovered for 17 times without deactivation. It was proved that the presence of GVL inhibited the deposition of insoluble by-products on the surface and cavity of TS-1. The proposed reaction mechanism is shown in Scheme 3.16.

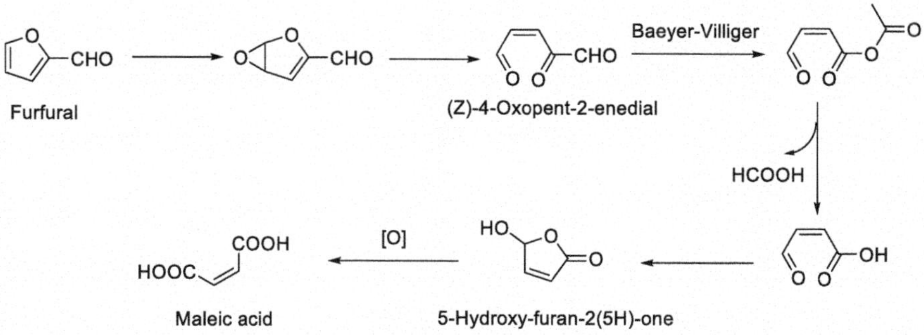

Scheme 3.16 Proposed Mechanism of Furfural Transformation into MA over TS-1 Catalyst

In 2017, Sotak et al. reported Cu-phosphate catalysts for aerobic oxidation of furfural to MAc [57]. $CaCuP_2O_7$ catalyst possessed the highest reactivity, achieving a 68% furfural conversion and a 37.3% MAc yield at 388 K, 0.8 MPa O_2. Mechanistic study proposed that the reaction proceeded through an initial oxidation of furfural to furoic acid, which further decarboxylated into furan cation and attacked by a water molecule to form 2(5H)-Furanone intermediate. Finaly, this intermediate was oxidized into maleic acid in the presence of catalyst.

In 2021, Li et al. reported a bimetallic catalyst supported on the hierarchical porous carbon (Cu-Pd@HPC) for furfural oxidation to MAc with H_2O_2 as oxidant [58]. The reaction achieved a 97.8% yield at 323 K in 4 h. DFT calculation showed that the introduction of Pd shifts the partial density of states of Cu toward the Fermi level, which enhanced the chemical adsorption of furfural on catalyst surface to facilitate the reaction.

3.5.2.2 HMF Conversion

In 2015, Yin et al. reported the use of heteropoly acid $H_5PV_2Mo_{10}O_{40}$ for the aerobic oxidation of HMF to MAc [59]. In acetonitrile/acetic acid solvent, the total yield of MAc and maleic anhydride reached 64% at 363 K, 10 atm O_2 and 8 h. Mechanism study showed that the cleavage of C-C bond between hydroxyl methyl group and furan ring of HMF triggered the production of MAc.

In 2018, Parunlescu and Coman et al. developed a metal-free catalyst (Fe_3O_4@SiO_2-NH2) for the oxidation of HMF to Mac in water [60]. A 35.6% HMF conversion and an 85% MAc selectivity were obtained under the reaction conditions of 110 °C, 10 bar O_2 and 8 h. Results showed that the $-NH_2$ group replaced the homogeneous strong base for catalysis process. The reaction path diagram is shown in Scheme 3.17.

In 2020, Deng et al. successfully oxidized HMF to MAc using V doped graphene oxide (V-GO) catalyst [61]. The effect of solvent on the reaction was investigated, where the reaction with GVL solvent provided a 97.81% HMF conversion and a 53.72% MAc and MA yield at 90 °C, 2 MPa O_2 and 4 h. 5-Hydroxyl-2(5H)-

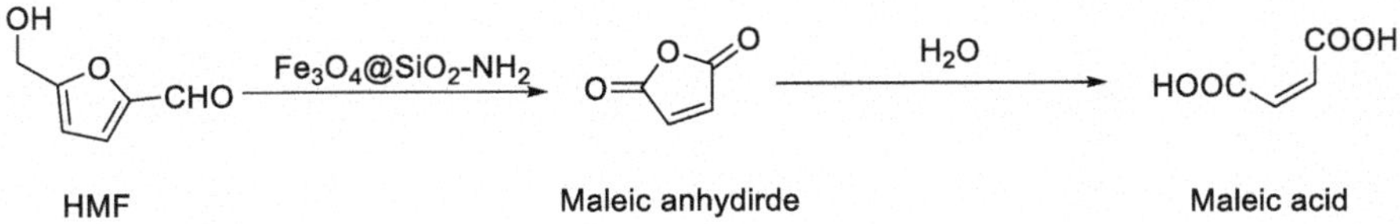

Scheme 3.17 Oxidation of HMF to Mac over basic catalyst

furanone was detected as the reaction intermediate by NMR characterization. Then, it was oxidized to MAc over vanadium catalyst under O_2 atmosphere. The solvent was proved to be important to 5-hydroxyl-2(5H)-furanone formation and its transformation.

3.5.2.3 Guaiacol Conversion

In 2014, Lin et al. reported a TS-1 catalyst for the catalytic oxidation of guaiacol to produce MAc under alkaline H_2O_2 solution [62]. The MAc yield reached 28% under the condition of 80 °C,.pH = 13.3, and 24 h. and Mechanism study showed that the synergistic effect of Brønsted base and TS-1 with H_2O_2 as oxidant leads to the ring-opening reaction of guaiacol to maleic acid, where the $^{\bullet}$OH radicals were the key species to proceed the reaction.

3.6 C5 Organic Acids

3.6.1 Furoic Acid Production

As an important downstream product of furfural, furoic acid has broad applications in food manufacturing, material preparation, optical technology, drug synthesis [63]. At present, furfural is the main raw material for furoic acid production.

In recent decades, several heterogeneous catalysts have been proposed for the conversion of furfural to furoic acid. Wojcieszak et al. prepared a core-shell structured SiO_2@Au@TiO_2 catalyst [64], which enabled the alkali-free air oxidation of furfural to furoic acid. Under optimal reaction condition with a 0.25% SiO_2@Au@TiO_2 catalyst, the furfural conversion reached 100%, and the selectivity of furoic acid exceeded 99%, which was one hundred times enhancement in catalytic activity as compared to that of traditional supported SiO_2@TiO_2@Au catalyst. The core-shell structure prevented the reversible adsorption of furoic acid on the active metal Au to avoid catalyst deactivation, and also promoted the metal-oxide interactions to form new catalytically active sites around Au-TiO_2 interface.

Xu et al. prepared the Co-NC catalyst for the conversion of furfuryl alcohol to furoic acid [65]. A 96.7% yield of furoic acid was obtained at 120 °C, 1 MPa O_2 in $NaHCO_3$ aqueous solution. The authors pointed out that the doping of N atoms in the

Scheme 3.18 Conversion of furfuryl alcohol or furfural to furoic acid with MnO_2 catalysts

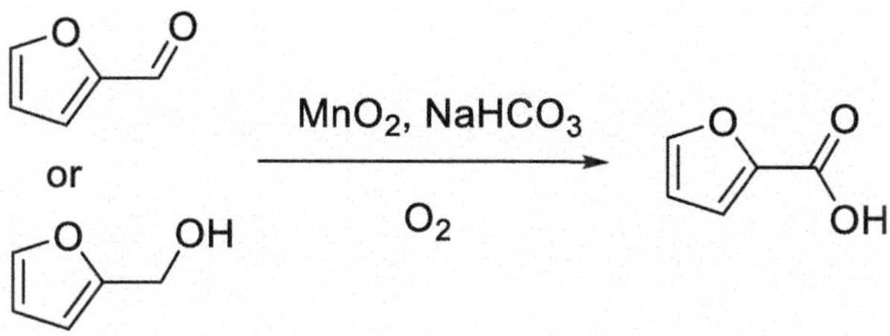

Scheme 3.19 Hydrogenation of furoic acid to tetrahydro-2-furoic acid

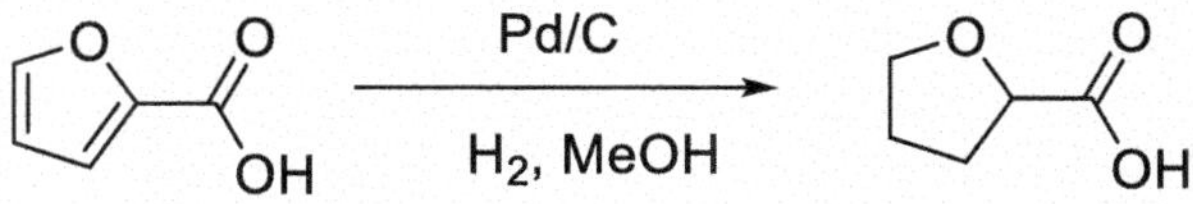

catalyst support could regulate the hydrogen binding standard Gibbs free energy, which would facilitate the activation of O-H bonds in the substrate.

Utilizing MnO_2 as catalyst [66], Hara et al. reported the conversion of furfural and furfuryl alcohol to furoic acid in $NaHCO_3$ aqueous solution under 1 MPa O_2, and achieved 95% and 99% product yield, respectively (Scheme 3.18). During the reaction, MnO_2 was partially reduced to $MnO_{2-\delta}$ due to its oxidation of substrate, but quickly reconverted to MnO_2 in the presence of O_2 for the next catalytic cycle.

3.6.2 Tetrahydro-2-Furoic Acid Production

Tetrahydro-2-furoic acid serves as a crucial chemical extensively employed in the synthesis of pharmaceuticals. One notable application is its application in the production of alfuzosin, a medication prescribed for the treatment of prostate enlargement [67].

Bhanu et al. presented a highly efficient system for the catalytic hydrogenation of furoic acid to tetrahydrofuroic acid with impressive yields ranging from 87 to 95% in water [68]. The reaction employed Pd/CoO_x catalyst and conducted under a mild reaction condition (30–60 °C, H_2 balloon). Experimental results indicated that the exceptional catalytic activity of Pd/CoO_x was attributed to the cooperative effect of Pd nanoparticles (approximately 3.5 nm) and the CoO_x support. It is worth noting that catalysts with either naked Pd nanoparticles or Pd immobilized on supports other than CoO_x exhibited lower activities.

Moreover, Nakagawa and Tomishige et al. reported the use of Pd/C catalysts for the hydrogenation of 2-furoic acid to tetrahydro-2-furoic acid in methanol solvent, and achieved 93% reaction conversion and 89% product selectivity at 100 °C and 40 bar H_2 pressure (Scheme 3.19) [69]. They also compared other carbon supported catalysts, such as Pt/C, Rh/C and Ru/C, which only offered moderate product yields (~60%).

3.6.3 LA Production

A lot of literatures have reported the synthesis of LA from real biomass feedstock. For example, Ga salt of polyoxometallate, known as GaHPMo, emerged as a potential substitute for traditional mineral acids in the rice straw conversion. A 56 wt% yield of LA could be obtained at 423 K in water [70]. GaHPMo exhibited superior catalytic performance in terms of activity for glucose conversion and selectivity for LA production relative to the parent HPMo. The reaction mainly proceeded through the hydrolysis of biomass to glucose, and further dehydrated to HMF. HMF was then hydrolyzed to generate LA and formic acid.

In addition, zeolites have been applied to the hydrolysis of biomass to LA. Ismadji et al. reported the conversion of corncob over acid-modified natural zeolite catalyst under subcritical water conditions [71]. The natural zeolite was modified by varied ratios of HCl, which changed the H^+ ions on the surface of zeolite. The highest yield of LA was 26 wt% at 200 °C in 60 min. A higher acidity of catalyst was found to promote the hydrolysis of cellulose and facilitat the dehydration of monomeric sugars into LA (Scheme 3.20).

3.6.4 D-Xylonic Acid Production

D-xylonic acid is a significant product derived from xylose or hemicellulose. It is renowned for good biocompatibility, biodegradability, and thermal stability, and can be used in diverse sectors such as cosmetics, agriculture, and pharmaceuticals [72]. Noble metal catalysts are reported for the oxidation of D-xylose to D-xylonic acid in the presence of alkaline reaction media. Liu and Peng et al. demonstrated that hierarchical porous carbon supported Au catalyst exhibited an outstanding reactivity in base-free aerobic oxidation of D-xylose [73]. Under the conditions of 100 °C, 3 bar O_2, the reaction achieved a 98% D-xylonic acid yield. Theoretical results indicated that Au NPs were anchored on pyridine N sites of catalyst support, and Au/NC catalyst could adsorb the reactants and intermediates formed during reaction, which facilitate their conversions (Scheme 3.21).

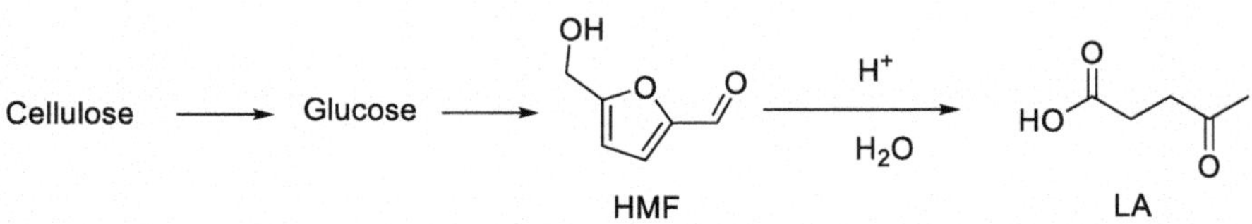

Scheme 3.20 Cellulose conversion to LA over acidic catalyst in water

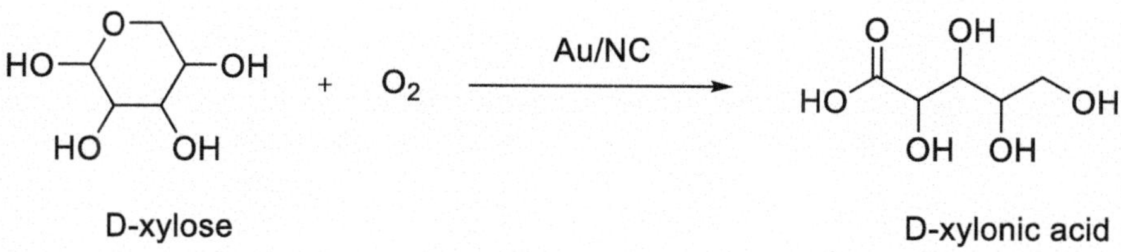

Scheme 3.21 Oxidation of D-xylose to D-xylonic acid over Au catalyst

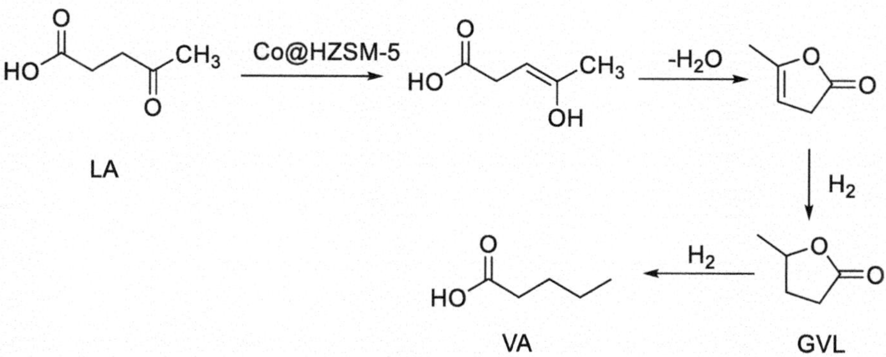

Scheme 3.22 Selective hydrogenation of LA to VA over Co catalyst

3.6.5 *Valeric Acid Production*

Valeric acid (VA) is an important platform chemical for synthesizing chemical intermediates and biofuels [74]. The esterification of VA provides valerate esters, which have been widely used as fuel additives. The synthesis of VA from LA encompasses the following steps (Scheme 3.22): (1) Hydro-cyclization of LA to GVL utilizing a hydrogenation catalyst and acid sites. (2) Ring-opening of GVL to form pentanoate ester by acid site catalysts. (3) Subsequent reduction of pentanoate ester isomers to VA. Sun et al. reported the synthesis of VA with a Co@HZSM-5 catalyst using both batch and fixed-bed reactors [75]. The reactions were conducted at a temperature of 240 °C under 3 MPa H_2, with either H_2O or EtOH as the solvent. The system with EtOH provided a complete LA conversion and a 97% VA yield. The Co NPs enclosed within HZSM-5 zeolite exhibited excellent stability without leaching or sintering in liquid-phase reaction (Scheme 3.22).

3.7 C6 Organic Acids

3.7.1 *Saccharic Acids Production*

Saccharic acids are important intermediates in pharmaceutical and food industries, which are traditionally produced by enzymatic oxidation of sugars. Glucose is one of

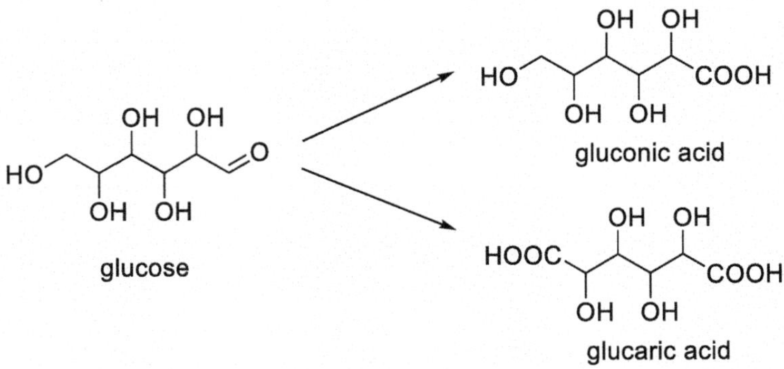

Scheme 3.23 Selective oxidation of glucose to gluconic acid and glucaric acid

Table 3.1 Catalysts for gluconic acid production from glucose

Entry	Substrate	Catalyst	Condition	Yield (%)	Ref.
1	Glucose	Pt/HT	50 °C, 1 bar O_2, Na_2CO_3 solution	83	[76]
2	Glucose	Au/ZrO_2	313 K, pH = 9	98	[77]
3	Glucose	Au/CMK-3	383 K, 0.3 MPa O_2	78	[78]
4	Glucose	Au/TiO_2	433 K, 3 bar O_2	84	[79]
5	Glucose	Pd/C	333 K	97	[80]
6	Glucose	Pd/C	323 K	50	[81]
7	Glucose	BiPd	313 K, pH = 9, air	99.3	[82]

the substrates to produce gluconic acid and glucaric acid in chemistry industry, both of which can be used as the food additives, drug additives, and biodegradable polymer monomer additives (Scheme 3.23).

3.7.1.1 Gluconic Acid Production

Noble metals are usually employed to catalyze the transformation of glucose to gluconic acid (Table 3.1). Anup Tathod et al. prepared Pt/hydrotalcite (HT) catalysts for glucose oxidation [76]. Under the reaction conditions of 50 °C, 1 bar O_2, and Na_2CO_3 solution, an 83% yield of gluconic acid was obtained after 12 h of reaction time. The use of basic support material facilitated desorption of intermediates and products from catalyst surface and suppressed the over-reaction to produced side products.

Au was also the well-known metal sites for oxidation reactions. Haruta et al. prepared a series of Au catalysts with ZrO_2, TiO_2, CeO_2 and Al_2O_3 supports by the solid grinding method, which offered samller Au sizes of 1.9–2.6 nm [77]. Au/ZrO_2 exhibited the highest catalytic activity in glucose oxidation, achieving a 98% yield of gluconic acid under the condition of 50 °C and pH 9.0. This study demonstrated that appropriate selection of support material was more important in the gas-phase reactions, whereas the control of the size of Au particles was more critical in liquid phase.

Apart from metal oxide supports, Yuan et al. prepared Au NPs supported on ordered mesoporous carbon (CMK-3) catalyst, and used it in the aerobic oxidation of glucose to gluconic acid under base-free conditions with molecular oxygen [78]. The reaction reached a 92% conversion with an 85% product selectivity at 383 K, 0.3 MPa O_2 in 2 h. The generation of acidic species in the reaction system suppressed the glucose isomerization to fructose, avoiding the dehydration route to generate side product, e.g. HMF, LA. H_2O_2 was generated during the reaction and acted as the oxidant for the oxidation.

The influence of catalyst preparation method was also crucial to the catalyst's reactivity. Hutchings et al. reported the base-free oxidation of glucose to gluconic acid with Au/TiO_2 catalyst prepared by the sol-immobilization method [79]. The preparation method and the amount of stabilizing ligand (PVA) had significantly impact on the final catalysts' activities. The maximum gluconic acid yield was 84% with the catalyst prepared with polyvinylalcohol: Au weight ratio of 0.1 at 160 °C, 3 bar O_2 in 1 h.

Other than Au catalyst, Pd is also one of the mostly studied metals used for glucose oxidation. Delidovich et al. compared the catalytic reactivities of Pd/C and Pt/C in the oxidation of glucose into gluconic acid with molecular oxygen [80]. Results showed that Pd/C was more reactive than Pt/C in the reaction, and achieved a 97% yield of gluconic acid at 60 °C. It was observed that the activities of prereduced and non-reduced Pt catalysts were identical, whereas the activity of Pd catalysts increased in twoolds when the catalyst was pre-reduced.

Haynes et al. synthesized two Pd/C catalysts by precipitation-reduction and urea-assisted sodium formate homogeneous deposition method, respectively [81]. The variation of thermal treatment temperature (573, 873 and 1173 K) resulted in different metal particle sizes of the catalysts. The yields of gluconic acid were strongly influenced by Pd particle size, with the catalyst of with an average particle size of 7 nm attaining the highest product yield of 50% at 50 °C.

The introduction of a second metal may promote the catalyst's reactivity. Besson et al. synthesized BiPd catalyst and subjected it to glucose oxidation. Under the condition of 313 K, pH = 9 and air atmosphere, a high gluconic acid yield of 99.3% was obtained [82]. The addition of bismuth element improved the stability of catalyst and prevented the oxidation of Pd NPs to loss their catalytic activities.

3.7.1.2 Glucaric Acid Production

Glucaric acid is a sugar-derived building block and used as a potential ingredient for deicing agents, corrosion inhibitors, and adipic acid precursors. Traditional processes for glucaric acid production relied on concentrated nitric acid, which generate significant amounts of toxic substances and waste products. Therefore, the use of green oxidant (air, molecular oxygen or hydrogen peroxide) attracts considerable research interests.

Chaudhari et al. synthesized a bimetallic PtCu catalyst supported on TiO_2 for the oxidation of glucose to glucaric acid [83]. Under the condition of 318 K and 0.1 MPa O_2, the reaction obtained a TOF of 3542 h^{-1} with a glucaric acid selectivity of 46% over Pt_1Cu_3/TiO_2. By comparing with the core-shell PtCu catalysts and monometallic Pt, Cu catalysts, the alloy structured PtCu generated well dispersed Pt and Cu particles of smaller sizes, resulted in enhanced oxidation activity and improved glucaric acid selectivity.

Later, the same research group designed bimetallic $PtPd/TiO_2$ catalyst for the direct oxidation of glucose to glucaric acid. [84]. The reaction achieved a 44% product yield at 318 K, 0.1 MPa O_2. Reaction pathway study indicated the reaction mainly proceed through the oxidation of glucose to gluconic acid, and further oxidation to glucaric acid. And glucose had a substrate inhibition effect for second step oxidation, resulting the accumulation of gluconic acid in the reaction mixture. Synergistic effect between the two metal was observed with improved product selectivity and yields.

3.7.2 *Furoic Acid Production*

HMF is a versatile substance that can be converted into a variety of chemicals used for the production of bioplastics, biofuels or macromolecular. The oxidation of HMF could produce several a series of furoic acids such as FFCA, HMFCA and FDCA depending on the catalytic system used (Scheme 3.24).

3.7.2.1 FDCA Production

FDCA is a representative biorefinery product with high potential as a replacement for the petroleum-based terephthalic acid for polymer synthesis. A lot of catalytic systems have been developed based on noble metal catalytic systems for FDCA production over the past few years.

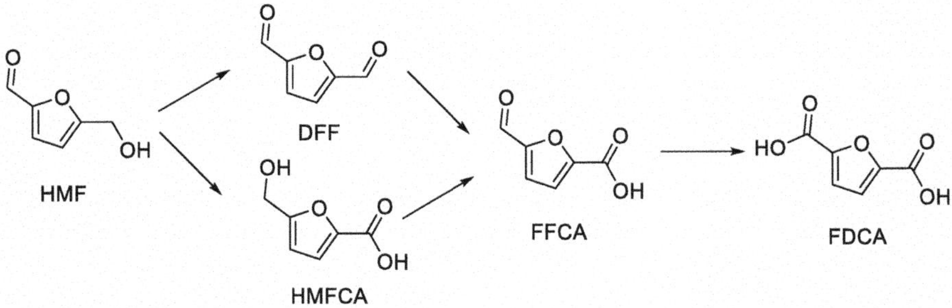

Scheme 3.24 Reaction pathways for selective oxidation of HMF

Casanova et al. reported the use of Au-CeO_2 and Au-TiO_2 catalysts for HMF oxidation [85]. High FDCA yields of 99% were obtained in NaOH solution under a mild condition of 65 °C and 10 bar air. HMFCA was quickly formed when the reaction started, which indicated a facile oxidation nature of aldehyde group as compared to $-CH_2OH$ group. HMFCA was gradually converted as the reaction proceeded. The base additive was proposed to desorb the acid product form Au surface to regenerate catalyst's activity.

Gorbanev et al. prepared a series of $Ru(OH)_x$ catalyst on different supports (TiO_2, Al_2O_3, Fe_3O_4, ZrO_2, CeO_2, MgO, La_2O_3, HT, etc.) and subjected them to the oxidation of HMF [86]. All the catalysts were found to be active in the oxidation reactions with distinct catalytic performances. For basic supports with MgO, HT, the yields of FDCA reached 95% under the condition of 2.5 bars O_2 and 140 °C, however, the dissolving of supports were observed. This also highlighted the importance of basic sites in HMF to FDCA oxidation.

Siyo et al. prepared a series of PVP stabilized Pd nanoparticles of different particle size for the oxidation of HMF to FDCA [87]. When the reaction proceed in an alkaline aqueous solution, the highest FDCA yield of 90% was obtained over Pd NPs with a mean diameter of 1.8 nm (prepared by a 4:1 molar ratio of NaOH: Pd solution). The work also revealed that smaller Pd particle sizes afforded higher FDCA yields.

3.7.2.2 FFCA Production

FFCA is an intermediate in HMF to FDCA conversion, but it may serve as a building block for chemical synthesis or the precursor of FDCA. Noble metal-based catalytic systems have been reported to efficiently transform HMF into FFCA. Saha et al. prepared two heterogeneous nanocatalysts (Au-TiO_2 and Au-CeO_2) for HMF oxidation into FFCA [88]. The reaction achieved a 79% FFCA yield over Au-TiO_2 catalyst in the presence of 1 wt% trifluoroacetic acid at 130 °C, 10 bar O_2. The reaction proceeded through a direct oxidation of $-CH_2OH$ group of HMF to form 2,5-diformylfuran, which was further oxidized to FFCA product. This was contrary to the traditional HMF oxidaiton with basic condition, where the aldehyde group was firstly oxidized to generate HMFCA.

Liu et al. reported the use of Ru/C catalyst for the oxidation of HMF in the presence of Mg-Al hydrotalcite (HT, Mg/Al molar ratio = 3/1) in water [89]. The reaction achieved 100% HMF conversion with 83% FFCA selectivity under the condition of 353 K, 2.0 MPa O_2. The addition of HT facilitated oxidation of the aldehyde group of HMF to FFCA. The experiments showed that the predominant species on Ru/C is metallic Ru^0 (the active sites) and water is the oxygen source for the oxidation of the formyl group.

Ventura et al. prepared mixed MgO·CeO_2 catalyst for oxidation of HMF [90]. The reaction was conducted in water at 403 K, 0.9 MPa O_2, and a 90% yield of FFCA was obtained. The use of mixed oxides allowed tuning of the basicity of the catalysts, avoiding the need for external bases for efficient and selective conversion of 5-HMF and waste formation.

3.7.2.3 HMFCA Production

HMFCA is also a crucial intermediate in HMF oxidation, which has been regarded as an essential indicator of the system's capacity to generate FDCA. Donoeva and Jongh et al. prepared carbon-supported Au catalysts for the oxidation of HMF [91]. The reaction provided a 93% yield of HMFCA under the condition of 90 °C, 10 bar O_2. The product selectivity strongly depend on the surface properties of the carbon support, with the acidic carbon supports being the most active support for the production of HMFCA.

3.7.3 *Adipic Acid Production*

Adipic acid is an aliphatic dicarboxylic acid mainly used in the production of nylon products or polyurethane products. It can be prepared from biomass-derived glucose, HMF and FDCA.

Zhang and Wang et al. prepared $Pt/Nb_2O_5 \cdot xH_2O$ catalyst for the conversion of FDCA to adipic acid [92]. Under the condition of 473 K and 3 MPa H_2, the reaction afforded a 38% yield of target product in water. The niobic acid participated in the cleavage of C-O-C bond inside the tetrahydrofuran ring and the C-OH bond in intermediates. Reaction pathway study indicated the hydrogenation of FDCA to THFDCA, whose C–O–C ether bond inside the furan ring was cleaved and the hydroxyl group were hydrogenolyzed to provide AA product (Scheme 3.25).

Later, the same research group reported a two-step conversion of glucose to adipic acid [93]. Pt/CNT effectively catalyzed the selective oxidation of glucose in basic solution, offering an 82% yield of glucaric acid. Then, Pd-ReOx/AC catalyst was used for the removal of four OH groups in glucaric acid to adipic acid in the form of methyl adipate in methanol and the yield of adipic acid reached 99%.

1,6-Hexanediol is one of the most relevant building blocks originating from biomass and it can transforms into AA. Robert Wojcieszak et al. prepared $Au/MgF_2–MgO$ for the base-free oxidation of 1,6-hexanediol in water [94]. The reaction achieved a 43% yield of AA at 110 °C and under 15 bar of air. The leaching of Mg was prevented by the introduction of MgF_2 in the oxides.

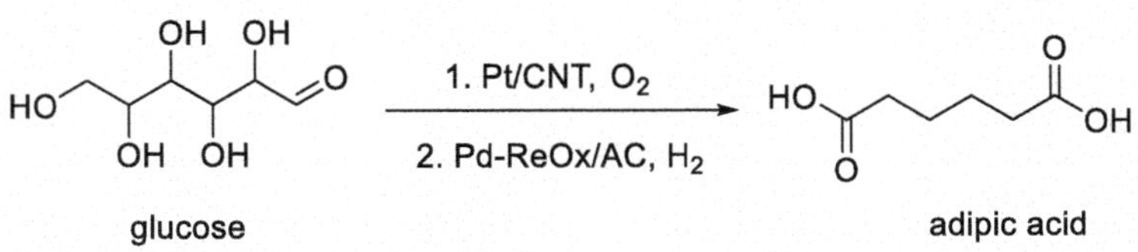

Scheme 3.25 A two step conversion of glucose to AA.

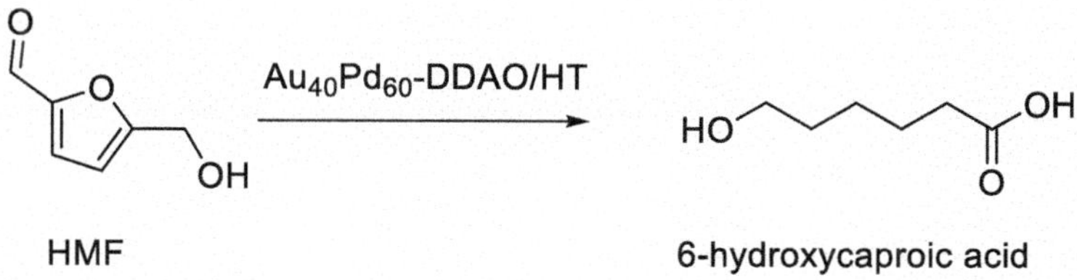

Scheme 3.26 Synthesis of 6-hydroxycaproic acid from HMF over AuPd catalyst

3.7.4 6-Hydroxycaproic Acid

6-Hydroxycaproic acid is a compound that has potential applications in the skin pharmaceutical, cosmetic and polymer industries. It can be prepared by the reduction of adipic acid or the oxidation of cyclohexane and cyclohexanone derived from fossil resources. Currently, 6-hydroxycaproic acid can be selectively prepared from biomass-derived HMF moleucles.

Tuteja et al. reported Au-Pd NPs supported on hydrotalcite as a heterogeneous catalyst for HMF conversion [95], and selective oxidation of 1,6-hexanediol into 6-hydroxycaproic acid was achieved using aqueous H_2O_2, and 1,6-hexanediol was prepared by biomass-derived HMF. $Au_{40}Pd_{60}$-DDAO/HT catalyst achieved a 93% yield of 6-hydroxycaproic acid under the condition of 353 K in 8 h. This work used N, N-dimethyldodecylamine N-oxide as an efficient capping agent, and achieved high selectivity (Scheme 3.26).

3.8 Conclusions and Future Outlook

Effective transformations of biomass resource into organic acids have been extensively studied over the past few years, and a lot of value-added compounds such as formic acid, acetic acid, glycolic acid, gluconic acid, glucaric acid, malonic acid and others, are produced by the catalytic oxidation process in the presence of heterogeneous catalysts. The oxidants used in these research work are O_2, H_2O_2, TMPO, nitric acid, and etc., which are activated to form the active species or directly oxidize the catalysts for oxidation reactions. The catalysts used in the oxidation reactions include supported metal nanoparticles, metal oxides, zeolites, among which the supported metal nanoparticles are the most frequently used types in the selective oxidations, featuring high reactivity high selectivity and good stability.

Albeit that, most of the above conversions mainly focused on the conversion of pure biomass derived compounds, e.g. glucose, furfural, which definitely require more complex procedures to obtain the starting materials and increase the cost of production. Thus, one pot conversion of real biomass into organic acids in over heterogeneous catalysts is an ultimate goal. However, the components of biomass are very complex, cascade reactions involving different reaction steps are required to ensure complete conversions. Composite catalysts integrating necessary catalytic sites are therefore the key to the above conversions. To facilitate industrial application, the stability of the catalysts should also be critically concerned. We anticipate that by developing new efficient catalyst and catalytic systems, the organic acid family from biomass resources will be further enriched and the procedures will be more simple and cost-efficient.

References

1. Huber GW, Iborra S, Corma A. Synthesis of transportation fuels from biomass: chemistry, catalysts, and engineering. Chem Rev. 2006;106:4044–55. https://doi.org/10.1021/cr068360d.
2. Zhang ZH, Huber GW. Catalytic oxidation of carbohydrates into organic acids and furan chemicals. Chem Soc Rev. 2018;47:1351–0. https://doi.org/10.1039/c7cs00213k.
3. Wang TF, Nolte MW, Shanks BH. Catalytic dehydration of C6 carbohydrates for the production of hydroxymethylfurfural (HMF) as a versatile platform chemical. Green Chem. 2014;16: 548–25. https://doi.org/10.1039/c3gc41365a.
4. Zakrzewska ME, Bogel LE, Bogel LR. Ionic liquid-mediated formation of 5-hydroxymethylfurfural-a promising biomass-derived building block. Chem Rev. 2011;111: 397–21. https://doi.org/10.1021/cr100171a.
5. Tong XL, Ma Y, Li YD. Biomass into chemicals: conversion of sugars to furan derivatives by catalytic processes. Appl Catal A Gen. 2010;385:1–13. https://doi.org/10.1016/j.apcata.2010.06.049.
6. Chen X, Liu Y, Wu JW. Sustainable production of formic acid from biomass and carbon dioxide. Mol Catal. 2020;483:1–16. https://doi.org/10.1016/j.mcat.2019.110716.
7. Shen F Jr, Li JL, Guo HX, Zhang X, Qi XH. Critical assessment of reaction pathways for conversion of agricultural waste biomass into formic acid. Green Chem. 2021;23:1536–26. https://doi.org/10.1039/d0gc04263c.
8. He RT, Ma T, Cheng J, Jin BB, Xu J. Formation of formic acid from glucose with simultaneous conversion of Ag2O to ag under mild hydrothermal conditions. ACS Omega. 2021;6:11260–6. https://doi.org/10.1021/acsomega.1c00106.
9. Wu LX, Yang Y, Cheng J, Shi XY, Zhong H, Jin FM. Highly efficient conversion of carbohydrates into formic acid with a heterogeneous MgO catalyst at near-ambient temperatures. ACS Sustain Chem Eng. 2022;10:15423–14. https://doi.org/10.1021/acssuschemeng.2c04502.
10. Li R, Lin QX, Wang YX, Yang WR, Liu XX, Li WY, Wang XH, Wang XY, Liu CF, Ren JL. Brønsted acid-driven conversion of glucose to xylose, arabinose and formic acid via selective C-C cleavage. Appl Catal B Environ. 2021;286:1–11. https://doi.org/10.1016/j.apcatb.2020.119862.
11. Atsushi T, Wataru O, Tatsumi I. Oxidative conversion of glucose to formic acid as a renewable hydrogen source using an abundant solid base catalyst. ChemistryOpen. 2021;10:954–6. https://doi.org/10.1002/open.202100074.
12. Chagas P, Thibau MA, Breder S, Souzab PP, Caldeira GS, Portilho MF, Castro CS, Oliveira LCA. Metal-free bifunctional silica for conversion of waste glycerol from biodiesel: sustainable production of formic acid. Chem Eng J. 2019;369:1102–7. https://doi.org/10.1016/j.cej.2019.03.068.
13. Xu JL, Zhao YF, Xu HJ, Zhang HY, Yu B, Hao LD, Liu ZM. Selective oxidation of glycerol to formic acid catalyzed by Ru(OH)4/r-GO in the presence of FeCl3. Appl Catal B Environ. 2014;155:267–73. https://doi.org/10.1016/j.apcatb.2014.02.034.

14. Lu T, Hou Y, Wu W, Niua M, Wang Y. Formic acid and acetic acid production from corn cob by catalytic oxidation using O2. Fuel Process Technol. 2018;171:133–7. https://doi.org/10.1016/j.fuproc.2017.11.010.
15. Huo Z, Fang Y, Yao G, Zeng X, Ren D, Jin F. Improved two-step hydrothermal process for acetic acid production from carbohydrate biomass. J Energy Chem. 2015;24(2):207–6. https://doi.org/10.1016/S2095-4956(15)60302-3.
16. Gao X, Chen X, Zhang J, Guo W, Jin F, Yan N. Transformation of chitin and waste shrimp shells into acetic acid and pyrrole. ACS Sustain Chem Eng. 2016;4(7):3912–9. https://doi.org/10.1021/acssuschemeng.6b00767.
17. Khallouk K, Solhy A, Khalfaouy RE, Kherbeche A, Barakat A. Mechanosynthezized Zn3V2O8 mixed oxide as efficient catalyst of xylose conversion to glycolic acid in water. Catal Lett. 2023;153(7):2210–3. https://doi.org/10.1007/s10562-022-04151-8.
18. Liao S, Tian Q, Xiao Y, Qin D, Li J, Hu C. Glycerol valorization towards glycolic acid production over Cu-based biochar catalyst. ChemSusChem. 2022;15(23):e202201537–12. https://doi.org/10.1002/cssc.202201537.
19. Xu S, Tian Q, Xiao Y, Zhang W, Liao S, Li J, Hu C. Regulating the competitive reaction pathway in glycerol conversion to lactic acid/glycolic acid selectively. J Catal. 2022;413:407–10. https://doi.org/10.1016/j.jcat.2022.07.003.
20. Dai X, Adomeit S, Rabeah J, Kreyenschulte C, Brgckner A, Wang H, Shi F. Sustainable Co-synthesis of glycolic acid, formamides and formates from 1,3-dihydroxyacetone by a Cu/Al2O3 catalyst with a single active sites. Angew. 2019;58(16):5251–5. https://doi.org/10.1002/anie.201814050.
21. Zhang Y, Cao J, Zhou Y, Li Y, Li L, Xie X. MoO3–catalyzed transformation of corn stalk cellulose to glycolic acid: an experimental and DFT study. Cellulose. 2023;30(6):3523–15. https://doi.org/10.1007/s10570-023-05139-2.
22. Jiang Z, Zhang Z, Song J, Meng Q, Zhou H, He Z, Han B. Metal-oxide-catalyzed efficient conversion of cellulose to oxalic acid in alkaline solution under low oxygen pressure. ACS Sustain Chem Eng. 2016;4(1):305–7. https://doi.org/10.1021/acssuschemeng.5b01212.
23. Cai J, Ma H, Zhang J, Du Z, Huang Y, Gao J, Xu J. Catalytic oxidation of glycerol to tartronic acid over Au/HY catalyst under mild conditions. Chin J Catal. 2014;35(10):1653–8. https://doi.org/10.1016/S1872-2067(14)60132-7.
24. Ventura M, Williamson D, Lobefaro F, Jones MD, Mattia D, Nocito F, Aresta M, Dibenedetto A. Sustainable synthesis of oxalic and succinic acid through aerobic oxidation of C6 polyols under mild conditions. ChemSusChem. 2018;11(6):1073–9. https://doi.org/10.1002/cssc.201702347.
25. Li H, Na J, Cong H, Wu L, Gao X. Thermodynamics foundation and separation process design for production of propionic acid from ethanol carbonylation catalyzed by iodide. Ind Eng Chem Res. 2020;59(13):6090–101. https://doi.org/10.1021/acs.iecr.9b06505.
26. Huo ZB, Xiao JF, Ren DZ, Jing FM, Wang T, Yao GD. Chemoselective synthesis of propionic acid from biomass and lactic acid over a cobalt catalyst in aqueous media. Green Chem. 2017;19(5):1308–14. https://doi.org/10.1039/c6gc03036j.
27. Li X, Pang J, Dai YS, Wang HQ, Tang CM, Chen Z, Yang CL. Sustainable production of propionic acid: catalytic deoxygenation of lactic acid over MoOx/Fe. New J Chem. 2021;45(35):15831–9. https://doi.org/10.1039/d1nj02050a.
28. Christopher DE, Jonathan KB, Stuart HT, Graham JH, Simon AK. Perovskite supported catalysts for the selective oxidation of glycerol to tartronic acid. Catal Lett. 2023;153:2026–35. https://doi.org/10.1007/s10562-022-04111-2.
29. Della Pina C, Falletta E, Rossi M. A green approach to chemical building blocks. The case of 3-hydroxypropanoic acid. Green Chem. 2011;13(7):1624–32. https://doi.org/10.1039/c1gc15052a.
30. Wu L, Dutta S, Mascal M. Efficient, chemical-catalytic approach to the production of 3-Hydroxypropanoic acid by oxidation of biomass-derived Levulinic acid with hydrogen peroxide. ChemSusChem. 2015;8(7):1167–9. https://doi.org/10.1002/cssc.201500025.

31. Mohammad M, Nishimura S, Ebitani K. Selective aerobic oxidation of 1,3-propanediol to 3-hydroxypropanoic acid using hydrotalcite supported bimetallic gold nanoparticle catalyst in water. AIP Conf Proc. 2015;1649:58–66.
32. Landi G, Lisi L, Russo G. Oxidation of propane and propylene to acrylic acid over vanadyl pyrophosphate. J Mol Catal A Chem. 2002;239(1–2):172–9.
33. Sun D, Yamada Y, Sato S, Ueda W. Glycerol as a potential renewable raw material for acrylic acid production. Green Chem. 2017;19(14):3186–213. https://doi.org/10.1039/c7gc00358g.
34. Omata K, Matsumoto K, Murayama T, Ueda W. Direct oxidative transformation of glycerol to acrylic acid over Nb-based complex metal oxide catalysts. Catal Today. 2016;259:205–12. https://doi.org/10.1016/j.cattod.2015.07.016.
35. Dos Santos MB, Andrade HMC, Mascarenhas AJS. Oxidative dehydration of glycerol over alternative H,Fe-MCM-22 catalysts: sustainable production of acrylic acid. Microporous Mesoporous Mater. 2019;278:366–77. https://doi.org/10.1016/j.micromeso.2019.01.016.
36. Peng J, Li X, Tang C, Bai W. Barium sulphate catalyzed dehydration of lactic acid to acrylic acid. Green Chem. 2014;16(1):108–11. https://doi.org/10.1039/c3gc42028k.
37. Matsuura Y, Onda A, Yanagisawa K. Selective conversion of lactic acid into acrylic acid over hydroxyapatite catalysts. Catal Commun. 2014;48:5–10. https://doi.org/10.1016/j.catcom.2014.01.009.
38. Dusselier M, Van Wouwe P, Dewaele A, Makshina E, Sels BF. Lactic acid as a platform chemical in the biobased economy: the role of chemocatalysis. Energy Environ Sci. 2013;6(5):1415–42. https://doi.org/10.1039/c3ee00069a.
39. Kupila R, Lappalainen K, Hu T, Romar H, Lassi U. Lignin-based activated carbon-supported metal oxide catalysts in lactic acid production from glucose. Appl Catal A-Gen. 2021;612:118011. https://doi.org/10.1016/j.apcata.2021.118011.
40. Xu H, Ye X, Shi X, Zhong H, He DP, Jin BB. ZnO as a simple and facile catalyst for acid-base coordination transformation of biomass-based monosaccharides into lactic acid. Mol Catal. 2022;522:112241. https://doi.org/10.1016/j.mcat.2022.112241.
41. Yang X, Yang L, Fan W, Lin HF. Effect of redox properties of LaCoO3 perovskite catalyst on production of lactic acid from cellulosic biomass. Catal Today. 2016;269:56–64. https://doi.org/10.1016/j.cattod.2015.12.003.
42. Zhang X, Xu S, Li Q, Zhang G, Xia H. Recent advances in the conversion of furfural into bio-chemicals through chemo- and bio-catalysis. RSC Adv. 2021;11(43):27042–16. https://doi.org/10.1039/d1ra04633k.
43. Verma M, Mandyal P, Singh D, Gupta N. Recent developments in heterogeneous catalytic routes for the sustainable production of succinic acid from biomass resources. ChemSusChem. 2020;13(16):4026–8. https://doi.org/10.1002/cssc.202000690.
44. Pembere AMS, Louis H, Wu H. Mechanism and dynamics of Baeyer–Villiger oxidation of furfural to maleic anhydride in presence of H2O2 and Au clusters. J Mol Model. 2023;29:359. https://doi.org/10.1007/s00894-023-05764-5.
45. Zhu J, Yin G. Catalytic transformation of the furfural platform into bifunctionalized monomers for polymer synthesis. ACS Catal. 2021;11(1):10058–25. https://doi.org/10.1021/acscatal.1c01989.
46. Tachibana Y, Masuda T, Funabashi M, Kunioka M. Chemical synthesis of fully biomass-based poly(butylene succinate) from inedible-biomass-based furfural and evaluation of its biomass carbon ratio. Biomacromolecules. 2010;11(10):2760–5. https://doi.org/10.1021/bm100820y.
47. Palai YN, Shrotri A, Fukuoka A. Selective oxidation of furfural to succinic acid over Lewis acidic Sn-Beta. ACS Catal. 2022;12(6):2760–9. https://doi.org/10.1021/acscatal.1c05348.
48. Dunlop AP, Smith S. Preparation of succinic acid. US patent, 2676186. 1954.
49. Stephan V, Frits V, Jacobus V. Succinic acid from biomass. WO 2012/044168 A1. 2012.
50. Podolean I, Kuncser V, Gheorghe N, Macovei D, Parvulescu VI, Coman SM. Ru-based magnetic nanoparticles (MNP) for succinic acid synthesis from levulinic acid. Green Chem. 2013;15(11):3077–5. https://doi.org/10.1039/c3gc41120f.

51. Podolean I, Rizescu C, Bala C, Rotariu L, Parvulescu VI, Coman SM, Garcia H. Unprecedented catalytic wet oxidation of glucose to succinic acid induced by the addition of n-Butylamine to a RuIII catalyst. ChemSusChem. 2016;9(17):2307–4. https://doi.org/10.1002/cssc.201600474.
52. Li J, Tian W, Shi L. Highly active and selective Nickel–Platinum catalyst for the low temperature hydrogenation of maleic anhydride to succinic anhydride and synthesis of succinic acid at 40 °C. Catal Lett. 2011;141(4):565–6. https://doi.org/10.1007/s10562-010-0533-7.
53. Nielsen ER. Vapor phase oxidation of furfural. Ind Eng Chem. 1949;41(2):365–3. https://doi.org/10.1021/ie50470a026.
54. Alonso-Fagundez N, Agirrezabal-Telleria I, Arias PL, Fierro JLG, Mariscal R, Granados ML. Aqueous-phase catalytic oxidation of furfural with H2O2: high yield of maleic acid by using titanium silicalite-1. RSC Adv. 2014;4(98):54960–12. https://doi.org/10.1039/c4ra11563e.
55. Alba-Rubio AC, Fierro JLG, Leon-Reina L, Mariscal R, Dumesic JA, Granados ML. Oxidation of furfural in aqueous H2O2 catalysed by titanium silicalite: deactivation processes and role of extraframework Ti oxides. Appl Catal B Environ. 2017;202:269–11. https://doi.org/10.1016/j.apcatb.2016.09.025.
56. Rodenas Y, Mariscal R, Fierro JLG, Alonso DM, Dumesic J, Granados ML. Improving the production of maleic acid from biomass: TS-1 catalysed aqueous phase oxidation of furfural in the presence of γ-valerolactone. Green Chem. 2018;20(12):2845–11. https://doi.org/10.1039/c8gc00857d.
57. Sotak T, Hronec M, Gal M, Dobrocka E, Skriniarova J. Aqueous-phase oxidation of furfural to maleic acid catalyzed by copper phosphate catalysts. Catal Lett. 2017;147(11):2714–9. https://doi.org/10.1007/s10562-017-2191-5.
58. Zhao X, Kong X, Wang F, Fang R, Li Y. Metal sub-nanoclusters confined within hierarchical porous carbons with high oxidation activity. Angew. 2021;60(19):10842–7. https://doi.org/10.1002/ange.202016591.
59. Lan J, Lin J, Chen Z, Yin G. Transformation of 5-hydroxymethylfurfural (HMF) to maleic anhydride by aerobic oxidation with heteropolyacid catalysts. ACS Catal. 2015;5(4):2035–6. https://doi.org/10.1021/cs501776n.
60. Tirsoaga A, Fergani ME, Parvulescu VI, Coman SM. Upgrade of 5-hydroxymethylfurfural to dicarboxylic acids onto multifunctional-based Fe3O4@SiO2 magnetic catalysts. ACS Sustain Chem Eng. 2018;6(11):14292–9. https://doi.org/10.1021/acssuschemeng.8b02962.
61. Chai L, Hou X, Cui X, Li H, Zhang N, Zhang H, Chen C, Wang Y, Deng T. 5-Hydroxymethylfurfural oxidation to maleic acid by O2 over graphene oxide supported vanadium: solvent effects and reaction mechanism. Chem Eng J. 2020;388:124187–8. https://doi.org/10.1016/j.cej.2020.124187.
62. Su J, Yang L, Liu RN, Lin H. Low-temperature oxidation of guaiacol to maleic acid over TS-1 catalyst in alkaline aqueous H2O2 solutions. Chin J Catal. 2014;35(5):622–8. https://doi.org/10.1016/S1872-2067(14)60039-5.
63. Douthwaite M, Huang X, Iqbal S, Miedziak PJ, Brett GL, Kondrat SA, Edwards JK, Sankar M, Knight DW, Bethell D, Hutchings GJ. The controlled catalytic oxidation of furfural to furoic acid using AuPd/Mg(OH)2. Cat Sci Technol. 2017;7(22):5284–10. https://doi.org/10.1039/c7cy01025g.
64. Ferraz CP, Navarro-Jaén S, Rossi LM, Dumeignil F, Ghazzal MN, Wojcieszak R. Enhancing the activity of gold supported catalysts by oxide coating: towards efficient oxidations. Green Chem. 2021;23(21):8453–5. https://doi.org/10.1039/d1gc02889h.
65. Liu X, Luo Y, Ma H, Zhang SJ, Che PH, Zhang MY, Gao J, Xu J. Hydrogen binding initiated activation of O−H bonds on a nitrogen-doped surface for the catalytic oxidation of biomass hydroxyl compounds. Angew Chem Int Ed. 2021;133(33):18251–8. https://doi.org/10.1002/ange.202103604.
66. Hayashi E, Komanoya T, Kamata K, Hara M. Heterogeneously-catalyzed aerobic oxidation of 5-hydroxymethylfurfural to 2,5-furandicarboxylic acid with MnO_2. ChemSusChem. 2017;10(4):654–5. https://doi.org/10.1002/cssc.201601443.

67. Bapat UR, Potams JP, Subramanian N, Valgeirsson J. U.S. Patent US2010/01740373A1. 2010.
68. Priya B, Kumar A, Dostagir SNHM, Shrotri A, Singh SK. Catalytic hydrogenation of biomass-derived furoic acid to tetrahydrofuroic acid derivatives over Pd/CoOx catalyst in water. ChemCatChem. 2022;14(16):e202200590–9. https://doi.org/10.1002/cctc.202200590.
69. Asano T, Takagi H, Nakagawa Y, Tamura M, Tomishige K. Selective hydrogenolysis of 2-furancarboxylic acid to 5-hydroxyvaleric acid derivatives over supported platinum catalysts. Green Chem. 2019;21(22):6133–13. https://doi.org/10.1039/C9GC03315G.
70. Kumar VB, Pulidindi IN, Mishra RK, Gedanken A. Development of Ga salt of molybdophosphoric acid for biomass conversion to levulinic acid. Energy Fuel. 2016;30(12): 10583–9. https://doi.org/10.1021/acs.energyfuels.6b02403.
71. Hartono CD, Marlie KJ, Putro JN, Soetardjo FE, Ju YH, Sirodj DAN, Ismadji S. Levulinic acid from corncob by subcritical water process. Int J Ind Chem. 2016;7:401–9. https://doi.org/10.1007/s40090-016-0086-8.
72. Ma J, Zhong L, Peng X, Sun R. D-Xylonic acid: a solvent and an effective biocatalyst for a three-component reaction. Green Chem. 2016;18(6):1738–13. https://doi.org/10.1039/c5gc01727k.
73. Meng X, Li Z, Li D, Huang Y, Ma J, Liu C, Peng X. Efficient base-free oxidation of monosaccharide into sugar acid under mild conditions using hierarchical porous carbon supported gold catalysts. Green Chem. 2020;22(8):2588–10. https://doi.org/10.1039/c9gc04333k.
74. Dayma G, Halter F, Foucher F, Mounaim-Rousselle C, Dagaut P. Laminar burning velocities of C4–C7 ethyl esters in a spherical combustion chamber: experimental and detailed kinetic modeling. Energy Fuel. 2012;26(11):6669–77. https://doi.org/10.1021/ef301254q.
75. Sun P, Gao G, Zhao Z, Xia C, Li F. Stabilization of cobalt catalysts by embedment for efficient production of valeric biofuel. ACS Catal. 2014;4(11):4136–7. https://doi.org/10.1021/cs501409s.
76. Tathod A, Kane T, Sanil ES, Dhepe PL. Solid base supported metal catalysts for the oxidation and hydrogenation of sugars. J Mol Catal A Chem. 2014;388-389:90–9. https://doi.org/10.1016/j.molcata.2013.09.014.
77. Ishida T, Kinoshita N, Okatsu H, Akita T, Takei T, Haruta M. Influence of the support and the size of gold clusters on catalytic activity for glucose oxidation. Angew Chem Int Ed. 2008;47: 9265–3. https://doi.org/10.1002/anie.200802845.
78. Qi P, Chen S, Chen J, Zheng J, Zheng X, Yuan Y. Catalysis and reactivation of ordered mesoporous carbon-supported gold nanoparticles for the base-free oxidation of glucose to gluconic acid. ACS Catal. 2015;5:2659–11. https://doi.org/10.1021/cs502093b.
79. Cao Y, Liu X, Iqbal S, Miedziak PJ, Edwards JK, Armstrong RD, Morgan DJ, Wang J, Hutchings GJ. Base-free oxidation of glucose to gluconic acidusing supported gold catalysts. Cat Sci Technol. 2016;6:10–117. https://doi.org/10.1039/C5CY00732A.
80. Delidovich IV, Taran OP, Matvienko LG, Simonov AN, Simakova IL, Bobrovskaya AN, Parmon VN. Selective oxidation of glucose over carbon-supported Pd and Pt catalysts. Catal Letters. 2010;140:14–7. https://doi.org/10.1007/s10562-010-0430-0.
81. Haynes T, Dubois V, Hermans S. Particle size effect in glucose oxidation with Pd/CB catalysts. Appl Catal A Gen. 2017;542:47–54. https://doi.org/10.1016/j.apcata.2017.05.008.
82. Besson M, Lahmer F, Gallezot P, Fuertes P, Fleche G. Catalytic oxidation of glucose on bismuth-promoted palladium catalysts. J Catal. 1995;152:116–5. https://doi.org/10.1006/jcat.1995.1065.
83. Jin X, Zhao M, Shen J, Yan W, He L, Thapa PS, Chaudhari RV. Exceptional performance of bimetallic Pt1Cu3/TiO2 nanocatalysts for oxidation of gluconic acid and glucose with O_2 to glucaric acid. J Catal. 2015;330:323–6. https://doi.org/10.1016/j.jcat.2015.05.018.
84. Jin X, Zhao M, Vora M, Shen J, Zeng C, Yan W, Chaudhari RV. Synergistic effects of bimetallic PtPd/TiO2 nanocatalysts in oxidation of glucose to glucaric acid: structure dependent activity and selectivity. Ind Eng Chem Res. 2016;55:2932–13. https://doi.org/10.1021/acs.iecr.5b04841.

85. Casanova O, Iborra S, Corma A. Biomass into chemicals: aerobic oxidation of 5-hydroxymethyl-2-furfural into 2,5-furandicarboxylic acid with gold nanoparticle catalysts. ChemSusChem. 2009;2:1138–6. https://doi.org/10.1002/cssc.200900137.
86. Gorbanev YY, Kegnæs S, Riisager A. Effect of support in heterogeneous ruthenium catalysts used for the selective aerobic oxidation of HMF in water. Top Catal. 2011;54:1318–6. https://doi.org/10.1007/s11244-011-9754-2.
87. Siyo B, Schneider M, Pohl MM, Langer P, Steinfeldt N. Characterization, and application of PVP-Pd NP in the aerobic oxidation of 5-hydroxymethylfurfural (HMF). Catal Letters. 2014;144:498–506. https://doi.org/10.1007/s10562-013-1186-0.
88. Saha B, Dutta S, Abu-Omar MM. Aerobic oxidation of 5-hydroxylmethylfurfural with homogeneous and nanoparticulate catalysts. Cat Sci Technol. 2012;2:79–3. https://doi.org/10.1039/c1cy00321f.
89. Xie J, Nie J, Liu H. Aqueous-phase selective aerobic oxidation of 5-hydroxymethylfurfural on Ru/C in the presence of base. Chin J Catal. 2014;35:937–44. https://doi.org/10.1016/S1872-2067(14)60136-4.
90. Ventura M, Lobefaro F, de Giglio E, Distaso M, Nocito F, Dibenedetto A. Selective aerobic oxidation of 5-hydroxymethylfurfural to 2,5-diformylfuran or 2-formyl-5-furancarboxylic acid in water by using MgO CeO2 mixed oxides as catalysts. ChemSusChem. 2018;11:1305–10. https://doi.org/10.1002/cssc.201800334.
91. Donoeva B, Masoud N, De Jongh PE. Carbon support surface effects in the gold-catalyzed oxidation of 5-hydroxymethylfurfural. ACS Catal. 2017;7:4581–10. https://doi.org/10.1021/acscatal.7b00829.
92. Wei L, Zhang J, Deng W, Xie S, Zhang Q, Wang Y. Catalytic transformation of 2,5-furandicarboxylic acid to adipic acid over niobic acid-supported Pt nanoparticles. Chem Commun. 2019;55:8013–6. https://doi.org/10.1039/c9cc02877c.
93. Deng W, Yan L, Wang B, Zhang Q, Song H, Wang S, Wang Y. Efficient catalysts for the green synthesis of adipic acid from biomass. Angew Chem Int Ed. 2021;60:4712–7. https://doi.org/10.1002/anie.202013843.
94. Capece N, Sadier A, Ferraz CP, Thuriot-Roukos J, Pietrowski M, Zieliński M, Wojcieszak R. Aerobic oxidation of 1,6-hexanediol to adipic acid over Au-based catalysts: the role of basic supports. Cat Sci Technol. 2020;10:2644–7. https://doi.org/10.1039/d0cy00183j.
95. Tuteja J, Nishimura S, Choudhary H, Ebitani K. Selective oxidation of 1,6-hexanediol to 6-hydroxycaproic acid over reusable hydrotalcite-supported Au-Pd bimetallic catalysts. ChemSusChem. 2015;8:1862–4. https://doi.org/10.1002/cssc.201500255.

Chapter 4
Organic Acids and Active Compounds from Hydrothermal Conversion of Biowaste and Its Application Potential

Yongdong Xu and Zhidan Liu

Abstract Hydrothermal conversion plays a crucial role in efficiently converting biomass s. There are organic acids and active compounds in aqueous phase product: Under certain reaction temperature and pressure hydrothermal treatment, biomass undergoes degradation and polymerization, yielding valuable small molecular acids and macromolecule humic acid, and other active compounds. The feedstock species and operational conditions influenced the yield and compounds distribution of hydrothermal aqueous products. These compounds showed various application potential including phytopathogen control, advanced material preparation, energy output and soil fertilizer, fostering comprehensive biomass utilization. Considering the effect of antimicroorganism and soil improvement of aqueous phase, the organic and inorganic compounds can be cyclically utilized in agriculture, serving as biofertilizers, soil amendments and phytopathogen control agents. In this chapter, the influence of compositions in feedstocks and operational conditions on organic acids and antimicrobial compounds in aqueous phase were discussed. Furthermore, the chapter addressed the potential application of hydrothermal aqueous product in agriculture for phytopathogen control and soil fertilizer. However, the challenge of agricultural application lies in the uncertainties of the beneficial components, environmental safety and crop safety.

Keywords Hydrothermal conversion · Biomass · Aqueous phase · Organic acids · Humic acid · Antimicrobial agent · Soil fertilizer

Y. Xu · Z. Liu (✉)
Laboratory of Environment-Enhancing Energy (E2E), Key Laboratory of Agricultural Engineering in Structure and Environment of Ministry of Agriculture and Rural Affairs, College of Water Resources and Civil Engineering, China Agricultural University, Beijing, China

State Key Laboratory of Efficient Utilization of Agricultural Water Resources, Beijing, China
e-mail: zdliu@cau.edu.cn

Z. Fang et al. (eds.), *Production of Organic Acids and Alcohols from Agricultural Residues and Food Wastes*, Biofuels and Biorefineries 13,
https://doi.org/10.1007/978-981-96-1222-2_4

4.1 Introduction

Hydrothermal conversion (HTC) is a method that involves chemical reactions under certain temperature and pressure, which converts various wet biomass into valuable products without drying process [1, 2]. In HTC process, the water environment triggers reactions such as depolymerization, condensation, and oxidation of organic substrates, resulting in the production of diverse substances [3]. Agricultural biomass such as poultry manure, straw forestry residues, is crucial feedstocks for HTC, making it particularly significant in reducing the amount of agricultural waste to produce valuable compounds [4, 5]. Compared to other technologies such as anaerobic fermentation, composting and landfill, it requires less operational time and has high efficiency on pathogenic microorganism reduction; and it cost less energy input than pyrolysis because there is no need for feedstock drying [2]. Through HTC, various high-value substances can be obtained, including hydrochar, bio-oil and energy gas [6–9]. Besides, aqueous by-products are generated [10, 11], where there are some organic acids and active compounds [12]. The aqueous can be generated via different hydrothermal processes, such as hydrothermal liquefaction, hydrothermal carbonization, hydrothermal humification or hydrothermal gasification. In this chapter, the aqueous product mainly refered to that from the former three processes.

These compounds in aqueous phase product from HTC have diverse structures and functions, exhibiting applications in chemical synthesis, biomass energy development, soil fertilizer and phytopathogen control [10, 13–15]. Under hydrothermal conditions, specific reaction pathways may lead to the formation of biologically active compounds from organic substrates: For example, nitrogen-containing substrates may generate compounds with fungicidal activity, while phenolic substrates may produce antioxidants via HTC [16, 17]. There are thousands of compounds in aqueous phase: The main components of aqueous phase include organic compounds such as small molecular acids, amides, esters, ketones, alcohols and phenols and some inorganic components such as ammonia nitrogen and potassium [4, 10, 18]. Apart from generating small-molecule acids in aqueous product, HTC can also produce large-molecule acids such as humic acid and fluvic acid, which play a crucial role in soil improvement and environmental remediation [19]. These compounds in aqueous products not only increases the value of HTC but also provides new chemical resources from renewable feedstocks [10].

Current research on HTC primarily focuses on the main products such as bio-oil, hydrochar and gas, with studies on the aqueous phase being more oriented towards wastewater treatment [20]. Utilizing the aqueous phase products has significant technical importance, enabling comprehensive application of hydrothermal conversion products. This chapter will introduce the organic chemicals (acids and active substances) in the aqueous phase products from biomass via HTC, including the conversion pathway of compositions in feedstocks and the influence of operational parameters on the characteristics of the aqueous phase. Given the numerous components in the aqueous phase products from HTC and their application potential, this chapter introduces the agricultural application prospects of biomass HTC aqueous phase products for phytopathogen control and soil fertilizer. The challenges faced in their agricultural application are also discussed.

4.2 Organic Chemicals in Aqueous Products from HTC

4.2.1 Small Molecular Organic Compounds in Aqueous Products

4.2.1.1 Small Molecular Acids in Aqueous Products from HTC

The HTC process yields complex mixtures of substances due to the intricate nature of raw material composition and reaction conditions [1, 21]. There are many kinds of small molecular acids, such as formic acid, acetic acid, and lactic acid in aqueous phase of HTC (Fig. 4.1) [12, 22]. The physicochemical properties of the reaction raw materials, the temperature and time of HTC and catalyst can affect the quality of hydrothermal acid production [1, 2].

A significant source of raw material for HTC to produce acid substances in aqueous phase is corn straw. Volatile fatty acids (VFAs) originate from the decomposition of cellulose and hemicellulose in straw. In a previous study using straw as feedstock, the production of VFAs is influenced by different hydrothermal temperatures and times. In order to compare different hydrothermal treatments by

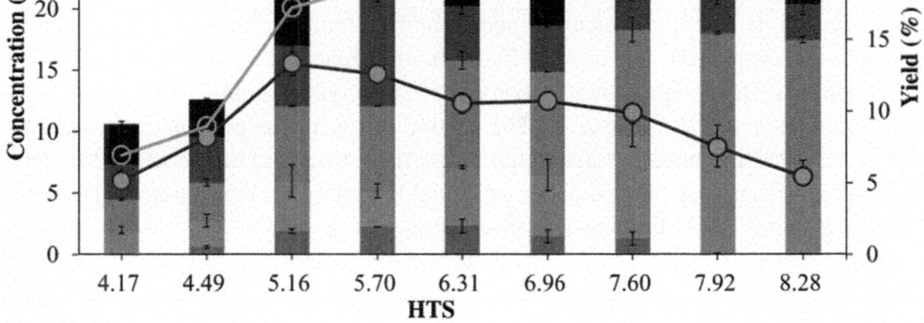

Fig. 4.1 Yields of reducing sugars and VFAs (Volatile fatty acids) at different HTS (Hydrothermal treatment severity). *Yield R* reducing sugars yield based on the dry biomass; *Yield V* VFAs yield based on the dry biomass. (Reprinted with permission from [13]. Copyright © 2016, Elsevier)

accounting for both temperature and time, hydrothermal treatment severity (HTS) is commonly quantified. The formula for HTS is:

$$\mathrm{HTS} = \int_{t1}^{t2} \exp\left(\frac{T - 100}{14.75}\right) \mathrm{d}t \tag{4.1}$$

In Eq. (4.1): t is the residence time (in minutes); $t1$ and $t2$ were the time at which the reactor temperature rose to and dropped to 100 °C, respectively (in degrees Celsius); The value 14.75 is a constant derived from the activation energy and gas constant for the degradation reactions involved.

The total yield of six VFAs (formic acid, lactic acid, acetic acid, propionic acid, butanoic acid and succinic acid) in aqueous phase generally increases with an increase in HTS. For instance, a prior study reported a significant increase in total VFAs when HTS increase from 4.17 to 6.31 (Fig. 4.1) [13]. As HTS increases, more compositions in feedstock transform into VFAs. The VFAs in aqueous phase are from the further decomposition of reducing sugars in reaction solution. The yield of reducing sugars during the hydrothermal process ranges between 5.15 and 13.31% (5.95–16.47 g/L), with the highest value observed at HTS 5.16. The rate of reducing sugar production initially increases with the rise in HTS from 4.17 to 5.16, followed by a decrease with continued HTS increase. Before reaching the maximum reducing sugar production, the accumulation is attributed to the hydrolysis of oligosaccharides initially derived from hemicellulose [2, 10, 13]. As hemicellulose is consumed, reducing sugars primarily stem from cellulose degradation. However, the degradation rate of reducing sugars surpasses that of generation, leading to a sustained decline in reducing sugar yield.

Under all HTS conditions, the concentration of acetic acid, as the primary VFA (23.67–36.24% of total VFAs, weight percentage), gradually increases from 2.50 to 10.09 g/L (Fig. 4.1). This is possibly due to the higher stability of acetic acid compared to other VFAs. Similar trends are observed for lactic acid and butanoic acid, with higher concentrations obtained at elevated HTS (6.96–8.28). Formic acid concentration decreases from 2.30 g/L (7.91% of total VFAs, at HTS 4.17) as HTS continues to rise [13], indicating the difficulty of accumulating formic acid owing to its active chemical properties at high temperature [22–25]. Additionally, a study employed a two-stage hydrothermal reaction to enhance acetic acid yield, where the first stage primarily converted carbohydrates into 5-hydroxymethyl-2-furfural, furfural and lactic acid. The second stage, in the presence of oxidants, further oxidized the products generated in the first stage into acetic acid [26]. This method significantly increased acetic acid yield, reaching a maximum of 16.3 and 17.5% for starch and cellulose, respectively.

Furthermore, the addition of oxidants in the solvent and control of the acid-base conditions in the reaction system facilitate the conversion of carbohydrate-based biomass into organic acids such as formic acid, acetic acid, and lactic acid. Catalysts like perovskite-type and solid heteropoly acids promote the production of lactic acid and formic acid from cellulose-based biomass [12, 27, 28]. The use of

$H_5PV_2Mo_{10}O_{40}$ solid acid catalyst with oxygen as the oxidant, under mild conditions (around 100 °C), converts natural carbohydrates into formic acid [29]. Under non-optimized conditions, soluble carbohydrates achieve a formic acid yield exceeding 40%, while more complex biomass, like wood chips, attains a 10% formic acid yield.

4.2.1.2 Compounds in Aqueous Products from Model Components

HTC aqueous products contain various substances. The composition of substances in raw materials can be classified as lipids, proteins, lignin, cellulose, hemicellulose and ash. The varied composition and content in raw materials contribute to the complex compounds in aqueous products.

The components in the raw materials are a crucial factor in determining the distribution of organic substances in the aqueous phase products. Five chemical components of hydrothermal feedstock namely lipids, proteins, lignin, cellulose and hemicellulose (and mixture of five compounds), were selected as model components to conduct reactions at five temperatures (200, 230, 260, 290 and 320 °C) for 30 min, resulting in 30 HTL aqueous products [10]. Results revealed significant differences in pH, oxidation-reduction potential (ORP), total organic carbon (TOC), inorganic carbon (IC), conductivity, total dissolved solid (TDS) and salinity among the 30 HTC aqueous phase samples, indicating the great influence of reaction temperature and feedstock composition on the content of chemicals in aqueous phase [2, 10, 30]. Generally, with increasing temperature, the diversity of compounds increased, while their relative concentrations of single compound decreased [10]. Small alkane compounds were consistently abundant in almost all HTC aqueous phase samples, with esters, phenols, aldehydes, phenols, and ketones being the top five most abundant substances (Fig. 4.2).

The non-uniform structure of hemicellulose makes it prone to decomposition in hydrothermal process [2]. During HTC, xylan (one kind of hemicellulose), initially breaks down into oligomers. And then, under mild conditions ($T < 200$ °C), oligomers further decomposed into monosaccharides. With increasing temperature, aldehydes, carboxylic acids, alkanes, lipids and alcohols are generated. For HTC aqueous phase from xylan sources at temperatures below 320 °C, a high proportion of aldehydes (>74.9%) was observed (Fig. 4.2e). Aldehydes were not detected at reaction temperature of 320 °C, suggesting the transformation of highly reactive aldehydes into alcohols, esters and ketones under high-temperature conditions. Cellulose initially decomposes into oligomers during HTC and then transforms into monosaccharide precursors through C-C bond cleavage and dehydration. Cellulosic materials undergo complete decomposition in subcritical water above 300 °C. At 200 °C and 230 °C, esters, aldehydes, and alcohols were the predominant compounds in cellulose HTC aqueous phase. At 260 and 290 °C, ketones increased in cellulose HTC aqueous phase, while at 320 °C, esters became the major compounds [10].

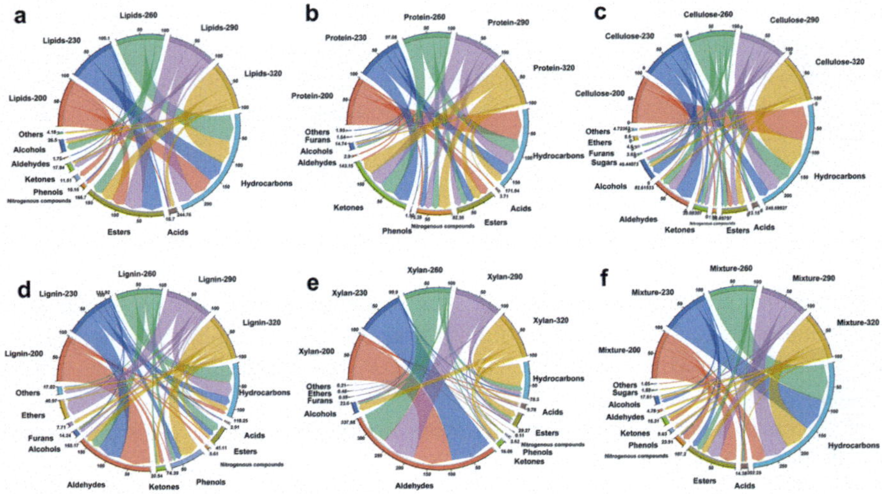

Fig. 4.2 Compounds distribution of HTC aqueous phase products from (**a**) lipids, (**b**) protein, (**c**) cellulose, (**d**) lignin, (**e**) xylan and (**f**) mixture feedstocks at 200, 230, 260, 290 and 320 °C (Relative peak areas based on the compounds detected by GC-MS). (Reprinted with permission from [10]. Copyright © 2022, Elsevier)

Lipids are essential components of HTC feedstock, found in livestock and poultry manure as well as algae [1, 31]. Studies indicate that the yield of biocrude from lipids during hydrothermal liquefaction surpasses the yields from proteins and lignocellulosic fibers [7, 32]. During HTC, lipids first decompose under mild conditions ($T < 260$ °C), and some decomposition products, such as short-chain esters, transfer to HTC aqueous phase. With further temperature increase, decarboxylation and decarbonylation reactions occur, leading to the formation of ketones, alcohols, aldehydes and acids. Esters are the major components in HTC aqueous phase at temperatures between 200 and 320 °C.

Protein is another important composition in raw materials, which influences the pH value of aqueous products. The high protein content will lead to alkaline aqueous, otherwise HTC aqueous phase from all other treatment groups exhibited acidity. Temperature plays a crucial role in protein hydrolysis in HTC [28]. Under mild temperatures (<230 °C), proteins and peptides undergo slight degradation and generate carboxylic acids and amino acids, with amines and pyrrole being the main nitrogen-containing compounds. But as the temperature increases, these components are further converted into amides through acylation reactions, resulting in an increase in the relative content and types of amides. Amino acid deamination is an important reaction that produces alpha keto acids and further converts them into ketones, resulting in abundant ketones in the protein HTC aqueous phase (Fig. 4.2b). The pH of mixed feedstock (mixture of five model compounds) HTC aqueous phase increased with reaction temperature from 3.95 to 5.36, related to the enhanced deamination reaction of proteins [3]. Amino acids in protein materials undergo deamination reactions during HTC, releasing ammonia nitrogen into HTC aqueous phase. With increasing hydrothermal reaction intensity, the deamination reaction strengthens, resulting in an increase in pH of the obtained HTC aqueous phase [20].

4.2.1.3 Compounds in Aqueous Products from Natural Biomass

There are complex interactions among the compositions in feedstocks and their intermediate products, leading to various compound in aqueous phase generation [2]. Currently, many studies have studied and compared the physicochemical properties of aqueous products obtained under different raw materials and reaction conditions [33–35]. Overall, 26–55% of the elements (by weight) in the raw materials transferred to the HTC aqueous phase [36]. The pH value ranges from 3.0 to 9.0, and the COD is concentrated in the range of 20–110 g/L. The components in the aqueous phase products mainly include organic acids, amides, esters, ketones, alcohols, and phenols [20].

The composition distribution in HTC aqueous phase (reaction temperature of 320 °C and reaction time of 30 min) from four kinds of natural biomass, microalgae, corn straw, cow manure, and swine manure are varied (Fig. 4.3) [1]. The ratio of lipids, proteins and carbohydrates leads to the difference of aqueous phase in compound concentration and kinds [1]. Microalgae, rich in proteins, produced nitrogen-containing ketones, peptides, pyrazoles, pyrimidines, and pyridines into

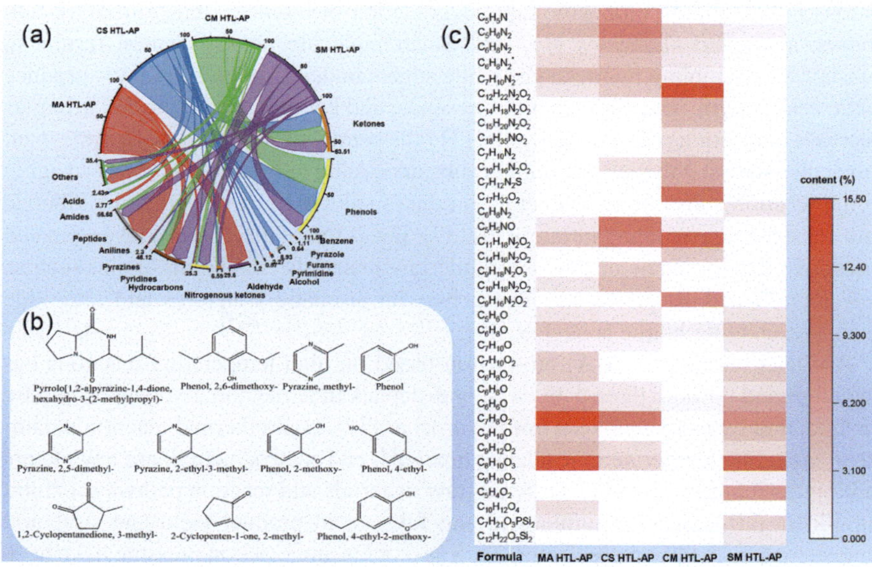

Fig. 4.3 Compound distribution (**a**), main compounds (**b**) and their relative contents (**c**) of four HTC aqueous phase products from microalgae, corn straw, cow manure and swine manure. (The relative peak areas were based on GC-MS analysis. The relative peak areas of listed compounds were higher than 2% in at least one HTC aqueous phase. *, Pyrazine, ethyl-, **, Pyrazine, 2,5-dimethyl-. (Reprinted with permission from [1]. Copyright © 2023, Elsevier)

aqueous phase. Deamination reactions and denitrification generate ammonia nitrogen, rendering aqueous phase product from microalgae alkaline (pH = 8.67). Swine manure, which also contains a significant amount of protein, resulted in the presence of pyridines and pyridines in aqueous phase product. During HTC, cellulose-based feedstocks can transform into phenols, ketones, and aldehydes. High cellulose content in corn straw and cow manure leads to higher levels of ketones and phenols in aqueous phase product. The low protein ratio in corn straw and cow manure resulted in acidic and oxidative aqueous phase product.

4.2.2 Macromolecule Humic Acid in Aqueous Products

In addition to the production of small molecular acids, adjusting the solvent in hydrothermal reactions can promote humification reactions to generate macromolecule humic acid and fulvic acid into aqueous phase [37]. Currently, various biomass including straw, animal manure, food waste and sludge, have been utilized for hydrothermal processes to obtain humic acids and fulvic acid in aqueous product and solid humin [38–40]. In the hydrothermal humification process, the composition in feedstock including cellulose, hemicellulose, lignin, starch, fats, and proteins undergo intricate processes of hydrolysis, dehydration, condensation, and polymerization to generate humic substances [41–43]. Alkaline and high-temperature conditions induce the decomposition and degradation of biomass, transforming it into sugars, acids and aldehydes [41, 44]. Small molecule polymerization results in substances resembling humic acids, while sugars under alkaline conditions produce lactic acid, butyric acid, various hydroxy acids, and ketones, leading to a continuous decrease in reaction pH [45]. When the pH falls below the neutralization point, small molecules start to polymerize, forming substances akin to fulvic acid [46, 47]. Lignin in the reaction substrate subsequently produces soluble phenols, which polymerize into substances resembling humic acids [37]. It was found that fulvic acid dominated the conversion products of glucose, while raw materials containing lignin, such as leaves and sawdust, produced substances with aromatic structures and fatty side chains resembling humic acids [19, 48].

Artificial humic acid (HA) production based on high-temperature reactions has been reported through three pathways: one-step alkaline method, two-step acid-base method, and two-step recovery from hydrochar [37]. Hydrothermal reaction parameters, including temperature, residence time, pH, and catalyst action, are key factors in biomass humification [37]. Different raw materials and reaction pathways exhibit significant differences in humic acid and fulvic acid product yield and structural characteristics. Additionally, factors such as oxygen and reaction pressure have been confirmed to promote humic acid production [41]. For biomass or waste liquids rich in lignin, increasing hydrogen peroxide, oxygen, or alkaline concentration can promote the generation of more molecules with phenolic, carboxyland quinone groups [49].

4.3 Application Potential of Aqueous Phase in Agriculture

4.3.1 Phytopathogen Control

These compounds in aqueous product from HTC have various applications in chemical synthesis, soil fertilizer and phytopathogen control [10, 13–15]. This chapter introduces the Application potential of aqueous phase in agriculture for Phytopathogen control and soil remediation.

Some compounds such as small molecular acids, aldehydes, phenols, and nitrogen-containing organic compounds in aqueous products obtained through HTC exhibit potent antimicrobial activity [10]. The antimicrobial potency of aqueous product from HTC is common and general, with experimental results demonstrating universal antimicrobial effects against *Escherichia coli* and *Staphylococcus aureus* for almost all aqueous products [1]. However, variations in antimicroorganism intensity exist among aqueous products due to differences in raw material sources and reaction conditions.

When conducting HTC experiments using model compounds as raw materials, aqueous product produced from xylan exhibits the largest inhibitory zone diameter, with the maximum inhibitory zone against *Escherichia coli* obtained at 230 °C in the xylan group (19.46 mm) and against *Staphylococcus aureus* at 200 °C in the xylan group (19.30 mm) [10], comparable to the inhibitory zone diameters produced by 100 μg/mL streptomycin (19.97 mm and 20.79 mm, respectively). As the reaction temperature increases, the diameter of the inhibitory zone on *Staphylococcus aureus* for xylan aqueous phase decreases from 19.30 mm to 13.29 mm. However, the dilutions of aqueous showed a different trend in antibacterial strength reduction (Fig. 4.4).

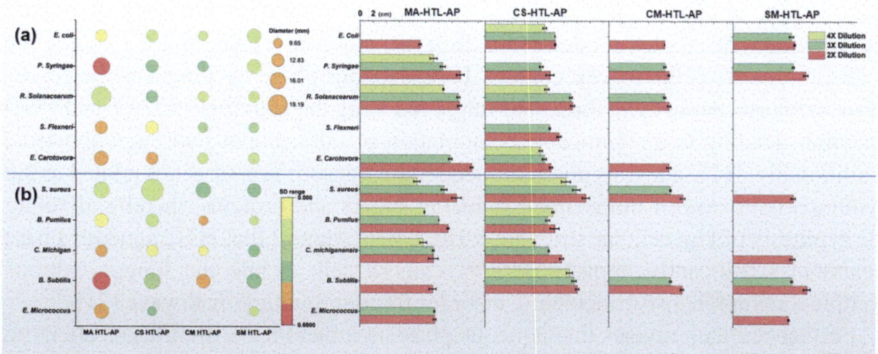

Fig. 4.4 Inhibition zone diameters of four HTC aqueous phase products from microalgae, cornstalk, cow manure and swine manure. The inhibition zone diameters of four HTC aqueous phase products on Gram-negative strains (**a**) and (**b**) Gram-positive strains. (**c**–**f**), the inhibition zone diameters on target strains of diluted HTC aqueous phase products from microalgae (**c**), cornstalk (**d**), cow manure (**e**) and swine manure (**f**). (Reprinted with permission from [1]. Copyright © 2023, Elsevier)

Aqueous products derived from four nature materials namely swine manure, cow manure, microalgae and corn straw were used to evaluate the antimicrobial strength on ten bacteria (5 g negative cell and 5 g positive cell) [1]. Experimental results showed that aqueous products derived from HTC have a broad antimicrobial spectrum, indicating the high potential of aqueous product for bacterial control. The compounds in aqueous can leading to structural changes, metabolic disruption, and even cell death in bacteria (Fig. 4.4) [1]. After HTC aqueous phase treatment, the surfaces of *Escherichia coli* and *Staphylococcus aureus* cells became wrinkled and deformed, and their volumes decreased. The treated *Staphylococcus aureus* exhibited irregular shrinkage, with some cells becoming oval-shaped. The antibacterial effect of HTC aqueous phase is likely related to the destruction of bacterial cell membranes. Hydrothermal aqueous treatment of *Escherichia coli* reduced intracellular ATP production by more than 16%. After HTC aqueous phase treatment, the accumulation of intracellular H_2O_2 significantly increased.

Aqueous products from swine manure and cow manure exhibit strong inhibitory effects against *Alternaria solani*, with their minimum inhibitory concentration (MIC) being 12% (v:v) under experimental conditions. Aqueous product from two kinds of manure significantly alter the mycelial morphology of *Alternaria solani*, causing metabolic disruption and disrupting cell membrane structure [4]. The antifungal strength and active compounds of HTC aqueous phase vary due to the distribution of compounds in the HTC raw material, with phenols and -containing compounds identified as the main active components in cow manure and swine manure HTC aqueous phase, respectively. Microalgae-derived HTC aqueous phase exhibits significant inhibition against *Botrytis cinerea*, with a MIC of approximately 14.8 mg/L [21]. HTC aqueous phase treatment induces metabolic disorders, morphological changes, and membrane permeability disruption in Botrytis cinerea. The experimental results showed that after HTC aqueous phase treatment, Botrytis cinerea experienced metabolic disorders and a decrease in intracellular ATP content. Transcriptome analysis indicated that HTC aqueous phase affects oxidative phosphorylation, mitochondrial respiratory chain complex synthesis, the TCA cycle, and carbohydrate metabolic processes in Botrytis cinerea, which might explain the reduction in intracellular ATP content. HTC aqueous phase also impacts the peroxidase synthesis pathway in the tested fungi, reducing the activities of CAT and POD enzymes, leading to a significant accumulation of intracellular hydrogen peroxide. Additionally, HTC aqueous phase influences the transcription of genes related to the synthesis pathways of intracellular reducing sugars and proteins, thereby affecting the synthesis of key cellular structures. These data suggest that HTC aqueous phase treatment significantly impacts *Botrytis cinerea* cells, while the fungus exhibits complex, comprehensive metabolic disorder through multiple pathways [21].

Further research reveals that aqueous phase treatment has a significant effect on seedborne fungi control, suggesting its potential application as a seed disinfectant (Fig. 4.5) [21]. In addition to its potential in inhibiting bacteria and fungi, the aqueous phase also exhibits significant inhibitory effects on viruses [50]. Compared to commercial antifungal agents such as carbendazim used in agriculture, the

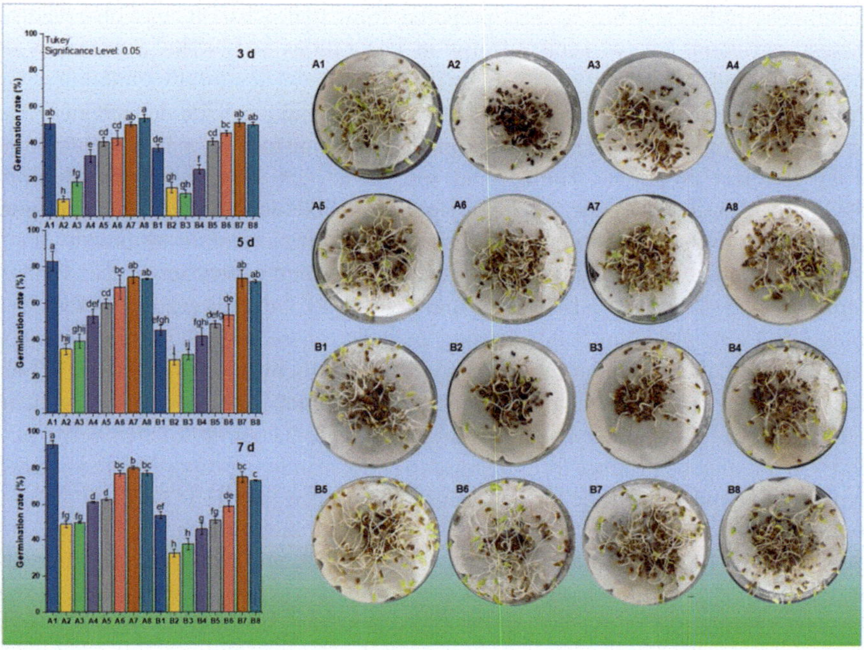

Fig. 4.5 Seed germination rates at 3d, 5d and 7d and images at 7d after treatment of aqueous products at different concentrations. (A1, control; A2–A8, 1–7× aqueous products treated; B1, *B. cinerea* infected seed, B2–B8, infected seed treated by 1–7× aqueous products.). (Reprinted with permission from [21]. Copyright © 2024, Elsevier)

aqueous product showed less toxicity to aquatic organisms when achieving the same disease prevention and control effect [42]. According to the phytopathogen control efficiency, a highly-efficient planting-breeding integration system based on hydrothermal resource utilization of animal wastes was proposed [51].

4.3.2 Soil Fertilizer

The application of humic acid and fluvic acid in agriculture demonstrates positive effects on soil quality and plant growth [52–54]. Firstly, humic acid and fluvic acid significantly improve soil structure by enhancing water retention, promoting soil aeration and permeability [55, 56]. This contributes to preventing soil drought and waterlogging, enhancing soil fertility. Secondly, humic acid and fluvic acid positively influences plant growth by providing essential nutrients and adjusting soil pH [57, 58]. Its rich organic content supports plant development, increases disease resistance, and improves overall crop yield [19, 41]. Furthermore, their chelating ability reduces metal toxicity, benefiting soil pH regulation and environmental friendliness [59]. In a humic acid prepared from hydrochar, it was observed to possess excellent heavy metal adsorption capabilities [41]. The application of humic acid and fluvic acid also supports sustainable agricultural development by promoting soil microbial activity, reducing fertilizer and pesticide dependence, and slowing soil degradation [56]. Additionally, the transformed substances exhibit strong fertilizer properties. Humic acid and fluvic acid in aqueous product from HTC hold promise as a crucial support for sustainable agriculture, contributing to efficient, environmentally friendly, and sustainable agricultural development.

Besides humic acid, hydrothermal liquefaction aqueous product contains many elements, N (5–20 g/L), P (1–10 g/L), and K (1–8 g/L), demonstrating its fertilizer potential. There are some studies using aqueous phase to provide nutrients for microalgae and these element can be absorbed efficiently by microalgae [18, 60]. Element enrichment can be achieved through membrane filtration: polypropylene membranes can concentrated 100% P in aqueous product from kitchen waste and 99.5% P in aqueous product from manure [61]. Researchers proposed a method for recovering N and P from the aqueous phase to produce $MgNH_4PO_4•6H_2O$ [62]. A previous study achieved 99% P and 40–100% ammonium nitrogen conversion to $MgNH_4PO_4•6H_2O$ precipitate by adding HCl, NaOH, and $MgCl_2$ to the aqueous phase. Firstly, an electrochemical leaching method was employed to extract the dissolved ammonium nitrogen from the aqueous phase, generating $(NH_4)_2SO_4$ with H_2SO_4. After acid washing of hydrothermal residues and pH adjustment to 9 with NaOH, addition of $MgCl_2$ resulted in the production of $MgNH_4PO_4•6H_2O$ [63]. This pathway allows 1000 kg of micalgae hydrothermal liquid phase to yield 78 kg of $MgNH_4PO_4•6H_2O$ and 158 kg of $(NH_4)_2SO_4$.

4.3.3 Challenge of Application

Although the aqueous phase products of hydrothermal liquefaction have certain agricultural application potential, there are still significant challenges in practical applications. While the direct application of the aqueous phase solution rich in humic acid and fluvic acid to the soil has been proven effective [55], the complexity of the HTC aqueous phase leads to uncertainties regarding the beneficial components, environmental safety, and crop safety.

Obtaining stable aqueous phase components is a significant challenge. The HTC aqueous phase contains various compounds, but their concentrations are generally low, and the types and contents of these compounds vary depending on the feedstock, reaction conditions, and separation pathways [64]. The biomass feedstocks for hydrothermal liquefaction are diverse, including microalgae, corn stalks, cow manure, and pig manure, with complex compositions. The reaction conditions of HTC are complex, and the separation methods of the aqueous phase also significantly affect the properties of the aqueous phase products. It is challenging to obtain aqueous phase products with stable physicochemical properties and functions.

Separating components from the aqueous phase products is highly challenging. Although current research has already explored the direct application of solutions rich in humic acid and fulvic acid to the soil, humic acid in the aqueous phase can be separated by adjusting the pH of the solution [19, 65]. Nitrogen and phosphorus elements in the aqueous phase can be recovered using techniques such as membrane separation and chemical precipitation [20]. However, the variety and low content of small molecular substances in the aqueous phase products pose significant cost challenges for the separation process [66]. From the perspective of providing plant nutrients, the application of mixtures might be an important approach for the aqueous phase. From the perspective of crop disease prevention and control using aqueous phase products, identifying the effective antibacterial components in the aqueous phase might be an important prerequisite for developing aqueous phase antibacterial agents. The HTC aqueous phase extracted from the aqueous phase has shown a broad antibacterial spectrum against both Gram-negative and Gram-positive bacteria. Previous studies have also observed the synergistic antibacterial activity among HTC aqueous phase components. However, the relationship between inhibition strength and compound distribution still needs further study, especially the contribution of specific components to antibacterial strength. Additionally, for the application of antibacterial agents, it is necessary to evaluate the stability and safety of HTC aqueous phase in terms of its chemical composition and physical properties.

4.4 Conclusions and Future Outlook

Hydrothermal technology, as a distinctive biomass conversion method, holds significant importance in the transformation of biomass into valuable acids and other active substances.

1. Hydrothermal technology provides an efficient and comprehensive utilization pathway for biomass. Through hydrothermal treatment under high temperature and pressure conditions, degradation and conversion of biomass into acids and other organic compounds can be achieved. This process not only reduces the volume of biomass but also releases organic materials suitable for energy production, thereby promoting the comprehensive utilization of biomass. The high temperature and pressure conditions in the hydrothermal environment accelerate the degradation and hydrolysis of biomass, transforming it into organic and inorganic molecules. In this process, complex organic compounds in biomass such as cellulose, hemicellulose, lignin, can be effectively decomposed into acids like acetic acid, propionic acid, lactic acid, humic acid and fluvic acid.
2. Hydrothermal technology facilitates the cyclic utilization of organic matter in biomass. The organic compounds, including small molecular acids obtained through hydrothermal treatment, can be utilized as biofertilizers, soil amendments and phytopathogen control in agricultural applications. This not only ensures the effective utilization of organic matter in biomass but also provides renewable

organic fertilizers for agricultural production. However, the complexity of the HTC aqueous phase owing to compositions in feedstocks and varied operational parameters leads to uncertainties regarding the beneficial components, environmental safety and compound separation.

In conclusion, the significance of hydrothermal technology in converting biomass into small molecular acids and active compounds is evident in aspects such as resource integration, efficient organic matter conversion, harmful substance degradation, and cyclic utilization. Through this technological approach, the sustainable management and treatment of biomass waste can be achieved, contributing to ecological environmental protection and resource cycling.

Acknowledgements The authors are grateful for the financial support from National Natural Science Foundation of China (U21A20162; 52261145701) and the 2115 Talent Development Program of China Agricultural University, China.

References

1. Xu Y, Wang Y, Lu J, Yuan C, Zhang L, Liu Z. Understand the antibacterial behavior and mechanism of hydrothermal wastewater. Water Res. 2022;226:119318. https://doi.org/10.1016/j.watres.2022.119318.
2. Usman M, Chen H, Chen K, Ren S, Clark JH, Fan J, Luo G, Zhang S. Characterization and utilization of aqueous products from hydrothermal conversion of biomass for bio-oil and hydrochar production: a review. Green Chem. 2019;21:1553–72. https://doi.org/10.1039/c8gc03957g.
3. Tian C, Li B, Liu Z, Zhang Y, Lu H. Hydrothermal liquefaction for algal biorefinery: a critical review. Renew Sust Energ Rev. 2014;38:933–50. https://doi.org/10.1016/j.rser.2014.07.030.
4. Xu Y, Wang Y, Liu Z, Yuan C, Lu J, Wang Z, Liu Z. Harnessing fungicide potential of hydrothermal liquefaction water from livestock manure: a new vision of integrated crop-livestock system. J Clean Prod. 2023;404:136971. https://doi.org/10.1016/j.jclepro.2023.136971.
5. Zhu Z, Si B, Lu J, Watson J, Zhang Y, Liu Z. Elemental migration and characterization of products during hydrothermal liquefaction of cornstalk. Bioresour Technol. 2017;243:9–16. https://doi.org/10.1016/j.biortech.2017.06.085.
6. Yu S, He J, Zhang Z, Sun Z, Xie M, Xu Y, Bie X, Li Q, Zhang Y, Sevilla M, Titirici M, Zhou H. Towards negative emissions: hydrothermal carbonization of biomass for sustainable carbon materials. Adv Mater. 2024;2:e2307412. https://doi.org/10.1002/adma.202307412.
7. Wang S, Wang Y, Shi Z, Sun K, Wen Y, Niedzwiecki L, Pan R, Xu Y, Zaini IN, Jagodzińska K, Aragon-Briceno C, Tang C, Onsree T, Tippayawong N, Pawlak-Kruczek H, Jönsson PG, Yang W, Jiang J, Kawi S, Wang CH. Van Krevelen diagrams based on machine learning visualize feedstock-product relationships in thermal conversion processes. Commun Chem. 2023;6:273. https://doi.org/10.1038/s42004-023-01077-z.
8. Liu Z, Wang Z, Tang S, Liu Z. Fabrication, characterization and sorption properties of activated biochar from livestock manure via three different approaches. Resour Conserv Recycl. 2021;168:105254. https://doi.org/10.1016/j.resconrec.2020.105254.
9. Liu Z, Wang Z, Chen H, Cai T, Liu Z. Hydrochar and pyrochar for sorption of pollutants in wastewater and exhaust gas: a critical review. Environ Pollut. 2021;268:115910. https://doi.org/10.1016/j.envpol.2020.115910.

10. Xu Y, Lu J, Wang Y, Yuan C, Liu Z. Construct a novel anti-bacteria pool from hydrothermal liquefaction aqueous family. J Hazard Mater. 2022;423:127162. https://doi.org/10.1016/j.jhazmat.2021.127162.
11. Shen R, Jiang Y, Ge Z, Lu J, Zhang Y, Liu Z, Ren ZJ. Microbial electrolysis treatment of post-hydrothermal liquefaction wastewater with hydrogen generation. Appl Energy. 2018;212:509–15.
12. Yan W, Guan Q, Jin F. Catalytic conversion of cellulosic biomass to harvest high-valued organic acids. iScience. 2023;26:107933. https://doi.org/10.1016/j.isci.2023.107933.
13. Zhu Z, Liu Z, Zhang Y, Li B, Lu H, Duan N, Si B, Shen R, Lu J. Recovery of reducing sugars and volatile fatty acids from cornstalk at different hydrothermal treatment severity. Bioresour Technol. 2016;199:220–7. https://doi.org/10.1016/j.biortech.2015.08.043.
14. Lu J, Zhang J, Zhu Z, Zhang Y, Zhao Y, Li R, Watson J, Li B, Liu Z. Simultaneous production of biocrude oil and recovery of nutrients and metals from human feces via hydrothermal liquefaction. Energy Convers Manag. 2017;134:340–6. https://doi.org/10.1016/j.enconman.2016.12.052.
15. Cao M, Wang Y, Li H, Jia J, Li R, Liu Z. Renewable biocrude blends boost tribological properties and maintain fuel performance of diesel. Chem Eng J. 2023;474:145696. https://doi.org/10.1016/j.cej.2023.145696.
16. Gil-Lalaguna N, Bautista A, Gonzalo A, Sánchez JL, Arauzo J. Obtaining biodiesel antioxidant additives by hydrothermal treatment of lignocellulosic bio-oil. Fuel Process Technol. 2017;166:1–7. https://doi.org/10.1016/j.fuproc.2017.05.020.
17. Zhang T, Zhou H, Fu Y, Zhao Y, Yuan Z, Shao Z, Wang Z, Qin M. Short-time hydrothermal treatment of poplar wood for the production of a lignin-derived polyphenol antioxidant. ChemSusChem. 2020;13:4478–86. https://doi.org/10.1002/cssc.202000534.
18. Leng L, Li J, Wen Z, Zhou W. Use of microalgae to recycle nutrients in aqueous phase derived from hydrothermal liquefaction process. Bioresour Technol. 2018;256:529–42. https://doi.org/10.1016/j.biortech.2018.01.121.
19. Shao Y, Li Z, Long Y, Zhao J, Huo W, Luo Z, Lu W. Direct humification of biowaste with hydrothermal technology: a review. Sci Total Environ. 2024;908:168232. https://doi.org/10.1016/j.scitotenv.2023.168232.
20. Xu Y, Liu Z. Formation mechanism and resource recovery perspectives of aqueous phase from hydrothermal liquefaction of biomass. Prog Chem. 2021;33:2150–62. https://doi.org/10.7536/PC210420.
21. Xu Y, Yuan C, Qu A, Wang Y, Zhang L, Zhang Y, Lu J, Du T, Liu Z. Massive components in renewable hydrothermal liquid trigger ultra-high differentially expressed genes and diverse pathways for pathogen control. Next Sustain. 2024;3:100022. https://doi.org/10.1016/j.nxsust.2023.100022.
22. Chen Y, Yang Y, Liu X, Shi X, Wang C, Zhong H, Jin F. Sustainable production of formic acid and acetic acid from biomass. Mol Catal. 2023;545:113199. https://doi.org/10.1016/j.mcat.2023.113199.
23. Jin F, Yun J, Li G, Kishita A, Tohji K, Enomoto H. Hydrothermal conversion of carbohydrate biomass into formic acid at mild temperatures. Green Chem. 2008;10:612–61. https://doi.org/10.1039/b802076k.
24. Wu L, Yang Y, Cheng J, Shi X, Zhong H, Jin F. Highly efficient conversion of carbohydrates into formic acid with a heterogeneous MgO catalyst at near-ambient temperatures. ACS Sustain Chem Eng. 2022;10:15423–36. https://doi.org/10.1021/acssuschemeng.2c04502.
25. Quitain AT, Faisal M, Kang K, Daimon H, Fujie K. Low-molecular-weight carboxylic acids produced from hydrothermal treatment of organic wastes. J Hazard Mater. 2002;93:209–20. https://doi.org/10.1016/S0304-3894(02)00024-9.
26. Jin F, Zhou Z, Moriya T, Kishida H, Higashijima H, Enomoto H. Controlling hydrothermal reaction pathways to improve acetic acid production from carbohydrate biomass. Environ Sci Technol. 2005;39:1893–902. https://doi.org/10.1021/es048867a.

27. Yang X, Yang L, Fan W, Lin H. Effect of redox properties of LaCoO3 perovskite catalyst on production of lactic acid from cellulosic biomass. Catal Today. 2016;269:56–64. https://doi.org/10.1016/j.cattod.2015.12.003.
28. Zhang J, Sun M, Liu X, Han Y. Catalytic oxidative conversion of cellulosic biomass to formic acid and acetic acid with exceptionally high yields. Catal Today. 2014;233:77–82. https://doi.org/10.1016/j.cattod.2013.12.010.
29. Wölfel R, Taccardi N, Bösmann A, Wasserscheid P. Selective catalytic conversion of biobased carbohydrates to formic acid using molecular oxygen. Green Chem. 2011;13:2759–63. https://doi.org/10.1039/c1gc15434f.
30. Basar IA, Liu H, Eskicioglu C. Effects of municipal sludge composition on hydrothermal liquefaction products: aqueous phase characterization and biodegradability assessment. Bioresour Technol. 2024;400:130671. https://doi.org/10.1016/j.biortech.2024.130671.
31. Li H, Lu J, Zhang Y, Liu Z. Hydrothermal liquefaction of typical livestock manures in China: biocrude oil production and migration of heavy metals. J Anal Appl Pyrolysis. 2018;135:133–40. https://doi.org/10.1016/j.jaap.2018.09.010.
32. Lu J, Liu Z, Zhang Y, Savage PE. Synergistic and antagonistic interactions during hydrothermal liquefaction of soybean oil, soy protein, cellulose, xylose, and lignin. ACS Sustain Chem Eng. 2018;6:14501–9. https://doi.org/10.1021/acssuschemeng.8b03156.
33. Rout PR, Goel M, Pandey DS, Briggs C, Sundramurthy VP, Halder N, Mohanty A, Mukherjee S, Varjani S. Technological advancements in valorisation of industrial effluents employing hydrothermal liquefaction of biomass: strategic innovations, barriers and perspectives. Environ Pollut. 2023;316:120667. https://doi.org/10.1016/j.envpol.2022.120667.
34. Prestigiacomo C, Scialdone O, Galia A. Hydrothermal liquefaction of wet biomass in batch reactors: critical assessment of the role of operating parameters as a function of the nature of the feedstock. J Supercrit Fluids. 2022;189:105689. https://doi.org/10.1016/j.supflu.2022.105689.
35. Gu Y, Zhang X, Deal B, Han L, Zheng J, Ben H. Advances in energy systems for valorization of aqueous byproducts generated from hydrothermal processing of biomass and systems thinking. Green Chem. 2019;21:2518–43. https://doi.org/10.1039/c8gc03611j.
36. Lu J, Li H, Zhang Y, Liu Z. Nitrogen migration and transformation during hydrothermal liquefaction of livestock manures. ACS Sustain Chem Eng. 2018;6:13570–8. https://doi.org/10.1021/acssuschemeng.8b03810.
37. Wang M, Li Y, Peng H, Wang J, Li Q, Li P, Fan J, Liu S, Zheng G. Review: biotic and abiotic approaches to artificial humic acids production. Renew Sust Energ Rev. 2023;187:113771. https://doi.org/10.1016/j.rser.2023.113771.
38. Hu ZT, Huo W, Chen Y, Zhang Q, Hu M, Zheng W, Shao Y, Pan Z, Li X, Zhao J. Humic substances derived from biomass waste during aerobic composting and hydrothermal treatment: a review. Front Bioeng Biotechnol. 2022;10:1–17. https://doi.org/10.3389/fbioe.2022.878686.
39. Han S-Z, Zhang B-T, Wang M-Q, Guo R-B, Fu S-F. Optimization of alkaline hydrothermal treatment for humic acids production from corn straw digestate using the response surface methodology. J Environ Chem Eng. 2024;112465:112465. https://doi.org/10.1016/J.JECE.2024.112465.
40. Chen P, Yang R, Pei Y, Yang Y, Cheng J, He D, Huang Q, Zhong H, Jin F. Hydrothermal synthesis of similar mineral-sourced humic acid from food waste and the role of protein. Sci Total Environ. 2022;828:154440. https://doi.org/10.1016/J.SCITOTENV.2022.154440.
41. Shao Y, Geng Y, Li Z, Long Y, Ajmal M, Lu W, Zhao J. Unlocking the potential of humic acid production through oxygen-assisted hydrothermal humification of hydrochar. Chem Eng J. 2023;472:145098. https://doi.org/10.1016/j.cej.2023.145098.
42. Wang R, Zheng X, Feng Z, Feng Y, Ying Z, Wang B, Dou B. Hydrothermal carbonization of Chinese medicine residues: formation of humic acids and combustion performance of extracted hydrochar. Sci Total Environ. 2024;925:171792. https://doi.org/10.1016/J.SCITOTENV.2024.171792.

43. Wang L, Chi Y, Du K, Zhou Z, Wang F, Huang Q. Hydrothermal treatment of food waste for bio-fertilizer production: formation and regulation of humus substances in hydrochar. Sci Total Environ. 2022;838:155900. https://doi.org/10.1016/J.SCITOTENV.2022.155900.
44. Shao Y, Luo Z, Bao M, Huo W, Ye R, Ajmal M, Lu W. Enhanced production of hydrothermal humic acid in a two-step hydrothermal process with acid hydrothermal solution recycling. Chem Eng J. 2023;474:145634. https://doi.org/10.1016/J.CEJ.2023.145634.
45. Huo Z, Fang Y, Ren D, Zhang S, Yao G, Zeng X, Jin F. Selective conversion of glucose into lactic acid with transition metal ions in diluted aqueous NaOH solution. ACS Sustain Chem Eng. 2014;2:2765–71. https://doi.org/10.1021/sc500507b.
46. Sutradhar S, Yasin Arafat KM, Nayeem J, Jahan MS. ORGANIC ACID LIGNIN from RICE STRAW in PHENOL-FORMALDEHYDE RESIN PREPARATION for PLYWOOD. Cellul Chem Technol. 2020;54:463–71. https://doi.org/10.35812/CelluloseChemTechnol.2020.54.47.
47. Kumar A, Chauhan AS, Bains R, Das P. Rice straw (Oryza sativa L.) biomass conversion to furfural, 5-hydroxymethylfurfural, lignin and bio-char: a comprehensive solution. J Ind Eng Chem. 2021;104:286–94. https://doi.org/10.1016/j.jiec.2021.08.025.
48. Shao Y, Bao M, Huo W, Ye R, Liu Y, Lu W. Production of artificial humic acid from biomass residues by a non-catalytic hydrothermal process. J Clean Prod. 2022;335:130302. https://doi.org/10.1016/J.JCLEPRO.2021.130302.
49. Wang X, Muhmood A, Dong R, Wu S. Synthesis of humic-like acid from biomass pretreatment liquor: quantitative appraisal of electron transferring capacity and metal-binding potential. J Clean Prod. 2020;255:120243. https://doi.org/10.1016/j.jclepro.2020.120243.
50. Yu F, Zhao W, Qin T, Zhao W, Chen Y, Miao X, Lin L, Shang H, Sui G, Peng D, Yang Y, Zhu Y, Zhang S, Zhu X. Biosafety of human environments can be supported by effective use of renewable biomass. Proc Natl Acad Sci USA. 2022;119:1–3. https://doi.org/10.1073/pnas.2106843119.
51. Xu Y, Liu Z. Analysis and perspective of a highly-efficient planting-breeding integration system based on hydrothermal resource utilization of animal wastes. Nongye Gongcheng Xuebao. 2024;40:1–12. https://doi.org/10.11975/j.issn.1002-6819.202312206.
52. Wang R, Li D, Deng F, Cao Z, Zheng G. Production of artificial humic acid from rice straw for fertilizer production and soil improvement. Sci Total Environ. 2024;906:167548. https://doi.org/10.1016/j.scitotenv.2023.167548.
53. Yang S, Wang K, Yu X, Xu Y, Ye H, Bai M, Zhao L, Sun Y, Li X, Li Y. Fulvic acid more facilitated the soil electron transfer than humic acid. J Hazard Mater. 2024;469:134080. https://doi.org/10.1016/j.jhazmat.2024.134080.
54. Gao Y, Zhang Y, Ge X, Gong Y, Chen H, Su J, Xi B, Tan W. Differential responses of the electron transfer capacities of soil humic acid and fulvic acid to long-term wastewater irrigation. Sci Total Environ. 2024;933:173114. https://doi.org/10.1016/j.scitotenv.2024.173114.
55. Yang F, Zhang S, Cheng K, Antonietti M. A hydrothermal process to turn waste biomass into artificial fulvic and humic acids for soil remediation. Sci Total Environ. 2019;686:1140–51. https://doi.org/10.1016/j.scitotenv.2019.06.045.
56. Gai S, Liu B, Lan Y, Han L, Hu Y, Dongye G, Cheng K, Liu Z, Yang F. Artificial humic acid coated ferrihydrite strengthens the adsorption of phosphate and increases soil phosphate retention. Sci Total Environ. 2024;915:169870. https://doi.org/10.1016/j.scitotenv.2024.169870.
57. Cui S, Li P, Ji L, Wang T, Liu Y, Lan Y, Jiang J. Superabsorbent quaternary ammonium guar gum hydrogel with controlled release of humic acid for soil improvement and plant growth. Carbohydr Polym. 2024;337:122188. https://doi.org/10.1016/j.carbpol.2024.122188.
58. Maji D, Misra P, Singh S, Kalra A. Humic acid rich vermicompost promotes plant growth by improving microbial community structure of soil as well as root nodulation and mycorrhizal colonization in the roots of Pisum sativum. Appl Soil Ecol. 2017;110:97–108. https://doi.org/10.1016/j.apsoil.2016.10.008.

59. Yang F, Du Q, Sui L, Cheng K. One-step fabrication of artificial humic acid-functionalized colloid-like magnetic biochar for rapid heavy metal removal. Bioresour Technol. 2021;328: 124825. https://doi.org/10.1016/j.biortech.2021.124825.
60. Nguyen TAH, Bui TH, Guo WS, Ngo HH. Valorization of the aqueous phase from hydrothermal carbonization of different feedstocks: challenges and perspectives. Chem Eng J. 2023;472: 144802. https://doi.org/10.1016/j.cej.2023.144802.
61. Rao U, Posmanik R, Hatch LE, Tester JW, Walker SL, Barsanti KC, Jassby D. Coupling hydrothermal liquefaction and membrane distillation to treat anaerobic digestate from food and dairy farm waste. Bioresour Technol. 2018;267:408–15. https://doi.org/10.1016/j.biortech.2018.07.064.
62. Shanmugam SR, Adhikari S, Shakya R. Nutrient removal and energy production from aqueous phase of bio-oil generated via hydrothermal liquefaction of algae. Bioresour Technol. 2017;230: 43–8. https://doi.org/10.1016/j.biortech.2017.01.031.
63. Li Y, Tarpeh WA, Nelson KL, Strathmann TJ. Quantitative evaluation of an integrated system for valorization of wastewater algae as bio-oil, fuel gas, and fertilizer products. Environ Sci Technol. 2018;52:12717–27. https://doi.org/10.1021/acs.est.8b04035.
64. Watson J, Lu J, de Souza R, Si B, Zhang Y, Liu Z. Effects of the extraction solvents in hydrothermal liquefaction processes: biocrude oil quality and energy conversion efficiency. Energy. 2019;167:189–97. https://doi.org/10.1016/j.energy.2018.11.003.
65. Shao Y, Huo W, Ye R, Liu Y, Ajmal M, Lu W. Hydrothermal humification of lignocellulosic components: who is doing what? Chem Eng J. 2023;457:141180. https://doi.org/10.1016/J.CEJ.2022.141180.
66. Lyu H, Chen K, Yang X, Younas R, Zhu X, Luo G, Zhang S, Chen J. Two-stage nanofiltration process for high-value chemical production from hydrolysates of lignocellulosic biomass through hydrothermal liquefaction. Sep Purif Technol. 2015;147:276–83. https://doi.org/10.1016/j.seppur.2015.04.032.

Chapter 5
Acetic Acid Through Microwave Assisted Pyrolysis (MAP) of Residual/Waste Biomass as a Contribution to Circular Economy

Piero Frediani and Marco Frediani

Abstract Acetic acid may be obtained by fermentation of ethanol, but industrial processes for the production of acetic acid, are mainly using mineral oil as starting compound, the most importants through methanol carbonylation. A possible alternative to these processes, using natural and bio-renewable raw materials, may be the microwave assisted pyrolysis of residual/waste biomass. A bio-oil is formed, from which acetic acid together with other important bio-organic compounds are the raw materials for a bio-refinery. In this chapter, a review of some applications of microwave assisted pyrolysis for the production of acetic acid, using biomasses such as cellulose, lignin, wood pellets, wood residue from municipal solid waste, biomass residues from agricultural productions, or biomass removed as invasive products present in fields were tested. All these products, when pyrolyzed through a microwave assisted pyrolysis process gave a gas, a bio-oil, and a bio-char In these chapter the bio-oils formed from different biomasses as well as from different reaction conditions using the same biomass are reported. The acetic acid present in the bio-oil may be separated and utilized, and an environmental friendly scheme for this separation is reported. Other interesting products may be separated from the bio-oil and finally the residue may be employed as fuel.

Obviously this acid has the same characteristics of the product obtained by mineral oil so it may be directly employed for the production of new materials thus reducing both the request of mineral oil and the carbon dioxide formed during the processing and giving a strong contribution to the circular economy.

Keywords Microwave · Pyrolysis · Acetic acid · Residual biomass · Contaminated biomass · Waste biomass · Crop residues

---

P. Frediani · M. Frediani (✉)
Department of Chemistry "Ugo Schiff", University of Florence, Florence, Italy
e-mail: marco.frediani@unifi.it

Z. Fang et al. (eds.), *Production of Organic Acids and Alcohols from Agricultural Residues and Food Wastes*, Biofuels and Biorefineries 13,
https://doi.org/10.1007/978-981-96-1222-2_5

5.1 Introduction

The energy consumption in the world is continuously growing and it is estimated to achieve 760 exajoule in 2050 (Fig. 5.1) [1]. The countries out of the Organization for Economic Cooperation and Development (non-OECD), such as China and India, will require a large part of this energy due to their rapid economic growth and large population.

The actual global warming, associated with pollution and other environmental problems has been attributed mainly to the use of fossil fuels for energy production and many efforts are devoted to mitigate the environmental threats without depleting energy consumption. Alternative energy will be produced using different approaches such as solar energy, wind energy, new ways of nuclear energy, and energy from natural and renewable resources. The use of different and well-coordinated approaches could lead to a reduction of fossil-derived energy sources. Among these approaches, the use of lignocellulosic biomasses is one of the most valuable and sustainable feedstock for energy production because it is a renewable re source

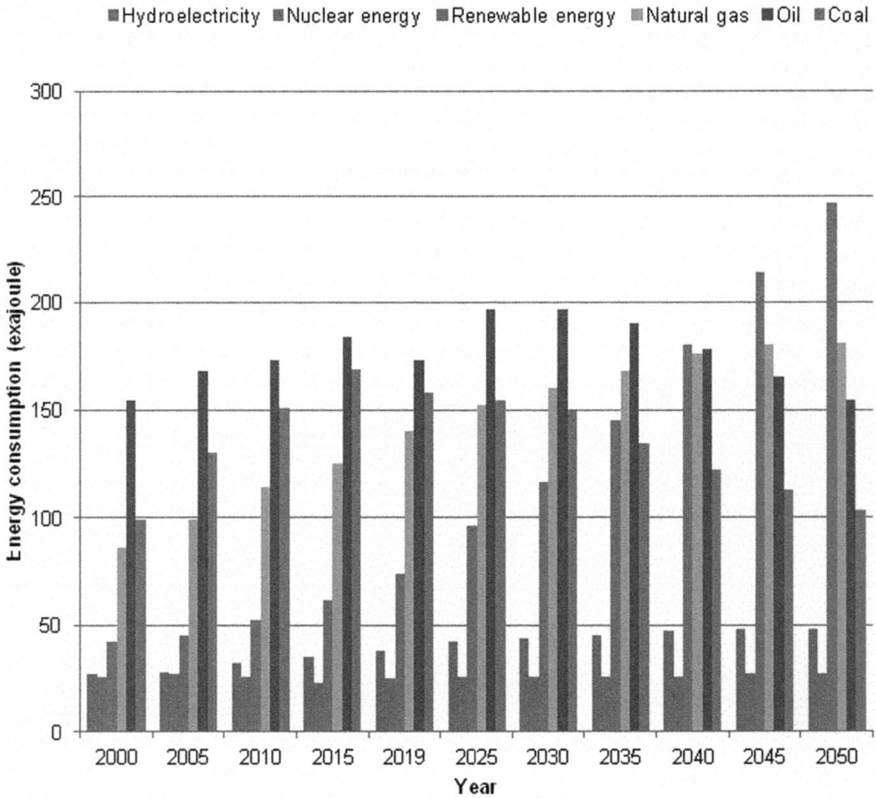

Fig. 5.1 Energy consumption worldwide from 2000 to 2019, with a forecast until 2050 [1]

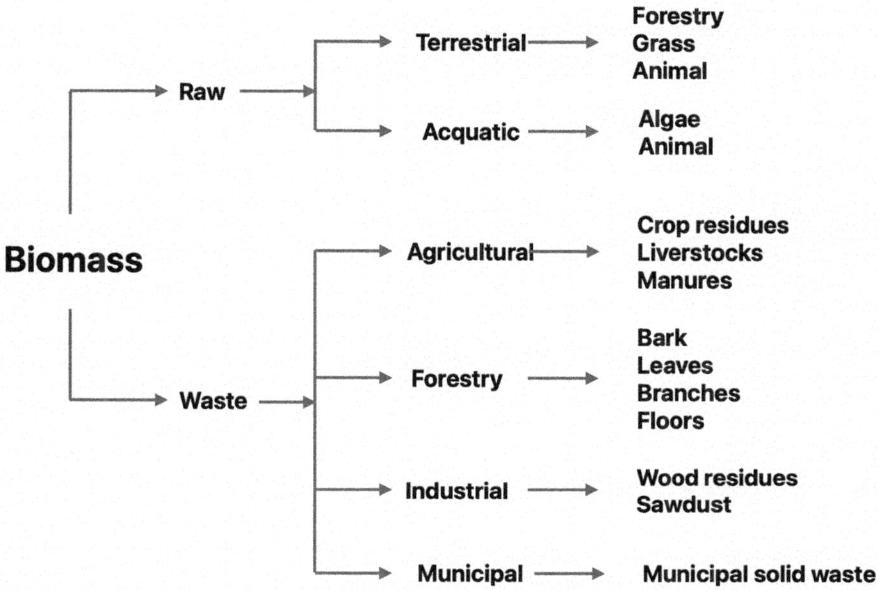

Fig. 5.2 Biomass availability

and approximately 550 × 10^9 tons of carbon fixed into biomass is present on the planet [2].

Lignocellulosic biomasses are formed through reactions [3] driven by solar energy., so they are renewable organic materials present in vegetables, algae, and animals (Fig. 5.2). They are employed as a relevant fuel in many countries, especially in the developing zones, for activities such as cooking and heating. Their use may be considered to have zero carbon emission because the carbon dioxide formed during their combustion is equivalent to that used by the plant itself for its grow. However, uses of biomass may cause pollution due to the release of the ash present in the biomass (5–10%) during combustion as a fine powder suspended in the gas and dispersed in the atmosphere. Today the use of biomass as a fuel for transportation and production of electric energy is increasing to limit the increasing amount of carbon dioxide in the atmosphere. In 2020, biomass provided only up to 5% of the total primary energy used in the United States [4] but it is estimated in 2040 renewable materials as the main source of energy [1].

Biomass employed as a raw material for the production of energy include wood and wood processing wastes (i.e., firewood, wood pellets, and wood chips, lumber and furniture mill sawdust and waste residue from pulp and paper mills [5]). Furthermore crops and waste materials (i.e., corn, soybeans, sugar cane, switchgrass, woody plants, algae, crop and food processing residues) are used. Finally biogenic materials (i.e., paper, cotton, wool products, food residue, yard, and wood wastes), animal manure and human sewage [6] must be considered.

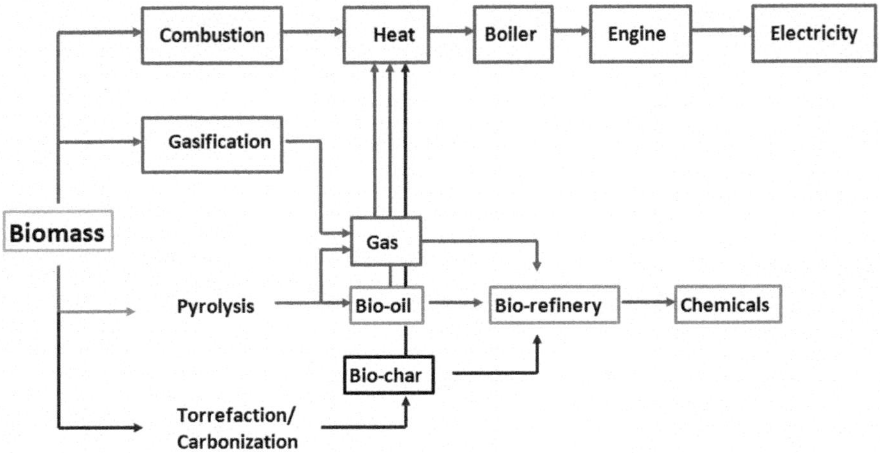

Fig. 5.3 Thermochemical utilization of biomass

Wood is one of the most important lignocellulosic biomass and the three main natural biopolymers are: cellulose, hemicellulose, and lignin together with a minor amount of other organic and inorganic compounds [7]. A synthetic chemical description of these three major components is reported in Sect. 5.5.1.

Wood is the starting material for the pulp and paper industry, it is employed as structural timber and sawn wood, to realize panels, furniture, and other purposes. Wood is largely available around all the world but there are large differences among the various kinds of wood. The chemical complexity of the three main biopolymers present lets wood as a potential source of valuable chemicals such as carbohydrate or their derivatives among which bio-ethanol may by formed (through first and second generation processes). Other products may be obtained from lignin such as phenols or products formed by complete decomposition of all biopolymers as carbon. It may be employed as a solid fuel, carbon black (for tire production, filler for plastic derivatives, ink for printer), catalyst support or used for energy storage and CO_2 sequestration processes.

Neat and/or residual/waste lignocellulosic biomass may be used as raw materials in several processes devoted to the production of chemical derivatives, energy, or fuels. Among these the thermochemical treatments of wood such as combustion, gasification, pyrolysis, pyro-gasification [8], torrefaction, and carbonization (Fig. 5.3) are employed to convert wood into products useful as a source of chemical compounds or fuel.

The first use of the wood as a source of energy is the combustion, however, this process presents several drawbacks because starting biomass has a high moisture content, low energy density, low heating value, presence of ash (sometimes also in high amounts). There are also some logistics issues connected with its production, collection and transport to destination. Biomass can be burned directly to produce heat or it may be converted into a liquid fuel through thermochemical processes such as pyrolysis because the use of liquid fuel is usually more easy than solid fuel.

Alternatively the liquid fuel may be employed as an interesting source of chemicals able to reduce the dependency on fossil fuel and the remaining bio-fuel may be the source of the energy required for the process.

The thermal decomposition of biomass may be carried out in an anoxic atmosphere, at a relatively high temperature realizing a pyrolysis process. The cracking of biopolymers gives products having lower molecular weight that is gas, bio-liquid and bio-solid with a higher heating value than the starting material. These products may be directly used as fuels to generate energy or, more conveniently, employed in a bio-refinery to produce raw materials for the synthesis of new products [3]. Pyrolysis has been considered an efficient, cost-effective, and significant process to convert organic materials into energy-rich products [9–11]. Pyrolysis may be run using classical heating technologies but alternative techniques such as solar energy, laser energy, or microwave heating have been successfully proposed for the thermal cracking of several kinds of biomasses [12, 13].

Different types of lignocellulosic biomass specifically produced, residue of biomass cultivation, waste, or contaminated biomass may be employed to produce bio-oil, bio-char, and pyrolytic gas [9, 14]. The thermal degradation of cellulose [9] give a bio-oil and it may be employed as a fuel or employed to produce valuable chemicals, such as acetic acid and levoglucosan. The yield of each fraction is mainly related to the kind of biomass used and the reaction variables (such as temperature, reaction time, and so on) employed during the pyrolysis process. Generally, the pyrolysis of woody biomass at 773–823 K results in bio-oil yield in the range of 50–70 wt%, bio-char yield of 20–30 wt%, and pyrolytic gas with a yield up to 10–25 wt% [15]. Among all the pyrolytic products, bio-oil is considered of great importance and foreseen as the future drop-in fuel, gas is a mixture of several compounds (CO_2, CO, CH_4, H_2, etc.) and it may be directly used for energy applications or employed as a syngas for production of organic compounds. Biochar can be used as a solid fuel, as a soil improver in agriculture, or in many other applications [16, 17].

Bio-oil can be considered a clean and renewable fuel compared to conventional fossil fuels since its combustion releases a very low amount of acid emissions (SO_x and NO_x). It may be directly used in turbines and boilers for power and heat generation, however, bio-oil is considered an unsuitable drop-in fuel due to its high oxygen content, low carbon/hydrogen ratio, acid pH, high instability, and low HHVs. The bio-oil properties can be substantially improved by various physical, thermal, chemical, and catalytic treatments. Several strategies may be applied at different stages of the pyrolysis, such as pre-pyrolysis, in-situ or ex-situ pyrolysis, and post-pyrolysis to improve the quality of the bio-oil formed. Finally, the bio-oil may be upgraded [18] and utilized for transportation or energy generation. Nevertheless, bio-oils can be used as a promising resource for obtaining various chemicals with high-added values improving its value, while the residue may be used as a fuel.

The microwave-assisted pyrolysis (MAP) which is the pyrolysis where the heating source is given by microwave is an alternative process concerning classical thermal heating. A correlation between classical and microwave heating with advantages and detriments of the two methodologies is reported in Table 5.1 [19, 20].

Table 5.1 Correlation between microwave and conventional heating, (from [19, 20] with modification)

Microwave heating		Conventional heating	
Advantages	Volumetric heating	Detriments	Surface heating
	Short reaction time (min)		Long reaction time (h)
	High heating efficiency		Low heating efficiency
	The low thermal conductivity of biomass may be overwhelmed by the presence of a microwave absorber		The biomass is hardly heated due to its low thermal conductivity
	The process is carried out in an anoxic atmosphere: The gas formed does not contain combustion gas		If direct heating is employed the gas formed is contaminated by combustion gas
Detriments	Electrical power is the only energy source: It is required to produce microwaves	Advantages	Every fuel source may be employed
	A microwave absorber is required for a fast process		Additives are not required, some heating carriers are sometimes used

An absorber is required in pyrolysis when the heating supplier is microwave if the materials are not able to convert the microwave into heat. Biomass can absorb microwave due to the humidity usually present in the substrate or compounds eventually present able to absorb microwave. However, the process proceeds with a low rate, at least in the first step of the pyrolysis, and may be increased over time due to the biochar formed during pyrolysis. The rate of the process may be strongly improved by the presence of an absorber that is compounds showing a high efficiency to convert microwave into heat. This property is attributed to the dissipation factor, called tan δ, which is the tangent of the rate between the imaginary (ε') and the real component of the permittivity (ε''). These values are in the range of 0.40 for silica [21], to 0.57–0.80 for carbon [22], to 0.86 for alumina [23], to 2.7 for iron [24] up to 5.15 for silicon carbide [18] as reported in Table 5.2 even if different values are reported in other literature source [25].

Pyrolysis of biomass using microwave as heating technology (microwave assisted pyrolysis: MAP) has been recently reviewed to produce bio-char, bio-oil and gas. A generic biomass [19, 26–71], lignin [72, 73], lignocellulosic biomass [74–80], macro and micro algae [81, 82], waste from agroindustrial productions [73, 83–85], animal breeding [86] sewage sludge [87, 88], separate municipal waste collection [89–91] were employed. These references are reported but they are not exhaustive and must be continuously updated due to the large scientific production.

Table 5.2 Bio-oil from MAP of α-cellulose [115]

Run	Set-up	Microwave absorber	Time (min)	T (K)	Bio-oil yield (wt %)	Bio-oil d_4^{20} (g/mL)	Acetic acid in bio-oil (mg/mL)	Acetic acid yield (g/kg)
1	A	Carbon	20	723	15.7	1.09	47.4	6.8
2	A	Iron	33	702	27.2	1.09	69.9	17.4
3	B	Carbon	20	745	16.2	1.09	3.4	0.5
4	B	Iron	37	681	23.6	1.07	51.8	11.4
5	A	Silicon carbide	40	659	58.3	1.05	89.2	49.5
6	A	Silica	33	481	62.9	1.41	120.6	53.8
7	A	Alumina	32	455	64.1	1.18	101.2	55.0
8	A	Graphite	18	532	26.7	1.07	43.1	10.8

Cellulose 100 g; cellulose/absorber ratio 2.0 (wt/wt); microwave power 3 kW

5.2 Industrial Processes for the Synthesis of Acetic Acid

Acetic acid may be obtained by fermentation of ethanol using acetic acid bacteria. It was reported just in the third century BC, so it is a very old process. This process is presently followed to produce the acetic acid employed for many foods because the law related to food production in several countries requires the use of vinegar obtained through a biological process.

The processes developed in the course of time for industrial production of acetic acid are reported in the following figure (Fig. 5.4). Over time different starting materials were used, initially fossil carbon for the synthesis of calcium carbide followed by the production of acetylene, hydration to acetaldehyde and oxidation to acetic acid. Many other processes were employed using acetylene or ethylene obtained from mineral oil. Alternatively acetic acid was produced, always using materials from fossil fuels, through the initial gasification of carbon to syngas followed by methanol synthesis and the direct carboxylation to acetic acid. This process covers today large part of the industrial production of acetic acid (75%) and it is realized using several catalytic processes from the old BASF process using cobalt catalyst through the Monsanto process using Rhodium catalyst to the new Cativa process using an Iridium catalyst.

5.3 Uses of Acetic Acid

Acetic acid is one of the commodities used in the Roman age to produce lead white and verdigris pigments. It is also used in the food industry and it is one of the components of the Mediterranean diet. In the year it is a raw chemical largely employed in industrial chemistry with a world production of around 17.48 × 10^6 ton/year (2022) and increased with an extrapolated value up to

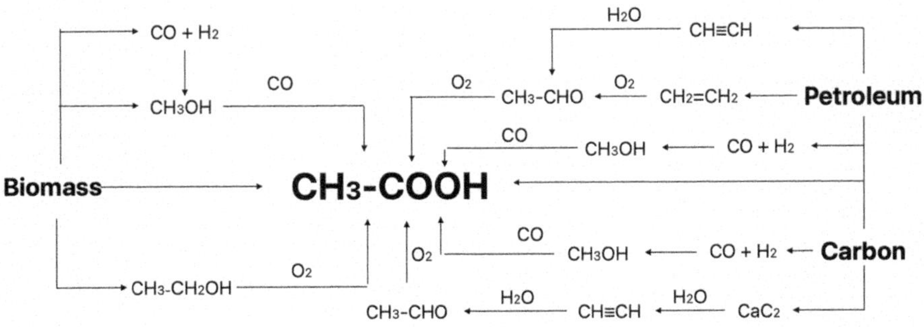

Fig. 5.4 Possible ways for industrial acetic acid synthesis

23.26×10^6 ton/year (2030) [92]. It is used as a solvent, for instance in the purification of terephthalic acid (one of the top chemical raw materials), as a starting material for the synthesis of vinyl acetate, followed by its polymerization to obtain one of the most used adhesives (poly(vinyl acetate)), or several copolymers employed as resins, adhesive paints, and coating materials. It is also employed as such or using the acetic anhydride (one of its derivative), to modify the natural polymer cellulose obtaining rayon acetate, a substitute of natural silk, or the base of photographic films, lenses for eyes glass, X-ray films, pharmaceutical products such as aspirin, dyes, foods, and so on (Fig. 5.5) [93].

5.4 Acetic Acid Identification and Quantification in Bio-oil

Bio-oil may be analyzed through several methods such as gas-chromatography, usually coupled with a flame ionization detector (GC-FID) [94], gas chromatography coupled with mass spectroscopy (GC-MS) [95–98] high performance liquid chromatography (HPLC) [99, 100], ion chromatography (IC) [101, 102], as reviewed by Sarcami et al. [103]. In the GC-FID method proposed by Undri et al. [94] bio-oil was dissolved in acetonitrile as a solvent and the solution was initially analyzed through thin-layer chromatography to evidence the high boiling products present. The same solution was also analyzed by GC-FID and several compounds were identified with a large range of retention factors (Rfs = 0–1). The GC let to separate each volatilizable component that were identified through its mass spectra while the peak area of each compound was employed to estimate their amount. In this way, acetic acid was identified in large part of the bio-oils tested and estimated as the major component present while other compounds such as limonen, cyclohexancarboxylic acid, 1,2,4-trimethoxybenzene and 2,6-dimethylphenol in all bio-oils, while 4-hydroxyacetophenone and (3,4,5-trimethoxy)acetophenone were present in bio-oil from MAP of wood. Cyclohexanone was identified in bio-oils obtained from pyrolysis using thermal heating or MAP without any absorber. The compounds having a high molecular weight were identified through HPLC.

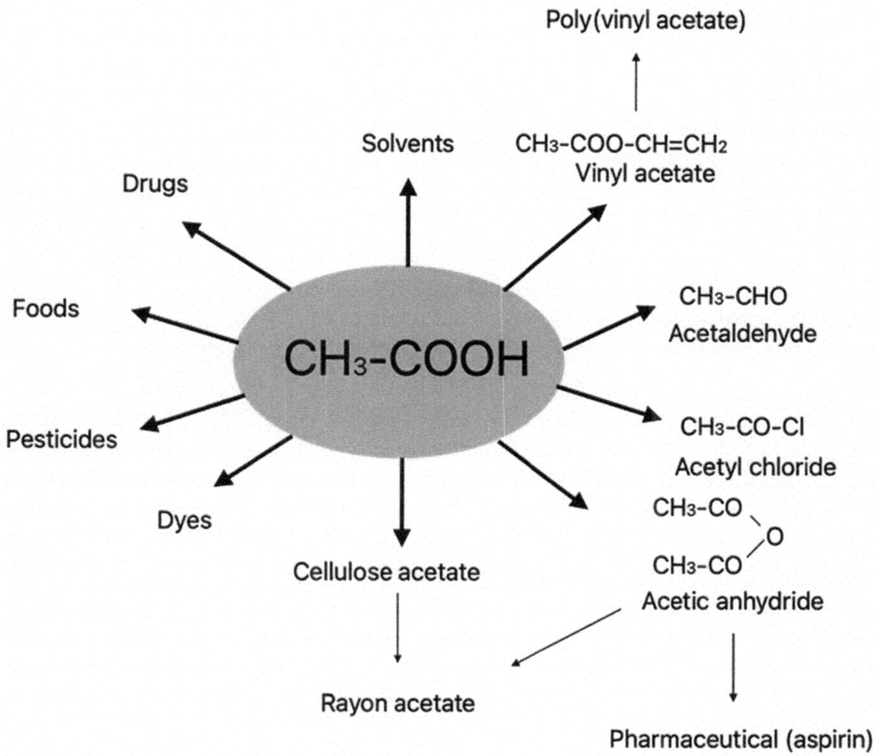

Fig. 5.5 Uses of acetic acid [93]

An implementation of the protocol was reported by Bartoli et al. [95] using the GC-MS method as a modified release of the method previously suggested for GC/MS-FID for identification and quantitative determination of the composition of the bio-oil. The compounds were identified through the corresponding spectra and they were quantified using peak area and response factor (Rf). A method to calculate Rfs of compounds was proposed using diphenyl as the internal standard. In this way, all the products were identified and quantified using GC-MS. For instance, the yields of identified products from poplar were in the range of 50 and 99% while those from MAP of cellulose were among 46 and 58%. This method, using an internal standard, let to evaluate the yield of acetic acid formed (g/kg) from the biomass employed for the MAP process. Another method is recently reported by Serras-Matilas et al. [98] using GC-MS close to that one reported by Bartoli et al. [95]. In Serras-Matilas method a certified saturated alkane mixture was employed together with an internal standard and 14 model compounds.

Acetic acid was the main product identified in all bio-oils from MAP of the biomass tested. It was obtained in large yields from poplar while it was formed in lower yield from MAP of cellulose suggesting its formation from several compounds present in the poplar or the inorganic compounds present may catalyze its formation.

5.5 MAP of Biomass

Many processes are described for the pyrolysis of biomass, and acetic acid was present in the bio-oil, but in this chapter our results of the MAP of biomass together with few data on other processes are reported. Unfortunately large parts of the paper devoted to the MAP of biomass put large attention to the yield of bio-oil. It was analyzed by GC or GC-MS to identify the compounds present, but a quantitative determination of these compounds was not carried out, as reported also by Serras-Malillos [98]. This lack of quantitative determination of acetic acid ruled out their utilization in this work.

The MAP of biomasses has been carried out in a multimode Microwave oven having a power of 6 kW (supplied by Bi.Elle s.r.l.—Italy), with a microwave power of 6 kW. A flask containing the biomass was inserted inside the oven and connected with two condensing systems, the first one cooled at 298 K and the second at 263 K. The condensed liquid was collected in a flask at the end of the two condensers while the un-condensable gas was collected in a gas holder. The biochar formed in the MAP was recovered at the end of the experiment in the flask where the biomass was initially inserted (Set-up A). A different set-up (Set-up B) was sometimes used where a fractionating column having a length of 0.2 m, internal diameter of 30.0 mm, filled with glass spheres having a diameter of 4.0 mm, was inserted between the oven and the condensing system to improve the residence time of high boiling products inside the microwave oven (Set-up B). A scheme of the two different set-ups is reported in Fig. 5.6 [104].

Temperature of the MAP was detected according to Undri et al. [105], however, several problems are connected with its detection as reported by Rosi et al. [106] and recently by Mohanty et al. [107].

MAP of biomass has been, sometimes, carried out without an external microwave absorber because the water present may convert microwave into heat, however, the process is run at a low rate. On the contrary, if a microwave absorber was added the rate of the process was very high. Different microwave power has been used in some experiments and the pyrolysis behavior has been tuned to achieve appropriate reaction conditions.

The typical three products of a MAP are obtained: a gas containing CO_2, CO, H_2, CH_4 and some other low molecular weight hydrocarbons, a bio-oil containing several organic compounds, among which acetic acid is frequently the main component, and a bio-char containing inorganic materials (ash) and a large amount of carbon (Fig. 5.7). The bio-oil collected was analyzed to evaluate the amount of acetic acid present as reported in Sect. 5.4.

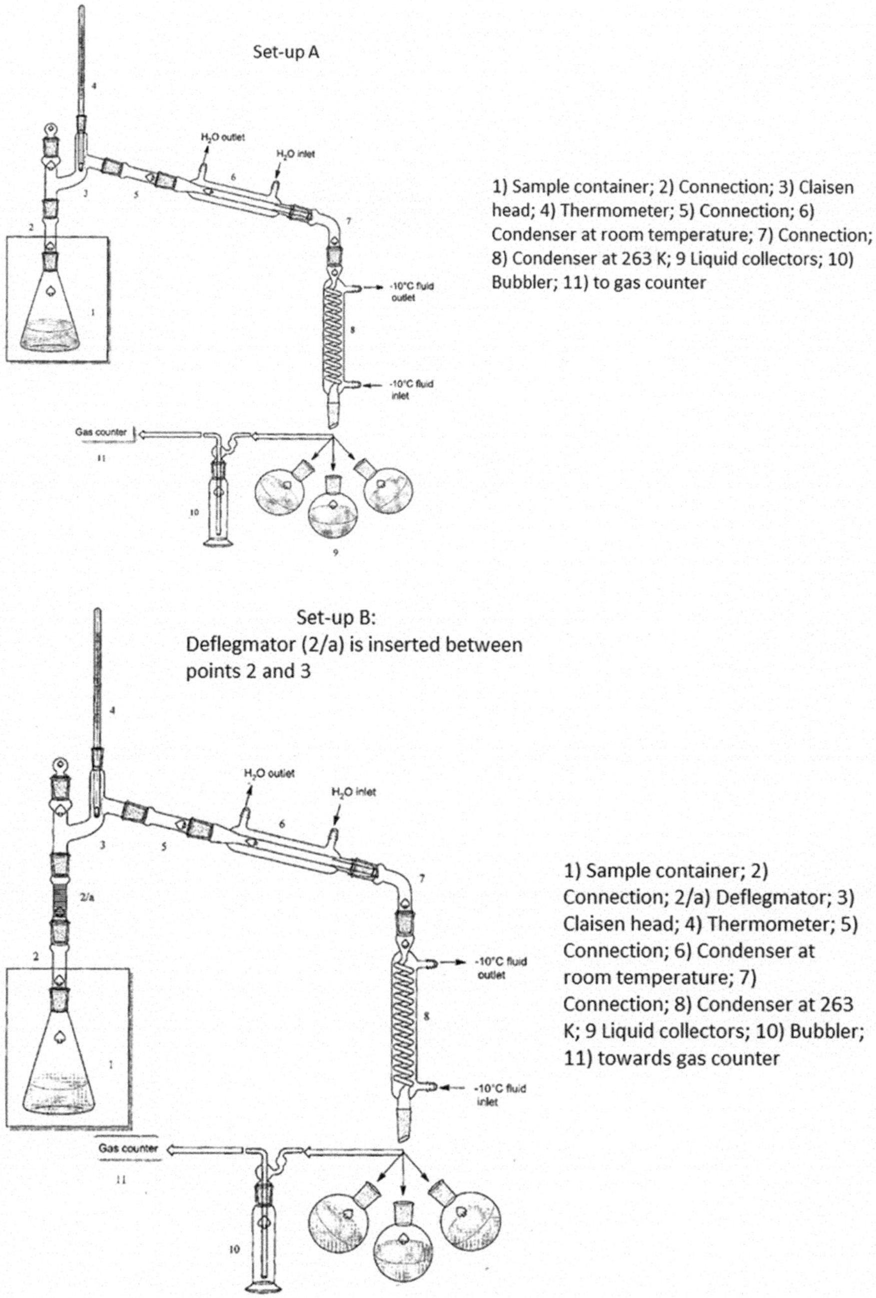

Fig. 5.6 Set-ups employed in microwave assisted pyrolysis [104]

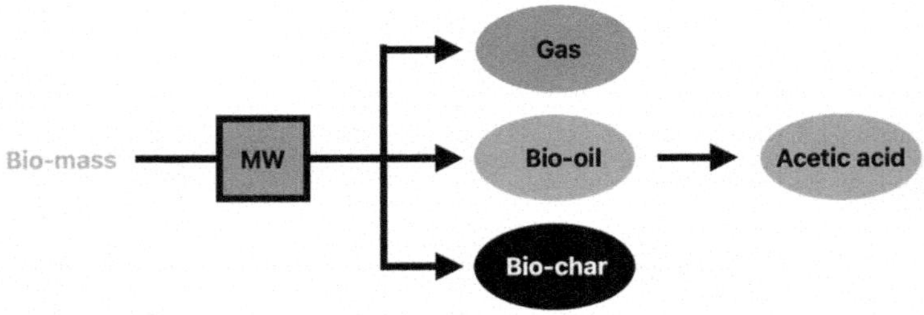

Fig. 5.7 Acetic acid from MAP of biomass

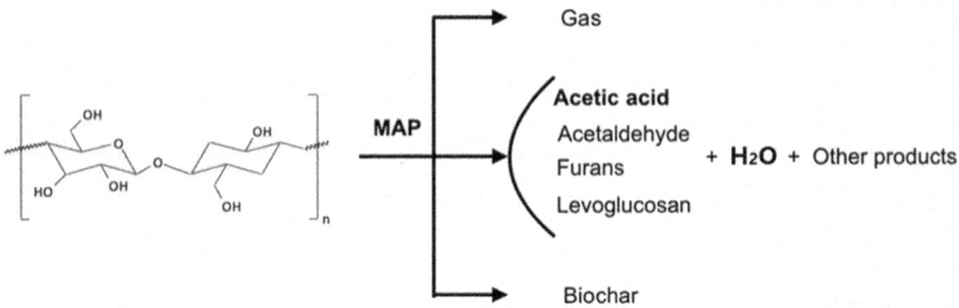

Fig. 5.8 Microwave assisted pyrolysis of α-cellulose [115]

5.5.1 Acetic Acid from MAP of Characteristic Components of Biomass

Every biomass may be employed for the production of acetic acid through MAP and wood as a biomass largely produced in the world. The three main components present in many biomasses are cellulose, hemicellulose, and lignin together with variable amounts of inorganic compounds. These three components were separately pyrolyzed to evaluate their contribution to the amount of acetic acid present in the bio-oil and to estimate the behavior of other biomasses having a known composition of the main components.

5.5.1.1 Acetic Acid Through MAP of α-Cellulose

α-Cellulose is one of the main component of wood and largely present in the earth [108, 109]. It is a biopolymer with formula minima $(C_6H_{10}O_5)_n$, obtained by linear polymerization of α-D-glucose through β-(1–4)-bonds (Fig. 5.8). The molecular weights of the biopolymer have a large dispersion from several hundred up to 4,000,000 amu [110], depending on the biomass in which it is formed. Many biomass among which green plants, many forms of algae, oomycetes contain α-cellulose in their cell walls, also biofilms secreted by some bacteria contain α-cellulose [111]. The amount of α-cellulose in cotton fiber is 90%, in wood it is 40–50%, while in dried hemp it is around 57% [112–114].

Neat α-cellulose was pyrolyzed through a MAP process using a multimode microwave oven, different microwave absorbers, and experimental setups. The performance of the MAP of α-cellulose in the production of acetic acid was reported in Table 5.2 [115]. The microwave absorber showed a strong influence on the MAP, especially for what concerns the rate of the process with a short reaction time in the range of 20 and 40 min, so the process must be considered a fast pyrolysis. Furthermore, the presence of iron as a microwave absorber led to a dark brown bio-oil with a yield of up to 37.6%, especially when a condenser is inserted in the apparatus between the oven and the collecting system (Set-up B in Fig. 5.6). Carbon as a microwave absorber induced the formation of a large amount of biogas from cellulose (yield 53.8%) while alumina (Al_2O_3) as an microwave absorber caused the formation of a high yield of biochar (64.1%) and low gas production (3.0%). Bio-oils had a low density and viscosity due to the presence of a large amount of furanosidic compounds. Bio-oils were characterized through quantitative GC–MS analysis using the above reported internal standard method and calculated or experimentally detected retention factors. A high yield of bio-oil and a high concentration of acetic acid was shown in a sample obtained through MAP of α-cellulose when silica or alumina were used as absorbers with a low dissipation factor (run 6 and 7). Even if the temperature can not be taken into account due to the problem connected with its determination, as above reported, the samples obtained in runs 6 and 7 show the lowest value. Higher yields of acetic acid formed in these samples (6 and 7) were an indication of a connection among the low efficiency of the microwave absorbers with the high yield of bio-oil. The low efficiency of the microwave absorber connected with the low temperature of the process caused a less extensive decomposition of the biomass towards gaseous products. The formation of water was affected by all parameters of the process, mainly by the microwave absorber employed because silica as microwave absorber let to produce a bio-oil with low water content (13%), while the presence of highly efficient microwave absorber such as carbon improve dehydration processes and a large amount of water (46%) was present. Surprising, silicon carbide, the most efficient microwave absorber among those employed, gave a yield of acetic acid very close to that one obtained with silica and alumina and very different from yields obtained using carbon or iron even if the temperature detected in the process (659 K) was very close to that one detected with iron as microwave absorber (681 K).

The best yield of acetic acid was 55.0 g/kg of the starting biomass and even if these values are not very high, however, the process may be interesting taking into account the following points: a) the starting materials are waste biomass; b) Other very interesting products, such as levoglucosan, furans and so on, are formed together with acetic acid and they may be isolated in a bio-refinery and they may contribute to the economy of the process; c) Bio-char and gas are formed in the MAP together with bio-oils and these products represent a further revenue of the MAP itself.

5.5.1.2 Kraft lignin as a Source for Acetic Acid Through MAP

Lignin is a rigid and very resistant bio-polymer containing large amounts of aromatic moieties forming a complex structure present as the support of tissues on most plants [116] or involved in cell walls, especially in wood and bark. The bio-polymer is formed by cross-linking of several different unities of lignol precursors such as coniferyl, sinapyl, and p-coumaryl alcohols [117]. The polymer is highly heterogeneous because these precursors are cross-linked in different ways (Fig. 5.9). Furthermore, small amounts of incomplete or modified mono-lignols are sometimes present in the structure, while other monomers are involved in the lignin of non-woody plants [118]. Lignins isolated from grass contain guaiacol unities, while sinapyl unities were identified in palm plants [119]. Different amounts of repetitive unitiy are present as a function of plant source and molecular weight. The degree of dispersion of the molecular weight is hardly detected due to the low solubility of the bio-polymer.

Usually, large amount of lignin are obtained as a by-product of the mill and paper industry as well as from bio-ethanol through the second generation process, and an availability of 50×10^6 ton/year is estimated worldwide. However, the utilization of lignin is very scarce and a large part of it (ca. 98%) is burned. Pyrolysis as well as gasification are considered promising utilization of lignin to convert this by-product into biochemicals, biomaterials, and biofuels. for a sustainable carbon economy.

MAP of Kraft lignin was reported using a low pressure (1 bar, 0.13 bar, 0.013 bar) and carbon as microwave absorbers. Interesting results were obtained at the lowest pressure employed (0.013 kPa) with a bio-oil yield of 37 wt% in 9 min (Table 5.3). The bio-oil was analyzed using a quantitative GC-MS method as above reported (Sect. 5.4) and ^{1}H-NMR or FT-IR ATR spectroscopy [120].

As expected the MAP of lignin gave a relatively low amount of acetic acid with respect to the MAP of α-cellulose. An acetic acid yield of 36.7 g/kg of lignin was obtained in run 1 (Table 5.3) in a short time, while using the set-up B the yield was

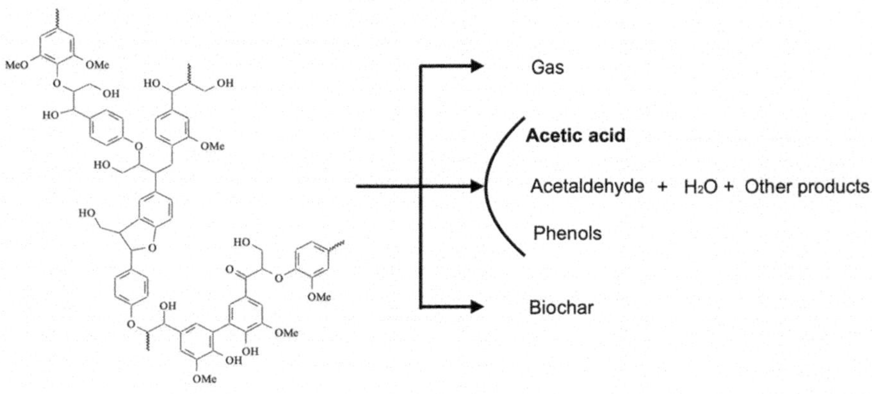

Fig. 5.9 Microwave assisted pyrolysis of lignin [120]

Table 5.3 Bio-oil from MAP of Kraft lignin using different set-up, in the presence of carbon as microwave absorber [120]

Run	Set-up	Time (min)	T (K)	Bio-oil yield (wt%)	Bio-oil d^4_{20} (g/mL)	Acetic acid in bio-oil (mg/mL)	Acetic acid yield (g/kg)
1	A	20	723	43.6	1.64	137.9	36.7
2	B	16	681	26.4	1.68	39.7	6.2
3	A	13	638	27.5	1.56	57.6	10.2
4	B	19	721	25.5	1.55	55.8	9.2
5	A	9	720	37.7	1.49	0.0	0.00
6	B	15	618	27.1	1.77	95.5	14.6

Lignin 100 g; lignin/absorber ratio 2.0 (wt/wt); microwave power 3 kW

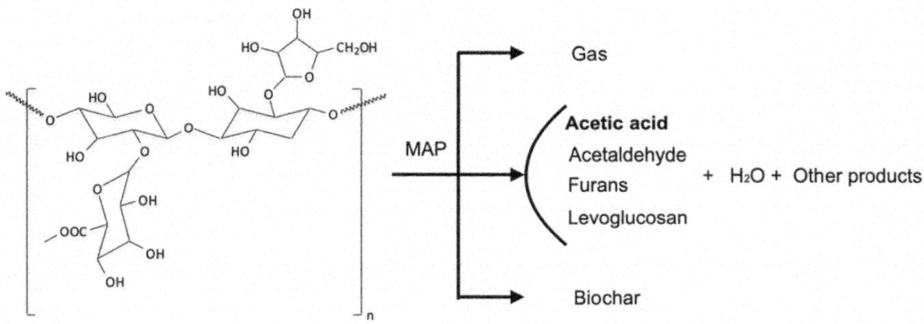

Fig. 5.10 Microwave assisted pyrolysis of hemicelluloses [123]

reduced presumably due to a pyrolysis of the acetic acid initially formed. The bio-oil contained a high amount of multi-substituted aromatic compounds and a lower amount of linear/cyclic organic products (C2–C5). These last products were formed through pyrolysis of side chains of lignin and they may be isolated in a bio-refinery.

5.5.1.3 Acetic Acid Through MAP of Hemicellulose

Hemicellulose is another polyose present in terrestrial plants [121] deriving from different sugars such as five-carbon sugars (xylose and arabinose), six-carbon sugars (glucose, mannose, and galactose), and six-carbon deoxy sugar (rhamnose). As a consequence the following groups are present in hemicellulose: xylan, glucuronoxylan, arabinoxylan, glucomannan, and xyloglucan. In most cases, groups deriving from xylose are present in largest amount, although in softwoods groups deriving from mannose are the most abundant (Fig. 5.10). Glucuronic and galacturonic acid, acid forms of regular sugars, can be also present [122]. Usually, hemicellulose and cellulose are present at the same time even if in different amount and they have distinct compositions and structures. Hemicellulose has a random, amorphous structure, easily hydrolyzed by dilute acid or base as well as degraded by a myriad of hemicellulase enzymes. On the contrary cellulose is crystalline and resistant to hydrolysis. Hemicellulose is not commercially available so pyrolysis of a model compound, xylan, was studied to evaluate the possible compounds formed from MAP of hemicellulose. The MAP of xylan was run by Li et al. without any microwave absorber and the classical three products: gas, bio-oil, and char were formed with 16–40%, 2–21%, and 40–82%, yield respectively [123]. The yield of products was affected by the amount of the starting products obtaining a high yield of bio-oil when a high amount of xylan (0.25–0.7 g) was used, while a higher yield of char was formed with a low amount of xylan ($<$0.25 g). This behavior was attributed to the different physical states of the volatiles present in the oven during the process because different pressures were present in the oven. This phenomenon may be employed for the selective production of liquid (bio-oil) or solid (biochar). Chemicals such as aldehydes, alkenes, phenols, polyaromatic hydrocarbons (PAHC), cyclic ketones, and furans, were present inside the bio-oil formed with a composition affected by the amount of xylan employed as starting material. However, acetic acid must be present, even if it was not reported while an energy efficiency of 26% was achieved for bio-oil production leading to an increase in the HHV of the products (concerning the original feedstock) of 19% for bio-oil.

5.5.2 MAP of Crop Residues from Biomasses Cultivation

The MAP of commercial biomass as wood pellets, and crop residue recovered from biomasses cultivation, such as *Vitis Vinifera*, olive pruning residue, and short rotation coppice, were performed and reviewed in the following paragraphs.

5.5.2.1 Acetic Acid Through MAP of Wood Pellets

Wood is one of the most available renewable biomass and is employed as a raw material for the pulp and paper industry, as structural timber in several constructions (sawn wood, panels, furniture, and many other uses) and it is also used to produce energy in stove pellets. The oldest way to produce energy using wood is its combustion, but this process is not very efficient and large amounts of ash is dispersed in the atmosphere dragged by the gas formed in the combustion. Now this use is strongly reduced and it remains largely employed only in developing countries. In recent years there has been a renewable interest in wood as a source of energy, especially in pelletized form, due to the possibility to use them in domestic automatic stoves with the aim to reduce both the use of fossil fuel and the CO_2 production. Taking into account the large availability and chemical complexity of wood, it is considered as a potential source of valuable chemicals for a bio-refinery. Bio-ethanol through the second generation process is one of the products obtained as well as the production of carbon-based char for activated carbon production, and gas for chemical synthesis. Pyrolysis of wood may be another way to supply bio-chemicals such as acetic acid in a process run at a relatively high temperature

Table 5.4 Bio-oil from MAP of wood pellets, using different absorbers and set-ups. In several runs only the upper fraction was analyzed [124]

Run	Microwave absorber	Set-up	Time (min)	Bio-oil yield (wt%)	Bio-oil d_4^{20} (g/mL)	Acetic acid yield (g/kg)
1[a]	–	A	10	0.0	1.049	0.4
2[a]	–	A	36	28.0	0.998	0.3
3[a]	–	A	45	23.3	1.244	0.5
4	Iron	A	26	35.4	n.d.	n.d.
5	Iron	A	26	39.3	n.d.	n.d.
6	Iron	A	26	41.5	n.d.	n.d.
7[b]	Carbon	A	25	47.7	n.d.	n.d.
8[a]	Carbon	A	23	41.9	0.753	0.3
9[c]	Carbon	A	67	46.1	n.d.	n.d.
10	Iron	B	29	34.6	n.d.	n.d.
11[a]	Iron	B	26	37.4	1.256	0.5
12[a]	Carbon	B	24	40.8	0.629	0.2
13[d]	Carbon	B	35	42.9	n.d.	n.d.
14[e]	Carbon	B	67	41.9	n.d.	n.d.

Sample 150 g; sample/absorber ratio 1.9 (wt/wt); microwave power 3 kW
[a] Only the upper phase of bio-oil was analyzed
[b] Sample/absorber ratio 4.7 (wt/wt)
[c] Microwave power 1.2–3 kW, sample/absorber ratio 2.8 (wt/wt)
[d] Sample/absorber ratio 3.0 (wt/wt)
[e] Microwave power 1.2–3 kW, sample/absorber ratio 2.8 (wt/wt)

(above 675 K) breaking of bonds of the components in an anoxic environment obtaining a bio-oil, a bio-char, and a gas [8, 124]. Acetic acid is present in the bio-oil fraction together with other valuable compounds such as carboxylic acids, phenols, and furanes. To check this use, wood pellets were employed as substrates in the MAP process using different microwave power, microwave absorber, mixing protocols between wood pellets and microwave absorbers, and apparatus set-up and results are reported in Table 5.4.

The MAP has been run without a microwave absorber but in this case, the process required a long time while in the presence of an absorber, the reaction time was strongly reduced.

Analysis of bio-oil showed the presence of compounds such as water, acetic acid, furans (such as furfural), carbohydrates, and their derivatives in the upper phase while compounds from pyrolysis of lignin such as phenols (phenol, m-cresol, guaiacol, veratrol, pyrocatechol, 4-methoxyphenol, resorcinol, and vanillin) and veratric acid (3,4-dimethoxybenzoic acid) were present in the lower phase.

Surprisingly the acetic acid was present in low amounts if correlated with results obtained using α-cellulose or lignin (Tables 5.3 and 5.4). Presumably, the acetic acid was present in the intermediate aqueous fraction, because it is soluble in water, but the analysis of this fraction was not reported.

5.5.2.2 Acetic Acid Through MAP of Crop Residues from Vitis Vinifera

Vitis vinifera is largely cultivated in almost all parts of the world among which the Mediterranean area, Central Europe, and southwestern Asia [125] are the main countries where it is present. Fresh or dried fruits are employed to produce raisins, sultanas, and currants and they are also employed to obtain juice or fermented to produce wine and vinegar. Wine is the starting bio-materials employed for acetic acid production through bacterial fermentation and this production covers almost 10% of the total production of acetic acid throughout the world, but this acid is mainly used for food production. Large amounts of pruning and grape residues are formed during the cultivation of Vitis vinifera with an estimated production of 1 ton/ha of cultivated field and a cost for their disposal of 50 €/y [126]. Several ways are available for their disposal but now the main followed procedure is burning. Other processes such as non-catalytic [127, 128] or catalytic [129] pyrolysis of pruning residue was proposed to obtain bio-oils.

MAP pyrolysis of crop residues of *Vitis vinifera* was run using several microwave absorbers (carbon, Fe powder, or SiC) in a multimode microwave oven (Table 5.5) [130]. The MAP may be directed towards the formation of bio-oil (34.9%) and gas (45.7%) or obtaining large amount of bio-char (up to 71.4%) using an appropriate combination of absorber/reactor arrangement. Liquids having dark brown color, low viscosity, and density were formed, and analytical and spectroscopic methods were employed for their characterization. Compounds present in the liquids were identified and quantified using the method reported in Sect. 5.4. A large concentration of acetic acid (up to 172.5 g/L) together with an appreciable amount of aromatics (up to 39.5 g/L) were identified in all bio-oils formed.

The water present in bio-oil was reduced when carbon was the microwave absorber and set-up B improving the residence time of the vapors inside the oven. It may be hypothesized a gasification reaction between the water present in the oven and the char followed by a water gas shift reaction causing a reduction of water and carbon and an increase of carbon monoxide and hydrogen.

Acetic acid was always obtained with yield higher than 22 g/kg of biomass and this value was strongly improved using carbon as microwave absorber and set-up B

Table 5.5 Bio-oil from MAP of Vitis vinifera, using different absorbers and set-up [130]

Run	Set-up	Microwave absorber	Time (min)	T (K)	Bio-oil yield (wt%)	Bio-oil d_4^{20} (g/mL)	Acetic acid yield (g/kg)
1	A	Carbon	16	499	33.1	1.14	22.0
2	B	Carbon	18	644	32.3	1.05	44.2
3	A	Iron	14	751	34.9	1.12	43.2
4	B	Iron	20	723	30.4	1.04	35.5
5	A	Silicon carbide	22	475	15.3	1.16	22.8
6	B	Silicon carbide	25	506	23.3	1.01	22.5

Sample 100 g; sample/absorber ratio 2.0 (wt/wt); microwave power 3 kW with a yield up to of 44.2 g/kg. On the contrary using iron as microwave absorber the use of set-up B does not improve the amount of acetic acid, presumably due to further reactions of the acetic acid with the products inside the oven, presumably catalyzed by the iron itself.

At the same time, other useful products such as aromatics and fuels may be obtained from bio-oils together with the gas and bio-char giving a contribution to the economy of the process.

5.5.2.3 Acetic Acid Through MAP from Olive Pruning Residue (OPR)

The cultivation of the Olive trees is largely followed in the Mediterranean regions because olives are directly employed as food. They are also largely employed for the production of olive oil, one of the key components of the Mediterranean diet [131], so they represent a commodity in southern European countries [132]. Several studies have been reported [133, 134] to improve plant productivity and the resistance of trees to environmental and biotic stress. These studies suggest pruning as important agronomic practice to improve the health of the plants andto give higher production. This practice causes large amounts of waste and an estimated cost for their disposal of 10–20 €/ha [135]. To reduce this cost pruning residue is burnt directly in the field where it is produced while other ways for its management are not developed up to now.

Thermochemical treatments of olive pruning residue (OPR), olive bagasse, and kernels are suggested for obtaining a bio-oil as proposed by several authors [136, 137].

A new and very appealing way to dispose OPR is the MAP process, that is using microwave heating and different microwave absorbers in a multimode batch reactor for an environmental friendly and economic disposal of these residues [138]. The MAP of these wastes allowed to obtain biochemicals using a short pyrolysis time, ranging from 15 to 36 min, a temperature in the range from 450 to 705 K and different microwave absorbers (Table 5.6). The MAP may be addressed towards the

Table 5.6 Bio-oil from MAP of Olive pruning residues, using different absorbers [138]

Run	Microwave absorber	Time (min)	T (K)	Bio-oil yield (wt%)	Bio-oil d_4^{20} (g/mL)	Acetic acid yield (g/kg)
1	Carbon	15	723	40.3	1.05	44.6
2	Iron	18	702	40.6	1.04	122.5
3	Silicon carbide	21	705	27.9	0.92	72.6
4	Silica	27	446	28.5	1.03	76.8
5	Alumina	36	450	25.5	1.04	46.1
6	Sodium hydroxide	17	568	56.2	1.10	131.8
7	Carbon/alumina	16	688	41.7	1.04	110.1

Sample 100 g; sample/absorber ratio 2.0 (wt/wt); microwave power 3 kW formation of large amounts of bio-oil (56.2%) and gas (41.7%) with a very low formation of bio-char (2.1%). However, in other conditions high amounts of bio-char (up to 61.2%) may be formed, while an intermediate amount of gas (26–30%) and bio-oil (40 wt%) were obtained using carbon or iron as microwave absorbers. In all experiments, the bio-oil was a dark-brown liquid having a low viscosity and density. A high concentration of acetic acid, even when NaOH was the microwave absorber suggesting the formation of acetic acid in the gas phase, was always present. Carbon or iron as the microwave absorber let to obtain a bio-oil with a low water concentration.

Surprisingly using sodium hydroxide as microwave absorber let to obtain the highest yield of acetic acid (131.8 g/kg), also iron or carbon/alumina as microwave absorber gave relatively high amounts of acetic acid (yield of 122.5 and 110.1 g/kg, respectively). On the contrary using carbon, silicon carbide, or silica, as microwave absorber relatively lower yield of acetic acid were formed.

Other products were present in the bio-oils such as aromatics, especially in the bio-oil formed when carbon was the microwave absorber.

Bio-chars having a calorific value close to commercial pellets (up to 25 MJ/kg) were obtained in all tests [139]. These biochar may be employed as a solid fuel or they may be converted into syngas and subsequently employed for the production of liquid fuels through the Fischer Tropsch process.

5.5.2.4 Acetic Acid Through MAP of Short Rotation Coppice of Poplar

In North Europe, especially in the UK and Sweden rotation coppice of poplar are employed to produce enegy or used in electric power stations. In the course of cultivation waste from stumps, roots, and leaves are formed and these residue were tested in the MAP to produce bio-oils available as a source of acetic acid. Bio-oil was obtained with a high yield (up to 32.0%) containing a small amount of water, having low density and viscosity (Table 5.7). Some spectroscopic and analytical techniques such as ^{1}H NMR, IR-ATR, and quantitative GC–MS method were employed for bio-oil characterization confirming the presence of acetic acid together with formic acid, acetic anhydride, furans, and some phenol-containing compounds.

Acetic acid was obtained with a yield in the range of 2.5–20.3 g/kg of poplar, the highest yield was obtained using a powder sample of poplar and set-up A. Lower amount of acetic acid was formed using iron as microwave absorber or the set-up B, that is increasing the residence time of products inside the oven. Presumably the acetic acid initially formed was decomposed in the course of the process improving the yield of gaseous products [140]. These results are analogous to those reported in Sect. 5.5.2.2 for the MAP of crop residue of Vitis vinifera where the role of the microwave absorber (iron) in the decomposition of acetic acid was hypothesized.

Table 5.7 Bio-oil from MAP of short rotation coppice of poplar, using different absorbers and set-up [140]

Run	Clone type	Clone shape	Microwave absorber	Set-up	Time (min)	T (K)	Bio-oil yield (wt%)	Bio-oil d_4^{20} (g/mL)	Acetic acid yield (g/kg)
1	I	Powder	Carbon	A	20	745	30.8	1.07	20.3
2	I	Powder	Iron	A	25	745	20.9	1.06	7.9
3	I	Powder	Carbon	B	21	745	27.7	1.04	14.7
4	I	Powder	Iron	B	26	745	12.2	1.04	2.5
5	I	Chips	Carbon	A	21	638	17.4	1.05	7.3
6	II	Chips	Carbon	A	16	606	29.6	1.07	16.7
7	I	Chips	Iron	A	21	702	32.0	1.17	9.7
8	II	Chips	Iron	A	22	617	26.6	1.23	8.5
9	III	Chips	Carbon	A	30	681	17.4	1.33	9.1

Sample 100 g; sample/absorber ratio 2.0 (wt/wt); microwave power 3 kW

5.5.3 *Acetic Acid from MAP of Non Edible Biomass*

Possible and very interesting starting materials available for MAP are non edible biomasses because they are not involved in the food cycle. Furthermore, they are a waste residue, sometimes classified as a special residue, and a cost is required for their disposal. MAP may contribute to this operation and in the same time a bio-oil may be formed useful as a fuel or employed to collect the acetic acid present together with other products. Another interesting source of biomass, for the production of acetic acid through a MAP, is the organic residue present in municipal solid waste (MSW).

5.5.3.1 Acetic Acid Through MAP of Arundo Donax

As an example of the disposal of a non edible biomass, the MAP of Around donax is reported in this paragraph. It is a perennial cane present in Eurasia and largely diffused in the Mediterranean area, it is also diffused in North America and Greater Middle East [141]. It grows rapidly in different soils and climatic conditions [142] so it is considered an invasive plant. The fast propagation rate cause an uncontrolled diffusion with fire and landslide risk in several regions and consequently, many ecosystems may be damaged in the future. To reduce these risks an environmental clearing was run with collection of large amount of residues of Arundo donax and they are usually disposed of through combustion. From an economical and environmental point of view, these residues may be also employed as starting materials for the production of bio-oil or bioethanol [143, 144]. Residue of Arundo donax may be employed as starting materials in the MAP using carbon as a microwave absorber and different reaction conditions (Table 5.8). Different organs of Arundo Donax were tested to evidence their contribution to the yield of the bio-oil formed. Also, the

Table 5.8 Bio-oil from MAP of Around Donax, using carbon as microwave absorber and different set-ups [145]

Run	Sample type	Set-up	Time (min)	T (K)	Bio-oil yield (wt%)	Bio-oil d_4^{20} (g/mL)	Acetic acid yield (g/kg)
1	Leaves	A	18	723	29.1	1.06	29.4
2	Stems	A	22	766	40.9	1.05	44.5
3	Rhizomes	A	21	681	35.5	1.03	38.9
4	Leaves	B	31	638	22.1	1.13	23.2
5	Stems	B	32	702	38.9	1.04	43.2
6	Rhizomes	B	40	707	28.7	1.03	32.9

influence of reaction parameters such as reaction temperature, microwave power, residence time in the oven, on the quality and quantity of bio-oil was tested [145]. In all experiments, a yield of bio-oil among 22.1 and 40.9% was obtained together with a small amount of gas (4.6%) and a large biochar production (up to 62.9%). The bio-oil was a one-phase dark brown liquid with low viscosity, it was characterized through several techniques such as FT-IR, ^{1}H-NMR, and a quantitative GC–MS analysis. Acetic acid was present in the bio-oil with yield among 23.2 and 44.5 g/kg together with other compounds such as aromatics, furans, and, sometimes, levoglucosan (up to 47.6 g/L). In conclusion the MAP from several organs of Arundo donax appears as an interesting way to dispose of these materials obtaining acetic acid and other biochemicals together with bio-char and gas through a fast pyrolysis process.

5.5.3.2 Acetic Acid Through MAP of Differentiated Waste Collection

Biomass is largely present in municipal solid waste collections as waste or contaminated materials and they may be disposed of through a bio-cell giving a biogas together with a soil improver but it may be contaminated by other residues eventually present and not destroyed during biodegradation such as inorganic materials, plasticizers, plastic filler, colorant, lubrificant and perfluoroalkyl compounds. These biomasses contain food residues, wood, paper, bio-plastics, and they may be disposed through a MAP obtaining a gas, a bio-oil, and a biochar. The bio-char is a source of carbon and may be devoted to the plethora of carbon uses reducing the production of Green House Gases (GHG) in the atmosphere [146]. The gas may be converted into syngas or directly employed to produce energy. The bio-oil contains acetic acid in agreement with the results reported in the MAP of cellulose, lignin, wood, and other biomasses as reported in the previous paragraphs. At the same time or after the separation of acetic acid the bio-oil residue may be employed to collect other interesting products such as furans or phenol-containing compounds and the remaining residue may be employed as a fuel.

5.5.3.3 Acetic Acid Through MAP of Corn-Derived Plastic Bags (CDP)

The European community with the aim to limit plastic pollution has forbidden the use of shoppers realized using synthetic plastics derived from mineral oils. These materials were substituted by "Bio-plastics", that is plastic materials obtained using bio-materials, that is compounds deriving from natural sources. The properties of these new materials are reported in the European Directive 94/62/EC and the norm EN13432:2000. The bio-plastics commercially available are mainly obtained using corn as starting materials and to a lesser extent from synthetic polymers obtained using a monomer from a natural source such as lactic acid (poly(lactic acid): PLA). These products are obtained through several different processes according to the performances required for the final polymer [147]. These bio-polymers are biodegradable plastics and are used to obtain shoppers because they are cheaper, have performances close to synthetic plastics, and may be disposed of using an anaerobic digestion. In the course of biodegradation their chemical content is completely lost as well as their energy, only gas may be sometimes recovered. Furthermore, the solid obtained as a residue of the biodegradation may be employed as soil improvement, but it may be contaminated by plasticizers, organic fillers and additives present in the starting bioplastic as above reported. These products may not be fully destroyed by the microorganisms present during biodegradation. Incineration may be also employed for their disposal and the energy content of bio-plastic was partially recovered, however, the heating value is relatively low due to the large amount of oxygen-containing compounds. In the course of biodegradation as well as in incineration the atom economy of the process is very poor.

Corn derived plastic bags (CDP) may be disposed of through a MAP recovering a gas, a bio-oil, and a bio-char avoiding the risks of contamination of the bio-char recovered due to fillers and additives present in CDP because in the course of the MAP process the organic products are completely destroyed. The process requires the use of an absorber because the plastic does not absorb microwave, so the MAP was realized using carbon or iron as microwave absorber [148]. In Table 5.9 are reported the results on the MAP of CDP. The bio-oil collected was always separated into three phases: upper, medium, and bottom, and these fractions were deeply characterized. Large amounts of aromatic acid, phthalates, and their derivatives were present in the upper fractions, while water was largely present in the middle phase (70 wt%), together with acids, alcohols, anhydro-sugars, and their pyrolysis products. The composition of the bottom fractions was very close to the upper fractions, the same classes of compounds were present in the upper and bottom fractions but in the last ones, oligomers of compounds present in the upper fractions were found. Using more drastic conditions the bottom fraction was reduced and the bio-char yield was improved. Surprisingly the acetic acid present in the medium fraction was relatively poor even if was the main component. Other chemicals (esters, other acids, aromatics) were present and these products may be separated or used as fuel avoiding their transformation into carbon dioxide, water, and a residue without any recovery of chemicals or energy.

Table 5.9 Bio-oil from MAP of Corn Derived Plastic Bag, using different microwave absorbers and set-ups. The intermediate fraction was analyzed and reported [148]

Run	Set-up	Microwave absorber	Microwave power (kW)	Time (min)	T (K)	Bio-oil yield (wt %)	Bio-oil d_4^{20} (g/mL)	Acetic acid yield (g/kg)
1	A	Carbon	3.0–1.2	58	808	47.6	1.0518	0.34
2	A	Carbon	1.5–3.0	32	723	48.4	1.0317	0.58
3	A	Iron	3.0	32	617	42.9	1.0467	0.20
4	B	Carbon	3.0–1.2	56	787	57.7	1.0435	1.02
5	B	Carbon	1.5–3.0	75	765	18.2	1.0394	0.39
6	B	Iron	3.0	41	765	44.2	1.0388	0.56

Sample 127 g; sample/absorber ratio 2.0–2.4 (wt/wt); microwave power 3 kW

5.5.3.4 Acetic Acid Through MAP of a Synthetic Bioplastic (PLA)

Lactic acid is an α-hydroxyacid [$CH_3CH(OH)COOH$] and the L form is produced by many tissues of the human body. It is also one of the high volume chemicals obtained by fermentation of many cheap compounds such as starchy or cellulosic material through yeast fermentation [149]. Several industries employ this acid as a synthetic intermediate in many chemical and biochemical syntheses. It is also used for the production of one of the most important synthetic bio-polymers (poly(lactic acid (PLA)) as a substitute for plastic from mineral oil. The starting material for the production of PLA is mainly obtained by industrial fermentation of carbohydrates (pentose and hexose sugars) and in minor amounts from acetic aldehyde. The L form of lactic acid is employed for the industrial production of PLA through a dehydro-condensation of lactic acid but the most interesting synthesis of high molecular weight PLA is the ring-opening polymerization of lactide, the cyclic dimer of lactic acid (Fig. 5.11).

PLA is a thermoplastic polymer and since 2010 it is one of the most used bio-plastic around the world even if it has not reached the classification of a commodity polymer due to its cost of production, higher than classical synthetic polymers obtained from mineral oil [150, 151]. PLA is the most widely used plastic for medical purposes such as prostheses, surgical sutures, [152] or in the form of anchors, screws, plates, pins, rods, and mesh because it can degrade into un-harmful lactic acid, so it is used as medical implants. Due to its biocompatibility and biodegradability, PLA has also found large interest as a polymeric scaffold for drug delivery purposes and to stimulate collagen synthesis in dermatological studies [153]. It is also used as material in 3D printing, but a considerable amount of PLA is used for a large variety of consumer products such as disposable tableware, cutlery, housings for kitchen appliances, electronics, shopper bags, and food packaging [154].

The MAP of waste PLA was studied to evaluate its recycling using different microwave absorbers (tire, carbon, or Fe), apparatus set-ups, and microwave power [155]. A liquid was obtained containing a large amount of the lactide (3,6-dimethyl-

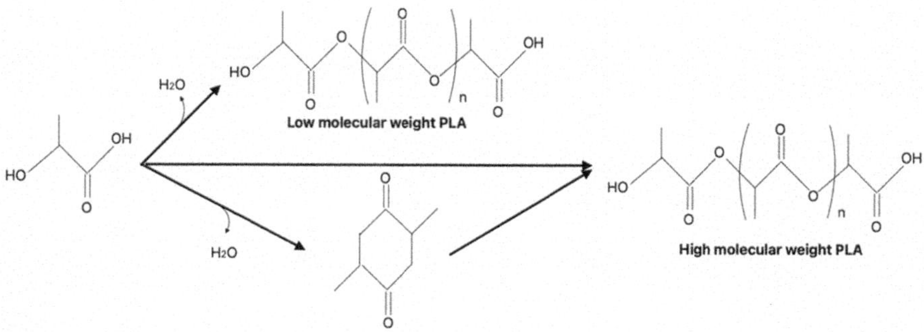

Fig. 5.11 Schematic synthesis of poly(lactic acid) (PLA)

Table 5.10 Bio-oil from MAP of poly(lactic acid) (PLA) a bio-plastic, using different microwave absorbers and set-ups [155]

Run	Substrate	Set-up	Microwave absorber	Sample/absorber ratio (wt/wt)	Time (min)	T (K)	Bio-oil yield (wt%)	Bio-oil d_4^{20} (mg/mL)	Acetic acid yield (g/kg)
1[a]	PLA	A	Carbon	2.1	61	723	28.4	1067	0.6
2[b]	PLA	A	Carbon	3.2	71	638	48.4	1087	2.5
3	PLA	A	Iron	2.0	41	659	42.9	1120	0.5
4[c]	PLA	B	Carbon	2.1	25	766	57.7	1134	0.7
5	PLA	B	Iron	2.0	70	659	18.2	953	0.6
6[d]	PLA	A	ELT	0.5	30	808	44.2	1079	0.2
7[e]	ELT	A	–	–	35	764	42.6	874	0.0
8	Lactide	A	Carbon	2.0	30	Nd	40.4	1000	0.8
9	Lactide	A	Iron	2.0	30	Nd	26.2	1.000	0.1

Sample 100 g; microwave power 3 kW
Nd not reported
[a] Sample 150 g, microwave power 1.5–3.0
[b] Microwave power 1.5–3 kW
[c] Sample 123 g
[d] End life tire (ELT) as microwave absorber
[e] End life tire (ELT) has starting material

1,4-dioxan-2,5-dione) the starting material employed for the production of high molecular weight PLA and other oxygenated organic compounds (Table 5.10). Pure L-lactide was separated from the collected liquids (up to 27.7%) while the meso-form of lactide was present in the liquid and identified through GC–MS analysis. As expected acetic acid was identified in the bio-oils in low amounts together with other organic compounds such as propionic acid (up to 17.1%), carbonyl compounds, and low molecular weight PLA. If the MAP was carried out in the presence of end life tire (ELF) as microwave absorber (which, in the same time, pyrolyze to aromatic compounds), cross-reactions take place in the course of the process affecting both yield and characteristics of the liquid, however, the amount of acetic acid remains lower and it was further diluted by the liquid oil formed by ELT pyrolysis.

5.6 MAP of Other Biomasses

Many other biomasses were pyrolyzed and several of them through a MAP and their results are reported in the literature as summarized at the end of Sect. 5.1, but these results are not described because they are only devoted to the production of bio-oil and they do not report a quantitative determination of the amount of acetic acid present.

5.7 Acetic Acid Separation from Bio-oil

Every bio-oil obtained from biomass pyrolysis contains a very large amount of compounds and many of them are thermolabile. Several methods are reported for the identification of acetic acid in bio-oil as reported in Sect. 5.4 however these methods can not be followed for an industrial separation. Acetic acid is, in fact, a commodity so it must be obtained in very high amounts using processes affordable by an economic and environmental point of view. Studies on the separation of acetic acid present in bio-oil from MAP of biomass are not reported.

The separation of acetic acid may be performed through several methods [156], such as distillation, reactive distillation that is conversion of acetic acid in its ester derivatives [157], fractional distillation followed by liquid-liquid extraction [158], membrane separation, pervaporation [159, 160], supercritical carbon dioxide extraction [161–163] as reviewed by Sarchami et al. [103]. This review is devoted to the production and separation of acetic acid, but surprisingly, it does not report the production of acetic acid through MAP of biomass and separation of the compounds present in the the bio-oils formed by an industrial point of view.

Some of the above reported procedures, however, may cause the degradation of thermolabile compounds present in bio-oil causing their lost or/and the formation of high boiling residues. In this way, several interesting compounds eventually present in bio-oil, such as limonene and so on, can not be collected or the residue of bio-oil can not be employed as a fuel due to the presence of large amounts of high boiling compounds.

A very interesting procedure for the separation of acetic acid and other interesting chemical compounds is reported by Pozo et al. [164] for acetic acid recovery from bio-oil of olive mill waste. This procedure consists on selective solvent extraction of aqueous and non-aqueous phases of bio oil at different pH values using hexane and ethyl acetate as solvents. In this way several bio-oil components were separated due to their polarity and solubility in different solvents but interesting thermolabile

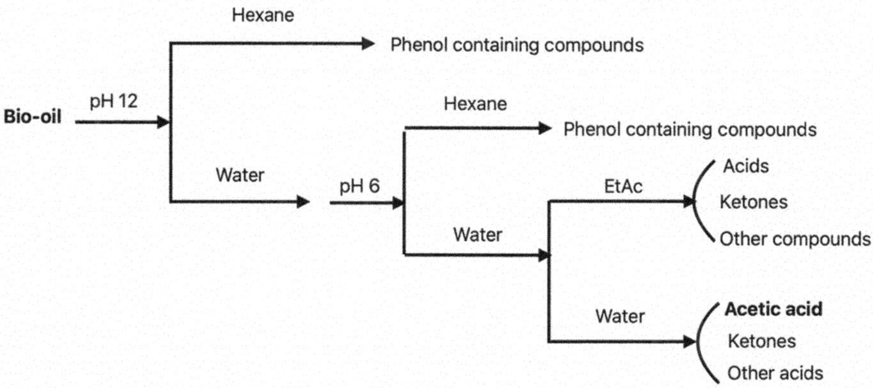

Fig. 5.12 Proposed solvent extraction for acetic acid separation

compounds were not decomposed. Furthermore, because acid components of bio-oil were eliminated the residue has more interesting properties if employed as a fuel (Fig. 5.12). These process has the potentiality of an industrial application because the Energy required for solvent extraction is low and solvents employed may be recycled avoiding the formation of large amount of waste and by-products.

5.8 Conclusion

Several applications of MAP for acetic acid production are described using different sources such as wood, non-edible biomass, crop residue from agricultural cultivation, contaminated biomass, or waste biopolymers. Also residue of municipal solid waste collection may be employed for the production of acetic acid through a MAP process. Contaminants eventually present in the starting materials are destroyed in the course of the pyrolysis as reported for the disposal of waste covid mask [20]. Furthermore Bambad et al. [165] reported a very interesting complete decomposition of per- and polyfluoroalkyl compounds (PFAS) present in the starting materials during a high temperature pyrolysis, that is the conditions usually present in the MAP process. This results is very interesting because PFAS are a very large and complex group of synthetic chemicals employed since 1950. They are persistent compounds due to their stability and multiple health effect are associated with PFAS exposure as reported in several scientific studies [166]. Finally, very recently, Du et al. [167] have reported the treatment of tannery sludge through MAP and the reduction of very toxic Cr(VI) compounds to low toxic Cr(III) compounds, while heavy metals were fixed in the residue. These results strongly improve the importance of the MAP process for the treatment of waste and removal of toxic substance.

Using appropriate biomass the MAP gives a bio-oil with an acetic acid yield up to 139.8 g/kg of biomass, and the acetic acid may be subsequently separated from the other components of bio-oil following the procedure reported in Sect. 5.7.

Taking into account the results above reported, MAP represents a friendly and economical way for the production of acetic acid reducing the carbon footprint due to the production from mineral oil. A commodity product such as acetic acid may be obtained in a friendly and economic way. Taking into account the presence of other useful products in the bio-oils such as furans, levoglucosan, and phenol-containing compounds, the residue of acid estraction may be sent to a bio-refinery, and the interesting compounds isolated and commercialized. Folllowing this way the economy of the process was improved and this process may represent a step towards the circular economy. Finally, other interesting compounds formed in the course of the MAP such as the gas [168] and mainly the bio-char [169–171] gave a further contribution to the validity of the process itself.

References

1. https://www.statista.com/statistics/222066/projected-global-energy-consumption-by-source. Accessed 2 July 2024.
2. Bar-On YM, Phillips R, Milo R. The biomass distribution on Earth. Proc Natl Acad Sci Unit States Am. 2018;115:6506–11. https://doi.org/10.1073/pnas.1711842115.
3. https://www.eia.gov/tools/glossary/index.php?id=Photosynthesis. Accessed 3 Mar 2022.
4. U.S. Energy Information and Administration. Biomass explained. https://www.eia.gov/energyexplained/biomass/. Accessed 28 June 2021.
5. https://www.eia.gov/tools/glossary/index.php?id=Black %20liquor. Accessed 3 Mar 2022.
6. https://www.eia.gov/energyexplained/biomass/waste-to-energy.php. Accessed 10 Nov 2023.
7. Nimz HH, Schmitt U, Schwab E, Wittmann O, Wood WF. Ullmann's encyclopedia of industrial chemistry, vol. 2000. Wiley-VCH Verlag GmbH & Co; 2000. https://doi.org/10.1002/14356007.a28_305.
8. Bridgwater AV. Renewable fuels and chemicals by thermal processing of biomass. Chem Eng J. 2003;91:87–102. https://doi.org/10.1016/S1385-8947(02)00142-0.
9. Kumar R, Strezov V, Lovell E, Kan T, Weldekidan H, He J, et al. Bio-oil upgrading with catalytic pyrolysis of biomass using copper/zeolite-nickel/zeolite and copper-nickel/zeolite catalysts. Bioresour Technol. 2019;279:404–9. https://doi.org/10.1016/j.biortech.2019.01.067.
10. Mohan D, Pittman CU, Steele PH. Pyrolysis of wood/biomass for bio-oil: a critical review. Energy Fuel. 2006;20:848–89. https://doi.org/10.1021/ef0502397.
11. Yang H, Yan R, Chen H, Lee DH, Zheng C. Characteristics of hemicellulose, cellulose and lignin pyrolysis. Fuel. 2007;86:1781–8. https://doi.org/10.1016/j.fuel.2006.12.013.
12. Miura M, Kaga H, Yoshida T, Ando K. Microwave pyrolysis of cellulosic materials for the production of anhydrosugars. J Wood Sci. 2001;47:502–6. https://doi.org/10.1007/BF00767906.
13. Robinson JP, Kingman SW, Barranco R, Snape CE, Al-Sayegh H. Microwave pyrolysis of wood pellets. Ind Eng Chem Res. 2010;49:459–63. https://doi.org/10.1021/ie901336k.
14. Kan T, Strezov V, Evans TJ. Lignocellulosic biomass pyrolysis: a review of product properties and effects of pyrolysis parameters. Renew Sust Energ Rev. 2016;57:1126–40. https://doi.org/10.1016/j.rser.2015.12.185.

15. Kumar R, Strezov V, Kan T, Weldekidan H, He J. Investigating the effect of cu/zeolite on deoxygenation of bio-oil from pyrolysis of pine wood. Energy Procedia. 2019;160:186–93. https://doi.org/10.1016/j.egypro.2019.02.135.
16. Weldekidan H, Strezov V, Town G. Review of solar energy for biofuel extraction. Renew Sust Energ Rev. 2018;88:184–92. https://doi.org/10.1016/j.rser.2018.02.027.
17. Frediani P, Frediani M. Bio-char and bio-attributed-carbon black from microwave assisted pyrolysis as a contribution to circular economy. Elsevier; 2024.
18. Vitolo S, Seggiani M, Frediani P, Ambrosini G, Politi L. Catalytic upgrading of pyrolytic oils to fuel over different zeolites. Fuel. 1999;78:1147–59. https://doi.org/10.1016/S0016-2361(99)00045-9.
19. Undri A, Rosi L, Frediani M, Frediani P. Microwave pyrolysis of polymeric materials. In: Chandra U, editor. Microwave heating. InTech; 2011. p. 207–26. https://doi.org/10.5772/24008.
20. Frediani M, Frediani P, Innocenti G, Mellone I, Simoni R, Oteri G. Microwave-assisted pyrolysis process: from a laboratory scale to an industrial plant. In: Kishk A, editor. Microwave technologies. London: IntechOpen; 2022. p. 104925. https://doi.org/10.5772/intechopen.104925.
21. Eser S, Jenkins RG, Derbyshire FJ, Malladi M. Carbonization of Coker feedstocks and their fractions. Carbon. 1986;24(1986):77–82. https://doi.org/10.1016/0008-6223(86)90213-7.
22. Mushtaq F, Mat R, Ani FN. A review on microwave assisted pyrolysis of coal and biomass for fuel production. Renew Sust Energ Rev. 2014;39:555–74. https://doi.org/10.1016/j.rser.2014.07.073.
23. Domenach AM, Moiroud A, Jocteur-Monrozier L. Leaf carbon and nitrogen constituents of some actonorhizal tree species. Soil Biol Biochem. 1994;26:649–53. https://doi.org/10.1016/0038-0717(94)90254-2.
24. Galema SA. Microwave chemistry. Chem Soc Rev. 1997;26:233–8. https://doi.org/10.1039/cs9972600233.
25. Saxena VK, Chandra U. Microwave synthesis: a physical concept. In: Chandra U, editor. Microwave heating. Intech; 2011. p. 14. https://doi.org/10.5772/22888.
26. Al-Qahtani AM. A comprehensive review in microwave pyrolysis of biomass, syngas production and utilisation. Energies. 2023;16:6876. https://doi.org/10.3390/en16196876.
27. Asomaning J, Haupt S, Chae M, Bressler DC. Recent developments in microwave-assisted thermal conversion of biomass for fuels and chemicals. Renew Sust Energ Rev. 2018;92:642–57. https://doi.org/10.1016/j.rser.2018.04.084.
28. Bridgwater AV, Peacocke GVC. Fast pyrolysis processes for biomass. Renew Sust Energy Rev. 2012;4:1–73. https://doi.org/10.1016/S1364-0321(99)00007-6.
29. Yin C. Microwave-assisted pyrolysis of biomass for liquid biofuels production. Bioresour Technol. 2012;120:273–84. https://doi.org/10.1016/j.biortech.2012.06.016.
30. Escalante J, Chen WH, Tabatabaei M, Hoang AT, Kwon EE, Lin KYA, Saravanakumar A. Pyrolysis of lignocellulosic, algal, plastic, and other biomass wastes for biofuel production and circular bioeconomy: a review of thermogravimetric analysis (TGA) approach. Renew Sust Energy Rev. 2022;169:112914. https://doi.org/10.1016/J.rser.2022.112914.
31. Ethaib S, Rozita Omar R, Kamal SMM, Biak DRA, Zubaidi SL. Microwave-assisted pyrolysis of biomass waste: a mini review. Processes. 2020;8:1190. https://doi.org/10.3390/pr8091190.
32. Fernández Y, Arenillas A, Menéndez JA. Chapter 31. Microwave heating applied to pyrolysis. In: Grundas SÅ, editor. Advances in induction and microwave heating of mineral and organic materials. InTech; 2011. p. 723–52.
33. Giorcelli M, Das O, Sas G, Forsth M, Bartoli M. A review of bio-oil production through microwave-assisted pyrolysis. PRO. 2021;9:561–77. https://doi.org/10.3390/pr9030561.
34. Hadiya V, Popat K, Vyas S, Varjani S, Vithanage M, Gupta VK, Delgado AN, Zhou Y, Show PL, Bilal M, Zhang Z, Sillan M, Mohanty SS, Patel Z. Biochar production with amelioration of microwave-assisted pyrolysis: current scenario, drawbacks and perspectives. Bioresour Technol. 2022;355:127303. https://doi.org/10.1016/j.biortech.2022.127303.

35. Hamzah HT, Veluru S, Hasan HM, Abed HY, Mahdi HS, Hamdullah NA, Tukarambai M, Poiba VR, Srikanth R. Microwave assisted pyrolysis of biomass feddstock fundamentals and the effect of process parameters. J Phys Conf Set. 2023;2550:012003. https://doi.org/10.1088/1742-6596/25509/1/012003.
36. Ingole PM, Ranveer AC, Deshmukh SM, Deshmukh DSM. Microwave assisted pyrolysis of biomass: a review. Int J Adv Techn Eng Sci. 2016;4:78–84.
37. Kim J-S. Production, separation and applications of phenolic-rich bio-oil—a review. Bioresour Technol. 2015;178:90–8. https://doi.org/10.1016/j.biortech.2014.08.121.
38. Li J, Dai J, Liu G, Zhang H, Gao Z, Fu J, He Y, Huang Y. Biochar from microwave pyrolysis of biomass: a review. Biomass Bioenerg. 2016;94:228–44. https://doi.org/10.1016/j.biombioe.2016.09.010.
39. Li S, Ji B, Zhang W. A review on the thermochemical treatments of biomass: implications for hydrochar production and rare earth element recovery from hyperaccumulators. Chemosphere. 2023;342:140140. https://doi.org/10.1016/j.chemosphere.2023.140140.
40. Liu J, Hou Q, Ju M, Ji P, Sun Q, Li W. Biomass pyrolysis technology by catalytic fast pyrolysis, catalytic co-pyrolysis and microwave-assisted pyrolysis. a review Catalysts. 2020;10:742. https://doi.org/10.3390/catal10070742.
41. Lu R-r, Shang H, Li J. Research progress on biomass pyrolysis technology for liquid oil production. Shengwuzhi Huaxue Gongcheng. 2010;44:54–9.
42. Luque R, Menendez JA, Arenillas A, Cot J. Microwave-assisted pyrolysis of biomass feedstocks: the way forward? Energ Environm Sci. 2012;5:5481–8. https://doi.org/10.1039/C1EE02450G.
43. Omoriyekomwan JE, Tahmasebi A, Dou J, Wang R, Yu J. A review on the recent advances in the production of carbon nanotubes and carbon nanofibers via microwave-assisted pyrolysis of biomass. Fuel Process Tech. 2021;2021(214):106686. https://doi.org/10.1016/j.fuproc.2020.106686.
44. Pritam K, Palla S, Puppala H, Srinivas BA, Luhaniwal J, Surya DV. Bibliometric analysis of research progress in microwave-assisted pyrolysis of biomass during 1979–2023. J Anal Appl Pyrol. 2024;2024:177106331. https://doi.org/10.1016/j.jaap2023.106331.
45. Singh A, Kushwaha A, Sen S, Goswami S, Katiyar S, Kumar A, Borah SN, Goswami L, Hussain CM. Chapter 7. Recent advancement in microwave-assisted pyrolysis for biooil production. In: Hussain C, Sunpreet Singh S, Goswami L, editors. Waste-to-energy approaches towards zero waste. 1st ed. Elsevier; 2021. https://doi.org/10.1016/B978-0-323-85387-3.00014-8.
46. State RN, Volceanov A, Muley P, Boldor D. A review of catalysts used in microwave assisted pyrolysis and gasification. Biores Techn. 2019;277:179–94. https://doi.org/10.1016/j.biortech.2019.01.036.
47. Suriapparao DV, Tejasvi R. A review on role of process parameters on pyrolysis of biomass and plastics: present scope and future opportunities in conventional and microwave-assisted pyrolysis technologies. Process Saf Environ Protect. 2022;162:435–62. https://doi.org/10.1016/j.psep.2022.04.024.
48. Yaashikaaa PR, Kumara PS, Sunita Varjanic S, Saravanand A. A critical review on the biochar production techniques, characterization, stability and applications for circular bioeconomy. Biotechn Rep. 2020;28:e00570. https://doi.org/10.1016/j.btre.2020.e00570.
49. Yao Y, He Y-F, Liu J-M, Ma X-X, Wang W, Liu G-Q. Progress of catalysts and absorbents for microwave assisted pyrolysis of biomass. Xiandai Huagong. 2016;36:12–6. https://doi.org/10.16606/j.cnki.issn0253-4320.2016.02.004.
50. Zeng Q-z, Yu F-w, Sheng J-f, Lu NY, J-b M-z X. Research progress in microwave drying pretreatment and microwave-assisted pyrolysis of biomass. Guocheng Gongcheng Xuebao. 2013;2013(13):1061–7.
51. Zhang X, Rajagopalan K, Lei H, Ruan R, Sharma BK. An overview of a novel concept in biomass pyrolysis: microwave irradiation. Sust Energ Fuels. 2017;2017(1):1664–99. https://doi.org/10.1039/C7SE00254H.

52. Zhang Y, Chen P, Liu S, Fan L, Zhou N, Min M, Cheng Y, Peng P, Anderson E, Wang Y, Wan Y, Liu Y, Li B, Ruan R. Chapter 6. Microwave-assisted pyrolysis of biomass for bio-oil production. In: Samer M, editor. Pyrolysis. Intech; 2017. p. 129–66. https://doi.org/10.5772/67442.
53. Zhang Y, Chen P, Liu S, Peng P, Min M, Cheng Y, Anderson E, Zhou N, Fan L, Liu C, et al. Effects of feedstock characteristics on microwave-assisted pyrolysis—a review. Biores Techn. 2017:230143. https://doi.org/10.1016/j.biortech.2017.01.046.
54. Zhang Y, Fan S, Liu T, Fu W, Li B. A review of biochar prepared by microwave-assisted pyrolysis of organic wastes. Sust Energy Techn Assess. 2021;50:101873. https://doi.org/10.1016/j.seta.2021.101873.
55. Ellison CR, Hoff R, Mărculescu C, Boldor D. Investigation of microwave-assisted pyrolysis of biomass with char in a rectangular waveguide applicator with built-in phase-shifting. Appl Ener. 2019;259:114217. https://doi.org/10.1016/j.apenergy.2019.114217.
56. Kumar R, Strezov V, Weldekidan H, He J, Singh S, Kan T, Dastjerdi B. Lignocellulose biomass pyrolysis for bio-oil production: a review of biomass pre-treatment methods for production of drop-in fuels. Renew Sust Energ Rev. 2020;123:109763. https://doi.org/10.1016/j.rser.2020.109763.
57. Kurian J, Raghavan GSV. Microwave-assisted pyrolysis of biomass: an overview. Biofuel Biorefin Product Biofuel Chem Pyrol. 2020:185–206. https://doi.org/10.1007/978-981-15-2732-6_7.
58. Mokhta ZM, Ong MY, Salman B, Nomanbhay S, Salleh SF, Wayne Chew KW, Show PL, Chen WH. Simulation studies on microwave-assisted pyrolysis of biomass for bioenergy production with special attention on waveguide number and location. Energy. 2020;190:116474. https://doi.org/10.1016/j.energy.2019.116474.
59. Luo H, Zhang Y, Zhu H, Zhao X, Zhu L, Liu W, Sun M, Miao G, Li S, Kong L. Microwave-assisted low-temperature biomass pyrolysis: from mechanistic insights to pilot scale. Green Chem. 2021;23:821–7. https://doi.org/10.1039/D0GC03348K.
60. Narde SR, Neelancherry Remya N. Biochar production from agricultural biomass through microwave-assisted pyrolysis: predictive modelling and experimental validation of biochar yield. Environ Develop Sust. 2022;24:11089–102. https://doi.org/10.1007/s10668-021-01898-9.
61. Waitongkham R, Pianroj Y, Punvichai T, Karrila S, Chumkaew P, Jumrat S. Characterization of bio-oil production by microwave pyrolysis from cashew nut shells and Cassia fistula pods. Green Process Synthesis. 2023;2023(12):20230084. https://doi.org/10.1515/gps-2023-0084.
62. Allende S, Brodie G, Jacob MV. Breakdown of biomass for energy applications using microwave pyrolysis: a technological review. J Environ Res. 2023;226:115619. https://doi.org/10.1016/j.envres.2023.115619.
63. Goong SY, Chan YH, Ye PNY, Lock SSM, Chin BLF, Yin CL, Lan JCW, Lam SS. Microwave-assisted pyrolysis in biomass and waste valorisation: insights into the life-cycle assessment (Lca) and techno-economic analysis (tea), SSRN; 2023. https://papers.ssrn.com/sol3/papers.cfm?abstract_id=4646091. Accessed 15 July 2024.
64. Hamzah HT, Veluru S, Hasan HM, Abed HY, Mahdi HS, Hamdullah NA, Tukarambai M, Venkata Rao Poiba VR, Srikanth R. Microwave assisted pyrolysis of biomass feedstock fundamentals and the effect of process parameters—a review. J Phys Conf Ser. 2023;2550:012003. https://doi.org/10.1088/1742-6596/2550/1/012003.
65. Maggiotto G, Colangelo G, Milanese M, de Risi A. Thermochemical technologies for the optimization of olive Wood biomass energy exploitation: a review. Energies. 2023;16:6772. https://doi.org/10.3390/en16196772.
66. Scarlett AL Microwave-assisted pyrolysis for biomass recovery and applications. PhD thesis, James Cook University; 2023. https://doi.org/10.25903/m7yy-nv90.
67. Selvam MS, Balasubramanian P. Emerging trends and research frontiers of biochar derived through microwave assisted pyrolysis: a scientometric review. Bior Technol Rep. 2023;24:101601. https://doi.org/10.1016/j.biteb.2023.101601.

68. Syed NR, Zhang B, Mwenya S, Aldeen AS. A systematic review on biomass treatment using microwave-assisted pyrolysis under PRISMA guidelines. Molecules. 2023;28:5551. https://doi.org/10.3390/molecules28145551.
69. Yin G, Zhang F, Gao Y, He W, Zhang Q, Yang S. Increase of bio-char yield by adding potassium salt during biomass pyrolysis. J Energ Inst. 2023;110:101342. https://doi.org/10.1016/J.joei.2023.101342.
70. Jiang Z, Liang Y, Guo F, Wang Y, Li R, Tang A, Tu Y, Zhang X, Wang J, Li S, Kong L. Microwave-assisted pyrolysis-a new way for the sustainable recycling and upgrading of plastic and biomass:a review. ChemSusChem. 2024;17:e202400129. https://doi.org/10.1002/cssc.202400129.
71. Rijo B, Soares Dias AP, de Jesus N, Pereira MF. Home trash biomass valorization by catalytic pyrolysis. Environm. 2023;10:186. https://doi.org/10.3390/environments10100186.
72. Wang Y, Dai L, Fan L, Shan S, Liu Y, Ruan R. Review of microwave-assisted lignin conversion for renewable fuels and chemicals. J Anal Appl Pyrol. 2016;119:104–13. https://doi.org/10.1016/j.jaap.2016.03.011.
73. Chen L, Hu J, He Y, Wang H, Deng Q, Mi B, Wu F. Microwave-assisted pyrolysis of waste lignin to prepare biochar for $Cu2^{+}$ highly-efficient adsorption: performance, kinetics and mechanism resolution. Separ Purif Techn. 2024;342:127070. https://doi.org/10.1016/j.seppur.2024.
74. Bu Q, Morgan HM Jr, Liang J, Lei H, Ruan R. Catalytic microwave pyrolysis of lignocellulosic biomass for fuels and chemicals. Adv Bioenerg. 2016;1:69–123. https://doi.org/10.1016/bs.aibe.2016.09.002.
75. Huang YF, Chiueh PT, Shang-Lien Lo SL. A review on microwave pyrolysis of lignocellulosic biomass. Sust Environ Res. 2016;26:103–9. https://doi.org/10.1016/j.serj.2016.04.012.
76. Jayakumar M, Hamda AS, Abo LD, Daba BJ, Prabhu SV, Rangaraju M, Jabesa A, Periyasamy S, Suresh S, Baskar G. Comprehensive review on lignocellulosic biomass derived biochar production, characterization, utilization and applications. Chemosphere. 2023;345:140515. https://doi.org/10.1016/j.chemosphere.2023.140515.
77. Morgan HM Jr, Bu Q, Liang J, Liu Y, Mao H, Shi A, Lei H, Ruan R. A review of catalytic microwave pyrolysis of lignocellulosic biomass for value-added fuel and chemicals. Biores Techn. 2017;230:112–21. https://doi.org/10.1016/j.biortech.2017.01.059.
78. Mutsengerere S, Chihobo CH, Musademba D, Nhapi I. A review of operating parameters affecting bio-oil yield in microwave pyrolysis of lignocellulosic biomass. Renew Sust Energ Rev. 2019;104:328–36. https://doi.org/10.1016/j.rser.2019.01.030.
79. Nomanbhay S, Salman B, Hussain R, Ong MY. Microwave pyrolysis of lignocellulosic biomass—a contribution to power Africa. Energy Sustain Soc. 2017;7:23. https://doi.org/10.1186/s13705-017-0126-z.
80. Lataf A, Khalil Awad AE, Joos B, Carleer R, Yperman J, Schreurs S, D'Haen J, Cuypers A, Vandamme D. Iron oxide-activated carbon composites for enhanced microwave-assisted pyrolysis of hardwood. Environments. 2024;11:102. https://doi.org/10.3390/environments11050102.
81. Lee XJ, Ong HC, Gan YY, Chen W-H, Mahlia T, Meurah I. State of art review on conventional and advanced pyrolysis of macroalgae and microalgae for biochar, bio-oil and bio-syngas production. Energ Conv Manag. 2020;210:112707. https://doi.org/10.1016/j.enconman.2020.112707.
82. Liu Y, Wang Y, Wang Y, Yang H, Wu X, Ruan R. Development of downstream processing technology for microalgae biofuel. Nongye Jixie Xuebao. 2014;45:158–67. https://doi.org/10.6041/j.issn.1000-1298.2014.03.027.
83. Reddy BR, Sridevi V, Kumar TH, Rao CS, Palla VCS, Suriapparao DV, Swami Undi GSNVKSN. Synthesis of renewable carbon biorefinery products from susceptor enhanced microwave-assisted pyrolysis of agro-residual waste: a review. Proc Saf Environ Protect. 2022;164:354–72. https://doi.org/10.1016/J.psep.2022.06.027.

84. Kuo LA, Tsai WT, Yang RY, Tsai JH. Production of high-porosity biochar from rice husk by the microwave pyrolysis process. PRO. 2023;11:3119. https://doi.org/10.3390/pr11113119.
85. Foong SY, Liew RK, Lee BHK, Lam SS. Microwave pyrolysis combined with CO2 and steam as potential approach for waste valorization. Key Eng Mat. 2022;914:187–92. https://doi.org/10.4028/p-q43662.
86. Tsai WT, Kuo LA, Tsai CH, Huang HL, Yang RY, Tsai JH. Production of porous biochar from cow dung using microwave process. Materials. 2023;16:7667. https://doi.org/10.3390/ma16247667.
87. Li S, Li C, Shao Z. Microwave pyrolysis of sludge: a review. Sust Environ Res. 2022;32:23. https://doi.org/10.1186/s42834-022-00132-z.
88. Zaker A, Chen Z, Wang X, Zhang Q. Microwave-assisted pyrolysis of sewage sludge: a review. Fuel Process Techn. 2019;2019(187):84–104. https://doi.org/10.1016/j.fuproc.2018.12.011.
89. Foong SY, Low F, Yek PNY, Yang Y, Meng L, Kee LR, Verma M, Rajamohan N, Lam SS. Production of value-added biochar and bio-oil from microwave CO_2 pyrolysis of medium-density fiberboard waste. J Sust Sci Manag. 2024;19:75–84. https://doi.org/10.46754/jssm.2024.06.006.
90. Neha S, Remya N. Optimization of biochar production from microwave co-pyrolysis of food waste and low-density polyethylene. Biomass Conv Bioref. 2023;13:9465–74. https://doi.org/10.1007/s13399-023-03948-w.
91. Allende S, Brodie G, Jacob MV. Microwave pyrolysis of various wastes and analysis of energy recovery. Biores Technol Rep. 2024;26:101821. https://doi.org/10.1016/j.biteb.2024.101821.
92. Market volume of worldwide in the year 2030. 2022. https://www.statista.com/statistics/1245203/acetic-acid-market-volume-worldwide/#:~:text=In. Accessed 14 Sept 2023.
93. Cavani F, Centi G, Di Serio M, Rossetti I, Salvini A, Strukul G. Fondamenti di chimica industriale: Materie prime - Prodotti - Processi - Sostenibilità, Zanichelli ed. SpA; 2022. p. 558.
94. Undri A, Abou-Zaid M, Briens C, Berruti F, Rosi L, Bartoli M, Frediani M, Frediani P. A simple procedure for chromatographic analysis of bio-oils from pyrolysis. J Anal Appl Pyrolysis. 2015;114:208–21. https://doi.org/10.1016/j.jaap.2015.05.019.
95. Bartoli M, Rosi L, Frediani M, Frediani P. Protocol: a simple protocol for quantitative analysis of bio-oils through gas-chromatography/mass spectrometry. Eur J Mass Spectrom. 2016;22:199–212. https://doi.org/10.1255/ejms.1432.
96. Liu Y, Shi Q, Zhang Y, He Y, Chung KH, Zhao S, et al. Characterization of red pine pyrolysis bio-oil by gas chromatography-mass spectrometry and negative-ion electrospray ionization Fourier transform ion cyclotron resonance mass spectrometry. Energy Fuel. 2012;26:4532–9. https://doi.org/10.1021/ef300501t.
97. Zheng A, Zhao Z, Chang S, Huang Z, He F, Li H. Effect of torrefaction temperature on product distribution from two-staged pyrolysis of biomass. Energy Fuel. 2012;26:2968–74. https://doi.org/10.1021/ef201872y.
98. Serras-Malillos A, Perez-Martinez BB, Iriondo A, Acha E, Lopez-Urionabarrenechea A, Caballero BM. Quantification of the composition of pyrolysis oils of complex plastic waste by gas chromatography coupled with mass spectrometer detector. RSC Adv. 2024;14:9892–911. https://doi.org/10.1039/d4ra00226a.
99. Ren S, Ye XP, Borole AP, Kim P, Labbe N. Analysis of switchgrass-derived bio-oil and associated aqueous phase generated in a semi-pilot scale auger pyrolyzer. J Anal Appl Pyrolysis. 2016;119:97–103. https://doi.org/10.1016/j.jaap.2016.03.013.
100. Liang S, Han Y, Wei L, McDonald AG. Production and characterization of bio-oil and bio-char from pyrolysis of potato peel wastes. Biomass Convers Biorefinery. 2015;5:237–46. https://doi.org/10.1007/s13399-014-0130-x.
101. Haverly MR, Okoren KV, Brown RC. Thermal stability of fractionated bio-oil from fast pyrolysis. Energy Fuel. 2016;30:9419–26. https://doi.org/10.1021/acs.energyfuels.6b01606.

102. Choi YS, Johnston PA, Brown RC, Shanks BH, Lee KH. Detailed characterization of red oak-derived pyrolysis oil: integrated use of GC, HPLC, IC, GPC and Karl-Fischer. J Anal Appl Pyrolysis. 2014;110:147–54. https://doi.org/10.1016/j.jaap.2014.08.016.
103. Sarcami T, Batta N, Berruti F. Production and separation of acetic acid from pyrolysis oil of lignocellulosic biomass: a review. Biofpr. 2021;15:1–26. https://doi.org/10.1002/bbb2273.
104. Frediani P, Rosi L, Frediani M, Undri A, Occhialini S, Meini S. Production of hydrocarbons from pyrolysis of tyres, WO2012110991, PCT/IB2012/050748. 2012.
105. Undri A, Meini S, Rosi L, Frediani M, Frediani P. Microwave pyrolysis of polymeric materials: waste tires treatment and characterization of the value-added products. J Anal Appl Pyrol. 2013;103:149–58. https://doi.org/10.1016/j.jaap.2012.11.011.
106. Bartoli M, Frediani M, Briens C, Berruti F, Rosi L. An overview of temperature issues in microwave-assisted pyrolysis. PRO. 2019;7:658. https://doi.org/10.3390/pr7100658.
107. Mohanty A, Patel DK, Panigrahi SK. A combined theoretical, computational, and experimental investigation to evaluate performance of suitable thermocouple for microwave processing. Measurement. 2024;230:114460. https://doi.org/10.1016/j.measurement.2024.114460.
108. Klemm D, Heublein B, Fink H-P, Bohn A. Cellulose: fascinating biopolymer and sustainable raw material. Angew Chem Int Ed. 2005;44:3358–93. https://doi.org/10.1002/anie.200460587.
109. https://en.wikipedia.org/wiki/Biopolymer; https://en.wikipedia.org/wiki/Organic_compound. Accessed 3 Mar 2022.
110. Nishiyama Y, Langan P, Chanzy H. Crystal structure and hydrogen-bonding system in cellulose Iβ from synchrotron X-ray and neutron fiber diffraction. J Am Chem Soc. 2002;124:9074–82. https://doi.org/10.1021/ja0257319.
111. Romeo T. Bacterial biofilms. Berlin: Springer; 2008. p. 258–63. ISBN: 978-3-540-75418-3.
112. https://www.britannica.com/science/cell-biology/Cellulose. Accessed 28 June 2024.
113. Pettersen RC. Chapter 2. Chemical composition of wood. In: The chemistry of solid wood. 1984. pp. 57–126. https://doi.org/10.1021/ba-1984-0207.ch002.
114. Carus S. Multi-criteria evaluation of lignocellulosic niche crops for use in biorefinery processes. 2011. https://api.semanticscholar.org/CorpusID:43993705. Accessed 27 June 2024.
115. Bartoli M, Rosi L, Giovannelli A, Frediani P, Frediani M. Pyrolisis of α-cellulose using a multimode microwave oven. J Anal Appl Pyrol. 2016;120:284–96. https://doi.org/10.1016/j.jaap.2016.05.016.
116. Saake B, Lehnen. Lignin in Ullmann's encyclopedia of industrial chemistry. Weinheim: Wiley-VCH; 2007. p. 21. https://doi.org/10.1002/14356007.a15_305.pub3.
117. Gargulak JD, Lebo SE, McNally TJ. Lignin. In: Kirk-Othmer, editor. Encyclopedia of chemical technology. Wiley; 2015. https://doi.org/10.1002/0471238961.12090714120914.a01.pub3.
118. Ralph J, et al. Elucidation of new structures in lignins of CAD- and COMT-deficient plants by NMR. Phytochemistry. 2001;57:993–1003. https://doi.org/10.1016/S0031-9422(01)00109-1.
119. Kuroda K, Ozawa T, Ueno T. Characterization of sago palm (Metroxylon sagu) lignin by analytical pyrolysis. J Agric Food Chem. 2001;49:1840–7. https://doi.org/10.1021/jf001126i.
120. Bartoli M, Rosi L, Frediani P, Frediani M. Bio-oils from microwave assisted pyrolysis of Kraft lignin operating at reduced residual pressure. Fuel. 2020;278:118175. https://doi.org/10.1016/j.fuel.2020.118175.
121. Scheller HV, Ulvskov P. Hemicellulose. Ann Rev Plant Biol. 2010;2010(61):263–89. https://doi.org/10.1146/annurev-arplant-042809-112315.
122. Ebringerová A, Hromádková Z, Heinze T. Hemicellulose in polysaccharides I: structure, characterization and use, advances. In: Heinze T, editor. Polymer science. Springer; 2005. p. 1–67. https://doi.org/10.1007/b136816.
123. Li T, Remón J, Shuttleworth PS, Jiang Z, Fan J, Clark JH, Budarin VL. Controllable production of liquid and solid biofuels by doping-free, microwave-assisted, pressurized pyrolysis of hemicelluloses. Energ Conv Manag. 2017;144:104–13. https://doi.org/10.1016/j.enconman.2017.04.055.

124. Undri A, Abou-Zaid M, Briens C, Berruti F, Rosi L, Bartoli M, Frediani M, Frediani P. Bio-oil from pyrolysis of wood pellets using a microwave multimode oven and different microwave absorbers. Fuel. 2015;153:464–82. https://doi.org/10.1016/j.fuel.2015.02.081.
125. https://en.wikipedia.org/wiki/Vitis_vinifera. Accessed 20 Sept 2023.
126. Spinelli R, Magagnotti N, Nati C. Harvesting vineyard pruning residues for energy use. Biosyst Eng. 2010;105:316–22. https://doi.org/10.1016/j.biosystemseng.2009.11.011.
127. Encinar JM, Beltrán FJ, Bernalte A, Ramiro A, González JF. Pyrolysis of two agricultural residues: olive and grape bagasse. Influence of particle size and temperature. Biomass Bioenerg. 2010;11:397–409. https://doi.org/10.1016/S0961-9534(96)00029-3.
128. Demiral I, Ayan EA. Pyrolysis of grape bagasse: effect of pyrolysis conditions on the product yields and characterization of the liquid product. Bioresour Technol. 2011;102:3946–51. https://doi.org/10.1016/j.biortech.2010.11.077.
129. Encinar JM, Beltrán FJ, Ramiro A, González JF. Catalyzed pyrolysis of grape and olive bagasse. Influence of catalyst type and chemical treatment. Ind Eng Chem Res. 1997;36:4176–83. https://doi.org/10.1021/ie960795b.
130. Bartoli M, Rosi L, Giovannelli A, Frediani P, Passaponti M, Frediani M. Microwave assisted pyrolysis of crop residues from Vitis vinifera. J Anal Appl Pyrolysis. 2018;130:305–13. https://doi.org/10.1016/j.jaap.2017.12.018.
131. https://nycancer.com/news/olive-oil-key-component-mediterranean-diet#, 2019. Accessed 23 Sept 2023.
132. Salomone R, Ioppolo G. Environmental impacts of olive oil production: a life cycle assessment case study in the province of Messina (Sicily). J Clean Prod. 2012;28:88–100. https://doi.org/10.1016/j.jclepro.2011.10.004.
133. Hamilakis Y. Food technologies/technologies of the body: the social context of wine and oil production and consumption in bronze age Crete. World Archaeol. 1999;31:38–54. https://doi.org/10.1080/00438243.1999.9980431.
134. Tognetti R, Giovannelli A, Lavini A, Morelli G, Fragnito F, d'Andria R. Assessing environmental controls over conductances through the soil–plant–atmosphere continuum in an experimental olive tree plantation of southern Italy. Agric Forest Meteor. 2009;149:1229–43. https://doi.org/10.1016/j.agrformet.2009.02.008.
135. Fratini R. Risorse forestali della Toscana e aspetti riguardanti il mercato del legno locale. Agri. 2009;87:80–3.
136. Şensoz S, Demiral İ, Gercel HF. Olive bagasse (Olea europea L.) pyrolysis. Bioresour Technol. 2006;97:429–36. https://doi.org/10.1016/j.biortech.2005.03.007.
137. Zabaniotou A, Kalogiannis G, Kappas E, Karabelas AJ. Olive residues (cuttings and kernels) rapid pyrolysis product yields and kinetics. Biomass Bioenergy. 2000;18:411–20. https://doi.org/10.1016/S0961-9534(00)00002-7.
138. Bartoli M, Rosi L, Giovannelli A, Frediani P, Frediani M. Characterization of bio-oil and bio-char produced by low-temperature microwave-assisted pyrolysis of olive pruning residue using various absorbers. Waste Manag Res. 2020;38:213–25. https://doi.org/10.1177/0734242X19865342.
139. Abdullah H, Wu H. Biochar as a fuel: 1. Properties and grindability of biochars produced from the pyrolysis of mallee wood under slow heating conditions. Energy Fuel. 2009;23:4174–81. https://doi.org/10.1021/ef900494t.
140. Bartoli M, Rosi L, Giovannelli A, Frediani P, Frediani M. Bio-oil from the residue of short rotation coppice of poplar using microwave-assisted pyrolysis. J Anal Appl Pyrolysis. 2016;119:224–32. https://doi.org/10.1016/j.jaap.2016.03.001.
141. CABI. Around donax (giant reed). In: Invasive species compendium. Wallingford, UK: CAB International; 2020. http://www.iucngisd.org/gisd/species.php?sc=112. Accessed 26 June 2024.
142. https://en.wikipedia.org/wiki/Arundo_donax. Accessed 28 June 2024.

143. Pirozzi D, Fiorentino N, Impagliazzo A, Sannino F, Yousuf A, Zuccaro G, Fagnano M. Lipid production from Arundo donax grown under different agronomical conditions. Renew Energy. 2015;77:456–62. https://doi.org/10.1016/j.renene.2014.12.046.
144. Corno L, Pilu R, Adani F. Arundo donax L.: a non-food crop for bioenergy and bio-compound production. Biotechnol Adv. 2014;32:1535–49. https://doi.org/10.1016/j.biotechadv.2014.10.006.
145. Bartoli M, Rosi L, Giovannelli A, Frediani P, Frediani M. Production of bio-oils and bio-char from Arundo donax through microwave-assisted pyrolysis in a multimode batch reactor. J Anal Appl Pyrolysis. 2016;122:479–89. https://doi.org/10.1016/j.jaap.2016.10.016.
146. Rosi L, Frediani M, Undri A, Bartoli M, Frediani P. Chapter 4. A friendly Management of Waste/contaminated polymeric materials from differentiated waste collection through microwave pyrolysis. In: Tzortzakis N, editor. Municipal solid waste: management strategies, challenges, and future directions. Nova Publisher; 2017. p. 83–125.
147. La Mantia FP, Morreale M. Green composites: a brief review. Compos A Appl. 2011;42:579–88. https://doi.org/10.1016/j.compositesa.2011.01.017.
148. Undri A, Rosi L, Frediani M, Frediani P. Microwave assisted pyrolysis of corn derived plastic bags. J Anal Appl Pyrolysis. 2014;108:86–97. https://doi.org/10.1016/j.jaap.2014.05.013.
149. Krishna BS, Nikhilesh GSS, Tarun B, Saibaba NKV, Gopinadh R. Industrial production of lactic acid and its applications. Int J Biotech Res. 2019;1:42–54.
150. Ceresana: bioplastics—study: market, analysis, trends. www.ceresana.com. Accessed 10 Nov 2023.
151. Nagarajan V, Mohanty AK, Misra M. Perspective on Polylactic acid (PLA) based sustainable materials for durable applications: focus on toughness and heat resistance. ACS Sust Chem Eng. 2016;4:2899–916. https://doi.org/10.1021/acssuschemeng.6b00321.
152. Wee YJ, Kim JN, Yun JS, Ryu HW. Utilization of sugar molasses for economical L(+)-lactic acid production by batch fermentation of enterococcus faecalis. Enzym Microb Technol. 2004;35:568–73. https://doi.org/10.1016/j.enzmictec.2004.08.0082004.
153. Ray S, Ta HT. Investigating the effect of biomaterials such as poly-(l-lactic acid) particles on collagen synthesis in vitro: method is matter. J Funct Biomater. 2020;11:51. https://doi.org/10.3390/jfb11030051.
154. Auras R, Lim L-T, Selke SEM, Tsuji H, editors. Poly(lactic acid): synthesis, structures, properties, processing, and applications. Wiley; 2010. https://doi.org/10.1002/9780470649848.
155. Undri A, Rosi L, Frediani M, Frediani P. Conversion of poly(lactic acid) to lactide via microwave-assisted pyrolysis. J Anal Appl Pyrolysis. 2014;110:55–65. https://doi.org/10.1016/j.jaap.2014.08.003.
156. Huang HJ, Ramaswamy S, Tschirner UW, Ramarao BV. A review of separation technologies in current and future biorefineries. Sep Purif Technol. 2008;62:1–21. https://doi.org/10.1016/j.seppur.2007.12.011.
157. Li Z, Zhang L, Chen G, Wu L, Liu B, Li Y, et al. A new method for comprehensive utilization of wood vinegar by distillation and liquid–liquid extraction. Process Biochem. 2018;75:194–201. https://doi.org/10.1016/j.procbio.2018.08.012.
158. Brueckner TM, Pickup PG, Hawboldt KA. Improvement of bark pyrolysis oil and value added chemical recovery by pervaporation. Fuel Process Technol. 2020;199:106292. https://doi.org/10.1016/j.fuproc.2019.106292.
159. Saha B, Chopade SP, Mahajani SM. Recovery of dilute acetic acid through esterification in a reactive distillation column. Catal Today. 2000;60:147–57. https://doi.org/10.1016/S0920-5861(00)00326-6.
160. Sagehashi M, Nomura T, Shishido H, Sakoda A. Separation of phenols and furfural by pervaporation and reverse osmosis membranes from biomass—superheated steam pyrolysis-derived aqueous solution. Bioresour Technol. 2018;98:2018–26. https://doi.org/10.1016/j.biortech.2006.08.022.

161. Wang J, Cui H, Wei S, Zhuo S, Wang L, Li Z, et al. Separation of biomass pyrolysis oil by supercritical CO_2 extraction. Smart Grid Renew Energy. 2010;1:98–107. https://doi.org/10.4236/sgre.2010.12015.
162. Feng Y, Meier D. Comparison of supercritical CO_2, liquid CO_2, and solvent extraction of chemicals from a commercial slow pyrolysis liquid of beech wood. Biomass Bioenergy. 2016;85:346–54. https://doi.org/10.4236/sgre.2010.12015.
163. Reyhanitash E, Zaalberg B, Ijmker HM, Kersten SRA, Schuur B. CO_2-enhanced extraction of acetic acid from fermented wastewater. Green Chem. 2015;17:4393–400.
164. Del Pozo C, Bartoli J, Puy N, Fabregas E. Separation of value added chemical groups from bio oil of olive mill waste. Ind Crop Prod. 2018:160–7. https://doi.org/10.1016/j.indcrop.2018.08.062.
165. Bamdad H, Papari S, Moreside E, Berruti F. High-temperature pyrolysis for elimination of per- and Polyfluoroalkyl substances (PFAS) from biosolids. PRO. 2022;10:2187. https://doi.org/10.3990/pr10112187.
166. PFAS, ingredients in various everyday products. https://www.niehs.nih.gov/health/topics/agents/pfc#:~:text=Per-%20and%20polyfluo-roalkyl%20substances. Accessed 15 July 2024.
167. Du H, Yuan D, Li W, Wang L, Li Y, Che L, Tian W, Salama E, Ossman M, Lin F. Efficient removal of toxic organics and reduction of Cr(VI) to Cr(III) from tannery sludge: a comparative study of microwave pyrolysis and conventional pyrolysis. Sep Pur Techn. 2024;354:128736. https://doi.org/10.1016/j.seppur.2024.128736.
168. Zhang Y, Liang Y, Li S, Yuan Y, Zhqang D, Wu Y, Xie H, Brindhadevi K, Pughazendhi A, Xia C. A review of biomass pyrolysis gas: forming mechanism, influencing parameters, and product application upgrades. Fuel. 2023;347:128461. https://doi.org/10.1016/j.fuel.2023.128461.
169. Jeyasubramanian K, Thankagiri B, Sakthivel A, Dhaveethu Raja J, Seenivasan S, Vallinayagam P, Madhavan D, Nalathi Devi S, Rathika B. A complete revierw on biochar: production, property, multifaced applications, interaction mechanism, and computational approach. Fuel. 2021;292:120243. https://doi.org/10.1016/j.fuel.2021.120243.
170. Bartoli M, Giorcelli M, Tagliaferro A. A comprehensive overview on biochar-based materials for catalytic applications. Catalysts. 2023;13:1336. https://doi.org/10.3390/catal13101336.
171. Bartoli M, Giorcelli M, Jagdale P, Rovere M, Tagliaferro A. A review of non-soil biochar applications. Materials. 2020;13:261. https://doi.org/10.3390/ma13020261.

Chapter 6
Synthesis of 5-Hydroxymethylfurfural (HMF) from Agricultural and Food Residues: An Intermediate for 2,5-Furandicarboxylic Acid (FDCA) Production

Surabhi Pandey, Valérie Orsat, and Marie-Josée Dumont

Abstract The diminishing fossil fuel reserves and greenhouse gas emissions have made it imperative for the chemical industry to produce bio-based alternatives as a replacement of petroleum-based platform chemicals. This chapter primarily focuses on 2,5-furandicarboxylic acid (FDCA), a bio-based precursor that can be further utilized in the synthesis of a various chemicals and polymers. The synthesis of FDCA involves chemical, electrocatalytic, and biological methods, wherein the chemical methods have proven most favourable in terms of productivity, purity, and reaction rate. The chemical pathway for FDCA synthesis involves carbohydrate dehydration into 5-hydroxymethylfurfural (HMF), which is oxidized into FDCA. To achieve maximum FDCA production, it is important to select an efficient catalytic system that can give high HMF yields and can also be scaled up in the future. Therefore, this chapter enlists different catalytic systems for HMF synthesis including homogeneous catalytic systems, biphasic catalytic systems, ionic liquids, and heterogenous catalytic systems. The substrate plays a major role in controlling the scalability, productivity, and economics to produce HMF and FDCA. Agricultural residues and food wastes represent significant, yet underutilized, sources of biomass that can be leveraged to produce valuable chemicals. Agricultural residues include the non-edible parts of crops, such as straw, husks, and stems, which are typically left in the field after harvesting. Food wastes encompass a wide range of materials discarded during the processing, distribution, and consumption of food, including

---

S. Pandey · V. Orsat
Department of Bioresource Engineering, McGill University, Sainte-Anne-de-Bellevue, QC, Canada
e-mail: surabhi.pandey@mail.mcgill.ca; valerie.orsat@mcgill.ca

M.-J. Dumont (✉)
Department of Chemical Engineering, Université Laval, Pavillon Adrien-Pouliot, Québec, QC, Canada
e-mail: marie-josee.dumont@gch.ulaval.ca

Z. Fang et al. (eds.), *Production of Organic Acids and Alcohols from Agricultural Residues and Food Wastes*, Biofuels and Biorefineries 13,
https://doi.org/10.1007/978-981-96-1222-2_6

peels, pulp, and spoiled or uneaten food. Both categories of biomass are rich in carbohydrates, such as cellulose, hemicellulose, and starch, which can be converted into HMF and FDCA as described in this chapter. These chemicals are crucial to produce bio-based plastics, resins, and other sustainable materials, offering a promising route to reduce reliance on fossil fuels and mitigate environmental impacts.

Keywords 2,5-Furandicarboxylic acid · Agro-food wastes · Heterogeneous catalysts · Homogeneous catalysts · Renewable chemicals · Thermochemical processes

6.1 Introduction

Currently, most fuels and chemical production rely on fossil fuels as the starting material. However, the dwindling supply of fossil fuels, environmental pollution, and greenhouse gas emissions have driven the interest in biomass valorization [1]. The replacement of petroleum chemicals with bio-based chemicals and their large-scale production is hampered due to their techno-economic challenges [2]. A recent study conducted by the US Department of Energy has identified that 2,5-furandicarboxylic acid (FDCA) is a potential substitute for a variety of petrochemicals like terephthalic acid (TPA) and adipic acid [3]. It has applications in making polyesters, polyurethanes, and polyamides and has superior thermomechanical and gas barrier properties compared to TPA [4]. FDCA can be synthesized from hydroxymethylfurfural (HMF) which itself has been termed a "sleeping giant" since it serves as a bridge between carbohydrate and petroleum-based chemistry [5]. The presence of aldehyde and hydroxyl groups on the furan ring makes it a highly reactive molecule and a key precursor for the synthesis of biochemicals, solvents, and polymers [6–8]. It also finds its application in the food industry as a food additive and food flavouring agent [9]. The current reaction for fructose to FDCA can be carried out through two different routes wherein fructose gets dehydrated to produce HMF in the first step, which later gets oxidized to FDCA [10]. Unfortunately, the sustainable synthesis of FDCA faces various techno-economic bottlenecks including (a) HMF instability in acidic or basic conditions, (b) low solubility of FDCA, (c) partial oxidation of HMF into FDCA-producing intermediates like 2,5-diformylfuran (DFF), 5-formylfuran-2-carboxylic acid (FFCA) and 5-(hydroxymethyl)furan-2-carboxylic acid (HMFCA). These intermediates hinders the chain growth during downstream polymerization process causing low quality polymers [11, 12]. To overcome these challenges, Avantium reported a process that involves the use of methoxymethylfurfural (fructose dehydrated using methanol) in presence of acetic acid as the solvent with homogeneous catalyst composed of $Co(OAc)_2$, $Mn(OAc)_2$, and HBr, producing 96% FDCA yield [12]. However, the drawback of this method is the presence of a small amount of FFCA is produced with FDCA; which needs further purification incurring additional costs, hence a cheaper renewable feedstock and heterogeneous catalyst are desired [13].

On the other hand, the industrial manufacturing of HMF is limited to using fructose extracted from energy crops as a substrate [14]. Other than fructose, different monosaccharides and polysaccharides have been tested for HMF synthesis, however, their scale-up is impeded due to the instability of HMF. Agricultural and food residues are potential sources of carbohydrates for large-scale production of platform chemicals, yet only a minor fraction of these is used by the chemical industry. Agricultural wastes (rice husk, straw, sugarcane bagasse, and corn stover) fall under the lignocellulosic category which is mainly comprised of hemicellulose (25–35%), cellulose (40–50%), and lignin (5–20%) [15]. Apart from fats and proteins that hinder HMF production, the main components of food wastes are sugars (fructose, glucose, galactose), starch, and other di/polysaccharides (lactose, sucrose, inulin, cellobiose, cellulose, xylose, etc). A techno-economic analysis (TEA) was performed to synthesize HMF from fructose in a biphasic water-butanol with HCl as the catalyst and NaCl as the additive; the system estimated the minimum selling price (MSP) of HMF to $0.88/kg; which can further drop down by using cellulosic feedstocks [16]. Aristizábal et al. reported glucose conversion into HMF from sugar bagasse and coffee stem, further lowering the HMF production cost to $0.35/kg [17]. To estimate the FDCA market fit, a study showed that using furfural as the starting material could produce FDCA with MSP of $2/kg [18]. Another study was conducted using cellulose as the feedstock that estimated the MSP of FDCA to be $1.36/kg [19]. Motagamwala et al. used fructose as the feedstock and showed that the MSP can be brought down to $1.31/kg in a base-free solvent system [20]. Furthermore, using cellulose as the feedstock the MSP of FDCA was dropped down to $1.52/kg [21], $1.25/kg [22], and $1.13/kg [16] in different solvent-catalytic systems which is lower than the MSP to TPA (1.59/kg). Indeed, agricultural and food residues have the potential to become a more sustainable alternative to substitute sugars for HMF synthesis, and subsequently for FDCA synthesis [23, 24].

Therefore, this chapter provides a compositional analysis of different agricultural and food residues and compares the recent developments in reaction systems that have been used for the conversion of agricultural and food residues into HMF. A detailed analysis of different catalytic systems involving monophasic, biphasic, ionic liquid (IL), and heterogeneous systems for HMF synthesis from biomass has been provided. Insights on FDCA synthesis with multi-step conversion using HMF as the feedstock and direct FDCA synthesis from biomass are focused upon. The performance of solvents and catalysts for high yield and HMF selectivity, side reactions, recyclability, and mass transfer limitation accounted for different catalytic-solvent systems, and the use of green solvents as a pathway toward sustainability in HMF production has been examined.

6.2 Overview of Agricultural and Food Residues

The agriculture and food processing sector generates a large quantity of waste (accounting for 1.3 billion tonnes) each year which causes serious economic and environmental management problems. The disposal of food waste has always been a critical challenge for industrialists because it requires a huge cost of treatment due to its high Chemical Oxygen Demand (COD) [25]. Food waste valorization is an approach to convert these wastes or byproducts into platform chemicals that not only offer economic benefits but also reduce the environmental degradation caused by conventional disposal techniques [26]. The main components present in food waste can be categorized into carbohydrates, cellulose, hemicellulose, and lignin. Tables 6.1 and 6.2 provide the compositional analysis for different agricultural and food residues, which can be a possible substrate for HMF. Carbohydrates are represented by the general formula of $C_m(H_2O)_n$, and these are categorized as mono-, di-, oligo, and polysaccharides (starch and non-starch). Monosaccharides have a general formula of $(CH_2O)_n$, where n is 3 or more. Natural saccharides, also known as sugars, include mono- and disaccharides which are obtained by simple hydrolysis of polysaccharides [47]. Monosaccharides are divided according to the presence of the aldehydic (aldose) or ketonic (ketose) group. Starch, a polysaccharide of amylose and amylopectin, has the same chemical formula of $(C_6H_{10}O_5)n$ as cellulose. The units of glucose are connected by 1,4-α glycosidic bond in amylose and 1,6-α glycosidic bond in amylopectin [48], while cellulose contains 1,4-β glycosidic linkages [49]. These 1,4-β glycosidic bonds cannot be easily attacked by enzymes and catalysts, hence the depolymerization of cellulose is much more difficult than starch [50]. Lignocelluloses is the most commonly available consisting of 38–50% cellulose, 23–32% hemicellulose, and 15–25% lignin in addition to small amount of ash and organic substances. The cellulose and hemicellulose portions of the lignocellulosic biomass are polysaccharides of C_6 sugars units such as mannose, glucose, and galactose which dehydrate to HMF. The difference in physicochemical properties of hemicellulose and cellulose allows their separation before processing. They are the most abundant and renewable source which is potentially suitable for the conversion to value-added chemicals [51–53]. The spaces between them are filled with a more complex polymer known as lignin [54]. Glucose is the main building block for cellulose whereas, hemicellulose is a heterogeneously branched polymer of pentoses and hexoses, like xylose, arabinose, mannose, galactose, and glucose.

Most of the plant material consists of cellulose (50–70%), hemicellulose (10–40%), and lignin (10–30%) as their structural components [50]. Food wastes from industries are generally liquid effluent. The wastewater coming from the food industry is generally considered safe as it contains no or few toxic or hazardous compounds [26]. Cheese processing industries produce whey as the by-product which has lactose as the main carbohydrate. For tofu processing industries, the wastewater us composed of complex mixture of polysaccharides [55]. A large portion of the food processing wastes comes from fruits and vegetables [56], which are major contributors to methane emissions and subsequently affect landfills and the environment [57]. In large-scale apple processing industries, 25% of the apple is wasted and this fraction is known as apple pomace. The pomace is a heterogeneous mixture of peel, pomace, and seeds, with a high-water content and contains insoluble carbohydrates, such as cellulose, hemicellulose, and lignin. It also contains simple sugars like fructose, glucose, sucrose, carbohydrates, crude fiber,

Table 6.1 Quantification of soluble carbohydrates in different fruit wastes (% dry matter)

Agro-food wastes	Rhamnose (%)	Arabinose (%)	Xylose (%)	Mannose (%)	Galactose (%)	Sucrose (%)	Glucose (%)	Fructose (%)	Ref.
Apple pomace	1.7	5.5	6.2	2.8	4.2	9.2	25.2	24.7	[27]
Banana peel	0.6	4.4	5.6	3.6	2.8	–	30.1	15.2	[27]
Grapefruit peel	3.4	4.8	2.3	2.2	3.5	1.4	30.6	8.2	[27]
Lemon peel	2.1	5.2	2.6	2.1	4.6	–	27.9	3.3	[27]
Lime peel	2.5	8.5	2.5	2.0	4.3	–	22.5	0.7	[27]
Mandarin peel	2.9	3.3	2.4	2.3	3.9	7.4	39.4	10.3	[27]
Mango kernels	1.0	9.8	2.2	1.6	8.6	–	23.0	–	[28]
Olive pulp	1.0–3.0	17.0–20.0	17.0–20.0	2.0	4.0–5.0	–	31.0–34.0	–	[29]
Orange peel	2.1	5.6	2.2	2.4	2.7	5.6	35.5	12.2	[27]
Pear pomace	1.3	6.0	20.2	2.4	4.5	1.9	21.1	14.1	[27]
Potato pulp	1.4	10.0	–	–	55.0	–	–	–	[30]
Sugar beet pulp	2.3	18.1	1.5	1.2	4.5	–	20.0	–	[29]
Sweet potato	2.1	3.7	3.3	1.3	10.4	–	56.1	–	[31]

Table 6.2 Fraction of polysaccharides (starch and non-starch) in different fruit wastes (% dry mass)

	Non-starch polysaccharides					
	Soluble	Insoluble				
Agro-food wastes	Pectin	Cellulose	Hemicellulose	Lignin	Starch	Ref.
Apple pomace	3.5–14.3	7.2–43.6	4.26–24.4	15.3–23.5	17.9	[32, 33]
Banana peel	8.1–21.7	6.4–9.6	0.6–8.4	6.0–16.8	0.1–39.3	[34]
Black currant	2.7	12.0	25.3	59.3	–	[35]
Carrot pomace	3.9	51.6	12.3	32.2	–	[35]
Cherry pomace	1.5	18.4	10.7	69.4	–	[35]
Citrus wastes	25.0	22.0	11.1	2.2	–	[36]
Citrus wastes	15.3 [27]	8.8	7.96 [28]	2.0	–	[37]
Grape seed	0.3	17.8	18.0	64.0	–	[38]
Grape skin	4.0	6.0	31.0	59.0	–	[38]
Grapefruit peel	16.1	19.8	18.3	8.2	–	[39]
Kinnow mandarin	22.6	10.1	4.3	0.6	–	[40]
Lemon peel	13.0	23.1	8.1	7.6	–	[41]
Lemon pulp	22.5	36.2	11.1	7.6	–	[41]
Lime peel	–	33.24	14.0	12.9	–	[42]
Mandarin peel	14.2	8.9	20.8	17.2	–	[41]
Mango kernel	2.8–3.2	37.0–39.0	7.0–9.0	17.0–19.0	58	[43, 44]
Orange peel	42.5	9.2	10.5	0.8	3.75	[45]
Orange pulp	12.1	24.5	7.6	7.5	–	[41]
Pear pomace	13.4	34.5	18.6	33.5	–	[35]
Sugar beet pulp	24.0–32.0	22.0–30.0	24.0–32.0	3.0–4.0	–	[46]
Sweet orange peel	23.0	37.1	11.0	7.5	–	[41]
Sweet potato	15.7	31.2	11.4	16.9	–	[31]

pectin, proteins, minerals, and vitamins that makes them valuable components worth recovering [58]. Citrus juice industries have a large amount of waste material such as peels and seeds which are discarded during juice processing [59]. The citrus processing wastes mainly consist of skin (flavedo and albedo), pulp, and seeds which account for 50% weight of the original fruit [60]. They consist of insoluble carbohydrates (pectin and cellulose), sugars (fructose, glucose, and sucrose), acids (malic, citric, succinic, and tartaric acid), and essential oils (D-limonene) [61, 62]. Tomato wastes are the most produced and least explored wastes [63, 64]. Pomace is the remaining solid after juice removal or wine production and consists of 0.003–1% lycopene, 0–10% monosaccharides, 5–45% pectin, and 15–50% cellulose [65]. Potato processing wastewater consists of slit water from the washing of potatoes and processing water from operations like blanching, peeling, cutting, cooking, grinding, and packing and is a rich source of starch [55].

6.3 Pre-treatment of Lignocellulosic

The most recalcitrant substrate for HMF production is lignocellulosic wastes (such as peels of fruits and vegetables, corn stover, rice straw, and palm bagasse) [66]. Two thirds of the dry matter in cell walls are composed of cellulose and hemicellulose, and can be hydrolyzed to sugars, and further dehydrate into HMF. During hydrolysis, the polymers are broken down by the addition of water, and the process is known as saccharification [67]. The orientation of linkages and additional hydrogen bonding in cellulose makes these polysaccharides recalcitrant to hydrolysis. In contrast, hemicellulose can be readily hydrolysed to pentoses and hexoses because of its branched and amorphous nature [68]. The presence of lignin is a major setback in using lignocellulosic wastes as a feedstock for biochemicals because it makes them resistant to chemical and biological degradation [69]. Many reasons have been put forward for the effect of lignin on the hydrolysis of the substrate, but it has been established that the lignin must be removed to achieve complete hydrolysis of the substrate [27, 70]. These lignocellulosic wastes are highly recalcitrant to enzymatic hydrolysis due to the presence of lignin, crystallinity of cellulose, and particle size [71, 72] Some of the food wastes like cherry pomace and black currant have a high amount of lignin percentage (Table 6.2). Hence, they require additional processing before conversion into HMF. The pre-treatment process softens the material and makes it liable to hydrolysis as shown in Table 6.3. The approach to remove this recalcitrance barrier is through pre-treatment which makes the cellulose more accessible to hydrolytic enzymes for conversion to glucose. For an ideal pre-treatment process, the release of sugars should be maximum with limited degradation compounds and should be energy efficient [67, 69, 73]. The first necessary step in the chemistry of conversion involves depolymerization and hydrolysis of cellulose to glucose. Glucose is further catalytically degraded to chemicals. Pre-treatment of cellulose-like impregnation can lead to partial degradation of cellulose before hydrolytic reaction. According to the reaction mechanism, a catalyst that favours the splitting of water molecules will favour the hydrolysis of cellulose.

6.4 The Chemistry Behind HMF Production

Generally, HMF is absent in fresh produce and is developed in sugar-containing foods during the Maillard and caramelization reactions. Foods heavy in protein and carbohydrates are more likely to exhibit the former reaction, which occurs at lower temperatures (>50 °C) and an acidic pH (4–7) wherein the reaction between the carbonyl group of sugar and the amino group of amino acid occurs. Caramelization, on the other hand, involves the oxidation of sugars and occurs at elevated temperatures (>120 °C), dry conditions, and extreme pH (<3 or >9) [74]. These reactions occur during baking, heating of milk, roasting coffee, aging of wine, etc. At the onset of Maillard reaction, carbohydrates including maltose, fructose, and lactose react

Table 6.3 Different pre-treatments on food wastes

Agro-food waste	Pre-treatment	Reaction conditions	Soluble sugars %	Comments (adapted from Ref. [73])
Artichoke	Concentrated acid hydrolysis	0.5% H_2SO_4, 130 °C, 5 min	35.2	Presence of acid did not affect the sugar released
				Temperature did not significantly affect sugar yield
Cardoon	Concentrated acid hydrolysis	0.5% H_2SO_4, 130 °C, 5 min	78.2	Presence of acid did not affect the sugar released
				Sugar yield increased with the rise in temperature
				Time negatively impacted the release of soluble sugars
Pulse food	Concentrated acid hydrolysis	1% H_2SO_4, 130 °C, 15 min	36.6	Harsh pre-treatment was required because of higher starch
				Enzymatic hydrolysis was required to break starch into glucose
				Presence of acid improved sugar yield
				Temperature did not significantly affect the sugar yield
Red pepper	No acid pretreatment	0% H_2SO_4, 110 °C, 5 min	50.2	Presence of acid did not affect soluble sugar recovery
				Temperature did not significantly affect sugar yield
Tomato	No acid pretreatment	0% H_2SO_4, 110 °C, 5 min	35.5	Presence of acid did not affect soluble sugar recovery
				Temperature showed a positive effect on soluble sugar yield

with an amino group of lysine to form maltulosyl-lysine, fructosyl-lysine, and lactulosyl-lysine, respectively (Amadori products). In addition, heating milk in acidic conditions accounts for the transformation of lactulosyl-lysine through 1,2-enolization to form bound HMF. The pH is an important factor while deciding the product formed; low pH leads to 1,2-enolization whereas high pH gives rise to different side products through 2,3-enolization. Moreover, isomerization and degradation of lactose result in HMF formation through the Lobry de Bruyn–van Ekenstein mechanism [75]. Conversely, at elevated temperatures (usually >100 °C), lactose gets isomerized to lactulose with further degradation to galactose and other C_5/C_6 compounds. These reactions occur in the absence of amino acids and fall under the caramelization reaction [76]. Compared to starch and other polysaccharides, fructose gets easily dehydrated to HMF via acyclic or cyclic routes (Scheme 6.1). On the other hand, HMF synthesis from starch and cellulosic-rich biomass involves two additional steps of hydrolysis of glucan to glucose, followed by its isomerization to fructose [78]. The non-catalytic HMF production from polysaccharides requires high temperature. However, most of the studies presented in this

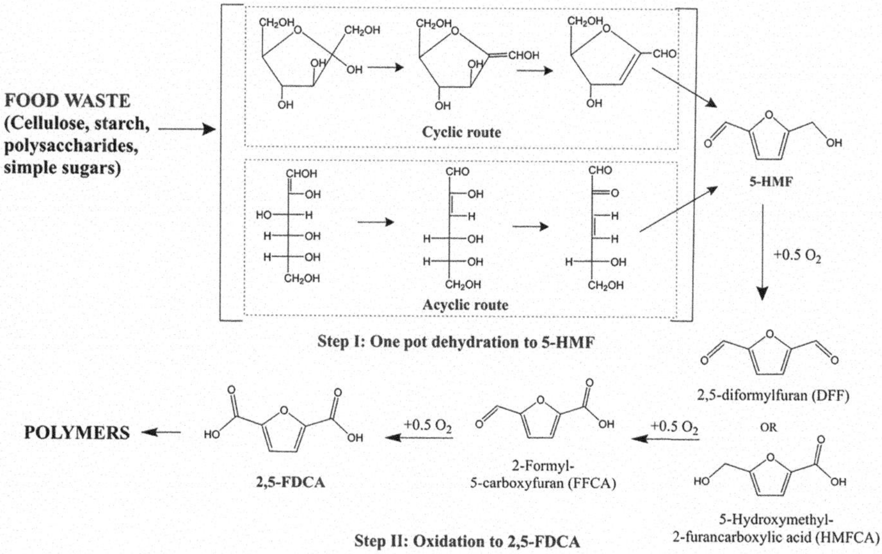

Scheme 6.1 Pathway for food wastes to HMF, and subsequently to FDCA conversion. (Adapted and modified with permission from [77], Copyright © 2016 American Chemical Society)

chapter involve acid-catalyzed reactions over Brønsted and/or Lewis acids. Brønsted acids favour the hydrolysis of C–O–C glycosidic bonds in polysaccharides and the dehydration of fructose to HMF by protonation at the hydroxyl group of sugar [79, 80]. Lewis acids, such as metal ions, promote glucose isomerization to fructose via 1,2 hydride shift [77]. It is important to note that excess Brønsted acidity might lead to undesirable side reactions such as the HMF rehydration to levulinic acid (LA) or formic acid (FA) [81], in addition to sugar polymerization reaction which can decrease the HMF yield [82] (Scheme 6.1).

6.5 Monophasic Catalytic Systems for HMF Synthesis

6.5.1 Mineral and Organic Acids

Mineral acids are widely used for the dehydration of carbohydrates into HMF because of their cheap cost and easy availability (refer to Table 6.4). Up to 91% HMF yield was obtained from fructose using 0.05 M HCl within a minute at 160 °C with 92% selectivity [81]. Wrigstedt et al. showed that HCl, H_2SO_4, HNO_3, and trifluoroacetic acid (TFA) had similar catalytic yields for fructose dehydration to HMF, where the yield was in the range of 85–91% [91]. Nevertheless, H_3PO_4 noticeably produced a lower HMF yield (60%) due to its low Bronsted acidity owing to its acid dissociation constants ($pK_a = 2.12$), which is lesser in comparison to other strong mineral acids (e.g., pK_a of $HNO_3 = -1.3$). The same mineral acids

Table 6.4 Conversion of biomass residues and their model compounds using monophasic reaction systems

Agro-food wastes	Carbohydrate (%)	Solvent	Catalyst	Heating mode	Temperature (°C)	Reaction time	HMF yield (%)	Ref.
Bread crust (5%)	Starch (67.6)	DMSO-water (1:1 v/v)	$SnCl_4$ (55.5 mM)	Oil bath	140	20 min	9.5	[83]
Bread crust (5%)	Starch (67.6)	ACN-water (1:1 v/v) (20 mL)	$SnCl_4 \cdot 5H_2O$ (55.5 mM)	Microwave reactor	140	10 min	26	[84]
Bread crust (5%)	Starch (67.6)	DMSO-water (1:1 v/v) (20 mL)	$SnCl_4 \cdot 5H_2O$ (55.5 mM)	Microwave reactor	140	60 min	27.4	[84]
Bread crust (5%)	Starch (67.6)	Acetone-water (1:1 v/v) (20 mL)	$SnCl_4 \cdot 5H_2O$ (55.5 mM)	Microwave reactor	140	10 min	26	[84]
Bread crust (5%)	Starch (67.6)	ACN-water (1:1 v/v) (20 mL)	$SnCl_4 \cdot 5H_2O$ (55.5 mM)	Microwave reactor	140	10 min	26	[84]
Bread crust (5%)	Starch (50–60)	DMSO-water (1:1 v/v) (20 mL)	$SnCl_4$ (55.5 mM)	Oil bath	160	60 min	29.5	[85]
Bread crust (5%)	Starch (50–60)	DMSO-water (1:1 v/v) (20 mL)	$AlCl_3$ (55.5 mM)	Oil bath	160	60 min	25.6	[85]
Bread crust (5%)	Starch (50–60)	DMSO-water (1:1 v/v) (20 mL)	$FeCl_3$ (55.5 mM)	Oil bath	160	20 min	9.8	[85]
Cooked penne (5%)	Starch (66.8)	DMSO-water (1:1 v/v) (20 mL)	$SnCl_4$ (55.5 mM)	Oil bath	140	40 min	8.1	[86]

Cooked rice	Starch (67.6)	DMSO-water (1:1 v/v)	$SnCl_4$ (55.5 mM)	Oil bath	140	20 min	8.1	[83]
Cooked rice (5%)	Starch (85.5)	DMSO-water (1:1 v/v) (20 mL)	$SnCl_4$ (55.5 mM)	Oil bath	140	40 min	4	[86]
Flax straw	–	Water, substrate:water (1:100)	HCl (pH 1.12)	Microwave reactor	180	20 min	0.8 g/100 g straw	[87]
Jerusalem artichoke	Inulin (50)	2-Butanol-water (1:1 v/v)	$FeCl_2/H_2O_2$ (pH 3.0–3.5) (1.3 mM)	Oil bath	180	90 min	46	[88]
Pretreated rice straw (1.53%)	Glucan (37.5), xylan (20.6), and lignin (12)	Water (500 mL)	(0.5%) sulfuric acid	Batch reactor with pressurized oil bath heating	180	3 h	15.3	[89]
Raw vegetable wastes (2.5%)	Cellulose (41.3)	DMSO-water (1:1 v/v) (20 mL)	$SnCl_4$ (55.5 mM)	Oil bath	140	20 min	4.9	[86]
Sugarcane bagasse (4%)	Cellulose	DMA-LiCl (0.5 g)	$Zr(O)Cl_2/CrCl_3$ (20 mol %) and [BMIM]Cl (38 wt. % as additive)	Microwave reactor	120	5 min	37	[78]
Triticale straw	–	Water, substrate:water (1:100)	HCl (pH 1.12)	Microwave reactor	180	20 min	0.98 g/100 g straw	[87]
Unskinned kiwifruit (5%)	Fructose (27), glucose (22), cellulose (3.4)	DMSO-water (1:1 v/v) (20 mL)	$SnCl_4$ (55.5 mM)	Oil bath	140	20 min	13.7	[86]
Watermelon flesh (5 wt/v%)	Fructose (11.8), glucose (4.1)	DMSO-water (1:1 v/v) (20 mL)	$SnCl_4$ (55.5 mM)	Oil bath	140	20 min	13.2	[86]

(continued)

Table 6.4 (continued)

Agro-food wastes	Carbohydrate (%)	Solvent	Catalyst	Heating mode	Temperature (°C)	Reaction time	HMF yield (%)	Ref.
Wheat straw	–	Water, substrate:water (1:100)	HCl (pH 1.12)	Microwave reactor	180	20 min	0.89 g/100 g straw	[87]
Wheat straw	Cellulose (30–35), hemicellulose (15–25), lignin (10–25)	Water	HCl (pH 1.12)	Microwave reactor	180	20 min	3.4	[15]
Carbohydrate as substrate (%)	**Substrate amount**	**Solvent**	**Catalyst**	**Heating mode**	**Temperature (°C)**	**Reaction time**	**HMF yield (%)**	**Ref.**
Cellulose (1 g)	–	Water (30 mL)	HCl (pH 3)	Muffle furnace	300	0 min	20.7	[90]
Cellulose (4%)	–	DMA-LiCl (0.5 g)	$Zr(O)Cl_2/CrCl_3$(20 mol%) and [BMIM]Cl (44 wt.% as additive)	Microwave reactor	120	20 min	62	[78]
Fructose	0.15 g	Water, substrate:water (1:100)	HCl (0.05 M)	Microwave reactor	160	1 min	91	[81]
Fructose	0.15 g	Water	HCl (pH 1.5–2.15)	Autoclave	150	2 h	38–43	[81]
Fructose	0.15 g	Water	HCl (0.05 M)	Microwave	160	1 min	91	[91]
Fructose	0.15 g	Water	HCl (0.05 M)	Oil bath	160	40 min	79	[91]
Fructose	0.15 g	Water	H_2SO_4 (0.05 M)	Microwave	160	1 min	85	[91]
Fructose	0.15 g	4.5 mL sat. KBr/MeCN (1:2 v/v)	HNO_3 (0.05 M)	Microwave	160	1 min	87	[91]

Fructose	0.15 g	4.5 mL sat. KBr/MeCN (1:2 v/v)	H_3PO_4 (0.05 M)	Microwave	160	1 min	60	[91]
Fructose	0.15 g	4.5 mL sat. KBr/MeCN (1:2 v/v)	TFA (0.05 M)	Microwave	160	1 min	86	[91]
Fructose	1%	Water	Acetic acid (10%)	Autoclave	150	2 h	29	[81]
Fructose	1%	Water	Acetic acid (20%)	Autoclave	150	2 h	43	[81]
Fructose	1%	Water	Acetic acid (50%)	Autoclave	150	2 h	48	[81]
Fructose	1%	Water	Formic acid (10%)	Autoclave	150	2 h	51	[81]
Fructose	1%	Water	Formic acid (20%)	Autoclave	150	2 h	53	[81]
Fructose	1%	Water	Lactic acid (20%)	Autoclave	150	2 h	50	[81]
Fructose	1%	Water	Lactic acid (50%)	Autoclave	150	2 h	64	[81]
Glucose	0.78 mmol	Water (30 mL)	$ZnCl_2$ (0.3 mmol) and HCl (0.3 mmol)	Stirrer batch reactor	180	60 min	0.26	[92]

catalyze the cellulose conversion into HMF reaction, for instance, HCl aqueous solution yield 21% HMF [90]. Owing to lower toxicity of organic acids, they were also tested for HMF synthesis. A study reported the effect of different organic acids including FA, lactic acid, and formic acid, producing HMF yield in the range of 48–64% at optimal reaction concentrations. It was observed that an increase in acid concentration (acetic and FA) resulted in a drop in HMF selectivity while the HMF yield increased [81]. However, these liquid Brønsted acids were not very efficient for the dehydration of complex polysaccharides as they require an additional isomerization step. For instance, the conversion of rice husk to HMF in the presence of sulphuric acid (0.5%) could reach only up to 15.3 g/kg straw at 180 °C and decreased further with an increase in reaction time [89].

6.5.2 Metal Salts

Most of the studies presented in Table 6.4 have used metal chloride as catalysts owing to their dual Brønsted and Lewis acidities. The catalytic activity of a metal chloride depends upon the type of metal. For instance, Yu et al. evaluated the effect of different metal chlorides ($SnCl_4$, $AlCl_3$, $FeCl_3$, and $CrCl_3$) for HMF production from bread wastes [85]. The highest HMF yield (29.5%) was obtained using $SnCl_4$ due to its strong Brønsted acidity, which facilitated the hydrolysis and dehydration of starch to HMF. The HMF yield (25.6%) obtained by using $AlCl_3$ as the catalyst for bread wastes was comparable to what was obtained by the conversion of pure starch (30.6%), under similar reaction conditions [93]. The Brønsted acidity of Sn metal ions can be attributed to their partial hydrolysis in the presence of an aqueous medium [85]. The study demonstrated better catalytic activity for $SnCl_4$ in comparison to the rest. The lower catalytic activity was attributed to the less hydrated trivalent metal ions (Al^{3+}, Fe^{3+}, and Cr^{3+}) than the tetravalent ions (Sn^{4+}) due to their lower negative hydration enthalpy [83]. The Fenton reaction, using $FeCl_2$ and hydrogen peroxide as catalysts, yielded 46% HMF from Jerusalem artichoke. Despite the fair yield, the catalyst used was not selective for HMF production and might have led to the oxidation of sugars [88].

Similar to the previous work, the catalytic effect of $AlCl_3$, $CrCl_3$, $FeCl_3$, $Zr(O)Cl_2$, and $SnCl_4$ were compared during the one-pot conversion of food wastes to HMF [86]. For sugar-rich food wastes like kiwifruit and watermelon, the presence of in situ organic acids competed with the Lewis acidity of metal ions, thus imposing an inhibitory effect on metal-catalysed dehydration. They also achieved 4.0% and 8.1% HMF yield from cooked rice and penne pasta in a microwave reactor (140 °C, 20 min) using $SnCl_4$, under similar reaction conditions. The study corroborated with the prior study, where tetravalent metal ions (Zr and Sn) showed superior performance over trivalent ions (Al, Cr, and Fe) in HMF synthesis. Notably, the authors observed that a major part of the product formed was glucose, while traces of fructose and HMF were detected. Another study demonstrated the conversion of whey permeate into HMF with lactose as the model carbohydrate with 59% and 74%

HMF yield from lactose and whey permeate respectively. The catalytic system was composed of $AlCl_3 \cdot 6H_2O$ with acetic acid acting synergistically to enhance the HMF yield. The increased HMF yield for whey permeate was accounted to the residual protein present which can initiate HMF formation through the Amadori pathway [94].

6.5.3 Polar Aprotic Solvents

This solvent category is characterized by the absence if acidic hydrogen and a moderate polarity, thereby proves promising for the conversion of biomass residues to HMF. Fructose dehydration to HMF in pure water has a high reaction rate and good selectivity, while glucose dehydration is not non-selective in aqueous media (about <6%). For binary mixtures, HMF yield from fructose was improved by adding aprotic solvents such as dimethyl sulfoxide (DMSO) as a modifier in water; for glucose however, it is impossible to achieve the same yield due to its inefficiency in higher boiling solvents. DMSO favourably shifts the equilibrium of the reaction by inhibiting acyclic reaction which might lead to undesirable products, agreeing with the results of other authors [95, 96]. The efficacy of different solvents such as acetonitrile (ACN), DMSO, and acetone was investigated for the conversion of bread wastes using $SnCl_4$ as a catalyst; where the highest yield (27.4% HMF) was observed using DMSO/H_2O [84] (Table 6.4). Despite the high yield obtained by DMSO/H_2O, the reaction time (120 min) was much longer than ACN/H_2O and acetone/H_2O (10 min). This result concurred with what was obtained by Yu et al. where a slight improvement in HMF yield (29.5%) using similar solvent-catalytic conditions was obtained [85]. For microwave heating, the dielectric constant and the dipole moment played a vital role in controlling the selectivity and yield of the reaction. For instance, HMF yield was much higher in DMSO/H_2O (17%) mixture than in aqueous media (5%), in the given temperature range of 120–150 °C [97].

Moreover, the addition of DMSO for cellulosic vegetable wastes increased the protonation of glucose for rapid HMF formation and further prevented its dampening to LA by preventing the carbonyl group of HMF from nucleophilic attack [86, 98]. However, Quin & Liu suggested that DMSO lowered the activation energy for the reaction involving glucose dehydration and humins formation via C2-OH and C1-OH protonation, respectively [99]. Hence, the study suggests that formation of humins is a major bottleneck at high temperature reaction. Another water-miscible solvent, *N,N*-dimethylacetamide (DMA) when used with LiCl, showed the unique property of dissolving cellulose and gave 37% and 62% HMF yield with sugarcane bagasse and cellulose, respectively, over $Zr(O)Cl_2/CrCl_3$ catalyst under microwave heating at 120 °C for 5 min. The DMA-Li^+ macrocations generated chloride ions which could hydrolyse cellulose by forming H-bonds with the hydroxyl groups of cellulose. The system could be used 3 times with minimal loss in catalytic activity [78]. The solvent used had the intrinsic property of dissolving cellulose and the catalyst facilitated the isomerization of glucopyranose to fructopyranose. Also, the dissociation of metal chloride to Cl^- ions and HCl formation promoted the dehydration reaction further.

6.6 Ionic liquid

Principally, IL is defined as molten salts existing in liquid form below 100 °C. They are characterized by their high thermal stability, low vapor pressure, and non-flammability [100]. Based on their properties, IL are classified as acidic (Lewis and Brønsted acids), basic, or neutral [101]. Brønsted acidic IL is generally grouped into acid-functionalized IL having acidic moieties ($-SO_3H$ or –COOH) and has been most widely used for the processing of agri-food biomass. Recently, –SO_3H functionalized Brønsted acidic IL, [DMA][CH_3SO_3] and [BBIM-SO_3][NTf_2] demonstrated excellent catalytic performances on four varieties of mushroom wastes owing to their proton donating ability [102] (Table 6.5). However, Brønsted acidic IL showed a very low yield for fructose to HMF conversion in comparison to protic IL, due to the promotion of side reactions with increased Brønsted acidity.

Protic IL is another class that is cheaper and easy to synthesize by a one-step reaction between Brønsted acid and Brønsted base. These have been used for the pretreatment of cellulosic agro-food wastes like cassava pulp [110], corn stalk [111], sugarcane bagasse [112], and also for their conversion to HMF affording a maximum HMF yield of 47.2% for corn stalk [107].

Immidazolium-based IL are well known for their remarkable solvation effect which enables the dissolution of complex polymers like cellulose [100]. Amongst different chlorine-based IL ([EMIM]Cl, [HMIM]Cl, and [OMIM]Cl), the highest HMF yield (69.1 mg/mL) was achieved using [OMIM]Cl with 0.3 M HCl at 100 °C for pretreated grape berry extract [108]. The hydrogen bonding between the chloride anions and the hydroxyl groups disrupted the carbohydrate structure [100]. Using [C_4mim]Cl/$CrCl_3$ catalytic system, rice straw, and corn stalk produced 45% and 47% HMF at 100 °C in 3 min under microwave radiation, which was close to that obtained by the mixture of cellulose and xylan (53%), under same reaction conditions. The formation of $[CrCl_{3+n}]^{n-}$ complex promoted glucose mutarotation and acyclic route for enolate formation with subsequent dehydration to HMF [103]. Yi et al. achieved a yield of 70% HMF in [OMIM]Cl for raw tapioca which was slightly less than that obtained by pure tapioca starch (73%), under similar reaction conditions [105, 109]. The HMF yield from sugarcane bagasse was increased from 37% to 42% by replacing DMA-LiCl solvent with [BMIM]Cl for the same reaction conditions [78]. The sole catalytic effect of IL ([BMIM]Br) was demonstrated by Li et al. [113], where they were able to achieve HMF yields of up to 95% under optimized conditions from fructose. The recyclability of the IL was tested and could be used 6 times without significant loss of activity. Br^- anion played the role of a base responsible for the hydrogen bond interaction and dehydration and acted as a nucleophile which resulted in the formation of by-products.

Table 6.5 Conversion of biomass residues and their model compounds in ionic liquid

Agro-food wastes	Carbohydrate (%)	Solvent	Catalyst	Heating	Temperature (°C)	Reaction time	HMF yield (%)	Ref.
Cellulose + xylan (0.1 g)	Hexose (36.1)	$[C_4mim]Cl$ (2.0 g)	$CrCl_3{\cdot}6H_2O$ (0.2 g)	Microwave reactor	100	2.5 min	53.0	[103]
Chicory root extract (10%, w/v)	Inulin (70–75)	[OMIM]Cl (4 g)-ethyl acetate (1 mL)	HCl (0.3 M), $CrCl_2$ (0.2 g)	NA	120	60 min	50.9	[104]
Corn stalk (0.1 g)	Hexose (36.1), pentoses (21.4)	$[C_4mim]Cl$ (2.0 g)	$CrCl_3{\cdot}6H_2O$ (0.2 g)	Microwave reactor	100	3 min	45	[103]
Corn starch (1 g)	–	[OMIM]Cl (4 g)-ethyl acetate (1 mL)	HCl (0.3 M), $CrCl_2$ (0.2 g)	Heating mantle	120	60 min	49.5	[105]
Kudzu root extract (10% w/v)	–	–	HCl (0.3 M), CrF_3 (0.1 g)	NA	80	2.5 h	34.8	[106]
Pretreated corn cob (1.5 g)	Cellulose (31.6), hemicellulose (33.3), lignin (17.6)	$TEAHSO_4$:H_2O (4:1), 10 g		Oven	120	2 h	38.9	[107]
Pretreated corn husk (1.5 g)	Cellulose (31.8), hemicellulose (19.7), lignin (12.2)	$TEAHSO_4$:H_2O (4:1), 10 g		Oven	120	2 h	47.2	[107]
Pretreated wheat bran (1.5 g)	Cellulose (40.2), hemicellulose (20.5), lignin (16.5)	$TEAHSO_4$:H_2O (4:1), 10 g		Oven	120	2 h	7.2	[107]
Pretreated grape berry extract (5 mL)	Fructose, glucose, sucrose	[OMIM]Cl (5 g)	$CrCl_2$ (0.2 g)	Heated mantle with stirrer	100	3 h	99.8 mg/mL	[108]
Pretreated grape berry extract (5 mL)	Fructose, glucose, sucrose	[HMIM]Cl (5 g)	$CrCl_2$ (0.2 g)	Heated mantle with stirrer	100	3 h	87.2 mg/mL	[108]
Pretreated grape berry extract (5 mL)	Fructose, glucose, sucrose	[EMIM]Cl (5 g)	$CrCl_2$ (0.2 g)	Heated mantle with stirrer	100	3 h	76.8 mg/mL	[108]
	Starch (70–75)		CrF_3 (0.1 g)		120	90 min	52.6	[109]

(continued)

Table 6.5 (continued)

Agro-food wastes	Carbohydrate (%)	Solvent	Catalyst	Heating	Temperature (°C)	Reaction time	HMF yield (%)	Ref.
Pretreated raw tapioca (10 wt.%)		[OMIM]Cl (4 g)-ethyl acetate (1 mL)		Heating mantle				
Pretreated raw tapioca (10 wt.%)	Starch (70–75)	[OMIM]Cl (4 g)-ethyl acetate (1 mL)	CrF_3 (0.1 g), $CrBr_3$ (0.1 g)	Heating mantle	120	90 min	69.7	[109]
Rice straw (0.1 g)	Hexose (37.5), Pentose (21.2)	[C_4mim]Cl (2.0 g)	$CrCl_3 \cdot 6H_2O$ (0.2 g)	Microwave reactor	100	3 min	47	[103]
Sugarcane bagasse (4 wt.%)	Cellulose	[BMIM]Cl (0.5 g)	$Zr(O)Cl_2/CrCl_3$ (20 mol%)	Microwave reactor	120	5 min	42	[78]
Tapioca starch (1 g)	–	[OMIM]Cl (4 g)-ethyl acetate (1 mL)	HCl (0.5 M), $CrCl_2$ (0.2 g)	Heating mantle	120	60 min	73.0	[105]
Wood ear mushroom (5 wt.%)	–	DMA-LiCl (2.0 g)	[DMA] [CH_3SO_3]	Microwave reactor	140	10 min	44.0	[102]
Wood ear mushroom (5 wt.%)	–	DMA-LiCl (2.0 g)	[BBIM-SO_3] [NTf_2] (20 wt.%)	Microwave reactor	140	2 min	47.0	[102]
α-Cellulose (0.1 g)	Hexose (36.1)	[C_4mim]Cl (2.0 g)	$CrCl_3 \cdot 6H_2O$ (0.2 g)	Microwave reactor	140	2.5 min	0.6	[78]

Note: 1-Butyl-3-methylimidazole chloride ([C_4mim]Cl), 1-butyl-3-methylimidazole bromide ([C_4mim]Br), *N,N*-dimethyl acetamide $[DMA]^+$, *N*-methylpyrrolidone $[NMP]^+$ and 1-butyl-3-(4-sulfobutyl)imidazole [BBIM-$BuSO_3H$]

6.7 Biphasic Catalytic Systems

Biphasic systems consist of an additional extracting layer that prevents HMF rehydration, condensation, and polymerization reactions to form LA and humins. Compared to a monophasic system, HMF yield increased up to 6 times (15.3–92.1 g/kg straw) using 2-PrOH as extracting solvents for rice husk, while acetone gave a rise of two-fold (Table 6.6). The extraction of HMF from the aqueous phase is controlled by the partitioning coefficient. The addition of NaCl enhanced phase separation, employing a further 3.2 times increase in HMF yield for acetone [89]. Similarly, Román-Leshkov and Dumesic greatly improved the partitioning coefficient of the biphasic system by the salting out effect of inorganic salts (such as NaCl), which altered the intermolecular forces between liquids, thus reducing their mutual solubility [123]. Smaller hydrated ions have a better influence on the salting-out effect. For instance, K^+ and Na^+ both showed superior HMF selectivity (84% and 82%) than Li^+ (72%). Hence, HMF yields in biphasic systems depend mainly on the selectivity of the extracting layer, its volume, and the partitioning coefficient.

The HMF yields obtained from bread wastes were comparable to the ones achieved using pure starch (26%) in a biphasic system at 170 °C in 6.6 min [116]. Dehydration reaction in the presence of ACN and acetone took place via glucose to fructose isomerization, while DMSO promoted direct glucose to HMF dehydration. Furthermore, DMSO as a modifier improved the selectivity of the MIBK–2-butanol/DCM biphasic system. Compared to the selectivity obtained in pure water (11%), extracting solvents improved the selectivity (28%) and the recovery of HMF (82% HMF in the organic phase). Along with the advantages of improved dehydration rate and HMF extractability, the increased DMSO content can reduce the HMF extracting capacity of the organic phase [116]. It can be seen from Tables 6.4 and 6.6 that the HMF yield increased from 0.26% to 48.8% for glucose by converting the monophasic system to a biphasic system using NaCl as the salt [92]. Potato tuber and girasol extract gave impressive yields of 30.7% and 58.2%, respectively in [OMIM]Cl/ethyl acetate biphasic system using HCl/$CrCl_2$ as catalysts. Although $CrCl_2$ was effective as a catalyst for the synthesis of HMF from potato tuber, it showed no effect with girasol extract [115]. In addition, some biphasic reaction systems demonstrated catalytic activity. For instance, the sole effect of the DMSO/DCM biphasic system facilitated carbohydrate dehydration in water [116]. This system solved the corrosion problem that occurred by using mineral acids and can be employed for higher concentrations of substrate (30%) with a slight decrease in HMF yield. However, due to its prolonged reaction time, this system is not suitable for scaled-up production.

Besides preventing HMF rehydration, the catalytic systems used in biphasic reactors can be recycled. For instance, the recyclability of the DMSO/[BMIM]Cl biphasic system with $AlCl_3{\cdot}6H_2O$ was tested after the first dehydration reaction of corn starch to HMF. The system showed just 4% decrease in catalytic activity after five reaction runs, with yields varying from about 60% to 48% [118]. However,

Table 6.6 Biphasic reaction systems for conversion of biomass residues and their model compounds to HMF

Substrate	Phase I	Phase II	Catalyst	Heating	Temperature (°C)	Time	HMF yield (%)	Ref.
Agro-food wastes								
Corncob (2.4%)	Water (0.8 mL)	THF (8 mL)	$NaHSO_4$ (0.66 wt.%)	Batch reactor	190	60 min	49.3	[114]
Girasol tuber extract (10%)	[OMIM]Cl (4 g)	Ethyl acetate (4 g/mL)	HCl (0.3 M)	Oil bath	120	60 min	58.2	[115]
Potato tuber extract (10%)	[OMIM]Cl (4 g)	Ethyl acetate (4 g/mL)	HCl (0.5 M) $CrCl_2$ (0.2 g)	Oil bath	120	60 min	30.7	[115]
Pretreated rice straw (1.53%)	Water (250 mL) + NaCl (75 g)	1-BuOH (250 mL)	H_2SO_4 (0.5%)	High pressure oil bath reactor	180	3 h	60 g/kg straw	[89]
Pretreated rice straw (1.53%)	Water (250 mL) + NaCl (75 g)	2-PrOH (250 mL)	H_2SO_4 (0.5%)	High pressure oil bath reactor	180	3 h	60 g/kg straw	[89]
Pretreated rice straw (1.53%)	Water (250 mL) + NaCl (75 g)	MIBK (250 mL)	H_2SO_4 (0.5%)	High pressure oil bath reactor	180	3 h	65.7 g/kg straw	[89]
Pretreated rice straw (1.53%)	Water (250 mL) + NaCl (75 g)	Acetone (250 mL)	H_2SO_4 (0.5%)	High pressure oil bath reactor	180	5 h	44 g/kg straw	[89]
Sugarcane molasses (0.15 g)	Water-NaCl (1.05 g)	THF	$ZnCl_2$/HCl (0.2 mmol)	Stirrer batch reactor	180	60 min	48.8	[92]
Model carbohydrates								
Cellobiose (10%)	3:7 Water:DMSO (7 g)	DCM (7 g)	No catalyst	Parr reactor	140	9.30 h	38.2	[116]
Cellobiose (10%)	4:6 Water:DMSO (1.5 g)	7:3 (w/w) MIBK:2-butanol (3 g)	HCl (0.25 M)	Parr reactor	170	6 min	27	[116]

Fructose (10%)	3:7 Water:DMSO (7 g)	DCM (7 g)	No catalyst	Parr reactor	140	2 h	87	[116]
Fructose (10%)	5:5 Water:DMSO (1.5 g)	7:3 (w/w) MIBK:2-butanol (3 g)	HCl (0.25 M)	Parr reactor	170	2.4 min	84.6	[116]
Fructose (30%)	7:3 (8:2 Water: DMSO):PVP (1.5 g)	7:3 MIBK:2-butanol (3 g)	HCl (0.25 M)	Parr reactor	180	3 min	75.6	[117]
Glucose (0.135 g)	Water-NaCl (1.05 g)	THF	$ZnCl_2$/HCl (0.3 mmol)	Stirrer batch reactor	180	60 min	48.8	[92]
Glucose (0.135 g)	Water-NaCl (1.05 g)	THF	$ZnCl_2$/HCl (0.2 mmol)	Stirrer batch reactor	180	60 min	62.7	[92]
Glucose (0.135 g)	Water-NaCl (1.05 g)	THF	$AlCl_3$/HCl (0.2 mmol)	Stirrer batch reactor	180	55 min	69.6	[92]
Glucose (10%)		DCM (7 g)	No catalyst	Parr reactor	140	4.3 h	29.8	[116]
Glucose (10%)	3:7 Water:DMSO (7 g)	DCM (7 g)	No catalyst	Parr reactor	140	4.3 h	29.8	[116]
Glucose (10%)	5:5 Water:DMSO (1.5 g)	7:3 (w/w) MIBK:2-butanol (3 g)	HCl (0.25 M)	Parr reactor	170	10.2 min	23.5	[116]
Inulin (10%)	3:7 Water:DMSO (7 g)	DCM (7 g)	No catalyst	Parr reactor	140	2.3 h	70	[116]
Inulin (10%)	5:5 Water:DMSO (1.5 g)	7:3 (w/w) MIBK:2-butanol (3 g)	HCl (0.25 M)	Parr reactor	170	3 min	75.5	[116]
Starch (10%)	3:7 Water:DMSO (7 g)	DCM (7 g)	No catalyst	Parr reactor	140	11 h	36.4	[116]
Starch (10%)	4:6 Water:DMSO (1.5 g)	7:3 (w/w) MIBK:2-butanol (3 g)	HCl (0.25 M)	Parr reactor	170	6.6 min	26.23	[116]
Sucrose (0.257 g)	Water-NaCl (1.05 g)	THF	$ZnCl_2$/HCl (0.3 mmol)	Stirrer batch reactor	180	60 min	49.6	[92]
Sucrose (0.257 g)	Water-NaCl (1.05 g)	THF	$ZnCl_2$/HCl (0.3 mmol)	Stirrer batch reactor	180	60 min	65.6	[92]

(continued)

Table 6.6 (continued)

Substrate	Phase I	Phase II	Catalyst	Heating	Temperature (°C)	Time	HMF yield (%)	Ref.
Sucrose (0.257 g)	Water-NaCl (1.05 g)	THF	$AlCl_3$/HCl (0.1 mmol)	Stirrer batch reactor	180	55 min	54.4	[92]
Sucrose (10%)	3:7 Water:DMSO (7 g)	DCM (7 g)	No catalyst	Parr reactor	140	4.30 h	50.8	[116]
Sucrose (10%)	4:6 Water:DMSO (1.5 g)	7:3 (w/w) MIBK:2-butanol (3 g)	HCl (0.25 M)	Parr reactor	170	3 min	50	[116]
Substrate	**Phase I**	**Phase II**	**Catalyst (recyclability)**	**Heating**	**Temp**	**Time**	**HMF yield**	**Ref.**
Recyclable biphasic systems								
Corn starch (1%)	[BMIM]Cl (0.1 g) + water	DMSO	$AlCl_3 \cdot 6H_2O$ (5 times)	Microwave batch reactor	150	20 min	47.81	[118]
Corn starch (1%)	[Bmim]Cl (0.2 wt. %)-water	MIBK (5 mL)	$AlCl_3 \cdot 6H_2O$ (6 times)	Microwave batch reactor	140	20 min	53	[119]
D-Xylose (1.25 mmol)	Water (1 mL)-NaCl (2.5 mmol)	CPME (3 mL)	$FeCl_3$ (4 times)	Microwave batch reactor	170	20 min	Furfural yield 76–80	[120]
Fructose (0.05 g)	Water (0.1 g)-[MBCIm]SO_3Cl	MeCN (2 g)	HCl (3 times)	Glass reactor	80	3 h	70	[121]
Fructose (20%)	ChCl (3.6 g)	MeCN (30 mL)	HCl (5 times)	Oil bath	100	4 h	92	[122]

DMSO, being a high boiling point solvent, is difficult to separate from the reaction mixture by evaporation. Replacing DMSO with low boiling point solvents like MIBK or 2-butanol can make the system more energy efficient. Later, the same group tested the recyclability of the MIBK/[BMIM]Cl biphasic system. The aqueous catalytic system was reactive and selective (64% HMF) up to the sixth recycling test, where the HMF yield dropped to 53% due to the accumulation of humins. The dehydration of D-xylose in water/CPME biphasic media with $FeCl_3$ as the catalyst and in the presence of NaCl showed satisfactory catalytic activity after 4 consecutive tests. The furfural yield increased from 76% to 80% after the fourth run, which witnessed the presence of initial D-xylose from the former cycle [120].

Most of the IL do not form a biphasic medium with ACN. Nevertheless, the [MBCIm]SO_3Cl/MeCN system does not dissolve and can afford greater than 70% HMF yield for the first three cycles, with a reduction to 38% in the fourth cycle for acidic dehydration of fructose [121]. Despite the green label that the IL carry due to their very low vapor pressure and high thermal stability, their application is limited due to recyclability issues and economic inefficiency. To alleviate this, deep eutectic solvents (DES) have been used in the reaction system owing to their excellent physicochemical properties. The low Brønsted acidity of DES/MeCN system gave 90.3% HMF yield from the dehydration of concentrated fructose (20 wt.%) with excellent recyclability in the presence of HCl. The system maintained HMF yields in the range of 88–92% for five consecutive runs, with an abrupt decrease in the sixth cycle. However, recrystallization of ChCl with other polar aprotic solvents is an issue for this process [122].

6.8 HMF Production over Heterogeneous Catalysts

Recent advances towards heterogenous catalysis for HMF production have shown to improve the corrosion, reusability and disposal challenges associated with the homogenous catalysts. Immobilization of acidic sites such as $-SO_3H$, –OH and –COOH groups on the heterogenous support enhances the catalytic activity, which is influenced by the thermal and morphological characteristics of catalytic support [124].

6.8.1 Silica-Based Materials

Amongst all heterogeneous supports, mesoporous silica catalysts (MSC) have high thermal stability, large surface area, uniform pore size distribution, and economical synthesis procedure [125]. As summarised in Table 6.7, SBA-X (Santa Barbara Amorphous) is the commonly used mesoporous silica owing to its thermal, mechanical, and chemical stability. SBA-15 has an ordered structure with a pore size of

2–50 nm, large BET (Brunauer-Emmett-Teller theory) surface area, tunable pore diameter, and water tolerance characteristics [144]. However, they exhibit lower catalytic activity due to the absence of proton-containing groups for the hydrolysis of polysaccharides in biomass residues. The reactivity of these catalysts is modulated by the pore structure, pore dimensions, and the functional groups present on the surface. Therefore, acidic moieties such as $–SO_3H$, –OH, and –COOH are doped onto MSC to enable the hydrolysis of cellulose and oligosaccharides. The proton of the $–SO_3H$, –OH, and –COOH groups attack the cellulose and hemicellulose present, forming water-soluble oligosaccharides which undergo further hydrolysis to form xylose and glucose. The next step involves the conversion of these monosaccharides to furfurals, HMF along with other by-products. For instance, silica-supported polytungstic acid (PTA) was used to catalyse the conversion of glucose producing a maximum HMF yield of 78.3% [145].

A study has confirmed that in comparison to non-sulfonated catalysts, sulfonated catalysts favour the production of side reactions. For instance, the hydrothermal conversion of rice straw was investigated using non-/sulphonated mesoporous silica as the solid acid catalyst. Compared to the non-sulfonated catalyst (SPF-15), the sulfonated catalysts favoured by-product formation along with monosaccharides as the main product. The surface area, pore volume, and acid amounts have been found to increase with the aging time of the catalyst. However, XRD studies have confirmed that sulphonation does not affect the structural integrity of the silica-based catalyst. The catalysts having a larger surface area, pore volume, and acid amounts demonstrated lower by-product formation [126]. The elevation in temperature from 150 to 220 °C was responsible for the increase in yield of furfural, HMF, and total organic acids, where the yield of organic acids rapidly decreased above 180 °C [126].

Compared to mesoporous silica materials (SBA-15, SBA-16, MCM-41), mesocellular silica foams (MSF) have better mass transfer properties since the pores are uniformly connected in the three-dimensional lattice. Accordingly, Huang et al. achieved 89.3% HMF yield with 95% selectivity using Nafion (15/30/45%)/MSF as a catalyst to carry out fructose dehydration to HMF in DMSO, which was an improvement over the use of Amberlyst-15, SO_4^{2-}/ZrO_2, and SAC-13 under similar reaction conditions [146]. A higher concentration of Nafion (45%) was responsible for lowering the pore volume, while the surface area remained unchanged. The strong interaction between electron attractive groups in Nafion and silanol groups of MSF resulted in the efficacious mass transfer of the reactant to the acidic sites, which was attributed to the well-distributed nanoparticles in the MSF framework. The catalyst could be recycled 5 times without a significant change in its activity. However, a minor decrease in activity was attributed to the leaching of Nafion or the replacement of acidic sites with by-products. The catalyst demonstrated fair HMF yield (69.5%) even at high fructose concentration (20%), opening the possibility for its practical application.

Table 6.7 HMF production over heterogeneous catalysts from agro-food residues

Entry [Ref.]	Agro-food wastes	Carbohydrates	Solvent	Catalyst	Surface area (m^2/g)	Pore volume (cm^3/g)	Acid amount (mmol/g)	Heating mode	Temperature (°C)	Time	HMF yield (%)
Silica-based catalysts											
1 [126]	Rice husk (3 g)	Cellulose 23.3%, hemi-cellulose 32.7%, acid insoluble lignin 14.4%, acid soluble lignin 1.9%	Water (30 g)	SBA-15 (0.5 g)	668	0.41	0.08	Batch reactor	180	1 h	<2
2 [126]	Rice husk (3 g)	Cellulose 23.3%, hemi-cellulose 32.7%, acid insoluble lignin 14.4%, acid soluble lignin 1.9%	Water (30 g)	MPS-A (0.5 g)	652	0.26	1.3	Batch reactor	180	1 h	2.5
3 [126]	Rice husk (3 g)	Cellulose 23.3%, hemi-cellulose 32.7%, acid insoluble lignin 14.4%, acid soluble lignin 1.9%	Water (30 g)	MPS-B (0.5 g)	680	0.58	2.2	Batch reactor	180	1 h	2.5

(continued)

Table 6.7 (continued)

Entry [Ref.]	Agro-food wastes	Carbohydrates	Solvent	Catalyst	Surface area (m^2/g)	Pore volume (cm^3/g)	Acid amount (mmol/g)	Heating mode	Temperature (°C)	Time	HMF yield (%)
4 [126]	Rice husk (3 g)	Cellulose 23.3%, hemicellulose 32.7%, acid insoluble lignin 14.4%, acid soluble lignin 1.9%	Water (30 g)	MPS-C (0.5 g)	713	0.65	2.8	Batch reactor	180	1 h	< 2.5
Carbonaceous materials											
5 [127]	Bread waste (0.5 g)	Starch 73.3%	DMSO/water (3:1 v/v) (10 mL)	PAB-600 °C (0.2 g)	1547	0.66	–	Autoclave	180	20 min	24.5
6 [127]	Rice wastes (0.5 g)	Starch 90.3%	DMSO/water (3:1 v/v) (10 mL)	PAB-600 °C (0.2 g)	1547	0.66	–	Autoclave	180	20 min	22.2
7 [127]	Spaghetti wastes (0.5 g)	Starch (NA)	DMSO/water (3:1 v/v) (10 mL)	PAB-600 °C (0.2 g)	1547	0.66	–	Autoclave	180	20 min	14.7
8 [128]	Cassava pulp (5%)	Starch (79.45%)	Water	Activated carbon (1 g)	–	–	–	Autoclave	50% power level	10 min	27.35 mg/1 g of cassava pulp hydrolysate

10 [129]	Cassava pulp (1 g)	Starch (62–68%)	Water (20 mL)	Activated carbon (1.0 g/g solid sample)	–	–	–	Autoclave	210	18 min	63.08 mg/100 g
11 [130]	Bread waste (0.5 g db)	Starch	DMSO/water	Sulfonated biochar (0.5 g)	–	–	–	Autoclave	180	20 min	30.4
12 [127]	Rice starch (0.5 g)	–	Water (10 mL)	PAB-600 °C (0.2 g)	1547	0.66	–	Autoclave	180	20 min	3.7
13 [131]	Cassava wastes (2%)	Starch 60%, cellulose 16%, hemicellulose 5%, and lignin 19%	Acetone/DMSO (70:30 w/w %)	Sulphonated carbon-based catalyst (0.05 g/g sample)	–	–	–	–	250	1 min	10.8
Ion exchange catalysts											
14 [132]	Pretreated food and beverage waste (70%)	Fructose (89.0 g/L)	DMSO/high fructose syrup (1:1 v/v) (10 mL)	Amberlyst 36	–	–	–	Pilot scale microwave reactor	140	40 min	71
15 [133]	Pretreated melon rinds (100 mg)	Cellulose (28.2%)	THF/water (3:1) + NaCl (0.35 g)	M-KSF (100 mg/mL)	–	–	–	Biphasic microwave reactor	180	30 min	5.9
16 [97]	Vegetable wastes (5%)	Cellulose	DMSO/water (1:1 v/v)	Amberlyst 36	10.4 [134]	0.06 [134]	5.4 [134]	Microwave reactor	120	120 min	17
Zeolites											
17 [135]	Rice straw		DMSO/water	Cu-Cr/ZSM-5				Autoclave	140	4 h	49.5

(continued)

Table 6.7 (continued)

Entry [Ref.]	Agro-food wastes	Carbohydrates	Solvent	Catalyst	Surface area (m^2/g)	Pore volume (cm^3/g)	Acid amount (mmol/g)	Heating mode	Temperature (°C)	Time	HMF yield (%)
18 [136]	Rice straw		DMSO/water	HSO_3–ZSM-5				Autoclave	140	4 h	54.10
19 [137]	Corncob		THF/water	HSO_3–ZSM-5					160	5 h	49
Metal oxides											
20 [138]	Jerusalem artichoke juice (pretreated)	Inulin (>75 wt. %)	Water/2-butanol (2: 3 v/v)	H_3PO_4 treated $Nb_2O_5{\cdot}nH_2O$ (0.1 g) (7 times)	214.9	–	–	Biphasic reactor	160	120 min	65
21 [139]	Jerusalem artichoke (1.2 g)	Inulin	Water/2-butanol (2: 3 v/v)	H_3PO_4 treated $Ta_2O_5{\cdot}nH_2O$ (0.1 g) (15 times)	141.5	–	–	Biphasic reactor	160	150 min	50
22 [139]	Jerusalem artichoke (1.2 g) pretreated	Inulin	Water/2-butanol (2: 3 v/v)	H_3PO_4 treated $Ta_2O_5{\cdot}nH_2O$ (0.1 g) (15 times)	141.5	–	–	Biphasic reactor	160	120 min	79
23 [140]	Rice starch	–	100 mL solvent (water/THF = 1: 4 v/v) + NMP-4 g NaCl	Phosphated TiO_2 (4 cycles)	151	0.43	(B: 0.72, L: 3.81)	Biphasic/batch reactor	180	105 MIN	99.7

24 [140]	Potato starch	–	100 mL solvent (water/ THF = 1: 4 v/ v) + NMP- 4 g NaCl	Phosphated TiO_2 (4 cycles)	151	0.43	(B: 0.72, L: 3.81)	Biphasic/ batch reactor	180	105 MIN	99.8
Metal phosphates											
25 [141]	Mixed food wastes (4%)	C (45.5 wt.%) and O (44.6 wt. %)	Water	ZrPO-600 (2 wt.%) (3 cycles)	46.4	0.179	0.75	High pressure reactor	160	6 h	4.3
26 [142]	Sugarcane bagasse (50 mg)	50% cellulose, 25% hemicelluloses and 25% lignin	DMA–LiCl (10 wt.%) 0.5 g	Mesoporous TiPO (10 mg)	193	–	–	Microwave reactor	140	5 min	26
27 [143]	Corn stalk (0.2 g)		H_2O-NaCl (1 mL)/ THF (5 mL)	Sn-Mont (0.1 g) $NbOPO_4$ (0.1 g)	–	–	–	Biphasic oil bath	160	2 h	44.1
28 [143]	Sugarcane bagasse (0.2 g)		H_2O-NaCl (1 mL)/ THF (5 mL)	Sn-Mont (0.1 g) $NbOPO_4$ (0.1 g)	–	–	–	Biphasic oil bath	160	2 h	27.2

6.8.2 Carbonaceous Materials

Carbon-based catalysts have become a hot research topic for the catalysis of carbohydrates and biomass residues. The carbon-based precursor can be obtained from biomass (refer to Table 6.7) via different carbonization methods encompassing dry pyrolysis, gasification, hydrothermal carbonization, and flash carbonization, followed by subsequent sulfonation. Most of the former works used pyrolysis, yet hydrothermal carbonization has the unique advantage of using a low temperature and wet biomass [147]. Accordingly, Kang et al. developed biochar using lignin, cellulose, wood meal, and D-xylose, formed by sulphonation via hydrothermal carbonization for lignin conversion to HMF. The reaction was carried out in IL and the HMF yields were in the range of 47–65% [148]. The major drawback of the reaction system was that the catalyst got partly deactivated in IL, which could be explained by the loss of acidic sites during the ion-exchange process between the catalyst and the IL [149]. However, for the conversion of raw biomass to HMF, H_3PO_4 activated biochar stands superior over sulfonated biochar due to ambient impregnation, one-step activation and carbonization, eco-friendliness, and improved mesoporous properties than sulphonation. Recently, Cao et al. synthesized H_3PO_4 activated pinewood sawdust-derived biochar (PAB) using an activation temperature of 400–600 °C. The catalytic activity was tested for HMF synthesis from starch-rich food wastes (bread, rice, and spaghetti). It was found that a high activation temperature and impregnation ratio improved the porosity and total acidity (due to the presence of C-O-PO_3 and C-PO_3 surface groups) of the catalyst [127]. Rice has a higher starch content than bread wastes, yet the HMF yield of the latter was higher due to the well-ground mix during the roasting process which increased the starch accessibility. The addition of DMSO inhibited the side reactions and no fructose was detected, suggesting the direct dehydration pathway of glucose in the presence of strong Brønsted acid.

6.8.3 Ion Exchange Resins

They are the copolymers of styrene, divinylbenzene, and sulfonic acid groups grafted on the benzene rings of styrene. The acid ion exchange resins are spherical with sizes ranging from 0.3 to 1.2 mm, typically existing in two morphological orders, i.e., gel-type (microporous) and macro reticular (macroporous). The styrene rings can be conventionally sulfonated either with one sulfonic group per styrene ring or over-sulfonated with more than one sulfonic group per styrene ring [146]. The commercially available ion exchange resins include Amberlyst, Purolite, and Nafion; holding the advantage of strong Brønsted acidity, high catalytic activity, and hydrophobic structural characteristics. Table 6.7 shows the progress of HMF production using ion exchange resins as the catalyst. Amongst these, Amberlyst-36 demonstrates strong Brønsted acidity equivalent to 35% sulphuric acid, cation exchange capacity of 5.4 mmol/g, surface area of 10.4 m^2/g, and pore size of 27.6 mm. The catalytic activity of the same was studied for HMF production from vegetable wastes, wherein increasing the reaction temperatures (120–150 °C) increased the cellulose dissolution and the HMF yield (up to 17%) in the DMSO-water mixture. On the downside, high temperatures can lead to the leaching of sulfonic groups and the formation of by-products such as glucose and HMF-derived humins, thereby limiting multi-cycle usage [97]. Another study demonstrated the conversion of high fructose syrup (80 g/L) derived from food and beverage wastes over Amberlyst-36, achieving a maximum HMF yield of 71% in 40 min with DMSO as a co-solvent. Using DMSO favored the proton released from Amberlyst via ion exchange which synergistically impacted the fructose dehydration into HMF. Overall, this study reported gram-scale HMF production (30 g/kg) where the solid catalyst could be recycled up to 4 times [132]. Significant gram-scale HMF synthesis from fructose was achieved using Purolite CT275D in a biphasic solvent system composed of dimethyl carbonate (DMC)/tetraethyl ammonium bromide (TEAB). The study reported the production of 22.76 g of crude HMF from 40 g of fructose, which accounted for 72% of HMF yield. The crude HMF was then isolated using a crystallization procedure leading to 50% of the pure crystalline product; meanwhile, the residual HMF-rich oil was directly reduced to bis(hydroxymethyl)furan (BHMF) [150]. Nafion, on the other hand, has been majorly explored for furfural synthesis. For instance, a comparative study of Nafion NR50 and Purolite CT275 was carried out in the MIBK-water system for the conversion of corn-stover hydrolysate into furfural which resulted in 78% and 80% furfural yield, respectively. Despite its high acidic properties, Nafion showed inferior catalytic performance compared to Purolite due to its low surface area (0.02 m^2/g) which led to 8 times higher catalytic loading compared to Purolite [151]. Pioneering research for the conversion of chitosan into HMF over Nafion 50 was evaluated in a biphasic system of MIBK-water with a yield of 33%. It is noteworthy that the sulfonic acid and the perfluorinated skeleton enhanced the chemical conversion of chitosan into HMF, which was comparable to the HMF production in homogeneous acidic systems including triflic acid (CF_3SO_3H), HCl, H_2SO_4, and HNO_3 with the same acid equivalence to that of Nafion 50 [152].

6.8.4 *Zeolites*

These are tetrahedral units composed of SiO_4 and $[AlO_4]^-$, where the negative charge of the latter is neutralized by a cation. The charge compensated with H^+ ion and the microporous structure (3–15 Å) makes the zeolites an excellent catalyst to produce furanic compounds [153]. Various crystalline materials such as HZSM-5 (Zeolite Soconyl Mobil), HMOR (Mordenite), HY (Faujasite), Hβ (Beta), Del-Nu-6 are investigated in monophasic and biphasic systems for HMF synthesis as listed in Table 6.7. The catalytic activity of the zeolites is influenced by the concentration and strength of the acidic sites, which could be modulated by changing the aluminum content or incorporating other elements into the zeolite structure. For instance, glucose dehydration in the presence of ZSM-5 zeolite (doped with H, Fe, or Cu) was effective in H_2O/MIBK biphasic systems, where the yield reached 42% by NaCl addition at 195 °C, after 30 min only. Under these reaction conditions, HMF did not rehydrate to LA [154]. A study evaluated the performance of Cu and Cr-doped ZSM-5 zeolite for the transformation of biomass-derived glucose (from enzymatic hydrolysis of rice straw) into HMF. It was concluded that the metal loading significantly impacted the HMF yield with a maximum of 0.3% and 0.2% of Cu and Cr, respectively [135]. Zeolite molecular sieves are another important type of catalyst widely used because of their regular pore structure, adjustable acidity, thermal stability, and reusability. A Cr-doped Ultra-stabilized Y molecular sieve (USY) was used for the direct conversion of rice straw into FDCA, where the reported yield was up to 67% at 180 °C in the presence of $KHCO_3$ as the base [155]. To provide additional acidity, ZSM-5 has been modified with sulfonic groups for the conversion of lignocellulosic biomass into HMF. The enzymatic hydrolysate of rice straw was used as the substrate in DMSO/water, which afforded around 54% HMF yield with no significant decrement in HMF yield after 5 recyclability tests [136]. The activity of HSO_3-ZSM-5 on corncob biomass was evaluated in tetrahydrofuran (THF)/water solvent system. The result was promising showing an HMF yield of 49%, simultaneously producing 89% furfural and 11% LA [137].

In addition, a study described glucose dehydration to HMF using a Sn-beta zeolite. The synthesis of Sn-beta requires hydrofluoric acid (HF), which is corrosive and hence harmful to humans and the environment [156, 157]. Moreover, the catalyst was active in the isomerization of glucose to fructose and needed a contemporary catalyst (HCl or Amberlyst-15) to carry forward the dehydration step. The HMF yield using Sn-beta combined with $NbOPO_4$ from corn stalk (44.1%) exceeded the yield for pure cellulose (39.1%) in a water/THF biphasic system, yet the main product was humins [143]. A synergistic role of solvent and acidic sites of HY zeolite can be seen for fructose conversion where the maximum yield obtained was 69.3% [158]. Nb-beta zeolite gave an impressive HMF selectivity value of 84.3% for glucose conversion of 97.4% in a biphasic water/MIBK solvent, at 180 °C and after 12 h [159]. Furthermore, Wang et al. [160] investigated the activity of Sn-Mont catalysts in monophasic as well as biphasic media. Sn-Mont exhibited low catalytic activity (glucose conversion 40.2% and HMF yield 6.5%) when pure water was used as the solvent, as it favoured the off-path reactions. Also, high boiling point solvents such as DMSO favoured the furanoid form (43.6% HMF), while low boiling point solvents like THF gave poor HMF yield due to low glucose solubility. To underscore this, a fixed ratio of THF-DMSO was used when the HMF yield rose to 53.5%. They found that the catalytic system was sufficient and needed no further addition of Brønsted acid. At lower temperatures (100–120 °C), the authors observed lower glucose conversion (<60%), however further increasing the temperature (up to 160 °

C) gave a better yield. The effect of glucose concentration (5%, 7%, and 10%) on the conversion ratio and HMF yield was marginal. They also conducted a catalyst recyclability test (6 times) which showed a low variation in the HMF yield [161].

6.8.5 Metal Oxides

Recently, oxides of metals in groups IV and V of the periodic table have been extensively used as heterogeneous catalysts for HMF synthesis (Table 6.7). For instance, hydrated niobium pentoxide, also called niobic acid ($Nb_2O_5{\cdot}nH_2O$), is commonly used due to its low toxicity, affordability, and ease of handling and thus, was investigated for the dehydration of Jerusalem artichoke and its model compounds to HMF in the biphasic water/2-butanol system [138]. Phosphoric acid treated niobic acid (NA-p) showed improved catalytic activity due to the synergistic effect of metal Lewis acid and protonated phosphate groups. The surface area which is also an important factor for tuning the active sites also increased (from 69.6 to 214.9 m^2/g) after H_3PO_4 treatment. The Lewis acid sites present in $Nb_2O_5{\cdot}nH_2O$ are water tolerant and are mainly responsible for its catalytic activity. The same catalyst, but untreated was tested by a previous group [162]. They reported the dehydration of fructose in water at 100 °C, without any side reactions. Hydrated tantalum oxide with treatment ($Ta_2O_5{\cdot}nH_2O$, TA-p) has been used by Yang et al., achieving 90% and 58% HMF yield from fructose and glucose respectively, at 160 °C [139]. The surface area increased from 41.6 to 141.5 m^2/g after phosphoric acid treatment and calcination. The bifunctional nature of the TA-p catalyst was also confirmed by the authors, where the basic and acidic sites took part in isomerization and dehydration reactions, respectively. Comparing the overall HMF yield from Jerusalem artichoke catalysed over TA-p and NA-p (79% vs 65%), it is evident that the former is more efficient in a catalysing dehydration reaction. This was reasoned by the excessive acidity and large surface area of NA-p which triggers the side reaction, hence lowering the yield. Mesoporous tantalum oxide enhanced the diffusion of the reactant and the product, more than the conventional one due to its high specific surface area (79 m^2/g), porous structure, and the existence of both Lewis and Brønsted acid sites (0.353 mmol/g). The dehydration of glucose took place in the biphasic water/MIBK system at 175 °C and 90 min only, achieving a yield of 23%, without undesirable reactions. Lewis acidity dominated in this type of catalyst, leading to the glucose-fructose-HMF dehydration pathway. Although the catalyst had enough Brønsted acidity, it was not sufficient to produce LA from HMF, hence it was established as a selective catalyst. Moreover, phosphoric acid grafted on mesoporous tantalum oxide followed by calcination at 300 °C enhanced its activity [163].

This reaction system gave superior yields than the one where crystalline TiO_2 (anatase) and ZrO_2 were used for fructose/glucose dehydration in the presence of pressurized water affording 7.68% HMF yield from glucose [164]. The reaction pathway followed during the dehydration depends on the acidity or basicity of TiO_2 and ZrO_2. The acid and base site density for TiO_2 were 17 and 8.9 μmol/g, respectively, while that of the ZrO_2 were 6.1 and 5.0 μmol/g, respectively. Hence, metal oxide's bifunctional role as Lewis acid and Lewis base could promote glucose isomerization to fructose (favored by base catalyst: ZrO_2) as well as its subsequent dehydration to HMF (favored by acid catalyst: TiO_2) [165]. Phosphated TiO_2 in water/THF biphasic reaction system (batch and continuous process) with *N*-methyl-2-pyrrolidone (NMP) gave excellent yield for rice and potato starch and other carbohydrates [140]. Furthermore, the flow reactor system used for the pre-treatment of cellulose (ball milling) afforded HMF in yield of 53%, demonstrating its feasibility for industrial application.

6.8.6 Metal Phosphates

Phosphates of transition metals (Nb, Zr, Cr, Sn, etc.) are another class of heterogeneous catalyst that displays interesting catalytic activity due to the presence of metal cations and PO_4^{3-} anion which governs their Lewis/Brønsted acidity. Similar to metal oxides, the calcination temperature is an important factor when comparing the catalytic activity of regenerated metal phosphates [166]. Niobium phosphate (NbPO) and zirconium phosphate (ZrPO) exhibit good catalytic behavior at high reaction temperatures temperature [167]. Hence, Antonetti et al. [168] studied fructose and inulin dehydration over NbPO and ZrPO in water under microwave heating. Despite its low surface area (108 vs 133 m^2/g), ZrPO has more acidic sites (mainly Lewis-type sites) per unit weight than NbPO, which explains why higher HMF yields and selectivity are obtained over NbPO. Moreover, NbPO in hydrous conditions develops stronger Brønsted acid sites than ZrPO, promoting the formation of undesired heavy compounds. On the other hand, the medium-strength Brønsted acidity of ZrPO in both aqueous and non-aqueous conditions make it more selective for HMF formation. Congruently, zirconium phosphate (ZrPO) was investigated by Parshetti et al. on mixed food wastes for HMF synthesis in aqueous conditions (Table 6.7) [141]. The yield obtained (4.3 wt.% HMF) was rather low compared to the previous study where pure carbohydrates were used as the substrate due to limited mass transfer [169]. Compared to ZrPO, tin phosphate (SnPO) has been proven to be a better catalyst for the conversion of high-concentration glucose (up to 20%) affording 52.3% HMF yield (for ZrPO). High concentration of Lewis acid sites, low Brønsted sites, and better surface area (than its metal oxide, SnO_2) make SnPO/EMIMBr an efficient catalytic system for glucose to fructose isomerization and its dehydration to HMF [169]. However, the tetrahedrally coordinated Sn^{4+} responsible for glucose to fructose isomerization is only active in anhydrous conditions which limits its use in aqueous systems. The same group of authors indicated the poor catalytic performance of CrPO due to the presence of inactive PO_4^{3-} anion, as free Cr^{3+} is excellent for this conversion [169]. In general, the main drawbacks of metal phosphates are their low surface area and small pore size. To alleviate this, Dutta et al. designed a macro/mesoporous titanium phosphate catalyst that demonstrated promising catalytic activity for the conversion of cellulose and sugarcane bagasse to HMF, under microwave heating [142]. Sugarcane bagasse produced more HMF than pure cellulose (26% vs 17%), under the same reaction condition, which can be ascertained due to the presence of hemicellulose and the amorphous form of cellulose present in sugarcane bagasse. Compared to the mesoporous tantalum oxide [163], mesoporous tantalum phosphate has a slightly better catalytic activity leading to 71 wt.% glucose conversion (vs 69%) and 37.8% HMF yield (vs 23%) from glucose. This can be attributed to the higher surface acidity (both Brønsted and Lewis) in the case of phosphate (1.48 vs 0.353 mmol/g). The main difference lies in the selectivity of the catalyst where Brønsted acidity dominated for phosphate, thus favouring direct glucose dehydration to HMF without fructose as an intermediate [170].

6.9 Green Solvents

The majority of polar aprotic solvents, such as DMSO and DCM, have better extraction capacities and greater HMF selectivity, however they are harmful for the environment [134]. The main attributes of green solvents are their lower boiling point, reduced toxicity, and biobased composition. By reducing the need to buy solvents generated from petroleum, these solvents can increase the process' sustainability [134]. The reaction systems discussed above include monophasic reactions (like water, aprotic solvents, and IL) or biphasic reactions where organic solvents like MIBK, DCM, ethyl acetate, diethyl ether, and acetone have been used as an organic phase for the extraction of HMF due to their high polarity. However, the large consumption of organic solvents harms the environment. Moreover, these reaction systems suffer inefficient HMF extraction, complex plant design, difficult recyclability, and energy-intensive HMF separation (for the IL and homogeneous catalysts). Learning from concepts of green chemistry, the reaction system should make use of heterogeneous catalysts and environmentally benign solvents which offer the advantage of recyclability of the catalyst and/or solvents system, to make the industrial processes more economically viable. Henceforth, the reaction systems in this section have focused principally on using heterogeneous catalysts and environmentally benign green solvents for the conversion of biomass residues to HMF (Table 6.8).

Owing to its low boiling point (82.2 °C), good sugar solubility, renewable nature, and easy recyclability; isopropyl alcohol is a greener medium than high boiling point aprotic solvents for the conversion of carbohydrates to HMF [180]. Consequently, it was used as a solvent medium with two different catalysts, HCl and Amberlyst-15

Table 6.8 Conversion of residue biomass and their model compounds to HMF in the presence of green solvents

Entry [Ref.]	Substrate	Carbohydrates	Green solvent	Catalyst	Heating mode	Temperature (°C)	Time	HMF yield (%)
Biomass residues								
Monophasic reaction systems								
1 [134]	Bread wastes (5%)	Starch 67.6%	PC-H_2O (1:1 v/v)	$SnCl_4 \cdot 5H_2O$ (55.5 mM)	Microwave reactor	120	10 min	20
2 [134]	Bread wastes (5%)	Starch 67.6%	GVL-H_2O (1:1 v/v)	$SnCl_4 \cdot 5H_2O$ (55.5 mM)	Microwave reactor	120	10 min	13.5
3 [171]	Corn stover (20%)	Glucan (35.1%), xylan (22.2.%), lignin (16.2%)	GVL-water (80:20)	H_2SO_4 (0.015 mM)	Flow-through reactor	117	1 h	5.4
4 [172]	Corn stalk (75 mg)	Glucan (31.6.%), xylan (20.5 wt.%), lignin (22.8 wt.%)	GVL (10 mL)	Sulfonated graphene (35 mg)	Autoclave	180	50 min	19.5
5 [173]	Sugarcane bagasse (1.2 g)	Cellulose 43.4 wt.%, hemi-celluloses 21.7 wt.%, lignin 20.3 wt.%,	Subcritical water	–	Batch-type reactor	270	10 min	3.1
6 [174]	Corncob (25 mg)	Cellulose 30.2 wt.%, xylan 31.7 wt.%.	GVL (2 mL)	SPTPA (6 mg)	Oil bath	175	85 min	32.3
7 [84]	Bread waste (5 wt./v%)	Starch (67.6%)	THF/water (1:1 v/v) (20 mL)	$SnCl_4 \cdot 5H_2O$ (55.5 mM)	Microwave reactor	140	120 min	8

Biphasic reaction system									
Entry [Ref.]	**Substrate (conc)**	**Monomer**	**Phase I**	**Phase II**	**Catalyst**	**Heating**	**Temperature (°C)**	**Time**	**HMF yield (%)**
8 [175]	Corn stover (0.05 g)	Glucan (35.1%), xylan (22.2.%), lignin (16.2%)	Water (1 mL)-NaCl (0.35 g)	THF (3 mL)	$AlCl_3{\cdot}6H_2O$ (0.1 mmol)	Microwave reactor	180	30 min	19
Model carbohydrates									
Monophasic reaction systems									
Entry [Ref.]	**Substrate**	**Green solvent**			**Catalyst**	**Heating**	**Temperature (°C)**	**Time**	**HMF yield (%)**
9 [176]	Fructose (9%)	Isopropyl alcohol (5 mL)			HCl (10 mol %)	Oil bath	120	2 h	87
10 [176]	Fructose (9%)	Isopropyl alcohol (5 mL)			Amberlyst 15 (20 mol%)	Oil bath	120	4 h	45
11 [177]	Fructose (5%)	CO_2-water-isopropanol, 1:1 (60 mL)			–	Parr reactor	190	2 h	67.1
12 [177]	Fructose (5%)	CO_2-DMSO-isopropanol, 1:1 (60 mL)			–	Parr reactor	190	2 h	12.5
13 [177]	Glucose (5%)	CO_2-water-isopropanol, 1:1 (60 mL)			–	Parr reactor	200	2.5 h	35.7
14 [177]	Sucrose (5%)	CO_2-water-isopropanol, 1:1 (60 mL)			–	Parr reactor	200	2.5 h	50.5
15 [177]	Inulin (5%)	CO_2-water-isopropanol, 1:1 (60 mL)			–	Parr reactor	190	2.5 h	46.6
16 [178]	Fructose (1.88 g)	GVL (10 mL)			HCl (0.853 mM)	Microwave reactor	130	10 min	66.5

(continued)

Table 6.8 (continued)

17 [172]	Xylose (60 mg)	GVL (10 mL)	Sulfonated graphene (28 mg)	Autoclave	190	40 min	9600%
18 [179]	Glucose (2%)	Water-GVL (1.5 g): (1:9)	$AlCl_3$ (5.0 mmol/L) HCl (3.17 mmol/L)		170	20 min	58.7
19 [179]	Glucose (2%)	Water-THF (1.5 g): (1:9)	$AlCl_3$ (5.0 mmol/L) HCl (3.17 mmol/L)		170	20 min	50.4
20 [179]	Fructose (2%)	GVL (1.5 g)-water (0.167 g)	Amb-70 (0.05 g)		130	9 min	71.2
21 [179]	Fructose (2%)	THF (1.5 g)-water (0.167 g)	Amb-70 (0.05 g)		130	10 min	77.3
22 [179]	Glucose (2%)	GVL (1.5 g)-water (0.167 g)	Amb-70 (0.1 g)		130	30 min	29.4
23 [179]	Glucose (2%)	THF (1.5 g)-water (0.167 g)	Amb-70 (0.1 g)		130	50 min	22.5
24 [179]	Glucose (2%)	GVL (1.5 g)-water (0.167 g)	Sn-β (0.05 g), Amb-70 (0.05 g)		130	20 min	58.9
25 [179]	Glucose (2%)	THF (1.5 g)-water (0.167 g)	Sn-β (0.05 g), Amb-70 (0.05 g)		130	30 min	63

Entry [Ref.]	Substrate (Conc)	Phase I	Phase II	Catalyst	Heating	Temperature (°C)	Time	HMF Yield (%)
Biphasic reaction system								
26 [179]	Fructose (5%)	Water-NaCl (1.5 g)	GVL (3.0 g)	$AlCl_3$ (5.0 mmol/L) HCl (3.17 mmol/L)	Biphasic reactor	150	20 min	94
27 [179]	Glucose (5%)	Water-NaCl (1.5 g)	GVL (3.0 g)	$AlCl_3$ (5.0 mmol/L) HCl (3.17 mmol/L)	Biphasic reactor	170	20 min	94
28 [175]	Glucose (0.25 mmol)	Water (1 mL)-NaCl (0.35 g)	THF (3 mL)	$AlCl_3 \cdot 6H_2O$ (0.1 mmol)	Microwave reactor	160	10 min	61
29 [175]	Glucose (0.25 mmol)	Water (1 mL)	THF (3 mL)	$AlCl_3 \cdot 6H_2O$ (0.1 mmol)	Microwave reactor	160	10 min	52
30 [175]	Cellobiose (0.25 mmol)	Water (1 mL)-NaCl (0.35 g)	THF (3 mL)	$AlCl_3 \cdot 6H_2O$ (0.1 mmol)	Microwave reactor	160	10 min	58

[176]. It can be inferred from Table 6.7 that the HMF yield was lowered from 87% to 45% when Amberlyst-15 was used as the catalyst instead of HCl. Being more acidic in nature, Amberlyst-15 resulted in the formation of more ether or acetylated by-products; thus, lowering the HMF yield and selectivity. The HMF yield was found to decrease with each recycling step due to the partial loss of HCl during the evaporation process. The use of Amberlyst-15 in place of HCl in the same reaction system accounted for better recyclability, yet the HMF yield and selectivity were low [176].

This reaction system was modified by Lin et al. [177] where they tested the use of CO_2-water-isopropanol on various carbohydrates. CO_2 imparts an acidic environment by forming carbonic acid and releasing H^+ ions when dissolved in water. Addition of isopropanol to the CO_2-water system increased the fructofuranose composition and greatly suppressed the pathway for LA and humins formation. Meanwhile, the hydroxyl group present in isopropanol binds with the H^+ ion, increasing the pH of the system is responsible for the increase in pH of the system. This hindrance in ionization activity of carbonic acid led to a decrease in HMF yield (87–67.14%) in CO_2-water-isopropanol system [177]. Replacing water with DMSO gave inferior HMF yields (12.54%) as the hydrogen ions could not be produced without water [177]. The HMF yield (67.14%) obtained from fructose was comparatively lower than for the previous study (87%).

Other biomass-derived solvents like tetrahydrofuran (THF) and γ-lactones have also been used for the conversion of carbohydrates and biomass residues to HMF. THF is the derivatized product of furfural, formed by dehydration of xylose [181]. Addition of salts like NaCl reduces the miscibility of THF/H_2O and makes the system biphasic in nature. Such a system generated a HMF yield of 61% from fructose (using NaCl) [175] and 43–53% using $NaHSO_4$ as salt [182]. However, having a low dipole moment and dielectric constant, THF witnessed low glucose conversion and favoured undesirable side reactions [84]. An improved HMF yield (49.3%) was obtained at 190 °C and 60 min with sodium bisulfate in water/THF for corn cob [114] [78]. Also, LA yield never exceeded more than 5%, which suggests that the reaction system is efficient in suppressing the rehydration of HMF. The H^+ ion released by $NaHSO_4$ in aqueous phase catalyse the hydrolysis of polysaccharides to monosaccharides and form HMF and furfurals. However, the main product obtained by Cao et al. was LA, which was the result of the inefficiency of H_2SO_4 to supress water-THF solubility, and hence it behaved as a monophasic system where HMF was rehydrated to LA [183]. Using water/THF-$AlCl_3{\cdot}6H_2O$, an HMF yield of 61% was achieved in 10 min at 160 °C using microwave heating [175].

γ-Valerolactone (GVL) is the hydrogenation product of LA which itself is derived from the rehydration of HMF. Being a platform chemical itself, it is non-toxic and can easily be separated by distillation, hence simplifying the downstream processing. Polypropylene carbonate (PC), in view of its low toxicity, high biodegradability and in situ CO_2 production, is also recognised as a green solvent [134]. The results evidenced a yield of 20% in 7.5 and 20 min for PC/H_2O and GVL/H_2O (1:1 v/v) solvent systems, respectively [134]. Recently, PC/H_2O and GVL/H_2O were examined for the conversion of bread waste to HMF over $SnCl_4$ using microwave heating (120 °C) [134]. A more recent study showed that the use of sulphonated graphene in GVL system at 180 °C for 50 min can yield 19.5% HMF from corn stalk [172]. Notably, a synergistic effect of Lewis ($AlCl_3$) and Brønsted (HCl) acids in a salt-free reaction system can produce high yields of HMF from glucose. The use of bio-based solvents in the monophasic system (GVL-water/THF-water) alleviates the problems that are encountered in upscaling biphasic reactors. However, the HMF yield of the system is reduced from 94% to 58.74% when no salt is added to the reaction system. GVL can extract HCl from the aqueous phase, bearing higher process cost as the organic layer will need an additional neutralization step [179]. The replacement of homogeneous catalyst with solid Brønsted acid (Amb-70) for direct dehydration of glucose in GVL, as well as THF, lowers the selectivity (<35%) (Gallo et al.) [177]. The combination of Sn-β/Amb-70 increased the selectivity of the system, facilitating isomerization of glucose to fructose. An additional advantage of using GVL as the solvent medium is its capability to dissolve humins, thus preventing the clogging due to deposition of humins on the surface of the solid catalyst. Water at low concentration (10%) is beneficial for the monophasic reaction system as it increases the solubility of sugar. Under similar reaction conditions, using GVL as the extracting solvent in a biphasic system gave comparable yields (61.6%) to those obtained from 2-sec-butylphenol (62%) [184].

However, the type of solvent used has a major impact on the catalytic activity of metal chlorides. For instance, the catalytic activity of $SnCl_4$ in the presence of PC/H_2O, GVL/H_2O, and acetone/H_2O was compared for the conversion of bread wastes into HMF [134]. The presence of Lewis acid sites (Sn^{4+}) was more in case of green solvents than acetone/water mixture wherein Sn^{4+} existed as inactive SnO_2 particles. Amongst all solvent systems, PC/H_2O had a superior solvent effect due to its pressure development via the liberation of CO_2. The study showed the synergistic effect of the solvent and the catalytic system on the HMF yield. Quantifiable amounts of fructose (2–13%) and glucose (9–62%) were detected in both solvent systems, which suggested that HMF was synthesized by the glucose to fructose isomerization pathway [134]. A modest improvement in HMF yield (61% versus 52%) has been observed by the addition of NaCl in the biphasic system. XRD studies have shown that GVL plays a role in disrupting the crystallinity of cellulose, thus availing more of the reactive amorphous part for the reaction [172].

6.10 Catalytic System for Oxidation of HMF into FDCA

The synthesis of FDCA was first carried out using mucic acid in the presence of a strong acid catalyst. Fittig and Heinzelman used 48% aqueous HBr as a catalyst and solvent for the dehydration of mucic acid to FDCA. However, the method used required severe reaction conditions (highly acidic environment, longer reaction time

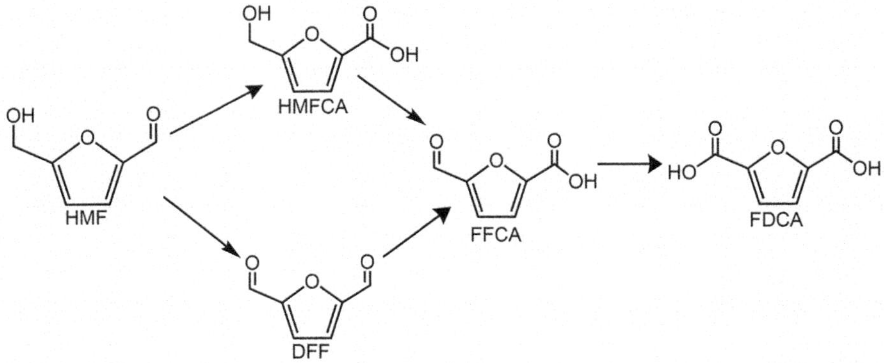

Scheme 6.2 Reaction mechanism of HMF oxidation to FDCA. (Adapted and modified with permission from [187], Copyright © 2024 Elsevier B.V.)

(>20 h), high temperature (120 °C), and nonselective yield <50%. Moreover, mucic acid itself is very costly, which makes the process implementation uncertain at a large scale [185]. The other methods involve the oxidation of furfurals or HMF in the presence of inorganic oxidants. The process of FDCA synthesis from furfural is quite intricate, and the yield is typically low. The simplest of all is the oxidation of HMF to FDCA using stoichiometric oxidants like $KMnO_4$ [186]. The drawbacks associated with the process are the high cost of oxidants, low selectivity, and generation of toxic wastes. Currently, most researchers follow the green chemistry approach to establish an environmentally benign synthesis route for FDCA. The oxidation of HMF followed the Mars-van Krevelen mechanism for the HMF oxidation into FDCA, as shown in Scheme 6.2. The intermediates formed during the reaction were HMFCA, DFF, and FFCA. The oxidation of HMF to FDCA conversion involves the simultaneous progression of the two competing reactions via (1) –CH_2OH oxidation to –CHO and (2) –CHO oxidation to –COOH [187].

6.10.1 Electrocatalytic Synthesis of FDCA from HMF

The electrochemical potential between electrodes is the main driving force for the electrochemical oxidation processes. Due to the electrode‘s electrochemical potential, electrons are used in place of oxidant in the electrochemical oxidation process. The process is initiated by electron transfer which eliminates the need for O_2 and other chemical oxidants [188]. The mechanism involves substrate oxidation by the electrode’s electrochemical potential. HMF is electrochemically oxidized to FDCA by two primary mechanisms: the “OH*-participated in oxidation mechanism” and the “electrochemical-chemical (E-C) participated in oxidation mechanism.” The high valence intermediate activated species are required for the E-C process to function. To be more precise, high valence species are formed when the catalyst molecules combine with OH^- (formed in presence of H_2O) at a specific potential. High valence intermediates then activate the substrate molecules to intermediate products such as DFF, HMFCA, FFCA, and finally into FDCA; after which they are reduced back to their initial state. On the other hand, the valence state transition of the catalyst is not included in the OH* process; instead, the provided voltage mainly promotes the adsorption and electron transfer of OH– and the consequent substrate dehydrogenation. At a specific potential, OH^- is adsorbed on the catalyst to generate OH* active species. Subsequently, OH* breaks the O–H and C–H bonds of the adsorbed HMF substrate molecule to form DFF. Following the stage of OH* activation, water molecules attack DFF and break the C–O bond, resulting in the production of an intermediate that is then transferred to FFCA. After repeating the OH* activation and water addition process, the FDCA product can finally be produced. Similarly, path 2 follows these next five steps: HMF addition leads to intermediate 1 activation, which then leads to HMFCA activation, after which FFCA addition occurs, leading to intermediate 2 activation, and ultimately FDCA is produced [188]. The selectivity of the surface reaction can also be monitored by changing the oxidation potential and faradaic current [189]. The first report on the electrochemical oxidation of HMF to FDCA came in 1995. They used an H-shaped cell and the oxidation of HMF occurred at an anode made up of nick oxide/ hydroxide. The yield obtained was 71% after 4 h in 1 M NaOH solution at the current density of 0.016 A/cm^2. Strasser and co-workers [190] studied the electrochemical oxidation of HMF using a Pt electrode in an alkaline solution (pH 10) and reported that HMF oxidation competed with the water oxidation leading to a very negligible yield of FDCA (<1%). Later, Li and co-workers used carbon-supported noble-metal catalysts such as Pd/C and Au/C. The reaction mechanism proposed by them is shown in Scheme 6.2. The oxidation of HMF in the presence of Pd/C followed two competitive routes depending on the electrode potential. The intermediates formed during the reaction were HMFCA, DFF, and FFCA. The oxidation of –CHO group occurred at a slower rate until the electrode potential was raised. However, no DFF was detected in Au/C catalytic system. The oxidation of –CHO was much easier than –OH group over Au/C catalyst which favoured HMFCA formation. However, the oxidation of the –OH group in HMFCA to the –COOH group in FDCA required a significantly greater potential. HMFCA was produced as the intermediate by the considerably quicker oxidation of the aldehyde group in HMF than the alcohol group over Au/C catalyst. However, high electrode voltage was needed to further oxidize the alcohol group in HMFCA to FDCA. In terms of HMF oxidation over Pd/C catalyst, two competitive pathways that were dependent on electrode voltage were found for HMF oxidation into the intermediate FFCA. They also studied the bimetallic catalyst $PdAu_2/C$ and interestingly obtained deeply oxidized products (83% yield of FDCA) at a lower electrode potential (0.9 V).

6.10.2 Biocatalytic Method

Contrary to the chemical oxidation processes, biocatalytic oxidation is carried out at relatively milder reaction conditions. Mara and co-workers reported the use of lipase as a biocatalyst for DFF oxidation into FDCA [191]. The system was catalyzed by the formation of peracid with the addition of hydrogen peroxide (30% v/v). The reaction afforded a high yield of FDCA (>99%) and excellent selectivity (100%). However, the method was ineffective in HMF oxidation. Another group of researchers used chloroperoxidase from *Caldariomyces fumago* to test the biocatalytic effect of HMF oxidation to FDCA. The method was restricted to its poor yields of FDCA (60–75%) as well as HMFCA (25–40%) [192]. Later, *Pseudomonas putida* S12 was used in a fermentative procedure that was investigated for HMF oxidation. The method resulted in 99.4% pure dry powder yield from the culture broth followed by acid precipitation which resulted in 76% recovery, with subsequent extraction using tetrahydrofuran. McKenna et al. created a one-pot tandem enzyme process to convert HMF to FDCA by employing aldehyde oxidase PaoABC and galactose oxidase M3-5 [193]. The study reported the highest substrate concentration (100 mM) for an enzyme-based process with 74% FDCA yield, at a neutral pH and ambient temperature. Apart from enzymes, whole-cell catalysis offers a promising pathway due to its environmentally friendly nature and the potential to operate under relatively mild conditions. Whole-cell catalysis refers to the use of entire living cells, usually microorganisms like bacteria, yeast, or fungi, as biocatalysts to convert substrates into desired products. One such study includes using filamentous fungus as biocatalysts in the oxidation process for the conversion of HMF into FDCA. Whole-cell biocatalysts, comprising five different species of fungi (*P. chrysosporium*, *P. purpurogenum*, *A. niger*, *T. reesei*, and *R. oryzae*), were employed in the oxidation process, with two species showing superior HMF conversion and product accumulation. The nitroxyl radical (2,2,6,6-tetramethylpiperidin-1-yl)oxidanyl (TEMPO), which favours the oxidation of –OH and –CHO functional groups, including the hydroxymethyl group of HMFCA, was combined with the fungal enzyme laccase in this chemo-biocatalytic system, which was further evaluated for use in relation to these two species (*A. niger* and *T. reesei*). Under optimal conditions, 88% FDCA was obtained with 1 g/L of *T. reesei* whole cells combined with 20% TEMPO and 2.5 mM laccase after 80 h [194]. Another pioneering study included the use of *Aspergillus flavus* for detoxification of HMF producing 0.83 g/L FDCA using HMF concentration of 1 g/L within 60 h. They found that using yeast extract media with mineral salt led to a more complete transformation of HMF into HMFCA and FDCA. It was observed that the fungus could grow and reach its reproductive stage on a solid medium with low HMF concentrations. Increased cell concentrations reduced the amount of oxygen present and deteriorated FDCA production, while HMF concentrations (>1 g/L) was shown to have hazardous impact on the organism [195].

6.10.3 Chemical Oxidants

Despite the advantages of electrochemical and biocatalytic conversion of HMF into FDCA, its synthesis at the industrial scale is mostly carried out by chemical oxidants classified into homogeneous and heterogeneous catalysts. A study used homogeneous catalyst system composed of $Co(OAc)_2$, $Mn(OAc)_2$, and HBr in an acetic acid environment for the aerobic HMF oxidation into FDCA. The yield of FDCA obtained was 60.9% in combination with another byproduct. A similar combination of catalyst $Co(OAc)_2/Zn(OAc)_2/NaBr$ was later tested for the aerobic HMF oxidation. That the authors reported that the method favored the oxidation of hydroxymethyl than aldehyde groups of HMF. Similar homogeneous catalysts were then employed for the aerobic HMF oxidation into FDCA [196]. In addition to oxygen, t-BuOOH was employed as an oxidant in the HMF-to-FDCA synthesis process. Riisanger and co-workers [197] used ACN with t-BuOOH, CuCl as catalysts, and LiBr as the additive in the system for HMF oxidation. The incorporation of additives gave a lower yield of FDCA from 45% to 43%. The recyclability of homogeneous catalysts and the separation of FDCA from metal salts present difficulties when employing them in the synthesis of FDCA. These problems can be solved by using heterogeneous catalysts, which are easily isolated from the reaction solution.

6.10.4 Noble Metal Catalysts

Noble metal catalysts are highly efficient for the oxidation of HMF to FDCA, including Au [198], Pd [199], and Pt [200]. The process typically necessitates the use of base additives and high temperatures. The use of bimetallic noble metals deposited onto carbon surfaces are also gaining impetus for the oxidation of HMF. For instance Pt, when used in combination with Pd, deposited onto the carbon surface and enhanced the catalytic activity of the system [201], as shown in Table 6.9. Under specific conditions (2.5 mL/s O_2 flow, 1.25 M NaOH, and 25 °C) a high FDCA yield of 99% was achieved in the short reaction time of 2 h with Pt-Pb/C catalyst, while the Pt/C catalyst resulted in an 81% FDCA yield at full HMF conversion. Besides Pt, Pd/C is also effective for the aerobic oxidation of HMF to FDCA (71% FDCA in 6 h), showing similar catalytic efficiency in comparison to Pt/C (79%) under specific conditions. Meanwhile, Au-containing bimetallic catalysts outperforms the monometallic Au catalysts for HMF oxidation to FDCA, due to the additional advantage of the tunable physical and chemical properties by varying the particle size, metal composition, and degree of mixing [215]. For instance, Au-Cu/TiO_2 exhibited greater catalytic activity and stability in comparison to Au/TiO_2 for HMF oxidation to FDCA, reason being that varying the molar ratios of Au/Cu via colloidal preparation increased the catalytic activity to almost the double of their monometallic Au counterparts [211].

Table 6.9 Results of the oxidation of HMF into FDCA over supported Pt catalysts

Catalyst	Oxidant	Base	Temperature (°C)	Time (h)	HMF concentration (%)	FDCA yield (%)	Ref.
Pt–Pb/C	1 bar O_2	1.25 M NaOH	25	2	100	99	[201]
Pt/C	1 bar O_2	1.25 M NaOH	25	2	100	81	[201]
Pt–Bi/C	40 bar air	2 equiv. of Na_2CO_3	100	6	100	>99	[200]
Pt/C	40 bar air	2 equiv. of Na_2CO_3	100	6	99	69	[200]
Pt–Bi/TiO_2	40 bar air	2 equiv. of Na_2CO_3	100	6	>99	99	[202]
Pt/TiO_2	40 bar air	2 equiv. of Na_2CO_3	100	6	90	84	[202]
Pt/RGO	1 atm O_2	5 equiv. of NaOH	25	24	100	84	[203]
Pt/C	690 kPa O_2	2 equiv. of NaOH	22	6	100	79	[198]
Pt/γ-$A1_2O_3$	0.2 bar partial O_2	pH = 9	60	6	100	99	[204]
Pt/γ-$A1_2O_3$	1 bar O_2	1 equiv. of Na_2CO_3	75	12	96	96	[205]
Pt/ZrO_2	1 bar O_2	1 equiv. of Na_2CO_3	75	12	100	94	[205]
Pt/C	1 bar O_2	1 equiv. of Na_2CO_3	75	12	100	89	[205]
Pt/TiO_2	1 bar O_2	1 equiv. of Na_2CO_3	75	12	96	2	[205]
Pt/CeO_2	1 bar O_2	1 equiv. of Na_2CO_3	75	12	100	8	[205]
Pt/$Ce_{0.8}Bi_{0.2}O_{2-\delta}$	1.0 MPa O_2	4 equiv. of NaOH	23	0.5	100	98	[206]
Pt/CeO_2	1.0 MPa O_2	4 equiv. of NaOH	23	0.5	100	20	[206]
Pt/PVP	1 bar O_2	No base	80	24	100	94	[207]
Pd/C	690 kPa O_2	2 equiv. of NaOH	23	6	100	71	[198]
Au/CeO_2	10 bar air	4 equiv. of NaOH	130	5	100	96	[199]
Au/TiO_2	10 bar air	4 equiv. of NaOH	130	8	100	84	[199]
Au/$Ce_{0.9}Bi_{0.1}O_{2-\delta}$	1 bar O_2	4 equiv. of NaOH	65	2	100	>99	[208]
Au/HY	0.3 atm O_2	5 equiv. of NaOH	60	6	>99	>99	[209]
Au/CeO_2	0.3 atm O_2	5 equiv. of NaOH	60	6	>99	73	[209]
Au/TiO_2	0.3 atm O_2	5 equiv. of NaOH	60	6	>99	85	[209]
Au/TiO_2	20 bar O_2	20 equiv. of NaOH	30	18	100	71	[210]
Au–Cu/TiO_2	10 bar O_2	4 equiv. of NaOH	95	4	100	99	[211]

Au_8–Pd_2/C	3 MPa O_2	2 equiv. of NaOH	60	2	>99	>99	[212]
Au/HT	1 atm O_2	No base	95	7	>99	>99	[212]
Au–Pd/CNT	0.5 MPa O_2	No base	100	12	100	94	[213]

In addition to active carbon, reduced graphene oxide (RGO) is also an effective support for metal deposition due to its abundance of functional sites. A study conducted with the use of Pt and Pd supported on RGO achieved complete HMF conversion, producing FDCA yields of 40.6% and 30.5%, respectively. Under identical conditions, Ru/RGO and Rh/RGO catalysts did not produce any FDCA, thus indicating that Pt was the most effective catalyst for converting HMF to FDCA [203]. Similar findings were reported in another study where Pt/C afforded 79% FDCA yield in 6 h in comparison to Pd/C catalyst with 71% FDCA yield [198]. Metal-oxide supports have also been extensively explored for HMF oxidation to FDCA. Several metal oxides including γ-Al_2O_3, ZrO_2, TiO_2, and CeO_2 have been used as support for the noble metals. Amongst all, the non-reducible oxides such as γ-Al_2O_3 and ZrO_2 performed better than reducible oxides (TiO_2 and CeO_2) due to the difference in oxygen storage capacities (OSC). For instance, Pt/Al_2O_3 and Pt/ZrO_2 exhibited higher catalytic performance resulting in 96% and 94% FDCA yield compared to Pt/TiO_2 and Pt/CeO_2 catalysts which resulted in low FDCA yield of 2% and 8%, respectively [205]. Additionally, supported Au catalysts have shown promising performance in the aerobic HMF oxidation to FDCA in aqueous medium. Au over TiO_2 and CeO_2 supports resulted in high FDCA yield of 99% in 8 h with 150 mol ratio of HMF/Au, 4 equiv. of NaOH and 10 bar of O_2. In contrast, Au supported on C and magnetic support (Fe_2O_3) gave lower FDCA yields of 44% and 15%, respectively. Subsequent studies confirmed that Au with CeO_2 outperforms the one with TiO_2 as the support as previously observed in catalytic alcohol oxidation [199]. Another study reported that using a mixed oxide support significantly increased the FDCA yield, for instance Au/$Ce_{1-x}Bi_xO_{2-\delta}$ improved the FDCA yield to 75% in comparison to Au/CeO_2 where the FDCA yield was limited to 39% [199]. It has been observed that basic environment significantly influences the HMF oxidation to FDCA. The addition of NaOH (20 equiv.) improved the catalytic performance of Au/TiO_2 catalyst to 71% FDCA, while a lower base concentration (<5 equiv.) resulted in drop in FDCA yield to 13% due to higher HMFCA formation which caused catalyst deactivation [212]. The absence of base leads to FDCA deposition on the surface of the catalyst due to low solubility, hence the reaction typically involves excess base which necessitates post-reaction acid neutralization, which is environmentally and economically disadvantageous [198]. Therefore, base-free oxidation of HMF to FDCA represents a more sustainable approach. A base free approach was developed using hydrotalcite-supported Au catalyst (Au/HT), achieving a 99% FDCA yield at 95 °C in 7 h under 1 bar O_2 in aqueous medium [212].

The studies on the oxidation of HMF to FDCA have reported the use of several metal catalysts and nanoparticles, however they require elevated temperature, high pressure and additives which reduces the sustainability of the process. Aerobic Pt nanoparticles [207] and Au catalytic systems [216] have been developed but they require a high pH environment which can cause the decomposition of HMF. The product obtained under the base environment is usually the salt form of FDCA and cannot be directly applied for polymer preparation by the polymer industry. Additionally, the conventional approach of producing FDCA by HMF oxidation often involves high temperatures and pressures, as well as a range of catalytic systems in high pH media. The disadvantage of employing homogenous bases is that the processes need additional acidification and purification before using FDCA, which increases the risk of reactor corrosion and produces a significant amount of mineral waste, thereby having negative impact on the environment [217]. To eliminate this, base free conversion of HMF to FDCA was investigated using bench flow reactor with Pt/ZrO_2 for the base free HMF to FDCA conversion, however very low HMF concentrations was needed to avoid FDCA accumulation on to the surface of the catalyst [218]. Yi et al. demonstrated an autolytic system using FDCA as the catalytic for fructose dehydration into HMF. The HMF produced was converted into FDCA over Ru/C catalyst in base-free aqueous environment affording 53% overall FDCA yield [219]. The reactions were conducted in Parr reactor, and very low HMF concentration (1 mmol) was used. Under optimized conditions (0.2 MPa O_2, 120 °C, and 10 h), 88% FDCA yield was achieved in a white precipitate form. The kinetic study indicated that the conversion of HMF to FDCA proceeds via formation of DFF and FFCA intermediates. Recently, carbon nanotubes supported Au-Pt alloy catalyst was used for the same purpose and were in corroboration to the one carried using Au/Pd-CNT system [213].

6.10.5 Non-noble Metal Catalyst

In general, non-noble metals have lower catalytic activity than noble metals, but they are more accessible and less costly. Reaction operational parameters are typically harsher operating conditions to make up for this lack of performance. Like their classical counterparts, non-noble metals frequently need a base addition. Recent studies, however, have showed that this performance difference was largely closed due to their remarkable catalytic activity for HMF oxidation [220]. Manganese-based catalysts proved to be the most effective across a range of catalytic systems because of the metal's various oxidation states, large oxygen storage capacity, and quicker oxygen absorption and reduction. When Mn was utilized alone or in combination with other metals, it produced outstanding results for both FDCA selectivity and HMF conversion. With a 93% yield of FDCA, Hayashi et al. proved that the most efficient crystalline form of manganese oxide was β-MnO_2 [221]. Moreover, the shape of β-MnO_2 can also have an impact on the catalytic activity. The template-free low-temperature crystallization process produced mesoporous β-MnO_2 nanoparticles with a high surface area that were found to have higher selectivity than β-MnO_2 made using the hydrothermal method [222]. Additionally, Bao et al. discovered that morphology is crucial, thereby they created holey MnO_2 nanoflakes to increase surface area and get remarkable outcomes in HMF conversion and FDCA selectivity [223]. To speed up the reoxidation of Mn, manganese oxides can be used in combination with other metal oxides. For instance, Han et al. combined Mn and Ce in 6:1 ratio which facilitated 98.5% HMF conversion and 91% FDCA selectivity under the reaction time of 12 h [224]. Cu in $CuMn_2O_4$ spinel oxide produced FDCA with a 92.1% yield, demonstrating the synergistic action of the two metals; while reducing the side reactions and accelerating the rate-limiting step in FDCA oxidation process [225]. Out of all the combinations that have been reported thus far, Mn and Co have proven to be the most beneficial. Liu et al. reported the solid-state grinding method used to prepare vitamin C-assisted mesoporous Mn–Co spinel oxides, which were then used in air to oxidize HMF. This strategy was highly inventive, completely sustainable, and can be scaled up to an industrial level. The technique states that vitamin C raised the Mn–Co oxide's oxygen vacancy (O_V) concentration, which in turn decreased the Mn–O bond intensity near O_V and increased lattice oxygen (O_L) reactivity. This led to an extremely effective HMF oxidation with air. Additionally, the catalyst was regenerated up to six times while maintaining its activity following an easy 2-day regeneration at 200 °C [226]. In a nutshell non-noble metal-based catalysts are being developed; the most promising ones are Mn and Co, which enable excellent HMF conversions and higher FDCA selectivity even under extremely mild reaction conditions.

6.11 Process Integration for Conversion of Agro-Food Waste to FDCA

The oxidation of HMF to FDCA has limitations for industrial applications due to the issues related to cost, availability, and stability of the substrate. The direct conversion of biomass to FDCA is always a promising and cost-effective option, and there have been several studies conducted on the direct conversion of carbohydrates for FDCA synthesis using feedstocks such as, cellulose, glucose, fructose, and lignocellulosic biomass [155, 213, 227–235]. This process is referred to as a "one-pot synthesis" (as shown in Scheme 6.3) as is often executed using a combination of acidic and metal site multifunctional catalysts.

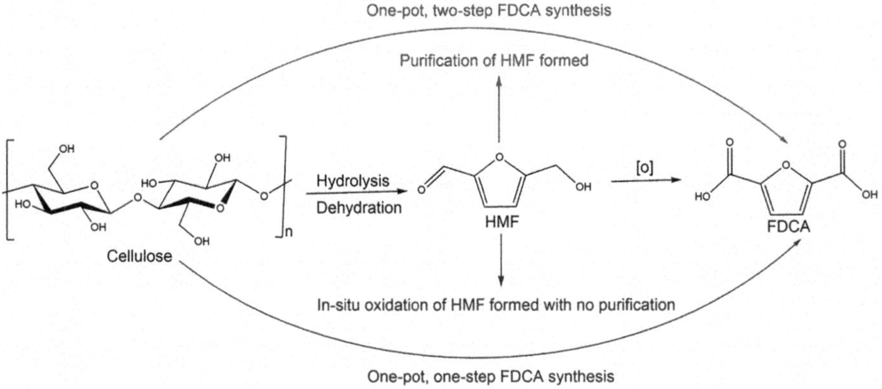

Scheme 6.3 Schematic representation of one-pot FDCA synthesis. (Adapted and modified with permission from [236], Copyright © 2023 American Chemical Society)

Few attempts have been made to convert fructose to FDCA in a one-pot, one-step reaction system. Kroger et al. carried out the direct conversion of fructose as a substrate to FDCA via acid-catalyzed dehydration and in situ oxidation. Two-system approach was applied where a membrane was used for phase separation and an oxidation catalyst was encapsulated in silicone beads. The biphasic reaction media was used constituting of water & solid acid catalyst and MIBK with solid oxidation catalyst. The maximum FDCA selectivity of 50% was achieved with 25% FDCA yield in a log reaction time of 7 days at 80 °C. This reaction system in presence of MIBK gave DFF as the main intermediate product during HMF oxidation. The rate of reaction was dependent on the membrane diffusion parameters, and water was added in the system to facilitate the oxidation of aldehyde to form FDCA. This system was limited for scaleup due to low FDCA yield and the difficulty to purify FDCA from the by-products formed [228]. Another study described the conversion of fructose into FDCA in a one-pot setup using t-BuOOH aqueous solution, obtaining a total FDCA yield of 59.8%. HMF was first produced from fructose over the Fe_3O_4@SiO_2-SO_3H catalyst, with HMF reaching a maximum yield of 93.1%; then, the Fe_3O_4@SiO_2-SO_3H catalyst was magnetically collected, and the nano-Fe_3O_4-CoOx was added in a solution of 70% aqueous t-BuOOH. The catalyst from the first step was recovered using an external magnet, and the catalyst for the oxidation step was added afterwards without the need for an additional purification step [231].

Ribeiro and Schuchardt used a bifunctional catalyst for redox reaction composed of cobalt acetylacetonate encapsulated in sol-gel silica. The study investigated the conversion of fructose to FDCA in one-pot, one-step process. The results demonstrated a 72% conversion of fructose with 99% selectivity for FDCA formation, achieved under conditions of 2.0 MPa air pressure and 160 °C [229]. Zhang et al. showed that the one-pot fructose and glucose conversion into FDCA within a tri-phasic system composed of tetraethylammonium bromide (TEAB) or water-methyl isobutyl ketone (MIBK)-water was possible. Phase I involved converting sugars into HMF using Amberlyst-15/$CrCl_3$ as the catalyst and either TEAB or water as the solvent. The HMF produced was extracted with MIBK (Phase II) and then transferred to the subsequent phase (water), where it was converted to FDCA using an Au_8Pd_2/HT catalyst. This process achieved FDCA yields of 78% from fructose-derived HMF and 50% from glucose-derived HMF [235].

Guangyu et al. reported a 97.1% HMF yield from fructose with 0.4 g Amberlyst-15, which was then separated by a straightforward filtration procedure. After collecting the crude HMF obtained in the form of filtrate, the second reaction was carried out in aqueous medium with Pt/C catalyst in DMSO. This reaction produced a 91% yield of FDCA [227]. Another study tested Amberlyst-15 as the solid catalyst for dehydration of fructose. The fructose-derived crude HMF was obtained with Amberlyst-15 in DMSO/H_2O solvent system, followed by direct HMF oxidation process over Ru/C without further purification, accomplishing an overall FDCA yield of 65% [237]. Yang et al. synthesized a magnetic catalyst with sulfonic acid sites (Fe_3O_4–RGO–SO_3H) for the dehydration step to achieve 81% HMF and a mixed oxide catalyst ($ZnFe_{1.65}Ru_{0.35}O_4$) for the HMF oxidation step to resulting in

70% FDCA. Besides, the use of magnetic catalysts has gained impetus due to its easy recovery from the reaction media using an external magnet [233]. Another study focused on one-pot, one step process wherein FDCA acted as the acid catalyst to facilitate the fructose dehydration into HMF. Subsequently, the crude HMF was oxidized in the presence of the Ru/C catalyst within the 1,4-dioxane/H_2O mixture at reaction conditions of 4 MPa O_2, 2 equiv. $NaHCO_3$, 3 h reaction time, and 140 °C temperature, yielding 79% FDCA. This system highlights the effectiveness of the catalytic system and its scale-up potential [232]. Base-free FDCA synthesis is carried out in another study using heterogeneous catalysts composed of ruthenium and manganese dioxide (Ru/MnO_2). This catalyst composition is notable for its high yield and selectivity for FDCA production. It outperforms conventional catalysts like Pt (Pt/C) and Ru (Ru/C) deposited onto carbon support when used under identical reaction conditions. The process starts with the fructose dehydration to produce 84% HMF in an acetone/water solvent system using a sulfonated porous organic polymer catalyst. Following the formation of HMF, the reaction proceeds with the oxidation over the Ru/MnO_2 catalyst yielding 92% FDCA. This reaction eliminates the use of base, which simplifies the process and avoids the use of mineral acids, including HCl and H_2SO_4, which are widely used for fructose acid-catalyzed dehydration reaction. Using acetone/H_2O solvent system improved the solubility of FDCA, thus eliminating the need of base such as KOH, NaOH, Na_2CO_3 or K_2CO_3. Moreover, scaling up the reaction with the fructose weight to 20 g was successfully demonstrated in this study [238]. The direct conversion of rice straw into FDCA was carried out using dilute HCl with ultrasonic pretreatment. This pretreatment method disrupts the lignocellulosic matrix, which makes the subsequent catalytic step more effective. The HMF was subsequently oxidized using MnO_2 catalyst with yield reaching up to 52.1% under optimal conditions: 1.9 g straw, 0.55 g, 34 min ultrasonication, 192.5 °C reaction temperature, and 56.6 mM HCl concentration. Moreover, the effect of heavy metals (Fe, Mn, Pd, and Cu) present in raw biomass was tested, showing that the presence of heavy metals favoured the FDCA production [239]. Another study focused on a bimetallic Cr-Mn catalyst supported on a zeolite (USY) [155]. Mn and Cr cooperation improved oxygen mobility, which raised FDCA production. Additionally, the addition of $KHCO_3$ was positively impacted the oxidation of HMF due to its ability to dissolve FDCA as a salt, thus preventing it from entering the catalyst's active sites. Under optimal conditions, 67% FDCA yield was obtained. Finally, raw biomass was examined for the detection of heavy metals, and it was shown that they might function as Lewis's acid sites, increasing the catalytic activity [155].

Recently, FDCA was produced in a one-pot, one step process using whey permeate powder (WPP) in presence of magnetic catalyst in a base-free system made of earth-abundant metals (Mn_1Fe_1). Ozone and t-BuOOH were used as oxidants in a GVL-water solvent solution to carry out the in-situ oxidation of HMF. When HMF was subjected to a 3 h ozonation, the FDCA yield increased from 39.7% (in the absence of ozonation) to 60% at 130 °C for 2 h. The same catalytic system when tested on WPP afforded an FDCA yield of 14.6% in a reaction time of 120 min at 130 °C [187]. Owing to the excellent catalytic activity of Au catalyst, selective oxidation of mixed substrate (HMF, glucose, fructose, LA, and FA) over Au/ZrO_2 was conducted in aqueous medium using air as oxidant. The by-products formed during FDCA synthesis has been examined in this study. The formation of LA adversely affected FDCA production by causing catalyst poisoning. The direct oxidation of impure HMF in presence of glucose gave high FDCA yield (74%) at the reaction condition of 125 °C, 8 equiv. of NaOH, 30 bar air within 5 h [240]. ILs have been investigated for FDCA synthesis by following two distinct procedures using either glucose or fructose as feedstocks. Since a prior isomerization step is required for glucose, the fructose reaction produced HMF yields higher than 85%, however, the glucose dehydration process achieved a yield 55% at low concentrations. $CrCl_3.6H_2O$ was employed as the catalyst (7% loading) for the reaction that began with glucose (2% loading) and conducted for 30 min at 120 °C. [Bmim]Cl was chosen as the solvent since it has been observed that this was suitable to be worked with substrates like cellulose and sugars. This process was efficient in production of dialkylimidazolium furan carboxylate ([dbmim]-FDCA) and can easily be separated as the final product by ethanol addition as an antisolvent. Pure FDCA can be formed by treating [dbmim]-FDCA with acid, and the ionic liquid can be recovered. The highest FDCA yield obtained in this process was >98% from HMF and 80% from sugars [241]. Sucrose conversion to FDCA was obtained using H_2SO_4 as catalyst in a tubular reactor running at 200 °C and 25 pressure. Herein, HMF was initially obtained by hydrolyzing and dehydrating sucrose using H_2SO_4 in a continuous reactor. This resulted in a 34% yield in 10 min of residence time. In the second stage, NaOH (2.5 M) was added to the reaction solution to achieve a total volume of 10 mL. The reaction solution was then directly oxidized to FDCA over an Au/ZrO_2 catalyst, achieving an FDCA yield of 84% from sucrose derived HMF. In terms of sucrose, this yield translates to an overall yield of 29% [230].

Furthermore, biphasic H_2O-dioxane solvent system was evaluated for FDCA synthesis using fructose as the starting substrate. Fructose was dehydrated to HMF in presence of TmaCl as a phase modifier and CH_3SO_3H as an acid catalyst. The application of TmaCl greatly increased the dehydration of sugars and HMF yield, affording 90.6% HMF. This biphasic-solvent system's demonstrated outstanding recyclability and facilitated 90% HMF separation in the organic phase. The isolated HMF dioxane solution from fructose dehydration was employed straight away as a starting solution to produce FDCA in presence of Ru/C with no HMF purification or alkali addition. This solvent system has high FDCA solubility, thereby, resulting in 90% FDCA with over 98% purity after distillation [242].

There are certain disadvantages to the one-pot, one step conversion system for carbohydrate to FDCA synthesis. Firstly, the acidic residues and impurities such as humins in the sugar derived HMF can hinder the catalytic activity during the oxidation process. Hence, the separation and purification of HMF becomes an essential step before the oxidation step. Secondly, the multistep process is capital and energy intensive which makes the price of FDCA less competitive than its petroleum counterparts [4]. The one-pot two-step process involves two reactions including fructose dehydration to HMF over an acid catalyst, followed by the removal of humins, catalyst or solvent. A study reported the conversion of fructose to FDCA, where dehydration was carried out using isopropanol as the solvent and HCl as the catalyst. Isopropanol was evaporated after the reaction, following by the purification and the oxidation over Au/HT catalyst, resulting in 98% FDCA yield in 7 h [234]. The same method when tested with Jerusalem artichoke as the feedstock afforded 52% FDCA yield. Another study reported FDCA production using rice straw, where the substrate was subjected to acidic DES mixture composed of ChCl and p-TSA. Post-reaction, the mixture was extracted with ethyl acetate to obtain HMF, followed by its oxidation to FDCA (98.8% in 5 h) using Ru@S-GO catalyst [243]. Inulin, an oligomer of fructose, with one terminal glucose unit and defined by β-(2–1) linkages between the fructose units, has also been explored for FDCA synthesis [36]. Using a Pt/C catalyst, inulin was oxidized to FDCA in a single pot in two stages, yielding a high 70% yield. The first stage involved the intermediate evaporation of the organic solvent (MeCN), while the second stage involved the addition of $NaHCO_3$. It should be noted that the FDCA yield enhanced from 52% to 70%, resulting in complete HMF and intermediates (HMFCA and FFCA) conversion [244].

Moreover, a study combined the chemo and bio-catalysed processes for direct conversion of biomass into FDCA. Fructose was converted to crude HMF by chemo catalysis employing recyclable Amberlyst IR-120 as a catalyst with >98% purity. This was followed by using bacterial strain *Klebsiella oxytoca* to convert 98% of HMF into 2667 mg/L of FDCA within 72 h under optimum conditions of pH 7, 37 °C, and 2 g whole cells [245]. Apart from using biomass as substrate, a study involves the carboxylation of 2-furoic acid derived from biomass with CO_2. Conventionally, this reaction has been catalysed by cesium carbonate (Cs_2CO_3), which is effective but costly due to the scarcity of cesium. To overcome this issue and reduce costs, the study explores the use of potassium carbonate (K_2CO_3) to facilitate carboxylation of furoic acid and CO_2. The usage of K_2CO_3 makes the process more economical yielding 61.6% FDCA at 330 °C of temperature, 2.0 MPa of pressure, 1.5 h of reaction time, 1.0 mmol of 2-furoic acid, and 1.25 mmol of K_2CO_3 [246].

6.12 Conclusions and Future Outlook

The utilization of biomass as feedstock has provided alternative routes for FDCA production. To date, there are limited studies on the biomass conversion into FDCA, while HMF has been extensively studied as the starting material. Interactions between biomass feedstock and solid heterogeneous catalysts pose challenges to the efficiency of the catalytic process for the direct conversion of biomass into FDCA. Hence, it is important to select a catalytic system that can overcome the mass transfer limitations between the substrate and the catalysts, additionally minimizing the catalyst deactivation and allowing catalyst recyclability. Even though the amount of catalyst needed for the reaction is small, its cost encompasses the most significant expenses in the biomass valorization process. Hence, it is imperative to carefully examine the stability and recyclability of catalysts in relation to its efficiency for HMF oxidation. Concerning reaction medium, green solvents are environmentally-friendly solvents for high product yields that can often be obtained in organic solvents and ILs. Although efforts have been made to produce FDCA using various feedstocks and a variety of catalysts, processes, and conditions, there is still no recognized FDCA market share in the polymer market despite the several lab-scale techniques that have been developed. Given the realities of industrialization and commercial economics, more advancements in oxidation techniques are still needed for mass manufacturing. The present chapter suggests the following points to be investigated in future:

(a) A thorough analysis and correlation of several catalysts for the HMF oxidation into FDCA will be beneficial in the process selection. Owing to their higher separation and recycling efficiency, heterogeneous catalysts are more advantageous than both homogeneous catalysts and autolytic (catalyst free) systems. Besides, non-noble metals are less explored than noble metal catalysts for HMF oxidation. The affordability, accessibility, and stability of transition metal catalysts make them a viable industrial substitute. Better than monometallic catalysts in terms of durability and catalytic ability, bimetallic catalysts are also viable alternatives. Expanded recyclability and stability of FDCA selectivity are achieved by alloying with other metals, such as palladium or copper. Further work in transition metal catalyst and bimetallic catalysts should be given more focus.
(b) The implementation of noble metal catalysts is still hampered by instability and degradation issues. Various methods, such as cage-encapsulated nanoparticles, carbon nanotube structure, and enhanced metallic particle-support contact, can prevent agglomeration and enhance the stability and reactivity of the catalyst. It is possible to increase the catalytic activity and stability even further by employing trimetallic or bimetallic alloying procedures. Molecular simulation in conjunction with catalyst and support design can offer novel ways to direct the procedure.
(c) In alkaline solution experiments, the procedure is more costly and less environmentally friendly. Furthermore, the neutralization of FDCA salt lengthens the operating time and costs while producing more byproducts. Base-free systems have not been thoroughly studied because of their poor yield and selectivity. As a result, creating a catalyst with an emphasis on base-free conversion systems with increased activity is much needed.
(d) The separation and instability issues with HMF can be resolved by using cellulose, glucose, and fructose in a one-pot FDCA synthesis process. Since these feedstocks are readily and affordably available, their direct use will increase the economic viability of optimized manufacturing systems. This objective will undoubtedly be accomplished by creating a multifunctional catalyst combining acid, base, and metal sites. It is recommended to place metal sites with hydrophobic nature support for HMF oxidation to FDCA and acid/base sites with hydrophilic support for carbohydrate dehydration to HMF. HMF will

have more affinity toward hydrophobic metal sites for its oxidation, improving carbohydrate absorption on acid/base sites. Both selectivity and overall yield will increase with this plan.

(e) The development of DES and ILs, with their various uses and acid-base characteristics, is ongoing. These solutions share the characteristics of reactive solvents, but before they can be used commercially, environmental issues must be resolved.

(f) Not much research has been performed on reaction kinetic study and mechanism for producing FDCA directly from sugars or via HMF oxidation. An in-depth knowledge of the reaction mechanism will make it easier to rationally build affordable, and effective catalysts with effective recycling.

(g) Lastly, large-scale FDCA production is imperative for the creation of a green environment. While many of the developed approaches are technically feasible, they are very expensive. Researchers have always faced the difficulty of designing optimized processes. This goal could be met, and the process improved by using an interdisciplinary research strategy that incorporates chemistry, material engineering, and process design.

References

1. Huber GW, Iborra S, Corma A. Synthesis of transportation fuels from biomass: chemistry, catalysts, and engineering. Chem Rev. 2006;106(9):4044–98. https://doi.org/10.1021/cr068360d.
2. Coyle WT. Next-generation biofuels: near-term challenges and implications for agriculture. DIANE Publishing; 2010.
3. Boussie TR, Dias EL, Fresco ZM, Murphy VJ. Production of adipic acid and derivatives from carbohydrate-containing materials. Google Patents; 2013.
4. Pandey S, Dumont MJ, Orsat V, Rodrigue D. Biobased 2,5-furandicarboxylic acid (FDCA) and its emerging copolyesters' properties for packaging applications. Eur Polym J. 2021;160: 110778. https://doi.org/10.1016/j.eurpolymj.2021.110778.
5. Bicker M, Hirth J, Vogel H. Dehydration of fructose to 5-hydroxymethylfurfural in sub- and supercritical acetone. Green Chem. 2003;5(2):280–4. https://doi.org/10.1039/b211468b.
6. Mukherjee A, Dumont MJ, Raghauan V. Review: sustainable production of hydroxymethylfurfural and levulinic acid: challenges and opportunities. Biomass Bioenergy. 2015;72:143–83. https://doi.org/10.1016/j.biombioe.2014.11.007.
7. Wang T, Nolte MW, Shanks BH. Catalytic dehydration of C 6 carbohydrates for the production of hydroxymethylfurfural (HMF) as a versatile platform chemical. Green Chem. 2014;16 (2):548–72. https://doi.org/10.1039/C3GC41365A.
8. Agirrezabal-Telleria I, Gandarias I, Arias PL. Heterogeneous acid-catalysts for the production of furan-derived compounds (furfural and hydroxymethylfurfural) from renewable carbohydrates: a review. Catal Today. 2014;234:42–58. https://doi.org/10.1016/j.cattod.2013.11.027.
9. National Center for Biotechnology Information. PubChem Database. 5-Hydroxymethylfurfural, CID = 237332. https://pubchem.ncbi.nlm.nih.gov/compound/237332. Accessed 20 Nov 2023.
10. Vinke P, van Dam HE, van Bekkum H, Centi G, TrifirÓ F, editors. New developments in selective oxidation, vol. 55. Amsterdam: Elsevier Science Publishers; 1990. p. 147–51.
11. Huang YT, Wong JJ, Huang CJ, Li CL, Jang GWB. Chemicals and fuels from bio-based building blocks. Wiley-VCH Verlag GmbH & Co. KGaA; 2016. p. 191–216.

12. van der Waal JC, Mazoyer E, Baars HJ, Gruter GJM. Liquid phase aerobic oxidation catalysis: industrial applications and academic perspectives. Wiley-VCH Verlag GmbH & Co. KGaA; 2016. p. 311–29.
13. Shaikh A, Parker KR, Janka ME, Partin LR. Process for purifying crude furan 2, 5-dicarboxylic acid using hydrogenation. Google Patents; 2014.
14. Kläusli T. AVA biochem: commercialising renewable platform chemical 5-HMF. Green Process Synth. 2014;3(3):235–6.
15. Yemis O, Mazza G. Optimization of furfural and 5-hydroxymethylfurfural production from wheat straw by a microwave-assisted process. Bioresour Technol. 2012;109:215–23. https://doi.org/10.1016/j.biortech.2012.01.031.
16. Kazi FK, Patel AD, Serrano-Ruiz JC, Dumesic JA, Anex RP. Techno-economic analysis of dimethylfuran (DMF) and hydroxymethylfurfural (HMF) production from pure fructose in catalytic processes. Chem Eng J. 2011;169(1–3):329–38. https://doi.org/10.1016/j.cej.2011.03.018.
17. Aristizábal V, Gómez Á. Biorefineries based on coffee cut-stems and sugarcane bagasse: furan-based compounds and alkanes as interesting products. Bioresour Technol. 2015;196: 480–9. https://doi.org/10.1016/j.biortech.2015.07.057.
18. Dubbink GHC, Geverink TRJ, Haar B, Koets HW, Kumar A, van den Berg H, van der Ham AGJ, Lange JP. Furfural to FDCA: systematic process design and techno-economic evaluation. Biofuels Bioproducts Biorefining. 2021;15(4):1021–30. https://doi.org/10.1002/bbb.2204.
19. Kim H, Choi J, Won W. Process synthesis and analysis of green plastic monomer production from cellulose. J Clean Prod. 2020;277:124072. https://doi.org/10.1016/j.jclepro.2020.124072.
20. Motagamwala AH, Won W, Sener C, Alonso DM, Maravelias CT, Dumesic JA. Toward biomass-derived renewable plastics: production of 2,5-furandicarboxylic acid from fructose. Sci Adv. 2018;4(1):eaap9722. https://doi.org/10.1126/sciadv.aap9722.
21. Kim H, Baek S, Won W. Integrative technical, economic, and environmental sustainability analysis for the development process of biomass-derived 2,5-furandicarboxylic acid. Renew Sustain Energy Rev. 2022;157:112059. https://doi.org/10.1016/j.rser.2021.112059.
22. Kim H, Lee S, Won W. Economical process for the co-production of renewable polymers and value-added chemicals from lignocellulosic biomass. J Clean Prod. 2020;276:124237. https://doi.org/10.1016/j.jclepro.2020.124237.
23. Ashok RPB, Oinas P, Forssell S. Techno-economic evaluation of a biorefinery to produce γ-valerolactone (GVL), 2-methyltetrahydrofuran (2-MTHF) and 5-hydroxymethylfurfural (5-HMF) from spruce. Renew Energy. 2022;190:396–407. https://doi.org/10.1016/j.renene.2022.03.128.
24. Davidson MG, Elgie S, Parsons S, Young TJ. Production of HMF, FDCA and their derived products: a review of life cycle assessment (LCA) and techno-economic analysis (TEA) studies. Green Chem. 2021;23(9):3154–71. https://doi.org/10.1039/d1gc00721a.
25. Gustafsson J, Cederberg C, Sonesson U, Emanuelsson A. The methodology of the FAO study: global food losses and food waste-extent, causes and prevention-FAO, 2011. SIK Institutet för livsmedel och bioteknik; 2013.
26. Hegde S, Lodge JS, Trabold TA. Characteristics of food processing wastes and their use in sustainable alcohol production. Renew Sustain Energy Rev. 2018;81:510–23. https://doi.org/10.1016/j.rser.2017.07.012.
27. Choi IS, Lee YG, Khanal SK, Park BJ, Bae H-J. A low-energy, cost-effective approach to fruit and citrus peel waste processing for bioethanol production. Appl Energy. 2015;140:65–74.
28. Gourgue CM, Champ MM, Lozano Y, Delort-Laval J. Dietary fiber from mango byproducts: characterization and hypoglycemic effects determined by in vitro methods. J Agric Food Chem. 1992;40(10):1864–8. https://doi.org/10.1021/JF00022A027.

29. Vierhuis E, Schols HA, Beldman G, Voragen AGJ. Isolation and characterisation of cell wall material from olive fruit (Olea europaea cv koroneiki) at different ripening stages. Carbohydr Polym. 2000;43(1):11–21. https://doi.org/10.1016/s0144-8617(99)00204-0.
30. Thomassen LV, Larsen DM, Mikkelsen JD, Meyer AS. Definition and characterization of enzymes for maximal biocatalytic solubilization of prebiotic polysaccharides from potato pulp. Enzym Microb Technol. 2011;49(3):289–97. https://doi.org/10.1016/j.enzmictec.2011.06.006.
31. Mei X, Mu TH, Han JJ. Composition and physicochemical properties of dietary fiber extracted from residues of 10 varieties of sweet potato by a sieving method. J Agric Food Chem. 2010;58(12):7305–10. https://doi.org/10.1021/jf101021s.
32. Singhal K, Thakur S, Sharma D. Nutritive value of dried and stored apple pomace and its further processing for improved utilization. Indian J Anim Nutr. 1991;8(3):213–6.
33. Dhillon GS, Kaur S, Brar SK. Perspective of apple processing wastes as low-cost substrates for bioproduction of high value products: a review. Renew Sustain Energ Rev. 2013;27:789–805. https://doi.org/10.1016/j.rser.2013.06.046.
34. Emaga TH, Robert C, Ronkart SN, Wathelet B, Paquot M. Dietary fibre components and pectin chemical features of peels during ripening in banana and plantain varieties. Bioresour Technol. 2008;99(10):4346–54. https://doi.org/10.1016/j.biortech.2007.08.030.
35. Nawirska A, Kwaśniewska M. Dietary fibre fractions from fruit and vegetable processing waste. Food Chem. 2005;91(2):221–5. https://doi.org/10.1016/j.foodchem.2003.10.005.
36. Pourbafrani M, Forgács G, Horváth IS, Niklasson C, Taherzadeh MJ. Production of biofuels, limonene and pectin from citrus wastes. Bioresour Technol. 2010;101(11):4246–50. https://doi.org/10.1016/j.biortech.2010.01.077.
37. Satari B, Karimi K, Taherzadeh MJ, Zamani A. Co-production of fungal biomass derived constituents and ethanol from citrus wastes free sugars without auxiliary nutrients in airlift bioreactor. Int J Mol Sci. 2016;17(3):302. https://doi.org/10.3390/ijms17030302.
38. Botella C, De Ory I, Webb C, Cantero D, Blandino A. Hydrolytic enzyme production by Aspergillus awamori on grape pomace. Biochem Eng J. 2005;26(2–3):100–6. https://doi.org/10.1016/j.bej.2005.04.020.
39. Boluda-Aguilar M, López-Gómez A. Production of bioethanol by fermentation of lemon (Citrus limon L.) peel wastes pretreated with steam explosion. Ind Crop Prod. 2013;41:188–97. https://doi.org/10.1016/j.indcrop.2012.04.031.
40. Oberoi HS, Vadlani PV, Nanjundaswamy A, Bansal S, Singh S, Kaur S, Babbar N. Enhanced ethanol production from Kinnow mandarin (Citrus reticulata) waste via a statistically optimized simultaneous saccharification and fermentation process. Bioresour Technol. 2011;102(2):1593–601. https://doi.org/10.1016/j.biortech.2010.08.111.
41. Marín FR, Soler-Rivas C, Benavente-García O, Castillo J, Pérez-Alvarez JA. By-products from different citrus processes as a source of customized functional fibres. Food Chem. 2007;100(2):736–41. https://doi.org/10.1016/j.foodchem.2005.04.040.
42. Shafie SA, Al-Attab KA, Zainal ZA. Effect of hydrothermal and vapothermal carbonization of wet biomass waste on bound moisture removal and combustion characteristics. Appl Therm Eng. 2018;139:187–95. https://doi.org/10.1016/j.applthermaleng.2018.02.073.
43. Banerjee J, Singh R, Vijayaraghavan R, MacFarlane D, Patti AF, Arora A. A hydrocolloid based biorefinery approach to the valorisation of mango peel waste. Food Hydrocoll. 2018;77:142–51. https://doi.org/10.1016/j.foodhyd.2017.09.029.
44. Kaur M, Singh N, Sandhu KS, Guraya HS. Physicochemical, morphological, thermal and rheological properties of starches separated from kernels of some Indian mango cultivars (Mangifera indica L.). Food Chem. 2004;85(1):131–40. https://doi.org/10.1016/j.foodchem.2003.06.013.
45. Taghizadeh-Alisaraei A, Hosseini SH, Ghobadian B, Motevali A. Biofuel production from citrus wastes: a feasibility study in Iran. Renew Sustain Energ Rev. 2017;69:1100–12. https://doi.org/10.1016/j.rser.2016.09.102.

46. Spagnuolo M, Crecchio C, Pizzigallo MDR, Ruggiero P. Synergistic effects of cellulolytic and pectinolytic enzymes in degrading sugar beet pulp. Bioresour Technol. 1997;60(3):215–22. https://doi.org/10.1016/S0960-8524(97)00013-8.
47. Ruppert AM, Weinberg K, Palkovits R. Hydrogenolysis goes bio: from carbohydrates and sugar alcohols to platform chemicals. Angew Chem Int Ed Engl. 2012;51(11):2564–601. https://doi.org/10.1002/anie.201105125.
48. Tester RF, Karkalas J, Qi X. Starch—composition, fine structure and architecture. J Cereal Sci. 2004;39(2):151–65. https://doi.org/10.1016/j.jcs.2003.12.001.
49. Nishiyama Y, Sugiyama J, Chanzy H, Langan P. Crystal structure and hydrogen bonding system in cellulose I(alpha) from synchrotron X-ray and neutron fiber diffraction. J Am Chem Soc. 2003;125(47):14300–6. https://doi.org/10.1021/ja037055w.
50. Zhou CH, Xia X, Lin CX, Tong DS, Beltramini J. Catalytic conversion of lignocellulosic biomass to fine chemicals and fuels. Chem Soc Rev. 2011;40(11):5588–617. https://doi.org/10.1039/c1cs15124j.
51. Taarning E, Osmundsen CM, Yang XB, Voss B, Andersen SI, Christensen CH. Zeolite-catalyzed biomass conversion to fuels and chemicals. Energy Environ Sci. 2011;4(3):793–804. https://doi.org/10.1039/c004518g.
52. Somerville C, Youngs H, Taylor C, Davis SC, Long SP. Feedstocks for lignocellulosic biofuels. Science. 2010;329(5993):790–2. https://doi.org/10.1126/science.1189268.
53. Shuai L, Pan XJ. Biohydrocarbons: next generation liquid fuels from lignocellulosic biomass. In: Research progress in paper industry and biorefinery (4th Isetpp), vol. 1–3; 2010. p. 1293–7.
54. Somerville C, Bauer S, Brininstool G, Facette M, Hamann T, Milne J, Osborne E, Paredez A, Persson S, Raab T, et al. Toward a systems approach to understanding plant cell walls. Science. 2004;306(5705):2206–11. https://doi.org/10.1126/science.1102765.
55. Wang LK, Lo HH, Hung Y-T, Yapijakis C. Waste treatment in the food processing industry. CRC Press; 2005.
56. Sagar NA, Pareek S, Sharma S, Yahia EM, Lobo MG. Fruit and vegetable waste: bioactive compounds, their extraction, and possible utilization. Compr Rev Food Sci Food Saf. 2018;17(3):512–31. https://doi.org/10.1111/1541-4337.12330.
57. Misi SN, Forster CF. Semi-continuous anaerobic co-digestion of agro-wastes. Environ Technol. 2002;23(4):445–51. https://doi.org/10.1080/09593332508618405.
58. Shalini R, Gupta DK. Utilization of pomace from apple processing industries: a review. J Food Sci Technol. 2010;47(4):365–71. https://doi.org/10.1007/s13197-010-0061-x.
59. Fernández-López J, Fernández-Ginés JM, Aleson-Carbonell L, Sendra E, Sayas-Barberá E, Pérez-Alvarez JA. Application of functional citrus by-products to meat products. Trends Food Sci Technol. 2004;15(3–4):176–85. https://doi.org/10.1016/j.tifs.2003.08.007.
60. Allaf T, Tomao V, Ruiz K, Chemat F. Instant controlled pressure drop technology and ultrasound assisted extraction for sequential extraction of essential oil and antioxidants. Ultrason Sonochem. 2013;20(1):239–46. https://doi.org/10.1016/j.ultsonch.2012.05.013.
61. Bagherian H, Ashtiani FZ, Fouladitajar A, Mohtashamy M. Comparisons between conventional, microwave- and ultrasound-assisted methods for extraction of pectin from grapefruit. Chem Eng Process. 2011;50(11–12):1237–43. https://doi.org/10.1016/j.cep.2011.08.002.
62. Bampidis VA, Robinson PH. Citrus by-products as ruminant feeds: a review. Anim Feed Sci Technol. 2006;128(3–4):175–217. https://doi.org/10.1016/j.anifeedsci.2005.12.002.
63. Del Valle M, Cámara M, Torija ME. Chemical characterization of tomato pomace. J Sci Food Agric. 2006;86(8):1232–6. https://doi.org/10.1002/jsfa.2474.
64. Chandrasekaran M. Valorization of food processing by-products. CRC Press; 2012.
65. Van der Hijden HTW, Suijker MJ, Tammes H. Tomato fibre composition and method for the preparation thereof. Google Patents; 2017.
66. Balat M. Production of bioethanol from lignocellulosic materials via the biochemical pathway: a review. Energy Convers Manag. 2011;52(2):858–75. https://doi.org/10.1016/j.enconman.2010.08.013.

67. Hamelinck CN, van Hooijdonk G, Faaij APC. Ethanol from lignocellulosic biomass: techno-economic performance in short-, middle- and long-term. Biomass Bioenergy. 2005;28(4): 384–410. https://doi.org/10.1016/j.biombioe.2004.09.002.
68. Lee D, Owens V, Boe A, Jeranyama P. Composition of herbaceous biomass feedstocks. Brookings, SD: South Dakota State University Publication; 2007.
69. Taherzadeh MJ, Karimi K. Pretreatment of lignocellulosic wastes to improve ethanol and biogas production: a review. Int J Mol Sci. 2008;9(9):1621–51. https://doi.org/10.3390/ijms9091621.
70. Angel Siles Lopez J, Li Q, Thompson IP. Biorefinery of waste orange peel. Crit Rev Biotechnol. 2010;30(1):63–9. https://doi.org/10.3109/07388550903425201.
71. Mosier N, Wyman C, Dale B, Elander R, Lee YY, Holtzapple M, Ladisch M. Features of promising technologies for pretreatment of lignocellulosic biomass. Bioresour Technol. 2005;96(6):673–86. https://doi.org/10.1016/j.biortech.2004.06.025.
72. Hendriks AT, Zeeman G. Pretreatments to enhance the digestibility of lignocellulosic biomass. Bioresour Technol. 2009;100(1):10–8. https://doi.org/10.1016/j.biortech.2008.05.027.
73. del Campo I, Alegría I, Zazpe M, Echeverría M, Echeverría I. Diluted acid hydrolysis pretreatment of agri-food wastes for bioethanol production. Ind Crop Prod. 2006;24(3): 214–21. https://doi.org/10.1016/j.indcrop.2006.06.014.
74. Wedzicha BL. Editorial. Food Chem. 1994;51(4):341. https://doi.org/10.1016/0308-8146(94)90183-x.
75. Van Boekel M. Effect of heating on Maillard reactions in milk. Food Chem. 1998;62(4): 403–14. https://doi.org/10.1016/S0308-8146(98)00075-2.
76. Tomasik P, Pałasiński M, Wiejak S. The thermal decomposition of carbohydrates. Part I. The decomposition of mono-, di-, and oligo-saccharides. In: Tipson RS, Horton D, editors. Advances in carbohydrate chemistry and biochemistry, Advances in carbohydrate chemistry and biochemistry, vol. 47. Academic Press; 1989. p. 203–78.
77. Nguyen H, Nikolakis V, Vlachos DG. Mechanistic insights into Lewis acid metal salt-catalyzed glucose chemistry in aqueous solution. ACS Catal. 2016;6(3):1497–504. https://doi.org/10.1021/acscatal.5b02698.
78. Dutta S, De S, Alam MI, Abu-Omar MM, Saha B. Direct conversion of cellulose and lignocellulosic biomass into chemicals and biofuel with metal chloride catalysts. J Catal. 2012;288:8–15. https://doi.org/10.1016/j.jcat.2011.12.017.
79. Zhou L, Liang R, Ma Z, Wu T, Wu Y. Conversion of cellulose to HMF in ionic liquid catalyzed by bifunctional ionic liquids. Bioresour Technol. 2013;129:450–5. https://doi.org/10.1016/j.biortech.2012.11.015.
80. Wrigstedt P, Keskiväli J, Leskelä M, Repo T. The role of salts and Bronsted acids in lewis acid-catalyzed aqueous-phase glucose dehydration to 5-hydroxymethylfurfural. ChemCatChem. 2015;7(3):501–7. https://doi.org/10.1002/cctc.201402941.
81. De Souza RL, Yu H, Rataboul F, Essayem N. 5-Hydroxymethylfurfural (5-HMF) production from hexoses: limits of heterogeneous catalysis in hydrothermal conditions and potential of concentrated aqueous organic acids as reactive solvent system. Challenges. 2012;3(2):212–32. https://doi.org/10.3390/challe3020212.
82. Hu X, Kadarwati S, Wang S, Song Y, Hasan MM, Li C-Z. Biomass-derived sugars and furans: which polymerize more during their hydrolysis? Fuel Process Technol. 2015;137:212–9. https://doi.org/10.1016/j.fuproc.2015.04.024.
83. Yu IKM, Tsang DCW, Yip ACK, Chen SS, Ok YS, Poon CS. Valorization of food waste into hydroxymethylfurfural: dual role of metal ions in successive conversion steps. Bioresour Technol. 2016;219:338–47. https://doi.org/10.1016/j.biortech.2016.08.002.
84. Yu IKM, Tsang DCW, Chen SS, Wang L, Hunt AJ, Sherwood J, De Oliveira Vigier K, Jerome F, Ok YS, Poon CS. Polar aprotic solvent-water mixture as the medium for catalytic production of hydroxymethylfurfural (HMF) from bread waste. Bioresour Technol. 2017;245 (Pt A):456–62. https://doi.org/10.1016/j.biortech.2017.08.170.

85. Yu IKM, Tsang DCW, Yip ACK, Chen SS, Wang L, Ok YS, Poon CS. Catalytic valorization of starch-rich food waste into hydroxymethylfurfural (HMF): controlling relative kinetics for high productivity. Bioresour Technol. 2017;237:222–30. https://doi.org/10.1016/j.biortech.2017.01.017.
86. Yu IKM, Tsang DCW, Yip ACK, Chen SS, Ok YS, Poon CS. Valorization of starchy, cellulosic, and sugary food waste into hydroxymethylfurfural by one-pot catalysis. Chemosphere. 2017;184:1099–107. https://doi.org/10.1016/j.chemosphere.2017.06.095.
87. Yemis O, Mazza G. Acid-catalyzed conversion of xylose, xylan and straw into furfural by microwave-assisted reaction. Bioresour Technol. 2011;102(15):7371–8. https://doi.org/10.1016/j.biortech.2011.04.050.
88. Seo YH, Han JI. Direct conversion from Jerusalem artichoke to hydroxymethylfurfural (HMF) using the Fenton reaction. Food Chem. 2014;151:207–11. https://doi.org/10.1016/j.foodchem.2013.11.067.
89. Amiri H, Karimi K, Roodpeyma S. Production of furans from rice straw by single-phase and biphasic systems. Carbohydr Res. 2010;345(15):2133–8. https://doi.org/10.1016/j.carres.2010.07.032.
90. Yin SD, Pan YL, Tan ZC. Hydrothermal conversion of cellulose to 5-hydroxymethyl furfural. Int J Green Energy. 2011;8(2):234–47. https://doi.org/10.1080/15435075.2010.548888.
91. Wrigstedt P, Keskiväli J, Repo T. Microwave-enhanced aqueous biphasic dehydration of carbohydrates to 5-hydroxymethylfurfural. RSC Adv. 2016;6(23):18973–9. https://doi.org/10.1039/c5ra25564c.
92. Gomes GR, Rampon DS, Ramos LP. Production of 5-(hydroxymethyl)-furfural from water-soluble carbohydrates and sugarcane molasses. Appl Catal A Gen. 2017;545:127–33. https://doi.org/10.1016/j.apcata.2017.07.049.
93. De S, Dutta S, Saha B. Microwave assisted conversion of carbohydrates and biopolymers to 5-hydroxymethylfurfural with aluminium chloride catalyst in water. Green Chem. 2011;13(10):2859–68. https://doi.org/10.1039/c1gc15550d.
94. Pandey S, Dumont MJ, Orsat V. Catalytic conversion of whey permeate into 5-hydroxymethylfurfural: optimization and kinetic modeling approach. Food Bioprod Process. 2021;129:134–43. https://doi.org/10.1016/j.fbp.2021.08.001.
95. Carlini C, Patrono P, Galletti AMR, Sbrana G. Heterogeneous catalysts based on vanadyl phosphate for fructose dehydration to 5-hydroxymethyl-2-furaldehyde. Appl Catal A Gen. 2004;275(1–2):111–8. https://doi.org/10.1016/j.apcata.2004.07.026.
96. Dias AS, Pillinger M, Valente AA. Dehydration of xylose into furfural over micro-mesoporous sulfonic acid catalysts. J Catal. 2005;229(2):414–23. https://doi.org/10.1016/j.jcat.2004.11.016.
97. Chen SS, Yu IKM, Tsang DCW, Yip ACK, Khan E, Wang L, Ok YS, Poon CS. Valorization of cellulosic food waste into levulinic acid catalyzed by heterogeneous Bronsted acids: temperature and solvent effects. Chem Eng J. 2017;327:328–35. https://doi.org/10.1016/j.cej.2017.06.108.
98. Tsilomelekis G, Orella MJ, Lin ZX, Cheng ZW, Zheng WQ, Nikolakis V, Vlachos DG. Molecular structure, morphology and growth mechanisms and rates of 5-hydroxymethyl furfural (HMF) derived humins. Green Chem. 2016;18(7):1983–93. https://doi.org/10.1039/c5gc01938a.
99. Qian X, Liu D. Free energy landscape for glucose condensation and dehydration reactions in dimethyl sulfoxide and the effects of solvent. Carbohydr Res. 2014;388:50–60. https://doi.org/10.1016/j.carres.2014.02.010.
100. Wasserscheid P, Welton T. Ionic liquids in synthesis. John Wiley & Sons; 2008.
101. Asim AM, Uroos M, Naz S, Sultan M, Griffin G, Muhammad N, Khan AS. Acidic ionic liquids: promising and cost-effective solvents for processing of lignocellulosic biomass. J Mol Liq. 2019;287:110943. https://doi.org/10.1016/j.molliq.2019.110943.

102. Alam MI, De S, Khan TS, Haider MA, Saha B. Acid functionalized ionic liquid catalyzed transformation of non-food biomass into platform chemical and fuel additive. Ind Crop Prod. 2018;123:629–37. https://doi.org/10.1016/j.indcrop.2018.07.036.
103. Zhang Z, Zhao ZK. Microwave-assisted conversion of lignocellulosic biomass into furans in ionic liquid. Bioresour Technol. 2010;101(3):1111–4. https://doi.org/10.1016/j.biortech.2009.09.010.
104. Yi YB, Lee JW, Hong SS, Choi YH, Chung CH. Acid-mediated production of hydroxymethylfurfural from raw plant biomass with high inulin in an ionic liquid. J Ind Eng Chem. 2011;17(1):6–9. https://doi.org/10.1016/j.jiec.2010.12.017.
105. Chun JA, Lee JW, Yi YB, Hong SS, Chung CH. Direct conversion of starch to hydroxymethylfurfural in the presence of an ionic liquid with metal chloride. Starch-Starke. 2010;62(6):326–30. https://doi.org/10.1002/star.201000012.
106. Yi YB, Ha MG, Lee JW, Chung CH. Effect of different halide types on HMF synthesis from kudzu extract in ionic liquid. J Clean Prod. 2013;41:244–50. https://doi.org/10.1016/j.jclepro.2012.10.023.
107. Semerci I, Güler F. Pretreatment of crop wastes from edible biomass with a protic ionic liquid. Int J Renew Energy Res. 2019;9(1):17–23. https://doi.org/10.20508/ijrer.v9i1.8596.g7559.
108. Yi YB, Lee JL, Choi YH, Park SM, Chung CH. Direct production of hydroxymethylfurfural from raw grape berry biomass using ionic liquids and metal chlorides. Environ Chem Lett. 2012;10(1):13–9. https://doi.org/10.1007/s10311-011-0322-6.
109. Yi YB, Ha MG, Lee JW, Chung CH. New role of chromium fluoride: its catalytic action on the synthesis of hydroxymethylfurfural in ionic liquid using raw plant biomass and characterization of biomass hydrolysis. Chem Eng J. 2012;180:370–5. https://doi.org/10.1016/j.cej.2011.10.055.
110. Weerachanchai P, Leong SS, Chang MW, Ching CB, Lee JM. Improvement of biomass properties by pretreatment with ionic liquids for bioconversion process. Bioresour Technol. 2012;111:453–9. https://doi.org/10.1016/j.biortech.2012.02.023.
111. Ma HH, Zhang BX, Zhang P, Li S, Gao YF, Hu XM. An efficient process for lignin extraction and enzymatic hydrolysis of corn stalk by pyrrolidonium ionic liquids. Fuel Process Technol. 2016;148:138–45. https://doi.org/10.1016/j.fuproc.2016.02.038.
112. Chambon CL, Mkhize TY, Reddy P, Brandt-Talbot A, Deenadayalu N, Fennell PS, Hallett JP. Pretreatment of South African sugarcane bagasse using a low-cost protic ionic liquid: a comparison of whole, depithed, fibrous and pith bagasse fractions. Biotechnol Biofuels. 2018;11(1):247. https://doi.org/10.1186/s13068-018-1247-0.
113. Li YN, Wang JQ, He LN, Yang ZZ, Liu AH, Yu B, Luan CR. Experimental and theoretical studies on imidazolium ionic liquid-promoted conversion of fructose to 5-hydroxymethylfurfural. Green Chem. 2012;14(10):2752–8. https://doi.org/10.1039/c2gc35845j.
114. Shi N, Liu QY, Wang TJ, Zhang Q, Ma LL, Cai CL. Production of 5-hydroxymethylfurfural and furfural from lignocellulosic biomass in water-tetrahydrofuran media with sodium bisulfate. Chin J Chem Phys. 2015;28(5):650–6. https://doi.org/10.1063/1674-0068/28/cjcp1501008.
115. Yi YB, Lee JW, Choi YH, Park SM, Chung CH. Simple process for production of hydroxymethylfurfural from raw biomasses of girasol and potato tubers. Biomass Bioenergy. 2012;39:484–8. https://doi.org/10.1016/j.biombioe.2012.01.011.
116. Chheda JN, Roman-Leshkov Y, Dumesic JA. Production of 5-hydroxymethylfurfural and furfural by dehydration of biomass-derived mono- and poly-saccharides. Green Chem. 2007;9(4):342–50. https://doi.org/10.1039/B611568C.
117. Román-Leshkov Y, Chheda JN, Dumesic JA. Phase modifiers promote efficient production of hydroxymethylfurfural from fructose. Science. 2006;312(5782):1933–7. https://doi.org/10.1126/science.112633.

118. Roy Goswami S, Dumont M-J, Raghavan V. Microwave assisted synthesis of 5-hydroxymethylfurfural from starch in AlCl3·6H2O/DMSO/[BMIM]Cl system. Ind Eng Chem Res. 2016;55(16):4473–81. https://doi.org/10.1021/acs.iecr.6b00201.
119. Roy Goswami S, Mukherjee A, Dumont M-J, Raghavan V. One-pot conversion of corn starch into 5-hydroxymethylfurfural in water-[Bmim]Cl/MIBK biphasic media. Energy Fuel. 2016;30(10):8349–56. https://doi.org/10.1021/acs.energyfuels.6b01699.
120. Guenic SL, Delbecq F, Ceballos C, Len C. Microwave-assisted dehydration of D-xylose into furfural by diluted inexpensive inorganic salts solution in a biphasic system. J Mol Catal A Chem. 2015;410:1–7. https://doi.org/10.1016/j.molcata.2015.08.019.
121. Okano T, Qiao K, Bao QX, Tomida D, Hagiwara H, Yokoyama C. Dehydration of fructose to 5-hydroxymethylfurfural (HMF) in an aqueous acetonitrile biphasic system in the presence of acidic ionic liquids. Appl Catal A Gen. 2013;451:1–5. https://doi.org/10.1016/j.apcata.2012.11.004.
122. Zuo M, Le K, Li Z, Jiang YT, Zeng XH, Tang X, Sun Y, Lin L. Green process for production of 5-hydroxymethylfurfural from carbohydrates with high purity in deep eutectic solvents. Ind Crop Prod. 2017;99:1–6. https://doi.org/10.1016/j.indcrop.2017.01.027.
123. Román-Leshkov Y, Dumesic JA. Solvent effects on fructose dehydration to 5-hydroxymethylfurfural in biphasic systems saturated with inorganic salts. Top Catal. 2009;52(3):297–303. https://doi.org/10.1007/s11244-008-9166-0.
124. Bhaumik P, Dhepe PL. Solid acid catalyzed synthesis of furans from carbohydrates. Catal Rev Sci Eng. 2016;58(1):36–112. https://doi.org/10.1080/01614940.2015.1099894.
125. Niakan M, Masteri-Farahani M, Seidi F. Catalytic fructose dehydration to 5-hydroxymethylfurfural on the surface of sulfonic acid modified ordered mesoporous SBA-16. Fuel. 2023;337:127242. https://doi.org/10.1016/j.fuel.2022.127242.
126. Li S, Qian EW, Shibata T, Hosomi M. Catalytic hydrothermal saccharification of rice straw using mesoporous silica-based solid acid catalysts. J Jpn Petrol Inst. 2012;55(4):250–60. https://doi.org/10.1627/jpi.55.250.
127. Cao L, Yu IKM, Tsang DCW, Zhang S, Ok YS, Kwon EE, Song H, Poon CS. Phosphoric acid-activated wood biochar for catalytic conversion of starch-rich food waste into glucose and 5-hydroxymethylfurfural. Bioresour Technol. 2018;267:242–8. https://doi.org/10.1016/j.biortech.2018.07.048.
128. Hermiati E, Mangunwidjaja D, Sunarti TC, Suparno O, Prasetya B. Microwave-assisted acid hydrolysis of starch polymer in cassava pulp in the presence of activated carbon. Proc Chem. 2012;4:238–44. https://doi.org/10.1016/j.proche.2012.06.033.
129. Hermiati E, Azuma JI, Tsubaki S, Mangunwidjaja D, Sunarti TC, Suparno O, Prasetya B. Improvement of microwave-assisted hydrolysis of cassava pulp and tapioca flour by addition of activated carbon. Carbohydr Polym. 2012;87(1):939–42. https://doi.org/10.1016/j.carbpol.2011.08.033.
130. Cao L, Yu IKM, Chen SS, Tsang DCW, Wang L, Xiong X, Zhang S, Ok YS, Kwon EE, Song H, et al. Production of 5-hydroxymethylfurfural from starch-rich food waste catalyzed by sulfonated biochar. Bioresour Technol. 2018;252:76–82. https://doi.org/10.1016/j.biortech.2017.12.098.
131. Daengprasert W, Boonnoun P, Laosiripojana N, Goto M, Shotipruk A. Application of sulfonated carbon-based catalyst for Solvothermal conversion of cassava waste to Hydroxymethylfurfural and furfural. Ind Eng Chem Res. 2011;50(13):7903–10. https://doi.org/10.1021/ie102487w.
132. Yu IKM, Ong KL, Tsang DCW, Haque MA, Kwan TH, Chen SS, Uisan K, Kulkarni S, Lin CSK. Chemical transformation of food and beverage waste-derived fructose to hydroxymethylfurfural as a value-added product. Catal Today. 2018;314:70–7. https://doi.org/10.1016/j.cattod.2018.01.011.
133. Lucas-Torres C, Lorente A, Cabañas B, Moreno A. Microwave heating for the catalytic conversion of melon rind waste into biofuel precursors. J Clean Prod. 2016;138:59–69. https://doi.org/10.1016/j.jclepro.2016.03.122.

134. Iris K, Tsang DC, Yip AC, Hunt AJ, Sherwood J, Shang J, Song H, Ok YS, Poon CS. Propylene carbonate and γ-valerolactone as green solvents enhance Sn (IV)-catalysed hydroxymethylfurfural (HMF) production from bread waste. Green Chem. 2018;20(9): 2064–74. https://doi.org/10.1039/C8GC00358K.
135. Hoang PH, Cuong TD. Preparation of metal-loaded ZSM-5 zeolite catalyst and its catalytic effect on HMF production from biomass. Appl Biochem Biotechnol. 2022;194(11):4985–98. https://doi.org/10.1007/s12010-022-03998-2.
136. Hoang PH, Dat NM, Cuong TD, Tung DT. Production of 5-hydroxymethylfurfural (HMF) from rice-straw biomass using a HSO 3–ZSM-5 zeolite catalyst under assistance of sonication. RSC Adv. 2020;10(23):13489–95. https://doi.org/10.1039/D0RA02037K.
137. Hoang PH, Cuong TD. Simultaneous direct production of 5-hydroxymethylfurfural (HMF) and furfural from corncob biomass using porous HSO3-ZSM-5 zeolite catalyst. Energy Fuel. 2020;35(1):546–51. https://doi.org/10.1039/D0RA02037K.
138. Yang F, Liu Q, Bai X, Du Y. Conversion of biomass into 5-hydroxymethylfurfural using solid acid catalyst. Bioresour Technol. 2011;102(3):3424–9. https://doi.org/10.1016/j.biortech.2010.10.023.
139. Yang F, Liu Q, Yue M, Bai X, Du Y. Tantalum compounds as heterogeneous catalysts for saccharide dehydration to 5-hydroxymethylfurfural. Chem Commun (Camb). 2011;47(15): 4469–71. Article. https://doi.org/10.1039/c0cc05138a.
140. Atanda L, Shrotri A, Mukundan S, Ma Q, Konarova M, Beltramini J. Direct production of 5-hydroxymethylfurfural via catalytic conversion of simple and complex sugars over phosphated TiO2. ChemSusChem. 2015;8(17):2907–16. https://doi.org/10.1002/cssc.201500395.
141. Parshetti GK, Suryadharma MS, Pham TPT, Mahmood R, Balasubramanian R. Heterogeneous catalyst-assisted thermochemical conversion of food waste biomass into 5-hydroxymethylfurfural. Bioresour Technol. 2015;178:19–27. https://doi.org/10.1016/j.biortech.2014.10.066.
142. Dutta A, Patra AK, Dutta S, Saha B, Bhaumik A. Hierarchically porous titanium phosphate nanoparticles: an efficient solid acid catalyst for microwave assisted conversion of biomass and carbohydrates into 5-hydroxymethylfurfural. J Mater Chem. 2012;22(28):14094–100. https://doi.org/10.1039/c2jm30623a.
143. Wang JJ, Liu XH, Hu BC, Lu GZ, Wang YQ. Efficient catalytic conversion of lignocellulosic biomass into renewable liquid biofuels furan derivatives. RSC Adv. 2014;4(59):31101–7. https://doi.org/10.1039/c4ra04900d.
144. Atchudan R, Cha BG, Lone N, Kim J, Joo J. Synthesis of high-quality carbon nanotubes by using monodisperse spherical mesoporous silica encapsulating iron oxide nanoparticles. Korean J Chem Eng. 2019;36(1):157–65. https://doi.org/10.1007/s11814-018-0200-z.
145. Huang FM, Su YW, Tao Y, Sun W, Wang WT. Preparation of 5-hydroxymethylfurfural from glucose catalyzed by silica-supported phosphotungstic acid heterogeneous catalyst. Fuel. 2018;226:417–22. https://doi.org/10.1016/j.fuel.2018.03.193.
146. Huang Z, Pan W, Zhou H, Qin F, Xu H, Shen W. Nafion-resin-modified mesocellular silica foam catalyst for 5-hydroxymethylfurfural production from D-fructose. ChemSusChem. 2013;6(6):1063–9. https://doi.org/10.1002/cssc.201200967.
147. Perez GP, Dumont MJ. Production of HMF in high yield using a low cost and recyclable carbonaceous catalyst. Chem Eng J. 2020;382:122766. https://doi.org/10.1016/j.cej.2019.122766.
148. Kang SM, Ye J, Zhang Y, Chang J. Preparation of biomass hydrochar derived sulfonated catalysts and their catalytic effects for 5-hydroxymethylfurfural production. RSC Adv. 2013;3(20):7360–6. https://doi.org/10.1039/c3ra23314f.
149. Rinaldi R, Meine N, vom Stein J, Palkovits R, Schüth F. Which controls the depolymerization of cellulose in ionic liquids: the solid acid catalyst or cellulose? ChemSusChem. 2010;3(2): 266–76. https://doi.org/10.1002/cssc.200900281.

150. Trapasso G, Mazzi G, Chicharo B, Annatelli M, Dalla Torre D, Arico F. Multigram synthesis of pure HMF and BHMF. Org Process Res Dev. 2022;26(10):2830–8. https://doi.org/10.1021/acs.oprd.2c00196.
151. Mittal A, Black SK, Vinzant TB, O'Brien M, Tucker MP, Johnson DK. Production of furfural from process-relevant biomass-derived pentoses in a biphasic reaction system. ACS Sustain Chem Eng. 2017;5(7):5694–701. https://doi.org/10.1021/acssuschemeng.7b00215.
152. Tzeng TW, Bhaumik P, Chung PW. Understanding the production of 5-hydroxymethylfurfural (HMF) from chitosan using solid acids. Mol Catal. 2019;479:164–9. https://doi.org/10.1016/j.mcat.2019.110627.
153. Ennaert T, Van Aelst J, Dijkmans J, De Clercq R, Schutyser W, Dusselier M, Verboekend D, Sels BF. Potential and challenges of zeolite chemistry in the catalytic conversion of biomass. Chem Soc Rev. 2016;45(3):584–611. https://doi.org/10.1039/c5cs00859j.
154. Moreno-Recio M, Santamaría-González J, Maireles-Torres P. Bronsted and Lewis acid ZSM-5 zeolites for the catalytic dehydration of glucose into 5-hydroxymethylfurfural. Chem Eng J. 2016;303:22–30. https://doi.org/10.1016/j.cej.2016.05.120.
155. Chai Y, Bai M, Chen A, Yuan J, Peng C, Zhao D, Yan B, Qin P. Cr-Mn bimetallic functionalized USY zeolite monolithic catalyst for direct production of 2,5-furandicarboxylic acid from raw biomass. Chem Eng J. 2022;429:132173. https://doi.org/10.1016/j.cej.2021.132173.
156. Nikolla E, Román-Leshkov Y, Moliner M, Davis ME. "One-pot" synthesis of 5-(hydroxymethyl)furfural from carbohydrates using tin-beta zeolite. ACS Catal. 2011;1(4):408–10. https://doi.org/10.1021/cs2000544.
157. Moliner M, Roman-Leshkov Y, Davis ME. Tin-containing zeolites are highly active catalysts for the isomerization of glucose in water. Proc Natl Acad Sci USA. 2010;107(14):6164–8. https://doi.org/10.1073/pnas.1002358107.
158. Wang LQ, Guo HQ, Xie QL, Wang JG, Hou B, Jia LT, Cui JL, Li DB. Conversion of fructose into furfural or 5-hydroxymethylfurfural over HY zeolites selectively in γ-butyrolactone. Appl Catal A Gen. 2019;572:51–60. https://doi.org/10.1016/j.apcata.2018.12.023.
159. Candu N, El Fergani M, Verziu M, Cojocaru B, Jurca B, Apostol N, Teodorescu C, Parvulescu VI, Coman SM. Efficient glucose dehydration to HMF onto Nb-BEA catalysts. Catal Today. 2019;325:109–16. https://doi.org/10.1016/j.cattod.2018.08.004.
160. Wang JJ, Ren JW, Liu XH, Xi JX, Xia QN, Zu YH, Lu GZ, Wang YQ. Direct conversion of carbohydrates to 5-hydroxymethylfurfural using Sn-Mont catalyst. Green Chem. 2012;14(9):2506–12. https://doi.org/10.1039/c2gc35699f.
161. Peela NR, Yedla SK, Velaga B, Kumar A, Golder AK. Choline chloride functionalized zeolites for the conversion of biomass derivatives to 5-hydroxymethylfurfural. Appl Catal A Gen. 2019;580:59–70. https://doi.org/10.1016/j.apcata.2019.05.005.
162. Carniti P, Gervasini A, Marzo M. Absence of expected side-reactions in the dehydration reaction of fructose to HMF in water over niobic acid catalyst. Catalysis Communications. 2011;12(12):1122–6. Article. https://doi.org/10.1016/j.catcom.2011.03.025.
163. Jiménez-Morales I, Moreno-Recio M, Santamaría-González J, Maireles-Torres P, Jiménez-López A. Mesoporous tantalum oxide as catalyst for dehydration of glucose to 5-hydroxymethylfurfural. Appl Catal B Environ. 2014;154–155:190–6. https://doi.org/10.1016/j.apcatb.2014.02.024.
164. Qi XH, Watanabe M, Aida TM, Smith RL. Catalytical conversion of fructose and glucose into 5-hydroxymethylfurfural in hot compressed water by microwave heating. Catal Commun. 2008;9(13):2244–9. Article. https://doi.org/10.1016/j.catcom.2008.04.025.
165. Watanabe M, Aizawa Y, Iida T, Aida TM, Levy C, Sue K, Inomata H. Glucose reactions with acid and base catalysts in hot compressed water at 473 K. Carbohydr Res. 2005;340(12):1925–30. Article. https://doi.org/10.1016/j.carres.2005.06.017.
166. Daorattanachai P, Khemthong P, Viriya-empikul N, Laosiripojana N, Faungnawakij K. Effect of calcination temperature on catalytic performance of alkaline earth phosphates in hydrolysis/

dehydration of glucose and cellulose. Chem Eng J. 2015;278:92–8. https://doi.org/10.1016/j.cej.2014.12.029.
167. Gliozzi G, Innorta A, Mancini A, Bortolo R, Perego C, Ricci M, Cavani F. Zr/P/O catalyst for the direct acid chemo-hydrolysis of non-pretreated microcrystalline cellulose and softwood sawdust. Appl Catal B Environ. 2014;145:24–33. https://doi.org/10.1016/j.apcatb.2012.12.035.
168. Antonetti C, Melloni M, Licursi D, Fulignati S, Ribechini E, Rivas S, Parajó JC, Cavani F, Galletti AMR. Microwave-assisted dehydration of fructose and inulin to HMF catalyzed by niobium and zirconium phosphate catalysts. Appl Catal B Environ. 2017;206:364–77. https://doi.org/10.1016/j.apcatb.2017.01.056.
169. Hou QD, Zhen MN, Liu L, Chen Y, Huang F, Zhang SQ, Li WZ, Ju MT. Tin phosphate as a heterogeneous catalyst for efficient dehydration of glucose into 5-hydroxymethylfurfural in ionic liquid. Appl Catal B Environ. 2018;224:183–93. https://doi.org/10.1016/j.apcatb.2017.09.049.
170. Jiménez-Morales I, Teckchandani-Ortiz A, Santamaría-González J, Maireles-Torres P, Jiménez-López A. Selective dehydration of glucose to 5-hydroxymethylfurfural on acidic mesoporous tantalum phosphate. Appl Catal B Environ. 2014;144:22–8. https://doi.org/10.1016/j.apcatb.2013.07.002.
171. Luterbacher JS, Rand JM, Alonso DM, Han J, Youngquist JT, Maravelias CT, Pfleger BF, Dumesic JA. Nonenzymatic sugar production from biomass using biomass-derived gamma-valerolactone. Science. 2014;343(6168):277–80. https://doi.org/10.1126/science.1246748.
172. Ma J, Li W, Guan S, Liu Q, Li Q, Zhu C, Yang T, Ogunbiyi AT, Ma L. Efficient catalytic conversion of corn stalk and xylose into furfural over sulfonated graphene in γ-valerolactone. RSC Adv. 2019;9(19):10569–77. https://doi.org/10.1039/C9RA01411J.
173. Iryani DA, Kumagai S, Nonaka M, Sasaki K, Hirajima T. Production of 5-hydroxymethyl furfural from sugarcane bagasse under hot compressed water. Proc Earth Planet Sci. 2013;6:441–7.
174. Zhang L, Xi G, Zhang J, Yu H, Wang X. Efficient catalytic system for the direct transformation of lignocellulosic biomass to furfural and 5-hydroxymethylfurfural. Bioresour Technol. 2017;224:656–61. https://doi.org/10.1016/j.biortech.2016.11.097.
175. Yang Y, Hu C-W, Abu-Omar MM. Conversion of carbohydrates and lignocellulosic biomass into 5-hydroxymethylfurfural using AlCl3·6H2O catalyst in a biphasic solvent system. Green Chem. 2012;14(2):509–13. https://doi.org/10.1039/C1GC15972K.
176. Lai L, Zhang Y. The production of 5-hydroxymethylfurfural from fructose in isopropyl alcohol: a green and efficient system. ChemSusChem. 2011;4(12):1745–8. https://doi.org/10.1002/cssc.201100489.
177. Lin HZ, Xiong QG, Zhao Y, Chen JP, Wang SR. Conversion of carbohydrates into 5-hydroxymethylfurfural in a green reaction system of CO-water-isopropanol. AICHE J. 2017;63(1):257–65. https://doi.org/10.1002/aic.15550.
178. Qi L, Mui YF, Lo SW, Lui MY, Akien GR, Horváth IT. Catalytic conversion of fructose, glucose, and sucrose to 5-(hydroxymethyl)furfural and levulinic and formic acids in γ-valerolactone as a green solvent. ACS Catal. 2014;4(5):1470–7. https://doi.org/10.1021/cs401160y.
179. Gallo JMR, Alonso DM, Mellmer MA, Dumesic JA. Production and upgrading of 5-hydroxymethylfurfural using heterogeneous catalysts and biomass-derived solvents. Green Chem. 2013;15(1):85–90. https://doi.org/10.1039/c2gc36536g.
180. Wang HL, Kong QQ, Wang YX, Deng TS, Chen CM, Hou XL, Zhu YL. Graphene oxide catalyzed dehydration of fructose into 5-hydroxymethylfurfural with isopropanol as cosolvent. ChemCatChem. 2014;6(3):728–32. https://doi.org/10.1002/cctc.201301067.
181. Alonso DM, Bond JQ, Dumesic JA. Catalytic conversion of biomass to biofuels. Green Chem. 2010;12(9):1493–513. https://doi.org/10.1039/c004654j.

182. Shi N, Liu QY, Zhang Q, Wang TJ, Ma LL. High yield production of 5-hydroxymethylfurfural from cellulose by high concentration of sulfates in biphasic system. Green Chem. 2013;15(7): 1967–74. https://doi.org/10.1039/c3gc40667a.
183. Cai CM, Zhang TY, Kumar R, Wyman CE. THF co-solvent enhances hydrocarbon fuel precursor yields from lignocellulosic biomass. Green Chem. 2013;15(11):3140–5. https://doi.org/10.1039/c3gc41214h.
184. Pagán-Torres YJ, Wang T, Gallo JMR, Shanks BH, Dumesic JA. Production of 5-hydroxymethylfurfural from glucose using a combination of Lewis and Brønsted acid catalysts in water in a biphasic reactor with an alkylphenol solvent. ACS Catal. 2012;2(6): 930–4. https://doi.org/10.1021/cs300192z.
185. Fittig R, Heinzelmann H. Production of 2, 5-furandicarboxylic acid by the reaction of fuming hydrobromic acid with mucic acid under pressure. Chem Ber. 1876;9:1198.
186. Ma ZM, Wang L, Li GY, Song T. Recent advances in electrocatalytic oxidation of 5-hydroxymethylfurfural to 2,5-furandicarboxylic acid by heterogeneous catalysts. Catalysts. 2024;14(2):157. https://doi.org/10.3390/catal14020157.
187. Pandey S, Mottoul M, Orsat V, Morin J-F, Dumont M-J. Base-free synthesis of renewable furan-2, 5-dicarboxylic acid (FDCA) over an earth-abundant magnetic catalyst. J Environ Chem Eng. 2024;12(3):112763. https://doi.org/10.1016/j.jece.2024.112763.
188. Guo L, Zhang X, Gan L, Pan L, Shi C, Huang ZF, Zhang X, Zou JJ. Advances in selective electrochemical oxidation of 5-hydroxymethylfurfural to produce high-value chemicals. Adv Sci (Weinh). 2023;10(4):e2205540. https://doi.org/10.1002/advs.202205540.
189. Holade Y, Morais C, Servat K, Napporn TW, Kokoh KB. Toward the electrochemical valorization of glycerol: Fourier transform infrared spectroscopic and chromatographic studies. ACS Catal. 2013;3(10):2403–11. https://doi.org/10.1021/cs400559d.
190. Vuyyuru KR, Strasser P. Oxidation of biomass derived 5-hydroxymethylfurfural using heterogeneous and electrochemical catalysis. Catal Today. 2012;195(1):144–54. https://doi.org/10.1016/j.cattod.2012.05.008.
191. Krystof M, Perez-Sanchez M, Dominguez de Maria P. Lipase-mediated selective oxidation of furfural and 5-hydroxymethylfurfural. ChemSusChem. 2013;6(5):826–30. https://doi.org/10.1002/cssc.201200954.
192. Hanke PD. U.S. Patent 2009/023174 A2, 2009.
193. McKenna SM, Leimkühler S, Herter S, Turner NJ, Carnell AJ. Enzyme cascade reactions: synthesis of furandicarboxylic acid (FDCA) and carboxylic acids using oxidases in tandem. Green Chem. 2015;17(6):3271–5. https://doi.org/10.1039/c5gc00707k.
194. Troiano D, Orsat V, Dumont M-J. Use of filamentous fungi as biocatalysts in the oxidation of 5-(hydroxymethyl) furfural (HMF). Bioresour Technol. 2022;344:126169. https://doi.org/10.1016/j.biortech.2021.126169.
195. Rajesh RO, Godan TK, Rai AK, Sahoo D, Pandey A, Binod P. Biosynthesis of 2,5-furan dicarboxylic acid by Aspergillus flavus APLS-1: process optimization and intermediate product analysis. Bioresour Technol. 2019;284:155–60. https://doi.org/10.1016/j.biortech.2019.03.105.
196. Saha B, Dutta S, Abu-Omar MM. Aerobic oxidation of 5-hydroxylmethylfurfural with homogeneous and nanoparticulate catalysts. Cat Sci Technol. 2012;2(1):79–81. https://doi.org/10.1039/c1cy00321f.
197. Hansen TS, Sádaba I, García-Suárez EJ, Riisager A. Cu catalyzed oxidation of 5-hydroxymethylfurfural to 2,5-diformylfuran and 2,5-furandicarboxylic acid under benign reaction conditions. Appl Catal A Gen. 2013;456:44–50. https://doi.org/10.1016/j.apcata.2013.01.042.
198. Davis SE, Houk LR, Tamargo EC, Datye AK, Davis RJ. Oxidation of 5-hydroxymethylfurfural over supported Pt, Pd and Au catalysts. Catal Today. 2011;160(1): 55–60. https://doi.org/10.1016/j.cattod.2010.06.004.

199. Casanova O, Iborra S, Corma A. Biomass into chemicals: aerobic oxidation of 5-hydroxymethyl-2-furfural into 2,5-furandicarboxylic acid with gold nanoparticle catalysts. ChemSusChem. 2009;2(12):1138–44. https://doi.org/10.1002/cssc.200900137.
200. Rass HA, Essayem N, Besson M. Selective aqueous phase oxidation of 5-hydroxymethylfurfural to 2,5-furandicarboxylic acid over Pt/C catalysts: influence of the base and effect of bismuth promotion. Green Chem. 2013;15(8):2240–51. https://doi.org/10.1039/c3gc40727f.
201. Verdeguer P, Merat N, Gaset A. Oxydation catalytique du HMF en acide 2, 5-furane dicarboxylique. J Mol Catal. 1993;85(3):327–44. https://doi.org/10.1016/0304-5102(93)80059-4.
202. Ait Rass H, Essayem N, Besson M. Selective aerobic oxidation of 5-HMF into 2,5-furandicarboxylic acid with Pt catalysts supported on TiO2-and ZrO2-based supports. ChemSusChem. 2015;8(7):1206–17. https://doi.org/10.1002/cssc.201403390.
203. Niu WQ, Wang D, Yang GH, Sun J, Wu MB, Yoneyama Y, Tsubaki N. Pt nanoparticles loaded on reduced graphene oxide as an effective catalyst for the direct oxidation of 5-hydroxymethylfurfural (HMF) to produce 2,5-furandicarboxylic acid (FDCA) under mild conditions. Bull Chem Soc Jpn. 2014;87(10):1124–9. https://doi.org/10.1246/bcsj.20140096.
204. Vinke P, van Dam HE, van Bekkum H. Platinum catalyzed oxidation of 5-hydroxymethylfurfural. In: Centi G, Trifiro F, editors. New developments in selective oxidation, Studies in surface science and catalysis, vol. 55; 1990. p. 147–58. https://doi.org/10.1016/S0167-2991(08)60144-5.
205. Sahu R, Dhepe PL. Synthesis of 2,5-furandicarboxylic acid by the aerobic oxidation of 5-hydroxymethyl furfural over supported metal catalysts. React Kinet Mech Catal. 2014;112(1):173–87. https://doi.org/10.1007/s11144-014-0689-z.
206. Miao Z, Wu T, Li J, Yi T, Zhang Y, Yang X. Aerobic oxidation of 5-hydroxymethylfurfural (HMF) effectively catalyzed by a Ce 0.8 Bi 0.2 O 2− δ supported Pt catalyst at room temperature. RSC Adv. 2015;5(26):19823–9. https://doi.org/10.1039/C4RA16968A.
207. Siankevich S, Savoglidis G, Fei ZF, Laurenczy G, Alexander DTL, Yan N, Dyson PJ. A novel platinum nanocatalyst for the oxidation of 5-hydroxymethylfurfural into 2,5-Furandicarboxylic acid under mild conditions. J Catal. 2014;315:67–74. https://doi.org/10.1016/j.jcat.2014.04.011.
208. Albonetti S, Lolli A, Morandi V, Migliori A, Lucarelli C, Cavani F. Conversion of 5-hydroxymethylfurfural to 2,5-furandicarboxylic acid over Au-based catalysts: optimization of active phase and metal-support interaction. Appl Catal B Environ. 2015;163:520–30. https://doi.org/10.1016/j.apcatb.2014.08.026.
209. Cai J, Ma H, Zhang J, Song Q, Du Z, Huang Y, Xu J. Gold nanoclusters confined in a supercage of Y zeolite for aerobic oxidation of HMF under mild conditions. Chemistry. 2013;19(42):14215–23. https://doi.org/10.1002/chem.201301735.
210. Gorbanev YY, Klitgaard SK, Woodley JM, Christensen CH, Riisager A. Gold-catalyzed aerobic oxidation of 5-hydroxymethylfurfural in water at ambient temperature. ChemSusChem. 2009;2(7):672–5. https://doi.org/10.1002/cssc.200900059.
211. Rasul S, Anjum DH, Jedidi A, Minenkov Y, Cavallo L, Takanabe K. A highly selective copper-indium bimetallic electrocatalyst for the electrochemical reduction of aqueous CO2 to CO. Angew Chem Int Ed Engl. 2015;54(7):2146–50. https://doi.org/10.1002/anie.201410233.
212. Villa A, Schiavoni M, Campisi S, Veith GM, Prati L. Pd-modified Au on carbon as an effective and durable catalyst for the direct oxidation of HMF to 2,5-furandicarboxylic acid. ChemSusChem. 2013;6(4):609–12. https://doi.org/10.1002/cssc.201200778.
213. Wan X, Zhou C, Chen J, Deng W, Zhang Q, Yang Y, Wang Y. Base-free aerobic oxidation of 5-hydroxymethyl-furfural to 2, 5-furandicarboxylic acid in water catalyzed by functionalized carbon nanotube-supported Au–Pd alloy nanoparticles. ACS Catal. 2014;4(7):2175–85. https://doi.org/10.1021/cs5003096.
214. Zhang Z, Deng D. Recent advances in the catalytic synthesis of 2,5-furandicarboxylic acid and its derivatives. ACS Catal. 2015;5(11):6529–44. https://doi.org/10.1021/acscatal.5b01491.

215. Wu Y, Wang D, Zhou G, Yu R, Chen C, Li Y. Sophisticated construction of Au islands on Pt-Ni: an ideal trimetallic nanoframe catalyst. J Am Chem Soc. 2014;136(33):11594–7. https://doi.org/10.1021/ja5058532.
216. Hashmi AK, Hutchings GJ. Angew Chem Int Ed. 2006;45:7896–936. https://doi.org/10.1002/anie.200602454.
217. Jiang M, Liu Q, Zhang Q, Ye C, Zhou G. A series of furan-aromatic polyesters synthesized via direct esterification method based on renewable resources. J Polym Sci A Polym Chem. 2012;50(5):1026–36. https://doi.org/10.1002/pola.25859.
218. Lilga MA, Hallen RT, Gray M. Production of oxidized derivatives of 5-hydroxymethylfurfural (HMF). Top Catal. 2010;53(15–18):1264–9. https://doi.org/10.1007/s11244-010-9579-4.
219. Yi GS, Teong SP, Zhang YG. Base-free conversion of 5-hydroxymethylfurfural to 2,5-furandicarboxylic acid over a Ru/C catalyst. Green Chem. 2016;18(4):979–83. https://doi.org/10.1039/c5gc01584g.
220. Totaro G, Sisti L, Marchese P, Colonna M, Romano A, Gioia C, Vannini M, Celli A. Current advances in the sustainable conversion of 5-hydroxymethylfurfural into 2,5-furandicarboxylic acid. ChemSusChem. 2022;15(13):e202200501. https://doi.org/10.1002/cssc.202200501.
221. Hayashi E, Yamaguchi Y, Kamata K, Tsunoda N, Kumagai Y, Oba F, Hara M. Effect of MnO2 crystal structure on aerobic oxidation of 5-hydroxymethylfurfural to 2, 5-furandicarboxylic acid. J Am Chem Soc. 2019;141(2):890–900. https://doi.org/10.1021/jacs.8b09917.
222. Yamaguchi Y, Aono R, Hayashi E, Kamata K, Hara M. Template-free synthesis of mesoporous β-MnO nanoparticles: structure, formation mechanism, and catalytic properties. ACS Appl Mater Interfaces. 2020;12(32):36004–13. https://doi.org/10.1021/acsami.0c08043.
223. Bao L, Sun FZ, Zhang GY, Hu TL. Aerobic oxidation of 5-hydroxymethylfurfural to 2, 5-furandicarboxylic acid over Holey 2 D Mn2O3 nanoflakes from a Mn-based MOF. ChemSusChem. 2020;13(3):548–55. https://doi.org/10.1002/cssc.201903018.
224. Han X, Li C, Liu X, Xia Q, Wang Y. Selective oxidation of 5-hydroxymethylfurfural to 2,5-furandicarboxylic acid over MnOx–CeO2 composite catalysts. Green Chem. 2017;19(4):996–1004. https://doi.org/10.1021/jacs.7b08861.
225. Wan X, Tang N, Xie Q, Zhao S, Zhou C, Dai Y, Yang Y. A CuMn 2 O 4 spinel oxide as a superior catalyst for the aerobic oxidation of 5-hydroxymethylfurfural toward 2, 5-furandicarboxylic acid in aqueous solvent. Cat Sci Technol. 2021;11(4):1497–509. https://doi.org/10.1039/D0CY01649G.
226. Liu H, Jia WL, Yu X, Tang X, Zeng XH, Sun Y, Lei TZ, Fang HY, Li TY, Lin L. Vitamin C-assisted synthesized Mn-Co oxides with improved oxygen vacancy concentration: boosting lattice oxygen activity for the air-oxidation of 5-(hydroxymethyl)furfural. ACS Catal. 2021;11(13):7828–44. https://doi.org/10.1021/acscatal.0c04503.
227. Chen GY, Wu LB, Fan H, Li BG. Highly efficient two-step synthesis of 2,5-furandicarboxylic acid from fructose without 5-hydroxymethylfurfural (HMF) separation: in Situ oxidation of HMF in alkaline aqueous H2O/DMSO mixed solvent under mild conditions. Ind Eng Chem Res. 2018;57(48):16172–81. https://doi.org/10.1021/acs.iecr.8b03589.
228. Kröger M, Prüße U, Vorlop K-D. A new approach for the production of 2, 5-furandicarboxylic acid by in situ oxidation of 5-hydroxymethylfurfural starting from fructose. Top Catal. 2000;13(3):237–42. https://doi.org/10.1023/A:1009017929727.
229. Ribeiro ML, Schuchardt U. Cooperative effect of cobalt acetylacetonate and silica in the catalytic cyclization and oxidation of fructose to 2,5-furandicarboxylic acid. Catal Commun. 2003;4(2):83–6. https://doi.org/10.1016/S1566-7367(02)00261-3.
230. Schade OR, Dannecker PK, Kalz KF, Steinbach D, Meier MAR, Grunwaldt JD. Direct catalytic route to biomass-derived 2,5-furandicarboxylic acid and its use as monomer in a multicomponent polymerization. ACS Omega. 2019;4(16):16972–9. https://doi.org/10.1021/acsomega.9b02373.

231. Wang SG, Zhang ZH, Liu B. Catalytic conversion of fructose and 5-hydroxymethylfurfural into 2,5-furandicarboxylic acid over a recyclable FeO-CoO magnetite nanocatalyst. ACS Sustain Chem Eng. 2015;3(3):406–12. https://doi.org/10.1021/sc500702q.
232. Xie WZ, Liu H, Tang X, Zeng XH, Sun Y, Ke XX, Li TY, Fang HY, Lin L. Efficient synthesis of 2,5-furandicarboxylic acid from biomass-derived 5-hydroxymethylfurfural in 1,4-dioxane/HO mixture. Appl Catal A Gen. 2022;630:118463. https://doi.org/10.1016/j.apcata.2021.118463.
233. Yang Z, Qi W, Su R, He Z. Selective synthesis of 2, 5-diformylfuran and 2, 5-furandicarboxylic acid from 5-hydroxymethylfurfural and fructose catalyzed by magnetically separable catalysts. Energy Fuel. 2017;31(1):533–41. https://doi.org/10.1021/acs.energyfuels.6b02012.
234. Yi G, Teong SP, Li X, Zhang Y. Purification of biomass-derived 5-hydroxymethylfurfural and its catalytic conversion to 2,5-furandicarboxylic acid. ChemSusChem. 2014;7(8):2131–5. https://doi.org/10.1002/cssc.201402105.
235. Yi GS, Teong SP, Zhang YG. The direct conversion of sugars into 2,5-furandicarboxylic acid in a triphasic system. ChemSusChem. 2015;8(7):1151–5. https://doi.org/10.1002/cssc.201500118.
236. Ban H, Cheng Y, Wang L, Li X. One-pot method for the synthesis of 2, 5-furandicarboxylic acid from fructose: in situ oxidation of 5-hydroxymethylfurfural and 5-acetoxymethylfurfural over Co/Mn/Br catalysts in acetic acid. Ind Eng Chem Res. 2022;62(1):291–301. https://doi.org/10.1021/acs.iecr.2c03815.
237. Liu H, Cao XJ, Wang T, Wei JN, Tang X, Zeng XH, Sun Y, Lei TZ, Liu SJ, Lin L. Efficient synthesis of bio-monomer 2,5-furandicarboxylic acid from concentrated 5-hydroxymethylfurfural or fructose in DMSO/HO mixed solvent. J Ind Eng Chem. 2019;77:209–14. https://doi.org/10.1016/j.jiec.2019.04.038.
238. Manjunathan P, Rao BS, Lee MM, Hidajat MJ, Yun GN, Hwang DW. Integrated process towards sustainable renewable plastics: production of 2,5-furandicarboxylic acid from fructose in a base-free environment. Appl Catal A Gen. 2023;667:119446. https://doi.org/10.1016/j.apcata.2023.119446.
239. Chai Y, Yang H, Bai M, Chen A, Peng L, Yan B, Zhao D, Qin P, Peng C, Wang X. Direct production of 2, 5-Furandicarboxylicacid from raw biomass by manganese dioxide catalysis cooperated with ultrasonic-assisted diluted acid pretreatment. Bioresour Technol. 2021;337:125421. https://doi.org/10.1016/j.biortech.2021.125421.
240. Naim W, Schade OR, Saraçi E, Wüst D, Kruse A, Grunwaldt JD. Toward an intensified process of biomass-derived monomers: the influence of 5-(hydroxymethyl)furfural byproducts on the gold-catalyzed synthesis of 2,5-furandicarboxylic acid Weiss. ACS Sustain Chem Eng. 2020;8(31):11512–21. https://doi.org/10.1021/acssuschemeng.0c01319.
241. Al Ghatta A, Hallett JP. High yield and isolation of 2, 5-furandicarboxylic acid from HMF and sugars in ionic liquids, a new prospective for the establishment of a scalable and efficient catalytic route. Green Chem. 2022;24(8):3309–13. https://doi.org/10.1039/D2GC00390B.
242. Chen G, Zheng L, Sun Q, Zong B. High-yield synthesis of 2, 5-Furandicarboxylic acid from fructose via a two-step strategy without 5-Hydroxymethylfurfural isolation using dioxane as a bridging solvent. Mol Catal. 2024;552:113677. https://doi.org/10.1016/j.mcat.2023.113677.
243. Mittal N, Kaur M, Singh V. Mild and selective catalytic oxidation of 5-HMF to 2,5-FDCA over metal loaded biomass derived graphene oxide using hydrogen peroxide in aqueous media. Mol Catal. 2023;546:113223. https://doi.org/10.1016/j.mcat.2023.113223.
244. Wrigstedt P, Keskiväli J, Perea-Buceta JE, Repo T. One-pot transformation of carbohydrates into valuable furan derivatives via 5-hydroxymethylfurfural. ChemCatChem. 2017;9(22):4244–55. https://doi.org/10.1002/cctc.201701106.

245. Parate RD, Dharne MS, Rode CV. Integrated chemo and bio-catalyzed synthesis of 2,5-furandicarboxylic acid from fructose derived 5-hydroxymethylfurfural. Biomass Bioenergy. 2022;161:106474. https://doi.org/10.1016/j.biombioe.2022.106474.
246. Han X, Wang Y, Liu G, Wang M, Guo C, Shen J. A sustainable and low-cost route to 2, 5-furandicarboxylic acid by carboxylation of biomass-based furoic acid and CO2. J CO2 Util. 2023;75:102572. https://doi.org/10.1016/j.jcou.2023.102572.

Chapter 7
Catalytic Conversion of Agricultural Residues and Food Wastes into 2,5-Furandicarboxylic Acid

Kui Wang, Mingyang Che, Taotao Lu, Yilun Luo, Jiaojiao Zhan, and Yuxiang Zhao

Abstract Agricultural and food wastes, rich in carbohydrates, are unavoidable biomass wastes that can be efficiently valorized into high-value molecules like 5-hydroxymethylfurfural (HMF) and subsequently 2,5-furandicarboxylic acid (FDCA). This chapter outlines the catalytic conversion process of carbohydrates from biomass into FDCA using hydrothermal processing and catalytic oxidation. FDCA, a valuable intermediate and substitute for terephthalic acid (TPA) in polyesters, can be converted into biofuels, plastics, and pharmaceuticals. Despite its potential, the conversion process is complex, yielding numerous by-products and requiring costly catalysts. Recent advancements in catalyst and solvent recycling, solvent system optimization, and process intensification through microreactors show promise for improving economic viability. Transitioning to continuous reactors with advanced electrocatalysts and innovative catalytic materials like nitrogen-doped porous carbon and bimetallic catalysts enhance adsorption performance and catalytic activity. Addressing the solubility issues of agricultural wastes and selecting appropriate media remain critical challenges. Future research should focus on cost-effective methods to improve selectivity and yield, leveraging interdisciplinary approaches and advanced technologies like machine learning. Integrating these advancements will enable scalable, sustainable FDCA production from biomass resources, paving the way for environmentally friendly chemical production.

Keywords Biomass wastes · Hydrothermal conversion · HMF · FDCA · Catalytic conversion

K. Wang (✉)
School of Chemistry and Chemical Engineering, Hefei University of Technology, Hefei, China

Smith School of Chemical and Biomolecular Engineering, Cornell University, Ithaca, NY, USA
e-mail: kuiwang@hfut.edu.cn

M. Che · T. Lu · Y. Luo · J. Zhan · Y. Zhao
School of Chemistry and Chemical Engineering, Hefei University of Technology, Hefei, China

Z. Fang et al. (eds.), *Production of Organic Acids and Alcohols from Agricultural Residues and Food Wastes*, Biofuels and Biorefineries 13,
https://doi.org/10.1007/978-981-96-1222-2_7

7.1 Introduction

7.1.1 Hydrothermal Conversion of Biomass

Many studies have shown that biomass is a viable feedstock for sustainable energy and the production of value-added products [1]. Biomass serves as the largest source for materials with added value, capable of substituting current fossil-based materials. However, the use of biomass in biorefineries is still in the early stages of development, due to the heterogeneity of feedstock properties and the intricacies of the conversion processes.

Biomass wastes, a crucial renewable resource, can also be converted into chemicals, animal feed, raw materials, and other valuable substances. Common biomass wastes include agricultural and forestry residues, food wastes, livestock manure, sludge, and algae, etc. [2, 3]. Agricultural and forestry wastes, such as straw, rice husk, edible fungi substrate, and various plant-derived materials, primarily comprises lignocellulose, consisting mainly of carbon (C), hydrogen (H), oxygen (O), nitrogen (N), and sulfur (S). Food wastes, originating from vegetables, fruits, grains, meat, dairy products, and eggs, is predominantly composed of carbohydrates, proteins, lipids, and trace elements like phosphorus (P), sodium (Na), calcium (Ca), and magnesium (Mg), essential for microorganisms.

Figure 7.1 illustrates the hydrothermal conversion of various biomass types into valuable products, highlighting its potential for sustainable energy production, chemical synthesis, and waste management. The left side presents diverse biomass

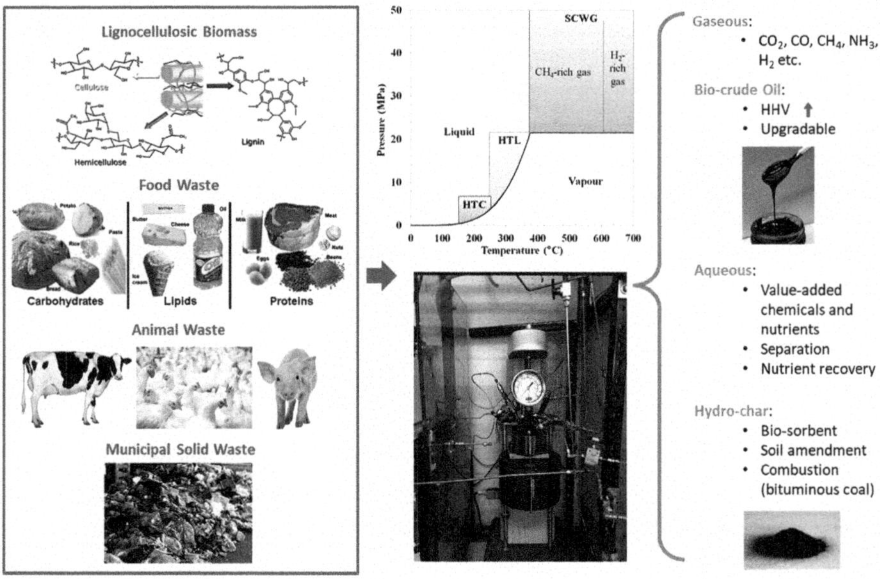

Fig. 7.1 Hydrothermal conversion of biomass wastes into useful products. (Adapted with permission from [3], Copyright © 2023 Elsevier)

sources: lignocellulosic biomass (43–47% cellulose, 25–35% hemicellulose, and 16–24% lignin) [4], food waste (carbohydrates, fats, proteins), animal waste (cow manure, chicken litter, pig manure), and municipal solid waste (household and business garbage, including plastic, paper, and food scraps) [3]. The right side showcases the hydrothermal treatment processes: hydrothermal carbonization (HTC), hydrothermal liquefaction (HTL), and supercritical water gasification (SCWG), each varying by temperature and pressure.

HTC typically operates at temperatures below 250 °C and yields hydrochar resembling bituminous coal. HTC, also known as wet torrefaction, represents an environmentally friendly thermochemical conversion technology with the potential to convert biomass wastes with high water contents, known as unavoidable biomass wastes, into hydrochar [3, 5]. The typical HTC process involves hydrolysis, decarboxylation, and re-polymerization reactions in either batch-type or continuous-type reactors [6, 7].

In the temperature range of 250–374 °C, HTL produces dense liquid products similar to petroleum crude oil, convertible into transportation fuel through hydrogenation. Compared to HTC, HTL is more conducive to biocrude production due to higher temperatures [6–8]. HTL can achieve full conversion of organics in biomass wastes, especially fats, proteins, and carbohydrates.

Above 374 °C, SCWG becomes predominant, leading to the production of synthesis gas, methane, or hydrogen, depending on specific temperature conditions [3, 9]. SCWG exhibits higher gasification efficiencies at lower temperatures compared to traditional gasification pathways. As the process occurs at supercritical pressure, smaller reactor volumes can be used, and the resulting pressurized gas products can be directly stored in pressurized tanks, leading to significant energy savings [10]. SCWG of biomass demonstrates high conversion rates to gases primarily composed of H_2, CO, CO_2, and C_1–C_4 hydrocarbons [11, 12].

Notably, biomass hydrothermal conversion eliminates the need for drying pretreatment of the feedstock, thus avoiding significant energy consumption [3, 13]. Water serves as both the solvent and the reaction medium during the process. As temperature and pressure increase, water transforms into a non-polar solvent with increasing ion product and reduced dielectric constant [14]. Hydrochar yield varies between 30% and 70%, influenced by biomass properties, preliminary treatment methods, operating conditions, reactor type, water-biomass ratio, and initial pH values [3]. Additionally, hydrothermal conversion is an exothermic process, which releases 25–38% of the energy value of dry biomass, enhancing the overall energy balance [15].

7.1.2 Roadmap from Biomass to FDCA

The escalating global demand for sustainable and environmentally friendly materials has positioned the development of synthetic pathways for green chemistry and bio-based platform molecules at the forefront of chemical research [16]. In this vein, the investigation of 5-hydroxymethylfurfural (HMF) and its derivative 2,5-furandicarboxylic acid (FDCA) has garnered extensive attention.

For the past decades, scholars have engaged in extensive research focused on converting agricultural and food wastes into furanic compounds, especially FDCA, which was considered a potential green replacement for TPA (terephthalic acid) and PET (polyethylene terephthalate) manufacturing. To achieve this objective, a range of methods and processes have been developed to economically and sustainably utilize such biomass, including hydrothermal conversion [17], infrared [18], ultrasonic and microwave radiation as energy media for the reactions [19, 20], etc. Lab-scale development of these methods has demonstrated outstanding results in converting biomass into furans. A general roadmap to produce FDCA from food and agricultural wastes can be summarized into two stages: (1) hydrothermal conversion of biomass wastes into HMF; and (2) catalytic conversion of HMF to FDCA.

HMF is recognized as a prominent platform molecule obtained from biomass resources like cellulose [21]. On the other hand, FDCA is identified as a promising bio-based precursor for polymers, potentially replacing petroleum-derived TPA. Notably, FDCA exhibits excellent performance in the synthesis of PET plastics [22], contributing to the development of more sustainable and bio-based materials.

Recent research has made significant progress in developing comprehensive technology for deriving FDCA from biomass and conducting an economic and environmental sustainability analysis [23–25]. The selection and optimization of the catalytic system are crucial in converting raw biomass and carbohydrates into HMF. Initially, cellulose is dehydrated to produce HMF, which is then oxidized to generate FDCA. The dehydration of cellulose to produce HMF is typically catalyzed by acids such as sulfuric, hydrochloric, and phosphoric acid. Studies show that using solid acid catalysts, such as ion exchange resins, can increase the HMF yield and simplify product separation, achieving a molar yield of 42% [25]. The oxidation of HMF to FDCA usually requires metal catalysts, with palladium (Pd)-based systems enabling an oxidation yield of 93.6%, enhancing FDCA yield and reducing by-products [26].

Multi-functional catalysts play a vital role in biomass conversion. Catalysts need high activity, selectivity, stability, and recyclability [27, 28]. Palladium catalysts can catalyze not only cellulose dehydration but also other dehydrogenation and hydrogenation reactions. Adding metals like copper or nickel to palladium catalysts allows precise control of activity and selectivity, improving the dehydration of cellulose to produce HMF. Nanocomposites with non-precious metals and magnetic core showed exceptional high conversion of 99.7% HMF conversion into FDCA [21].

Solvents also play a significant role in chemical reactions, where different types of solvents can markedly influence the reaction rate, selectivity, and yield. Various solvent systems were explored in the study to optimize the production process of FDCA. Water, as a solvent, boasts environmentally friendly, inexpensive, and non-toxic characteristics. However, in aqueous systems, HMF is prone to further degradation, necessitating the addition of salts or organic solvents to suppress side reactions. Organic solvents such as dimethyl sulfoxide (DMSO), dichloromethane, methanol, and ethanol can enhance reaction selectivity and yield. Other commonly used solvents such as γ-valerolactone (GVL)/H_2O solvent system [24, 25] and ionic liquids [29] represent a class of green solvents with low volatility and high thermal stability. Research indicates that ionic liquids can significantly improve the solubility of cellulose and reaction rates, thereby increasing the yield of HMF and FDCA [30].

Furthermore, by optimizing process parameters and catalyst prices, the minimum selling price of FDCA can be reduced to $1532/ton [127], highlighting the potential of biomass-derived FDCA as a substitute for petroleum-derived TPA. Life cycle assessments suggest that the production of biomass-derived FDCA is more environmentally favorable than petroleum-derived PTA production, exhibiting lower greenhouse gas emissions and energy consumption [23].

The conversion of HMF to FDCA involves a series of intricate chemical reactions, primarily consisting of multiple oxidation steps. Presently, the main pathways for this conversion include chemical oxidation, biocatalysis, and electrochemical oxidation. Each conversion pathway possesses distinct characteristics, and the selection of a particular method should consider factors such as reaction efficiency, cost, environmental impact, and product purity. Figure 7.2 shows the map of furan products from biomass that can serve as valuable monomers for the synthesis of new polymers, pharmaceutical intermediates, precursors for sustainable dyes, and platform compounds for various value-added chemicals [31, 32]. FDCA, characterized by its dicarboxylic acid structure, bears structural similarity to TPA derived from fossil fuels. Hence, FDCA is considered a renewable alternative to TPA in the manufacturing of PET polymer materials. The polymerization of FDCA with ethylene glycol yields polyethylene furanoate (PEF), which exhibits superior thermal performance and gas barrier properties (O_2, CO_2, H_2O) compared to PET. Consequently, PEF stands as a promising option for food and beverage packaging materials [31, 32]. For comparison, Table 7.1 lists the properties of FDCA, TPA, PEF, and PET.

Ongoing research has revealed that different reaction pathways can lead to diverse outcomes. The subsequent sections of this chapter will intricately explore the conversion steps described above. The intention is to offer a comprehensive overview of FDCA production through the catalytic conversion of biomass wastes, with a specific focus on carbohydrate-rich agricultural and food wastes.

7.2 Production of HMF from Biomass

7.2.1 Basic Principles

The main components of biomass are cellulose, lignin, hemicellulose, starch, protein and lipids [33]. There are many ways to produce HMF from raw biomass wastes. Studies have shown that HMF can be obtained not only from monosaccharides such as glucose [34] and fructose [35], but also from polysaccharides such as cellulose [36], inulin [37], chitin [38], etc. In recent years, the main feedstocks for the

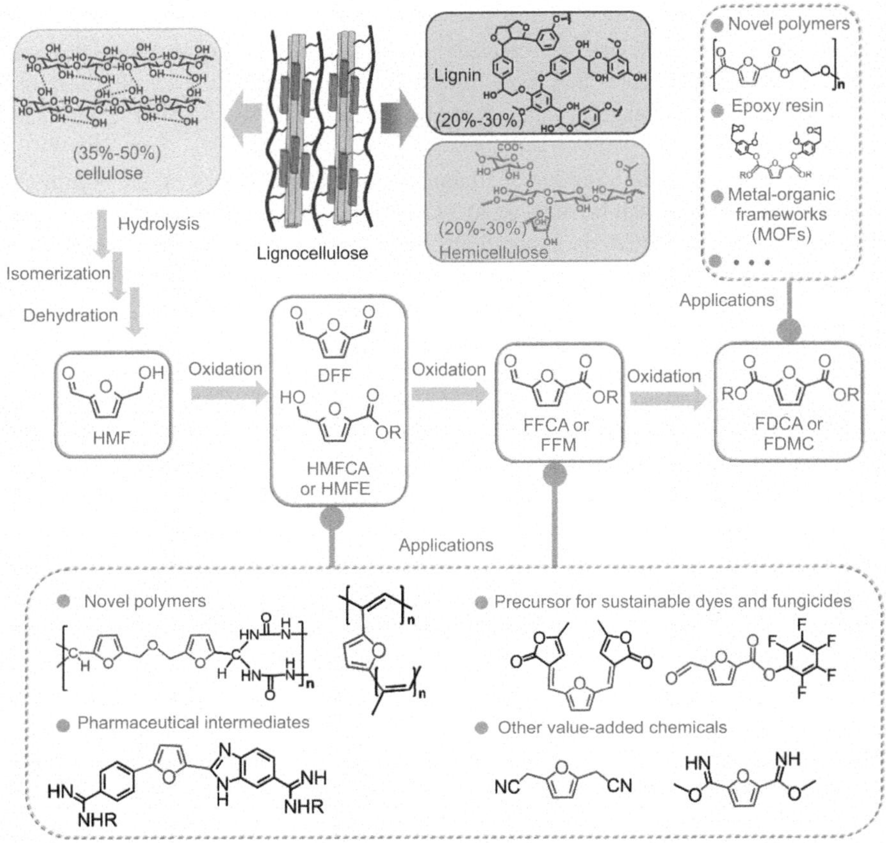

Fig. 7.2 Synthetic route for producing furanic chemicals from lignocellulose through selective oxidation of HMF and their applications. (Adapted with permission from [32], Copyright © 2022 Elsevier)

synthesis of biomass based HMF can be categorized as edible biomass, non-edible lignocellulosic biomass, and food waste biomass [39]. Due to their different chemical structures, the reaction networks also vary significantly. However, only biomass composed of monosaccharides or polysaccharides can be effectively converted into HMF in hydrothermal conversion reactions. Therefore, carbohydrate-rich agricultural and food wastes are the two primary biomass feedstocks for HMF production, including starchy root vegetable residues, cereal husks and bran, food processing residues, fruit peels and juice, etc.

HMF can be obtained through dehydration of hexose such as glucose and fructose, which are the two mostly studied compounds in the literature, but fructose usually results in a higher HMF yield than glucose due to the structural differences that fructose is more reactive than glucose in hydrothermal conversion process [40–42]. In addition, glucose can be isomerized into fructose [43], which can be catalyzed by bases, Lewis acids and enzymes.

Table 7.1 Comparison of the physical and chemical properties of FDCA, TPA, PEF, and PET

Property	FDCA	TPA	PEF	PET
Chemical formula	$C_6H_4O_5$	$C_8H_6O_4$	$(C_{10}H_8O_5)_n$	$(C_{10}H_8O_4)_n$
Molecular weight	156.09 g/mol	166.13 g/mol	192.17 g/mol	192.17 g/mol
Melting point	342 °C	300 °C	210–220 °C	260 °C
Density	1.61 g/cm^3	1.52 g/cm^3	1.37 g/cm^3	1.38 g/cm^3
Solubility	Soluble in alkaline solution	Slightly soluble in water	Poorly soluble in water	Poorly soluble in water
Thermal stability	High	High	High	High
Glass Transition temp (T_g)	–	–	86–92 °C	70–80 °C
Mechanical properties	–	–	High strength, high modulus	High strength, high modulus
Gas barrier properties	–	–	High	Low
Biodegradability	Biodegradable	Non-biodegradable	Biodegradable	Non-biodegradable
Main applications	Monomer for bio-based plastics	Monomer for synthetic polyesters	Bio-based plastic alternative to PET	Bottles, fibers, films, etc.
Molecular structure	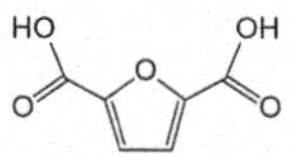	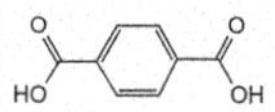	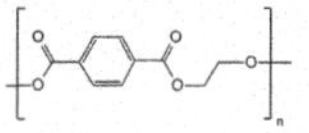	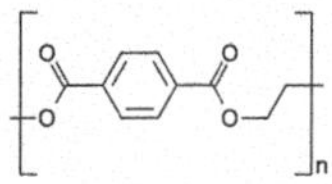

The main degradation pathways of glucose and fructose in hydrothermal conversion process is shown in Fig. 7.3. Yan et al. [44] investigated the isomerization of glucose to fructose and its further dehydration to produce HMF in subcritical water. They found that Brønsted base promotes glucose isomerization to fructose and Brønsted acid can significantly promote the dehydration of fructose to HMF. Glucose can also undergo retro-aldol condensation to glycolaldehyde and *d*-erythrose [45].

Cao et al. [46] found that the main product of the non-catalytic hydrothermal conversion of glucose was HMF at 160–200 °C. With the further increase of reaction temperature, the concentration of HMF began to decrease sharply due to degradation or polymerization into humic compounds. The reaction temperature and residence time are crucial factors for the yield of HMF. Levulinic acid (LA) and formic acid (FA) are two key byproducts formed by the rehydration of HMF, where a water molecule is added to the C_2–C_3 bond of the furan ring in HMF, forming LA and FA as the final products [47]. Rehydration of HMF is a key side reaction that lowers HMF to FDCA efficiency.

Cellulose is a polymer of glucose. The glucose monomer in cellulose is connected through β-1,4-glycosidic bonds [48]. Efficient hydrolysis of cellulose into glucose is

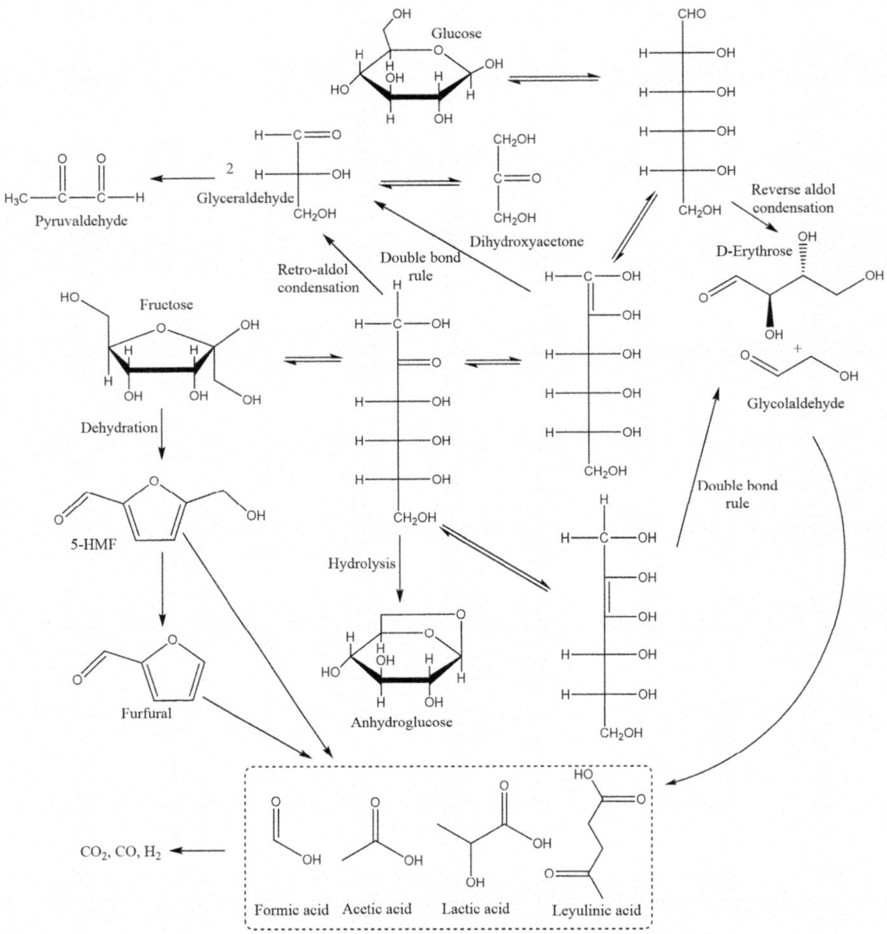

Fig. 7.3 Main reaction pathways of glucose and fructose in subcritical water. (Composed by authors based on [43–45, 54, 55])

a prerequisite for HMF production with cellulosic biomass feedstock. These glycosidic bonds can promote the formation of intra- and inter-molecular hydrogen bonds, which explains why cellulose is highly crystalline, will not swell in water and can resist the attack of enzymes [49]. The most crucial step for effectively converting cellulose is the breaking of β-1,4-glycosidic bonds through hydrolysis reaction.

Reaction temperature plays a great role on the product distribution and reaction path of cellulose. At lower temperature range (<180 °C), cellulose is mainly hydrolyzed into oligosaccharides; at moderate temperature range (180–220 °C), glucose is the main product; at higher temperature range (>220 °C), erythrose becomes the most stable compound in the product mixture [50]. In addition, hydrothermal conversion of cellulose changes its pathway with the initial pH of the reaction of solvent [51, 52]. Under acidic and neutral conditions, cellulose hydrothermal conversion products are mainly HMF and LA; under alkaline conditions, the product was converted to C_{2-5} carboxylic acids. Guo *et al.* [53] conducted an in-depth investigation into two pathways of cellulose hydrothermal conversion aimed at producing ethyl levulinate. In the first route, cellulose undergoes hydrolysis, leading to the formation of 1,6-anhydro-beta-*d*-glucopyranose (AGP), 1,6-anhydro-*β*-*d*-glucofuranose (AGF), 3-hydroxy-4-oxopentanoic acid (HOA), and ethyl 2-hydroxy-4-oxopentanoate (EHO). In the second route, cellulose is hydrolyzed to glucose, which undergoes a sequential transformation into 1,4:3,6-dianhydro-alpha-*d*-glucopyranose (DGP), HOA, and EHO, ultimately resulting in the production of ethyl levulinate.

7.2.2 *Effect of Operating Conditions*

7.2.2.1 Temperature and Holding Time

The increase in temperature results in the weakening of hydrogen bonds between water molecules, consequently altering the pathways through which hydrothermal conversion products are formed [56]. Temperature rise also has the potential to enhance both the solid yield and the distribution of diverse oxygen functional groups [57]. An extended reaction time can augment the solid yield, promoting the formation of hydrochar microspheres [58]. In this context, the effect of different temperatures and residence times on the hydrolysis of dextran was investigated [59], as shown in Table 7.2. HMF yield increased gradually with the increase of reaction temperature from 170 to 240 °C. The highest HMF yield of 36.56% was obtained when the holding time was 60 min. By increasing the reaction temperature and holding time from 170 °C and 60 min to 180 °C and 360 min, HMF yield increased from 0.02% to 8.85%, and the change of HMF yield was the same as that of monosaccharide conversion process.

7.2.2.2 Solvent Type

Due to the susceptibility of HMF to degradation through rehydration reactions, the dehydration of fructose in pure water with solid and inorganic acids typically lacks selectivity. Using solvents to extract HMF from the aqueous phase to the organic phase can reduce the rehydration of HMF to LA and FA and improve the HMF yield. Conversely, non-protonic solvents with high boiling points, such as DMSO (dimethyl sulfoxide), can prevent undesirable side reactions, thereby increasing the yield of HMF [60]. The solvent effect is crucial in the biomass dehydration and conversion to HMF. Kim *et al.* [61] summarized two widely recognized acyclic mechanisms (A and B) for glucose dehydration to form HMF, as shown in Fig. 7.4. Initially, hydroxyl protonation forms intermediate 1, a common step in both

Table 7.2 Yields of liquid products from hydrothermal conversion of dextran at various temperatures and holding times [59]

	Reaction conditions		Yield (%)		
Entry	Temperature (°C)	Time (min)	Polysaccharide	Monosaccharide	HMF
1	170	60	73.91	0.07	0.02
2	170	180	43.62	28.43	2.08
3	170	360	2.49	40.39	5.71
4	180	60	75.07	0.63	0.04
5	180	180	19.60	35.37	8.50
6	180	360	3.41	42.75	8.85
7	190	60	70.58	0.35	0.03
8	190	180	3.56	39.95	30.33
9	190	360	2.48	22.77	11.31
10	200	60	25.45	39.26	0.58
11	200	180	1.53	26.18	25.37
12	200	360	1.20	8.24	15.20
13	210	60	5.05	42.16	4.48
14	210	180	1.07	12.26	25.92
15	210	360	0.23	1.24	0.83
16	220	60	4.55	43.56	17.29
17	220	180	0.18	1.69	6.03
18	220	360	0.28	0.06	0.06
19	230	60	1.65	18.20	31.57
20	230	180	0.94	0.10	0.16
21	230	360	0.17	0.01	0.00
22	240	60	2.06	16.49	36.56
23	240	180	0.65	0.01	0.31
24	240	360	0.00	0.01	0.00
25	250	60	3.14	0.05	3.47
26	250	180	0.42	0.01	0.01
27	250	360	0.00	0.01	0.00

mechanisms. This leads to the formation of carbocation 2, releasing the first water molecule. From intermediate 2, dehydration proceeds via mechanism A or B.

In mechanism A, a second dehydration occurs from A2 to A3. The sequence A3 → A4 → A5 involves two hydride shifts, leading to dehydration to form A6, and finally deprotonation to form HMF. Mechanism B is more complex; hydrogen transfers directly to the hydroxyl group, forming enol B1, which dehydrates and protonates to form B3. B3 undergoes further protonation and dehydration steps, resulting in B5, which protonates to form HMF.

Solvent effects on HMF yield are primarily regulated by their influence on the protonation ability of B3, affecting the K_a value of B4 deprotonation. A higher K_a value facilitates deprotonation but lowers HMF yield. However, the separation of HMF from solvents with high boiling points poses considerable challenges and

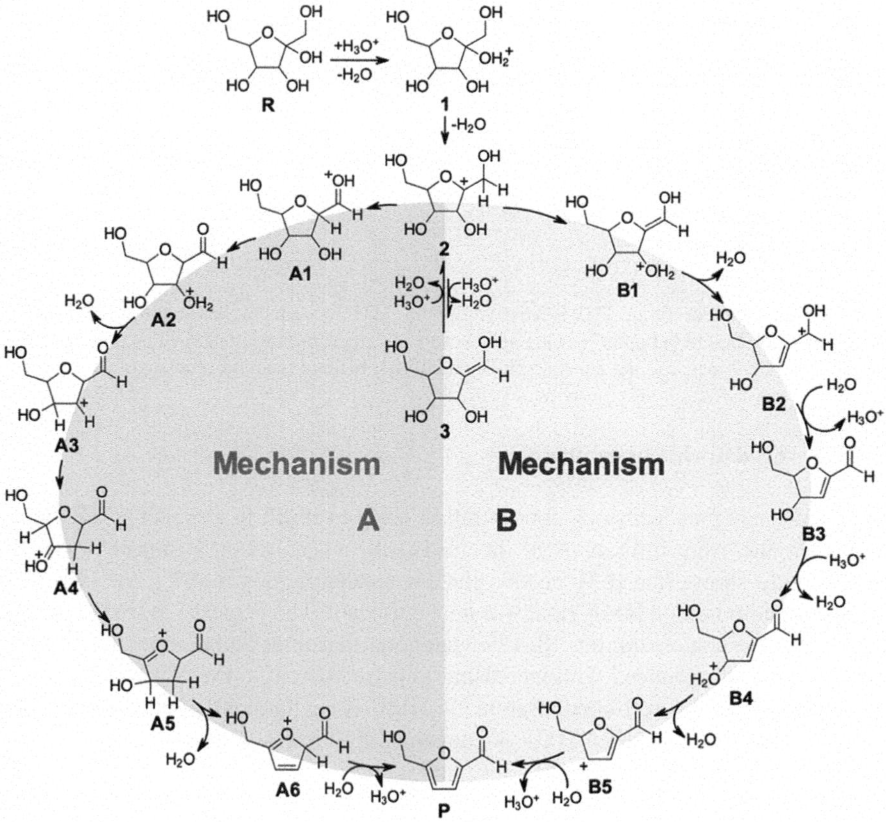

Fig. 7.4 Two mechanisms of fructose dehydration to HMF. (Adapted with permission from [61], Copyright © ACS)

demands energy-intensive separation processes, leading to a substantial increase in the overall experimental costs. To address this, the addition of salt (e.g., NaCl) to the aqueous phase enhances the distribution of HMF in the extractive phase through the "salting-out effect," thereby increasing the yield of HMF without relying on high boiling point solvents.

As indicated in Table 7.3, the effect of various solvents on the synthesis of HMF was evaluated in a salting-out system [62]. The aqueous phase was adjusted to pH = 0.6 with HCl, and the system operated at 150 °C with a solvent-to-water ratio of 3.2. The reaction time was 35 min, extended to 85 min when using tetrahydrofuran (THF). HMF selectivity and the distribution coefficient (R) were used to investigate the solvent's effects on HMF generation. A higher R value indicates a more significant salt precipitation effect of NaCl on the extraction solvent. Among the tested solvents, THF exhibited the highest HMF selectivity (83%) and extraction power (R = 7.1), indicating THF's strong affinity for HMF, and low water miscibility during this process.

Table 7.3 Dehydration results for 30 wt% fructose solutions in biphasic systems with different solvents [62]

Entry	Solvent	Conversion (%)	Selectivity (%)	R
1	1-Butanol	73	80	3.9
2	2-Butanol	67	85	3.7
3	2-Butanone	84	82	5.4
4	THF	53	83	7.1
5	THP	77	78	3.2
6	MTHF	87	76	3.7

Note: THF is tetrahydrofuran, THP is tetrahydropyran, MTHF is 2-methyltetrahydrofuran, R is the distribution coefficient (a higher R value indicates that a larger proportion of HMF is present in the organic phase compared to the aqueous phase, signifying better extraction efficiency)

7.2.2.3 Feedstock Concentration

In general, a higher feedstock concentration tends to result in a higher HMF yield. However, there are limitations to the feedstock concentration's impact on HMF yield. Study shows that [63], at low glucose concentrations (4–6%), the effect of sugar concentration on HMF yield was not significant. The decrease in HMF yield at higher glucose concentrations (8–12%) suggests that undesirable side effects may become more pronounced with increasing glucose content in the water. This could be attributed to the self-condensation of HMF (i.e., humins) or the reaction of glucose with HMF, wherein HMF is consumed by these side reactions, leading to a reduction in overall HMF yield.

7.2.3 Design of Catalysts

Catalyst design principles typically focus on high activity, selectivity, mechanical strength, thermal and chemical stability, and cost-effectiveness. These ensure that catalysts effectively facilitate reactions while being durable and economical.

For biomass hydrolysis, acids like HCl, H_2SO_4, HF and organic acids (oxalic, maleic, formic) are used [64]. However, these processes require precise water content control, are highly corrosive, and face issues with product/catalyst separation, reactor corrosion, catalyst recovery, and wastewater treatment.

Solid acids have gained attention for biomass hydrolysis due to their advantages over liquid acids, including easier separation, recycling, and environmental benefits. Examples include metal oxides, polymer-based acids, sulfonated carbon-based acids, heteropoly acids, H-shaped zeolites, and magnetic solid acids [64]. Additionally, advancements such as ionic liquids, microwave irradiation, supercritical water, and nanoparticles have facilitated biomass hydrolysis under milder conditions [65, 66].

Brønsted acid plays a crucial role in providing active acidic sites for the dehydration of fructose. However, the type of acid has minimal impact on the maximum HMF yield, primarily due to the sluggish conversion of glucose to HMF. By combining a solid acid catalyst, Amberlyst-15 (primarily composed of polymeric beads with strongly acidic sites), and a solid basic catalyst, Mg–Al-hydrotalcite (HT), in *N,N*-dimethylformamide, the conversion of HMF was 73% and the yield was 42% [67].

The use of $AlCl_3{\cdot}6H_2O$ as a catalyst in a biphasic reaction system resulted in a 61% yield of HMF from glucose [68]. Additionally, cellulose was obtained from mesoporous TiO_2 and ZrO_2 nanoparticles, yielding 18.2% and 29.2%, respectively [69]. However, it is noteworthy that the catalysts employed in these processes consist of toxic and expensive metal catalysts, and the HMF yield achieved may not be sufficiently high to offset the associated costs.

Ionic liquids are characterized by high thermal stability and a considerable solubilization capacity. Consequently, carbohydrates can be effectively dissolved in ionic liquids, even at elevated temperatures [70]. In recent years, extensive research has been conducted on the utilization of ionic liquids as catalysts. The incorporation of ionic liquids, such as [BMIm]Cl (1-Butyl-3-methylimidazolium chloride), has demonstrated remarkable results, with HMF yields reaching as high as 91% when $CrCl_3$ is employed as a catalyst [71].

The dehydration of carbohydrates facilitated by boric acid-supported surface catalysts serves to diminish product contamination and enables the establishment of environmentally friendly pathways for HMF synthesis. In this regard, silica-supported boric acid was synthesized and employed as a multiphase catalyst within ionic liquids. This innovative approach enables the direct conversion of economical and renewable mono- and polysaccharides into HMF under milder conditions [72].

Nonetheless, the high cost associated with ionic liquid solvent systems has constrained their application in industrial production. Exploiting the cost-effectiveness and water solubility of ionic liquids, HCl and $AlCl_3$ can serve as catalysts, allowing for the generation of high HMF yields and rendering them excellent catalytic raw materials [68]. It is crucial to note, however, that the hydrolysis products may contain inhibitors, which can significantly impact HMF formation. To address this, activated carbon proves to be a valuable tool for removing inhibitors. Activated carbon is chosen for its relative inexpensiveness, amenability to chemical modification, and the ability to recover metals after catalyst deactivation. Modified activated carbon catalysts, featuring Sn elements at 120 °C, enable the efficient conversion of carbohydrate-rich biomass wastes into high-value HMF molecules through localized transformation [73]. Additionally, the prepared catalysts exhibit both Brønsted and Lewis acid sites.

7.2.4 Catalyst Recycling

In the realm of green chemistry, the recycling and reutilization of catalysts are pivotal considerations, representing key factors for catalytic performance. Typically, after each catalytic experiment, the spent catalyst undergoes a process involving washing with acetone or water, filtration, and subsequent drying at an appropriate temperature for subsequent reaction cycles. Lima *et al.* [71] implemented a separation technique for the black residue obtained after catalysis experiments, involving centrifugation and membrane filtration. They then distilled off excess water and other potential volatile products from the ionic liquids under reduced pressure. Subsequently, the recovered ionic liquid catalyst was analyzed by infrared spectroscopy alongside the pure ionic liquid catalyst. The similarity in spectral peaks confirmed that the ionic liquid catalyst could be effectively recovered.

Kodatia *et al.* [69] conducted five dehydration reactions to recover the catalyst, showing good to moderate HMF yields after five catalytic runs. However, the HMF yields exhibited a decline of HMF yield from about 25% to around 10% with successive reactions. In a similar vein, Atanda *et al.* [74] executed six glucose dehydration reactions. After each reaction cycle, the catalyst was recovered through filtration, washed with acetone, dried, and subsequently utilized for the next reaction cycle. While the conversion of glucose remained relatively consistent across all experiments, the HMF yield showed a continuous decrease from 81% to 70% in the first three runs. Furthermore, the catalyst's color transitioned from white to dark brown, indicating a significant reduction in its activity during the recovery process.

The deactivation of the catalyst primarily stems from the adsorption of humic substances on the catalyst surface, which can be alleviated by burning them off with an oxygen stream, thereby removing the deposited material from the catalyst surface [75]. Yang *et al.* [37] adopted the decantation method to separate the catalyst at the end of the reaction, followed by washing and drying with deionized water. This process, conducted at a temperature of 160 °C with a reaction time of 180 min, was repeated seven times. The results indicated that the catalyst maintained good activity and stability in the system, with a final fructose conversion of about 90% and an HMF yield of about 85%.

In another study by Walia *et al.* [72], HMF was extracted with ethyl acetate, and the [bmim]HSO_4 + H_3BO_3–SiO_2 catalytic system was dried under vacuum to conduct a catalyst reusability test. The findings revealed that the catalyst system could be reused four times without a significant decrease in HMF conversion.

7.2.5 HMF Separation

The synthesis of HMF in a catalytic system can impose various side reactions. HMF and LA both result from glucose hydrolysis, with LA being the primary compound in the decomposition of HMF. Additionally, the dehydration of HMF can lead to the formation of FA. Traces of furfural and lactic acid may also be present. Consequently, the separation and purification of these products are crucial for enhancing the production and subsequent utilization efficiency of HMF. The detailed study of how to effectively separate these chemicals to increase the purity of HMF becomes imperative.

Liu *et al.* [63] used activated charcoal to purify the hydrolysate, effectively removing non-sugar components and decolorizing the solution. The mixture underwent gentle stirring for 30 min at room temperature, followed by filtration to separate solids and liquids. Analysis of the hydrolysate samples before and after activated carbon purification was conducted using high-performance liquid chromatography (HPLC) to determine the contents of non-sugar components in the hydrolysate. Subsequently, the organic solvent was removed from the organic layer of the reaction mixture through evaporation, yielding a highly viscous, deep brown liquid. This liquid was identified as an insoluble by-product known as humins, formed through a polymerization reaction triggered by the decomposition of glucose, intermediates, and HMF in the presence of water [76, 77]. The liquid was then dissolved in deionized water, filtered to eliminate the insoluble fraction, resulting in an HMF solution. Additionally, a different separation method was proposed by Liu *et al.* [63], one that bypasses the purification step and involves solid-liquid separation after auto-hydrolysis, ultimately achieving a higher yield of final HMF.

In a related study, Kang *et al.* [59] demonstrated that at the end stage of the catalytic reaction, the reaction solution was rapidly cooled in an ice water bath and centrifuged to separate the solid and liquid phases. The resulting solid phase underwent repeated washing with distilled water and ethanol, followed by vacuum filtration to achieve a pH of 7.0. Finally, after drying to a constant weight at 80 °C, HMF solid was obtained.

Lima *et al.* [71] outlined two methods for product separation. Firstly, they performed extraction with toluene or isobutyl methyl ketone. In this method, a pear-shaped bottom microreactor equipped with a purge gas valve was utilized, and the reaction was conducted in a nitrogen atmosphere. The reaction mixture was heated to 100 °C with the addition of a co-solvent, and extractive separations were carried out in a nitrogen atmosphere. Secondly, a decompression separation method was employed, where the reactants were placed in a round-bottom glass reactor connected to a condenser. The condenser, linked to a round-bottom flask for product collection via a vacuum adapter, was cooled with liquid nitrogen. The feed reactor was degassed under vacuum, placed in an ultrasonic bath filled with water at room temperature for approximately 15 min, and then immersed in an oil bath and heated to 100 °C for 4 h. This process allowed for the separation of products through evaporation under reduced pressure using a water inhaler, where HMF evaporated from the ionic liquid.

Research over recent decades has focused on HMF isolation through effective solvents, highly active catalysts, and new reaction processes [29, 78]. However, isolating HMF remains challenging, necessitating ongoing innovation in both process and intermediate separation methods.

7.3 Converting HMF to FDCA

7.3.1 *Basic Principles*

The conversion of HMF to FDCA involves a series of oxidation reactions. For instance, when utilizing metal oxidants, the hydroxyl functional group of HMF can undergo initial oxidation to aldehydes, yielding 2,5-diformylfuran (DFF), which is then further oxidized to 5-formyl-2-furancarboxylic acid (FFCA), culminating in the production of FDCA. Under electrochemical catalysis conditions, HMF is initially oxidized to 5-hydroxymethyl-2-furancarboxylic acid (HMFCA), followed by oxidation to FFCA, and ultimately resulting in the formation of FDCA [31, 32, 79], as shown in Fig. 7.5.

The adsorption effect of the catalyst on HMF, its oxidation intermediates, and products are crucial for the reaction's selectivity and efficiency. This adsorption effect influences the reaction by activating reactants, stabilizing intermediates, and facilitating product desorption. Effective catalyst adsorption of HMF increases its activation energy, accelerating the reaction rate. Selective adsorption of intermediates prevents their overreaction, improving product selectivity. Additionally, efficient desorption of products avoids accumulation on the catalyst surface, thereby preserving catalytic activity.

HMF can theoretically undergo conversion to FDCA through various oxidation pathways, namely chemical oxidation, biocatalysis, and electrochemical oxidation [31, 79]. Each of these methods comes with its own set of advantages and limitations, but they share a common objective of achieving higher product yield, better selectivity, and more environmentally friendly conversion processes. Catalyst design and the optimization of reaction conditions play pivotal roles in achieving efficient conversion in these pathways. Researchers have committed a lot of efforts to develop new catalysts and explore more optimized reaction conditions. Examples include the preparation of multifunctional magnetic nanocomposite catalysts [21], the utilization of photocatalysts [18], or the electrochemical oxidation of HMF to FDCA under acidic conditions [80], etc. These endeavors aim to reduce costs, increase yield and selectivity, while minimizing environmental impacts.

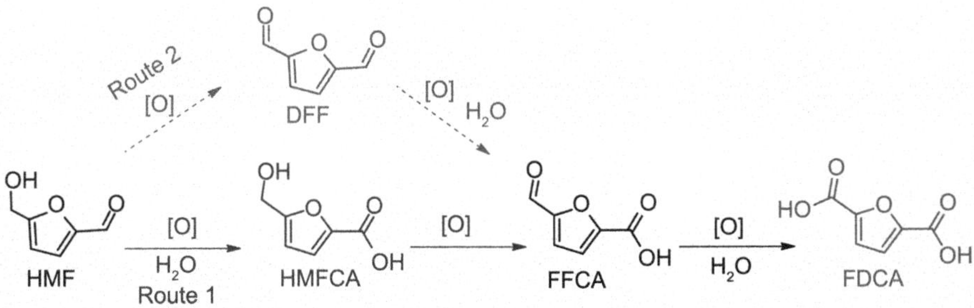

Fig. 7.5 Possible HMF oxidation pathways to produce FDCA. (Adapted with permission from [32], Copyright © 2022 Elsevier)

7.3.2 *Chemical Oxidation*

Chemical oxidation stands as one of the frequently employed methods for converting HMF to FDCA, wherein chemical oxidants are utilized to catalytically transform the hydroxyl group of HMF into a carboxyl group. This conversion generally unfolds in two stages: initially, HMF is oxidized to HMFCA, followed by the subsequent oxidation of HMFCA to FDCA. Chemical oxidation methods can be classified into noble metal oxidation, non-noble metal oxidation, and metal-free catalysis.

7.3.2.1 Noble Metal Oxidation

The utilization of noble metals in heterogeneous catalysis to facilitate the synthesis of FDCA from HMF stands as the current standard for FDCA production. This approach has garnered attention and interest from industrial leaders, signifying the initiation of commercial FDCA production. Catalysts play a pivotal role in chemical oxidation methods, and noble metal-supported heterogeneous catalysis, including platinum (Pt), gold (Au), ruthenium (Ru), and palladium (Pd), tends to favor the protonation and aldehyde hydration process under alkaline conditions, thereby promoting the oxidation process. Remarkable results in FDCA yield have been achieved through the interaction between catalytic species and their carriers. For example, Yu *et al.* [81] developed an efficient catalyst by combining polyvinylpyrrolidone-encapsulated Pt nanoparticles with chitosan-based high surface area carbon, leading to a 99% FDCA yield. The catalyst exhibited reusability for 10 cycles, maintaining activity for 5 cycles before a decline attributed to nanoparticle aggregation.

A Fe–N–C single-atom catalyst was prepared on β-Al_2O_3 support using a wet chemical method under mild alkali-free conditions [82]. The Fe–N_5 site generates reactive oxygen species, while the acidic and alkaline sites of γ-Al_2O_3 enhance the adsorption and conversion of HMF. This synergistic effect promotes the catalytic conversion of HMF to FDCA, achieving over 90% HMF conversion and a maximum FDCA yield of 99.8%. The catalyst can be reused up to five times.

Six MnO_2 catalysts were synthesized by hydrothermal and sol-gel method, which are more environmentally friendly than traditional approaches [83]. α-, β-, and γ-MnO_2 are rod-shaped with varying widths, δ-MnO_2 forms particle aggregates, ε-MnO_2 has a flower-like shape with a large specific surface area, and λ-MnO_2 features large pores. The FDCA yield effectiveness is ordered as $\alpha > \beta > \gamma \sim \lambda > \delta$-$MnO_2$, with the highest yield reaching 74%.

Limited work has been dedicated to studying the oxidation mechanism of HMF on Ru catalysts. Ru-based catalysts exhibit higher activity in organic solvents for the conversion of HMF to DFF compared to other catalysts [84]. Nie *et al.* [85] conducted research on the mechanism of HMF oxidation to DFF on Ru/C catalysts, as illustrated below:

$$R\text{-}CH_2OH + 2* \Leftrightarrow R\text{-}CH_2O^* + H^* \quad (7.1)$$

$$O_2 + 2* \Leftrightarrow 2O^* \quad (7.2)$$

$$R\text{-}CH_2O^* + O^* \rightarrow R\text{-}CHO^* + OH^* \quad (7.3)$$

$$H^* + O^* \Leftrightarrow OH^* + * \quad (7.4)$$

$$H^* + OH^* \rightarrow H_2O^* + * \quad (7.5)$$

$$R\text{-}CHO^* \Leftrightarrow R\text{-}CHO + * \quad (7.6)$$

$$H_2O^* \Leftrightarrow H_2O + * \quad (7.7)$$

The above proposed reaction mechanism for aerobic oxidation of HMF to DFF over Ru/C in DMF [26], where Ru-surface was denoted as * and adsorbed intermediate species denoted as X*. The reaction mechanism was a Langmuir-Hinshelwood mechanism. Initially, HMF and O_2 adsorb on the surface of Ru/C and then dissociate, forming various adsorbed intermediates. Subsequently, an adsorbed O abstracts a hydrogen atom in R-CH_2O, producing adsorbed DFF and OH*, where OH reacts with H to form adsorbed H_2O. Finally, DFF and H_2O desorb from the surface, completing a catalytic cycle. Isotope labeling experiments also confirmed that the activation of the C–H bond is the rate-limiting step in this process.

It is apparent that catalysts incorporating Au and Pd demonstrate heightened efficacy in oxidizing the aldehyde moiety of HMF as opposed to the hydroxymethyl side chain [86]. Consequently, even under non-alkaline conditions, the oxidation of HMF on Au and Pd-based catalysts predominantly follows the initial aldehyde oxidation pathway, as illustrated in Fig. 7.6, Route 1 [87, 88]. Conversely, Ru catalysts exhibit a propensity to oxidize the hydroxymethyl group rather than the aldehyde group in the primary oxidation step. Consequently, the process of oxidizing HMF to FDCA on Ru-based catalysts follows Route 2.

The oxidation mechanism of HMF on Au or Pt catalyst surfaces under excess alkaline conditions, utilizing H_2O as the oxidizing agent, is delineated in Fig. 7.6. Initially, the aldehyde group of HMF undergoes a reversible hydration reaction with H_2O in an alkaline environment, yielding a hemiacetal intermediate. Subsequently, this hemiacetal intermediate undergoes direct oxidation with H_2O, specifically labeled with ^{18}O, resulting in the production of HMFCA and FDCA with ^{18}O labeling. This labeling indicates the direct involvement of H_2O as the oxygen source in the conversion of the aldehyde group to the carboxylic group [89]. In contrast, when $^{18}O_2$ is employed as the oxygen source, no ^{18}O labeling is detected in the products, suggesting that O_2 merely plays an indirect role in the reaction by facilitating the removal of electrons adsorbed on the metal surface to complete the entire catalytic cycle. Therefore, the precise reaction pathway for the oxidation of the aldehyde group of HMF on Au or Pt catalyst surfaces under alkaline conditions entails the initial hydration reaction with H_2O, generating a hemiacetal intermediate, followed by the direct oxidation of this intermediate by H_2O to form the carboxylic group [60].

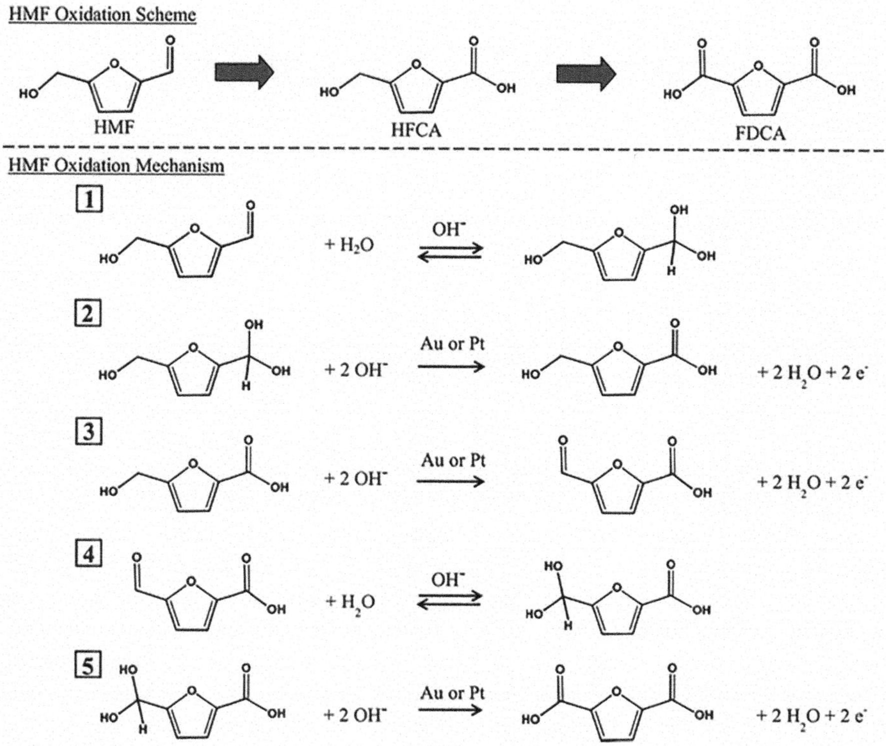

Fig. 7.6 Proposed reaction mechanism for the oxidation of HMF in an aqueous solution with the aid of excess base (OH^-) over either Pt or Au catalyst. (Adapted with permission from [89], Copyright © 2012 Royal Society of Chemistry)

7.3.2.2 Non-precious Metal Oxidation

However, the fundamental mechanisms that lead to the distinct reaction pathways in HMF oxidation with various noble metals remain unclear, which is also a critical aspect for comprehending the overall reaction mechanism. Exploring the adsorption behavior of intermediate products during HMF oxidation across diverse noble metal types holds the potential to yield valuable insights. Furthermore, a comprehensive examination of the oxygen activation process by different noble metals is warranted to enhance future understanding of these mechanisms.

Non-precious metals, while generally more cost-effective and readily available than their precious counterparts, often exhibit diminished catalytic activity. To address this performance disparity, more rigorous reaction conditions, including elevated temperature and prolonged reaction times, are frequently necessary. Similar to precious metal catalysts, the incorporation of alkaline substances is typically required. However, recent research indicates that this performance gap has been

Table 7.4 Oxygen hole formation energy in different MnO_2 phases [90]

The crystal structure of MnO_2	Reaction AE[a] (eV)
α-MnO_2	3.95 (O_A atom), 3.25 (O_B atom)
β-MnO_2	3.25 (O_A atom)
γ-MnO_2	3.84 (O_A atom), 3.15 (O_B atom)
λ-MnO_2	3.44 (O_B atom)

[a] Note: AE is activation energy

partially mitigated, with non-precious metal-based catalysts exhibiting noteworthy results. Among various catalytic systems, those based on manganese (Mn) or cobalt (Co) have demonstrated superior activity, attributed to the distinct oxidation states of these metals, their high oxygen storage capacity, and accelerated oxygen absorption and reduction kinetics. Mn and Co, either individually or in combination with other metals, have exhibited notable success in HMF conversion and FDCA selectivity, with the Mn and Co combination proving to be particularly effective.

Hayashi *et al.* [90] demonstrated the superior efficacy of β-MnO_2, a crystalline form of manganese oxide, achieving a notable 93% yield of FDCA. This improved performance is attributed to the higher vacancy formation energy of the oxygen atom site on the planar surfaces of α- and γ-manganese oxides compared to the bent oxygen atom site. In contrast, β- and λ-manganese oxides exclusively consist of planar and bent oxygen atom sites, resulting in lower vacancy formation energy and, consequently, exhibiting superior catalytic activity compared to α- and γ-manganese oxides. Table 7.4 presents the calculated oxygen hole formation energy of α-, β-, γ-, and λ-MnO_2 phases using Density Functional Theory (DFT).

The notation O_A refers to the planar μ_3-O (Planar $Mn_3(\mu_3\text{-O})$), and O_B refers to the bent μ_3-O (Bent $Mn_3(\mu_3\text{-O})$). The term "bent μ_3-O" designates the bent position of the oxygen atom in the MnO_6 octahedron, located between the faces of the octahedron, while "planar μ_3-O" refers to the oxygen atom situated within the face of the octahedron. Notably, in the case of β-MnO_2, it exclusively exhibits a planar μ_3-O structure, with an oxygen hole formation energy of 3.25 eV, the lowest among the phases examined.

Although the initial reaction rate R_0 of FFCA to FDCA is 16.4 μmol/h/m^2, the highest among all MnO_2 structures, however, β-MnO_2 only shows about 28% FDCA yield. This underscores the pivotal role of the surface area of β-MnO_2 in the final FDCA yield. Researchers have employed a template-free low-temperature crystallization method to synthesize high-surface-area β-MnO_2, enabling the attainment of a maximum FDCA yield up of 93% [60].

Tirsoaga *et al.* [21] synthesized a multifunctional magnetic nanocomposite catalyst, denoted as 10M@*x*Nb@MNP, where M = CoO_x, MnO_x, or FeO_x. Notably, the 10Co@22Nb@MNP catalyst exhibited an impressive 96.5% FDCA conversion rate, surpassing the catalytic efficiency of any reported non-precious metal-based system. The reaction pathway for this catalyst differs from those of non-precious metal oxide catalysts (e.g., MnO_x–CeO_2, Fe_3O_4–CoO_x) and Ru catalysts. Specifically, the proposed pathway is HMF → HMFCA → FFCA → FDCA, as depicted in Route 1 of

Fig. 7.5, with no identification of intermediate product DFF under the dual-function catalyst. This pathway aligns with those reported for precious metal catalysts such as Au and Pd.

Further investigation into the Mn@*x*Nb@MNP dual-function catalyst revealed a synergistic electron transfer mechanism between niobium and manganese oxides [21]. XPS results indicated that niobium, within the Nb-MCM framework, changed the electronic states of cobalt and manganese (note: MCM is short for "Mobil Composition of Matter", most commonly MCM-41, which is a type of mesoporous silica with a highly ordered structure, often used as a support for various active metals in catalytic reactions due to its large surface area and well-defined pore size distribution; essentially, it acts as a platform for other chemicals to react on). Niobium facilitated electron transfer to manganese oxide through its defect pores and NbO^-, leading to the reduction of manganese from the higher oxidation state Mn^{4+} to Mn^{3+} and the generation of basic centers on the manganese surface. This cooperative electron transfer between niobium and manganese enhanced the selectivity of HMF oxidation, improving the overall catalytic performance and facilitating more efficient conversion of HMF to FDCA.

Figure 7.7 elucidates the mechanism of synergistic effects between metal oxides and niobium in the decomposition of the oxidant *t*-BuOOH within the dual-function catalyst [21]. Specifically, Mn^{3+} engages with *t*-BuOOH, resulting in the generation of the higher oxidation state Mn^{4+} and OH^- ions. The OH^- ions, in turn, are partially adsorbed by Nb^+ sites, creating an active intermediate for the oxidation of HMF. Simultaneously, Mn^{4+} is reduced back to Mn^{3+}, maintaining the cyclic process of decomposing *t*-BuOOH and producing additional OH^- ions. This orchestrated electron transfer and synergistic interaction between the metal oxides and niobium enhance the dual-function catalyst‘s efficiency in decomposing the oxidant, surpassing the performance of a single-function catalyst.

The overall mechanism of the oxidation reaction of HMF catalyzed by non-precious metals are as follows: Initially, HMF adsorbs on to the catalyst surface through the CHO functional group, while *t*-BuOOH undergoes decomposition on the surface of metal oxides (such as Co, Mn, or Fe), yielding OH^- ions. Subsequently, these OH^- ions catalyze the generation of diol isomers and participate in dehydrogenation reactions through a combination of homogeneous and heterogeneous catalysis. This sequence leads to the formation of intermediate species, including carbonyl and carboxylic acid derivatives. The synergistic collaboration between the metal oxides and niobium enhances the production of OH^- ions. These OH^- ions, in turn, adsorb on the active Nb^+ centers, facilitating the more efficient oxidation of the intermediate species and ultimately culminating in the conversion of HMF to FDCA. This mechanistic insight elucidates the differences in yield and selectivity observed between single-function and dual-function catalysts [21, 91].

Metals significantly affect oxygen activation and catalytic defects. Kumar *et al.* used DFT simulations to reveal the crucial role of the Co–N_x bond in oxygen activation in Co–N/C catalysts [93]. They found that oxygen is most easily adsorbed where Co is bonded to two N atoms (Co–N/C–topCo), highlighting Co's role in oxygen adsorption and activation. The electron-rich region formed by Co and N

step 1 $Mn^{3+} + tBuOOH \rightarrow Mn^{4+} + tBuO\bullet + OH^{-}$

step 2 $Mn^{4+} + tBuOOH \rightarrow Mn^{3+} + tBuOO\bullet + H^{+}$

step 3 $2tBuOO\bullet \rightarrow 2tBuO\bullet + O_2$

step 4 $Nb\text{-}O^{-} + tBuO\bullet \rightarrow Nb\text{-}O\text{-}O\text{-}tBu$

step 5 $Nb\text{-}O\text{-}O\text{-}tBu + H^{+} \rightarrow Nb\text{-}O^{-}$

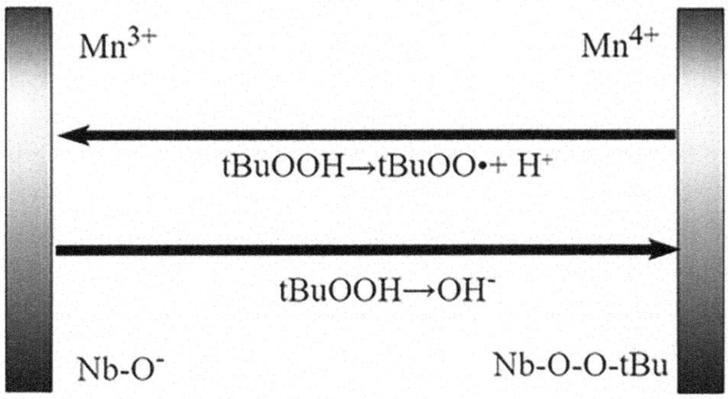

Fig. 7.7 Decomposition of *t*-BuOOH in the presence of the Mn@*x*Nb@MNP catalyst. (Modified by the authors based on [21])

atoms attracts oxygen molecules, enhancing catalyst activity. This study emphasized the influence of metals on oxygen activation in catalysts. High calcination temperatures can cause Co particle aggregation, reducing active sites and impairing catalyst performance. High temperatures induce Co atom migration and aggregation, forming larger Co particles and decreasing the density of Co–N_x bonds, thus affecting oxygen adsorption and activation. Additionally, high heating rates can create point defects in the carbon support, impeding electron transfer and reducing catalyst activity. Therefore, regulating calcination temperature and heating rate can effectively manage catalyst defects and enhance performance [93].

Oxygen vacancies also promote catalysis. Jiang *et al.* observed through O 1s spectrum analysis that the $CuCoO_x$ catalyst has a significantly higher ratio of oxygen vacancies to lattice oxygen compared to CuO_x and CoO_x, indicating a higher oxygen vacancy density in $CuCoO_x$ [94]. This is a key factor for its high catalytic activity. Oxygen vacancies enhance the catalyst's adsorption capacity for water and oxygen, promote lattice oxygen migration, and generate acid-base sites, improving catalytic efficiency. The synergistic effect between Cu and Co species further increases the oxygen vacancy rate, enhances oxygen species migration, and creates strong acid/base sites, enhancing $CuCoO_x$'s catalytic activity. This study provides a new direction for developing efficient and stable bimetallic catalysts for biomass conversion [94].

The adsorption effect of noble metal catalysts warrants further investigation. Jin et al. found that adsorption strength significantly influences reactant activation, product desorption, and side reactions in the selective oxidation of HMF to FDCA [92]. By adjusting the Ag/Pd ratio in the AgPd alloy catalyst, the electron density of Pd 3d orbitals can be regulated, affecting chemical adsorption strength and controlling catalytic activity and selectivity. DFT computational results showed that optimal interaction between bimetallic boundary sites and adsorbates enhances HMFCA oxidation and FDCA desorption, increasing FDCA yield [92].

Current research is prominently advancing non-precious metal catalysts centered on manganese and cobalt, due to their cost-effectiveness and environmental sustainability. These catalysts exhibit exceptional efficacy in catalytic processes, attaining elevated conversion rates and selectivity, particularly under mild oxidation conditions crucial for industrial chemical reactions. The ongoing development of these catalysts holds the promise of substituting conventional precious metal catalysts, thereby reducing costs and mitigating environmental impact.

7.3.2.3 Non-metal Catalysis

To circumvent the challenges associated with the direct oxidation of HMF to FDCA using expensive metal-based catalysts and the environmental concerns related to metal leaching, researchers are exploring sustainable and cost-effective non-metal alternatives. Carbon-based catalysts doped with heteroatoms (N, P, and S) have emerged as promising candidates due to their high catalytic activity, affordability, straightforward synthesis, and eco-friendliness. The presence of heteroatoms in carbon-based materials introduces surface defects, stemming from differences in atomic radius and bond length, thereby serving as active catalytic sites.

Among various doped carbon structures, nitrogen-doped carbon, particularly graphitic carbon structures doped with nitrogen atoms, has gained a wide attention as a non-metal dopant catalyst for diverse reactions including oxidation processes. Nguyen *et al.* [95] demonstrated that nitrogen-doped nanoporous graphitic carbon (NNC) catalysts can efficiently oxidize HMF to FDCA under aerobic conditions at 80 °C, achieving a yield of up to 80%. The significance of nitrogen presence, along with control over quantity and configuration within the nanoporous carbon framework, was highlighted. Through the synthesis of zeolitic imidazolate framework-8 (ZIF-8), NNC containing a substantial amount of nitrogen, particularly in the form of quaternary nitrogen (N-Q), is believed to act as the catalytic active site for the oxidation reaction. The aerobic oxidation mechanism is considered a radical process, wherein N-Q sites play a crucial role in generating oxygen radicals for oxidation.

This results in the conversion pathway from HMF to HMFCA (fast step), HMFCA to FFCA, and FFCA to FDCA.

Verma *et al.* [96] developed a solid catalyst based on chitosan for the air oxidation of HMF to produce FDCA, utilizing porous carbon nitride (PCN_x) derived from marine wastes chitosan. When PCN_x served as the catalyst, FDCA was obtained with observed yields of 8%, 46%, and 83% at 30 °C, 50 °C, and 70 °C, respectively. It's noteworthy that elevating the reaction temperature to 80 °C did not result in further improvement in the yield. Furthermore, to assess the impact of alkali on the reaction, various alkalis such as NaOH, KOH, and Na_2CO_3 were examined at 70 °C. However, these alkalis did not enhance the product yield under air conditions. Additionally, the PCN_x catalyst demonstrated stability over up to five cycles in the 5-HMF oxidation reaction without a loss of activity. XRD analysis confirmed that the catalyst morphology remained largely unchanged before and after the reaction, indicating its high stability.

Nitrogen-doped porous carbon significantly enhances the adsorption capacity of porous carbon for HMF, thereby improving the FDCA yield [97]. The presence of pyridinic nitrogen has minimal influence on HMF adsorption. Stronger HMF adsorption correlates with higher FDCA yields, highlighting the crucial role of adsorption performance in the oxidation conversion of HMF. This finding is consistent with Verma *et al.*'s research, who observed excellent stability and activity of the PCN_x catalyst in HMF oxidation [96].

The utilization of green and cost-effective non-metal catalysts, as demonstrated in the above-mentioned studies, holds substantial potential for the industrial-scale oxidation of HMF. The promising results obtained so far suggest that this avenue of research may bring about meaningful advancements. Further investigations in this field are anticipated, reflecting a growing interest in the development of sustainable and economically viable alternatives for the oxidation of HMF on an industrial scale.

7.3.3 Biocatalytic Methods

Biological methods for FDCA production, including whole-cell and enzyme catalysis, have attracted significant attention owing to their advantages such as mild reaction conditions, high selectivity, and environmentally friendly processes. These approaches utilize water solvents, benign oxidants, biodegradable catalysts, and involve low energy consumption, contributing to sustainable and eco-friendly manufacturing processes.

7.3.3.1 Whole-Cell Catalysis

Whole-cell biocatalysis leverages the natural detoxification mechanisms of microorganisms. Microorganisms, including bacteria and fungi, have evolved mechanisms to detoxify toxic molecules such as HMF, converting them into "non-toxic" compounds. This detoxification mechanism can be harnessed to produce FDCA. The use of microorganisms in this biotransformation process provides significant biocatalytic advantages, reducing environmental pollution and enabling the efficient conversion of harmful substances [98–101].

Whole-cell catalysts, in general, exhibit stability attributed to protective barriers, reactive substrates, endogenous coenzymes, and enzymatic pathways that sustain catalysis and inhibit by-products [102, 103]. For instance, certain bacteria like *Pseudomonas putida* and *Cupriavidus basilensis* can utilize HMF as a carbon source for growth and convert it into FDCA [104]. However, the FDCA yields obtained from these biocatalysts are relatively low, prompting the exploration of genetic engineering to enhance the conversion capability of microorganisms.

Researchers have employed genetic engineering methods to improve the conversion rates of microorganisms. This involves knocking out genes inhibiting FDCA production using gene editing techniques and enhancing the expression of specific enzymes to improve conversion rates. For example, the introduction of genes encoding HMF oxidase and HMF/furfural oxidoreductase into *Raoultella ornithinolytica* significantly increased the FDCA yield, with a molar conversion rate from HMF to FDCA increasing from 51.0% to 93.6% [105]. Additional modifications in batch reaction methods further increased the FDCA yield to 264.7 mM, with a yield of 96% [106]. *Escherichia coli* modifications using genes from *Cupriavidus basilensis* and *Comamonas testosteroni* resulted in a production of 144 mM of FDCA, with a yield of 96% [107].

While whole-cell catalysis holds promise, improvements are needed in areas such as product recovery rates and continuous carbon source supply. Recent research in whole-cell catalysis has expanded to filamentous fungi, which, with their unique characteristics, may offer advantages over bacteria in this biocatalytic process.

7.3.3.2 Enzyme Catalysis

Compared to whole-cell catalysis, the use of purified enzymes can offer higher reaction efficiency due to the absence of impurity products and reduced competitive reactions with other components. However, this approach has drawbacks, including the costly separation and purification of proteins, as well as the enzymes'sensitivity to temperature and pH changes, which may limit their applicability [108]. The oxidation of HMF to FDCA often necessitates multi-enzyme cascade reactions, cofactors, intermediates, or mutations. This is because a single enzyme typically cannot simultaneously catalyze the dual oxidation of alcohol and aldehyde functional groups; their activities are usually limited to the oxidation of either alcohols or aldehydes.

To address this challenge, researchers have extensively studied synthetic pathways facilitated by different enzymes, aiming to develop more effective methods for this transformation to contribute to sustainable chemical production. Determining the mechanism of action of specific enzymes requires their isolation and use in experiments. However, only a few enzymes exhibit activity towards HMF and its oxidation products like DFF, HMFA, and FFA. Most of these conversion reactions are slow, incomplete, and result in low FDCA yields, necessitating the combination of multiple enzymes for the reaction.

For example, aryl alcohol oxidase can convert HMF to FFA with a yield of 98% in 4 h but only produces 6% FDCA after 24 h [109]. A non-specific fungal peroxygenase can convert HMF to 97% HMFA within 24 h but only produces 10% FDCA. This enzyme can achieve a 90% FDCA yield by oxidizing FFA within 96 h. The selectivity of enzymes is evident in these examples. HMF oxidase stands out as an enzyme that can almost completely convert HMF to FDCA. Using this enzyme, HMF can be converted to 92% FFA and 8% FDCA within 5 h, reaching 95% FDCA after 24 h with a higher enzyme concentration. [110, 111]. Multi-enzyme cascade reactions have also shown promising results. The combined reaction of aryl alcohol oxidase, lactose oxidase, and non-specific peroxygenase can convert HMF to 25% DFF and 28% HMFA within 45 min, 70% FFA within 75 min, and 80% FDCA within 24 h [112].

Cajnko *et al.* [113] conducted tests on six different enzymes to investigate their effects on HMF and its oxidation products, aiming to establish a kinetic model and the mechanisms of enzyme actions. Enzymes exhibited activities included fungal aryl alcohol oxidases (AAOs), non-specific peroxygenases (UPOs), galactose oxidases (GAOs), laccases, and catalases. In multi-enzyme cascade reactions, such as those involving different aryl alcohol oxidases, the mechanisms become more complex due to the necessity to consider enzyme-enzyme interactions.

A multi-enzyme process was developed by Karich *et al.* [112], using three fungal enzymes, namely wild-type AAO, wild-type peroxygenase from *Agrocybe aegerita* (Aae UPO), and recombinant galactose oxidase (GAO), as shown in Fig. 7.8. AAOs, being flavoproteins active on benzyl alcohols and hydrate aldehydes, can oxidize

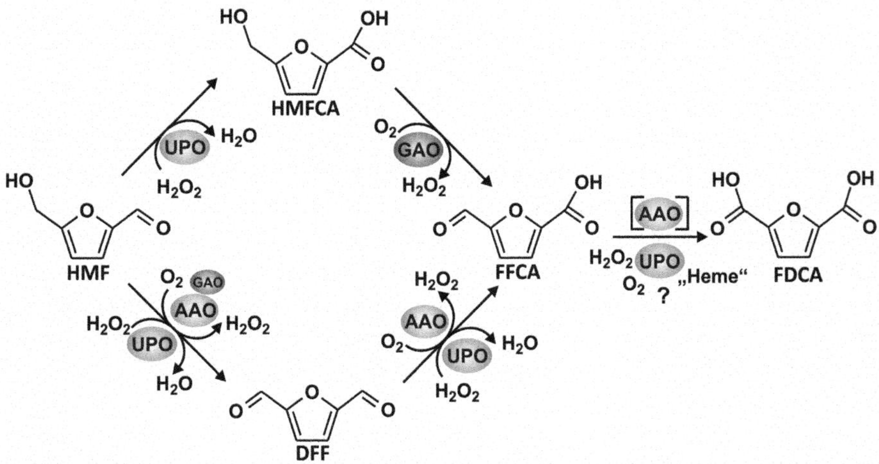

Fig. 7.8 Multi-enzyme cascade pathway for FDCA synthesis with fungal aryl alcohol oxidases (AAOs), non-specific peroxygenases (UPOs), and galactose oxidases (GAOs). (Adapted with permission from [112], Copyright © 2018 MDPI)

HMF and its derivative DFF to produce FFCA while reducing molecular oxygen to hydrogen peroxide. However, the excessive use of UPO needed adjustment to minimize the quantity of oxidase without compromising H_2O_2 levels. The combination of AAO from the Antarctic fungus *Moesziomyces antarcticus* with a non-specific peroxygenase achieved complete conversion of HMF to FDCA within 144 h. AAO oxidized HMF to DFF and FFCA, and UPO completely converted FFCA to FDCA. The inhibitory effect of H_2O_2 on AAO was addressed by adding catalase to remove H_2O_2 from the reaction mixture.

Despite the advantages of biotechnological conversion for FDCA production, challenges persist for large-scale applications. The complexity of the reaction environment requires precise control of pH, temperature, substrate concentration, and nutrients to maintain cell viability and enzyme stability [108]. Sensitivity to impurities and difficulties in separating and purifying FDCA from complex mixtures pose additional challenges. Traditional separation techniques may prove inefficient or costly in dealing with bio-based materials. Innovations in separation technologies are crucial for enhancing recovery yield and purity.

In short, while biotechnological conversion holds significant potential for FDCA production, addressing the complexities of the reaction environment and challenges in product separation and purification is essential for its industrial-scale application. Interdisciplinary research and technological innovations are expected to overcome these challenges, paving the way for the widespread adoption of biotechnological catalytic conversion in sustainable chemical production.

7.3.4 Electrochemical Oxidation

The electrocatalytic oxidation reaction of HMF (HMF-EOR) stands out as an environmentally friendly and clean method for converting HMF into FDCA, especially considering the increasing prevalence of renewable electricity resources. This oxidation process relies on the electrochemical potential generated through electron transfer under mild conditions, eliminating the necessity for external oxidizing agents. The significance of electrochemical oxidation of HMF is growing, primarily due to its reliance on non-precious metal catalysts and its operation at room temperature. The electrochemical oxidation of HMF can be represented by the following equations:

$$\text{Anode:}\quad \text{HMF} + 6\text{OH}^- \rightarrow \text{FDCA} + 4\text{H}_2\text{O} + 6\text{e}^- \tag{7.8}$$

$$\text{Cathode:}\quad 6\text{H}_2\text{O} + 6\text{e}^- \rightarrow 3\text{H}_2 + 6\text{OH}^- \tag{7.9}$$

$$\text{Overall:}\quad \text{HMF} + 2\text{H}_2\text{O} \rightarrow \text{FDCA} + 3\text{H}_2 \tag{7.10}$$

The hydrogen produced at the cathode in the electrocatalytic oxidation reaction of HMF (HMF-EOR) can serve as a clean energy source, aligning with current carbon-neutral policies. This further underscores the electrochemical oxidation method as a green and clean approach for oxidizing HMF. The reaction pathways of electrochemical catalytic reactions, illustrated in Routes 1 and 2 of Fig. 7.5, both have the potential to form FDCA. Ion-exclusion detection through HPLC reveals that the reaction pathway is influenced by the pH value. Under strong alkaline conditions ($pH \geq 13$), Route 1 is the primary pathway, while in non-strong alkaline environments ($pH < 13$), Route 2 is favored. However, HPLC, being an ion-exclusion detection method, cannot monitor intermediate products in real-time during the reaction process. Transient intermediate products may form briefly but quickly convert and remain undetected. To gain a more accurate understanding of the reaction pathway, advanced in-situ/real-time techniques and theoretical calculations should be combined with ion-exclusion detection.

For example, Wang *et al.* [114] developed in-situ and sum frequency generation (SFG) vibrational spectroscopy, a second-order nonlinear spectroscopy technique with intrinsic interfacial selectivity. This technique can capture interfacial molecular information that is challenging to obtain using other spectroscopic methods. It was employed to trace the reaction intermediates on the electrode surface. The study detected only HMFCA at different potentials and times, with no observation of the DFF signal throughout the entire oxidation process. This directly confirmed the electrochemical oxidation of HMF through the HMFCA pathway (i.e., Route 1) [80].

Figure 7.9 illustrates the specific steps of the electrochemical oxidation of HMF into FDCA through a direct oxidation mechanism in a strong alkaline environment: the aldehyde group of HMF preferentially adsorbs on the catalyst surface, undergoes hydration to form a diol, and then this diol reacts with the OH^- adsorbed on the catalyst surface, generating the intermediate product HMFCA through the activation of C–H/O–H bonds and deprotonation processes involving OH^-. Subsequently, another FFCA is generated on another alcohol chain through a similar reaction with OH^-. Finally, the aldehyde group of FFCA undergoes hydration again, and the resulting diol intermediate reacts with OH^- and is dehydrogenated, ultimately producing the final product FDCA. The application of potential promotes OH^- adsorption, driving the entire reaction [115].

In the electrochemical oxidation process, heterogeneous catalysts are commonly employed for achieving high conversion rates and yields. However, homogeneous catalysis can also be utilized. Homogeneous catalysts offer better contact with reactants, providing higher yields and Faradaic efficiency (FE) without the need to adjust the pH value. Choi *et al.* [118] demonstrated that using 2,2,6,6-tetramethylpiperidine-1-oxyl (TEMPO) as a medium significantly reduces the overpotential of HMF, suppressing water oxidation, leading to a 98% FDCA yield and FE exceeding 93%. Despite the high costs associated with using a precious metal electrode (Au) and an excess of TEMPO, making it unsuitable for industrial applications, Cardiel *et al.* [119] continued the research, replacing TEMPO with the more affordable 4-acetamido-TEMPO and using carbon felt instead of Au as the electrode. This modification achieved a 99.8% HMF conversion rate and 91.5% FDCA yield, making it more economically viable.

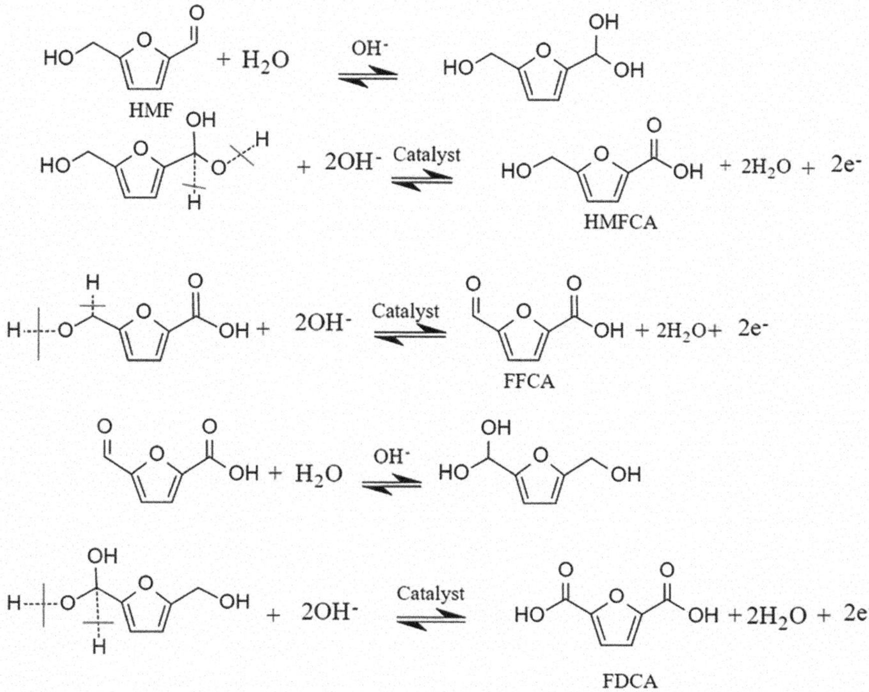

Fig. 7.9 Electrochemical reaction mechanism of HMF direct oxidation under strong alkaline environment (pH ≥ 13). (Modified based on [80, 116, 117])

In recent years, researchers have primarily focused on developing non-precious metal catalysts due to their lower costs and better catalytic performances. For instance, Song *et al.* [120] modified a NiB_x catalyst, and NiBx-P007 achieved the highest FDCA yield (90.6%) and FE (92.5%) by doping with different amounts of phosphorus. Chen *et al.*[121] prepared a Cu-Ni bimetallic catalyst, achieving 100% HMF conversion and a 93.3% yield. Researchers have also explored electrochemical oxidation of HMF under acidic conditions, with Stephen *et al.* [79] using a manganese oxide (MnO_x) anode in a pH = 1 H_2SO_4 solution to achieve a 53.8% FDCA yield. The acidic conditions allow for the direct separation of FDCA, reducing operating costs and offering potential for industrial production. Moreover, suppressing water oxidation is essential in electrochemical catalytic oxidation. Comparative analysis of the electrochemical oxidation reactions of HMF and its oxidation intermediates on MnO_x and Pt electrodes in a pH = 1 solution demonstrates that MnO_x exhibits higher selectivity than Pt for the oxidation of HMF to FDCA [79].

In conclusion, electrochemical oxidation has emerged as a promising strategy for the conversion of HMF to FDCA, demonstrating significant potential in the fields of green chemistry and sustainable technology. Continued research and technological advancements are expected to broaden its applications, providing an environmentally friendly and economically efficient pathway for the production of bio-based polymer materials.

7.3.5 Catalysts Design

Addressing the challenges in FDCA synthesis involves optimizing catalyst design and exploring alternatives to noble metals, with a focus on achieving higher yields and competitive production methods under non-alkaline conditions. To compete with TPA, developing efficient methods for the synthesis of FDCA is crucial. Noble metal catalysts like Pt, Au, Pd, or Ru supported on substrates typically operate in the presence of alkaline aqueous solutions. However, the use of alkaline substances in homogeneous phases can pose environmental challenges. Therefore, a current challenge is to develop reaction systems with proper catalysts that operate under non-alkaline conditions. The structural similarity between FDCA and its counterpart FFCA raises the issue of potential undesired combination, limiting further oxidation of FDCA. FFCA may also act as chain terminators in the polymerization process, affecting the production of PEF. Thus, removing FFCA from the final oxidation product is crucial for FDCA to remain competitive.

Efforts in FDCA production from HMF have focused on two main aspects: (1) improving catalyst design strategies for higher FDCA yields and (2) replacing noble metals with lower cost, more readily available non-noble metals such as Mn, Co, and Fe [91]. Tirsoaga *et al.* [21] developed a 10M@*x*Nb@MNP catalyst, achieving a 96.5% FDCA conversion rate. This catalyst combines molybdenum doping in a mesoporous silica-based MCM-41 shell with manganese, cobalt, and iron oxides as magnetic nanoparticle cores. The incorporation of Nb into the mesoporous silica-based MCM-41 shell enhances catalyst activity, stability, and reusability. The uneven layer of silicate formed on the magnetic iron oxide particles' surface, along with the well-dispersed distribution of molybdenum and other metal oxides, indicates that this synthesis method produces multifunctional catalysts effectively.

The novel dual-function system design offers advantages over using noble metals by operating without the need for a homogeneous base to facilitate the reaction mechanism. Experimental results demonstrate the synergistic effects of the catalyst's texture and structural characteristics on the selectivity of the product FDCA. The degree of FDCA selectivity follows the order of surface area and pore volume: Co > Fe > Mn. Higher surface area and pore volume correspond to greater FDCA selectivity.

Moreover, the combination of metal oxides (MO_x–CoO_x, MnO_x, or FeO_x) with Nb generates MO_x as a co-catalyst for Nb sites, providing a synergistic effect through the decomposition of *t*-BuOOH to produce hydroxyl groups. Once formed, these hydroxyl groups are captured by Nb cationic species, generating active sites for the dehydrogenation and oxidation of intermediate products to FDCA. This highlights the intricate interplay between different components in the catalyst system, contributing to its efficiency in the conversion of HMF to FDCA.

In another study, the synthesis of Fe nanoparticles doped into NH_2-functionalized SBA-15 (FeNPs@NH_2-SBA-15) and its use as a catalyst for the selective oxidation of HMF to FDCA has been investigated [122]. Operating under a pressure of 0.6 MPa and a temperature of 120 °C, this catalyst achieved a 100% HMF conversion rate and an 89.4% FDCA yield. The surface area and mesoporous structure of the catalyst play a crucial role in its performance. The preparation process involves several steps, starting with the synthesis of the SBA-15 mesoporous silica scaffold material. Subsequent modifications include surface functionalization with amino groups to obtain NH_2-SBA-15. Iron ions are then uniformly added to the surface of NH_2-SBA-15, and a reducing agent is introduced to convert the iron ions into iron nanoparticles (FeNPs) fixed on the surface. The resulting composite solid basic-oxidative catalyst, FeNPs@NH_2-SBA-15, is obtained through filtration and other treatments.

The SEM images reveal a fibrous structure with lengths of 20–30 μm and diameters of 3–5 μm, composed of aggregates of rod-shaped sub-particles with diameters of 1–2 μm and lengths of 0.5 μm [122]. HRTEM images show the regular hexagonal pore structure of the SBA-15 scaffold with long-range ordered channels [122]. Fe nanoparticles are embedded within these channels, maintaining an ordered hexagonal arrangement even after functionalization. Following ultrasonic treatment, uniformly distributed spherical Fe nanoparticles with an average size of approximately 6 nm are observed, indicating the SBA-15 scaffold structure remains intact. This structural stability enhances its effectiveness as a carrier for Fe nanoparticles in catalyzing the oxidation of HMF to FDCA [123–125].

The above research findings on non-noble metal catalysts showcase their significant potential in industrial chemistry, offering valuable tools for advancing sustainable chemistry goals. Catalyst design plays a crucial role in efficiently converting renewable resources into high-value chemicals, contributing to the progress of green chemistry and sustainable development. The innovations in materials science and catalytic chemistry highlighted in these studies demonstrate that it is feasible to reduce environmental impact without compromising performance. These advancements set the stage for future industrial applications with a focus on sustainability and environmental responsibility.

7.4 Other Considerations

7.4.1 Process Economy

In the context of FDCA production, researchers have considered the market price of purified PTA as an indicator [126]. The use of biomass wastes as a raw material for FDCA production, being a green and sustainable approach, enhances its attractiveness in the market [1]. Comparisons with market prices of PTA, a conventional chemical, indicate the potential competitiveness of FDCA. The choice of biomass wastes, such as food waste, agricultural and forestry wastes, as a feedstock offers advantages in terms of sustainability and ease of availability.

Several studies have investigated the techno-economic feasibility of converting biomass wastes into FDCA [127–129]. For instance, the conversion of lignocellulose to FDCA through an electrochemical method resulted in an estimated minimum selling price (MSP) of about 3.67€/kg [126]. However, challenges related to high production costs and low process yields currently limit the industrial-scale feasibility of such processes.

Specific types of food wastes, like waste from tomato, potato, orange, and olive processing, have been analyzed for their potential as feedstocks [130]. Techno-economic analyses indicate varying profitability based on the type of food wastes. Centralizing production and leveraging economies of scale have been identified as strategies to enhance profitability.

The choice of raw materials, such as starch, glucose, or high fructose corn syrup, can significantly impact the commercial viability of FDCA production [126, 131]. Using non-precious metal catalysts and diverse raw materials, researchers have explored different pathways with promising results. The FDCA final selling price is influenced by factors like raw material costs, reaction routes, equipment, and catalysts [126, 131].

In addition to chemical production, there is a growing emphasis on utilizing biomass wastes for biofuels. HTL is highlighted as a method to convert biomass wastes into transportation fuels. The MSP is estimated at $5.59/gal of gasoline-equivalent using an integrated production pathway involving fast pyrolysis and biocrude gasification [132].

Furthermore, the conversion of kraft lignin into depolymerized kraft lignin (DKL) and oxypropylated depolymerized kraft lignin (Oxy-DKL) demonstrates potential as bio-substitutes for petroleum-based polyols. The MSPs of these bio-substitutes are estimated at $1440/ton and $1623/ton, respectively [133].

In summary, the conversion of biomass wastes into high-value bio-based products, including chemicals, materials, and biofuels, holds great promise for reducing dependence on fossil fuels, contributing to environmental sustainability, and providing new revenue streams for the forestry and agricultural sectors. Researchers are actively exploring diverse pathways and conducting techno-economic analyses to determine the economic viability of these processes on a large scale [134].

7.4.2 Process Scale-Up

While lab-scale developments have shown promising results, scale-up and economically viable production, including catalyst and solvent recycling and product purification, remain significant challenges for large-scale implementation.

Hydrothermal reactions applied to biomass have led to the generation of many by-products, such as humins, posing challenges for commercial development.

Solvation has been proposed as an effective way to prevent harmful side reactions in the conversion of biomass into furans. For instance, a two-step process for the conversion of fructose to FDCA at a high fructose concentration (19 wt%) in a DMSO/H_2O mixture demonstrated improved stability of HMF and yielded up to 65% FDCA from crude HMF [135]. Other solvent systems, such as THF/concentrated seawater [136] and dimethyl carbonate: tetraethyl ammonium bromide (DMC:TEAB), have also shown promise in stabilizing HMF during conversion processes [137].

Process intensification technologies, microreactors, and adsorption beds have been evaluated through process modeling tools to reduce the MSPs and emissions in HMF production [138–140]. Continuous reactors are considered necessary for scale-up, with electrocatalysts like self-supported $Mn_{0.2}NiS$/GF nanosheets demonstrating remarkable activity and stability in continuous-flow electrolyzers [141].

The development of flow-based methods is proposed as an ideal future direction for the industry, offering advantages such as shorter reaction times, easier handling of large feed loads, greater stability for furans, and improved extraction and purification. The reduced retention time in a flow-based framework is expected to decrease the formation of humins and increase process yields. Addressing challenges related to the solubility of agricultural wastes and selecting appropriate acid and solvent media are crucial aspects in the development of these flow-based methods.

Looking ahead, scaling up FDCA production requires concentrated efforts in several key research areas. Firstly, advancements in catalyst and solvent recycling are crucial for economic viability. Overcoming by-product generation, particularly humins, remains a significant hurdle. Research should focus on optimizing solvent systems to stabilize intermediates like HMF effectively. Promising solvent systems such as THF/concentrated seawater and DMC/TEAB need further exploration and refinement.

Process intensification through microreactors and adsorption beds shows potential for reducing costs and emissions. Transitioning to continuous reactors, particularly those utilizing advanced electrocatalysts, is essential for industrial scalability. Integrating flow-based methods could revolutionize the industry by enhancing stability, reducing reaction times, and improving extraction processes. However, addressing solubility issues of agricultural wastes and selecting appropriate acid and solvent media remain critical challenges [142, 143].

In conclusion, while significant progress has been made, large-scale FDCA production still faces challenges that require focused research and innovative solutions. The future of FDCA production depends on successfully integrating these advancements into a cohesive, scalable process.

7.5 Conclusions and Future Outlook

The future of catalytic conversion of agricultural residues and food wastes into FDCA holds great promise for creating sustainable and valuable chemicals from biomass. Addressing the key challenges through focused research and innovative solutions will pave the way for efficient, scalable, and economically viable production processes. By leveraging advanced technologies, optimizing processes, and fostering interdisciplinary collaboration, the vision of large-scale FDCA production from renewable resources can become a reality.

The catalytic conversion of agricultural residues and food wastes into FDCA presents a promising avenue for sustainable chemical production, leveraging abundant biomass resources to create value-added products. Looking ahead, several key areas of research and development will be critical to advancing this field:

1. Development of cost-effective catalysts: Innovations in catalyst design are essential to enhance the efficiency and selectivity of the conversion processes. Future research should focus on developing catalysts that are not only highly active and selective but also economically viable. This includes exploring cheaper and more abundant materials as catalyst supports and active sites.
2. Process optimization and integration: Optimizing the conversion process to maximize yield and selectivity while minimizing by-product formation is crucial. Advanced process integration strategies, such as combining hydrothermal processing with catalytic electrochemical oxidation in a single, continuous system, could significantly improve efficiency. Process intensification techniques, including the use of microreactors and adsorption beds, should be further explored.
3. Enhanced solvent systems: Solvent systems play a pivotal role in stabilizing intermediates like HMF. Research should focus on identifying and optimizing solvent systems that can effectively stabilize these intermediates and facilitate their conversion to FDCA. The potential of green solvents and solvent recycling methods should also be investigated.
4. Innovative technologies and methodologies: Incorporating advanced technologies such as machine learning and artificial neural networks (ANN) can provide deeper insights into the conversion process and help optimize reaction conditions. These tools can assist in modeling and predicting catalyst performance, process conditions, and potential bottlenecks.
5. Scale-up and commercialization: Transitioning from laboratory-scale research to industrial-scale production requires addressing scalability issues. Developing continuous reactor systems, particularly those utilizing advanced electrocatalysts, is essential for achieving industrial-scale FDCA production. Ensuring the stability and robustness of these systems under continuous operation is a key challenge.
6. Environmental and economic sustainability: Assessing the environmental impact and economic viability of the entire conversion process is vital for commercial success. Life cycle assessments (LCA) and techno-economic analyses (TEA) should be conducted to evaluate the sustainability and cost-effectiveness of the proposed methods.
7. Interdisciplinary collaboration: Advancing the field of biomass conversion to FDCA will require interdisciplinary collaboration among chemists, chemical engineers, material scientists, and environmental scientists. Such collaboration

can foster innovation and lead to the development of holistic solutions that address both technical and environmental challenges.

8. Policy and regulatory support: Supportive policies and regulations can accelerate the development and adoption of biomass-based chemical production. Incentives for research and development, as well as for the commercialization of sustainable technologies, will be crucial in driving progress in this field.

References

1. Saqib NU, Sharma HB, Baroutian S, Dubey B, Sarmah AK. Valorisation of food waste via hydrothermal carbonisation and techno-economic feasibility assessment. Sci Total Environ. 2019;690:261–76. https://doi.org/10.1016/j.scitotenv.2019.06.484.
2. Zhu Y, Biddy MJ, Jones SB, Elliott DC, Schmidt AJ. Techno-economic analysis of liquid fuel production from woody biomass via hydrothermal liquefaction (HTL) and upgrading. Appl Energy. 2014;129:384–94. https://doi.org/10.1016/j.apenergy.2014.03.053.
3. Wang K, Tester JW. Sustainable management of unavoidable biomass wastes. Green Energy Resourc. 2023;1:100005. https://doi.org/10.1016/j.gerr.2023.100005.
4. Burhenne L, Messmer J, Aicher T, Laborie MP. The effect of the biomass components lignin, cellulose and hemicellulose on TGA and fixed bed pyrolysis. J Anal Appl Pyrolysis. 2013;101: 177–84. https://doi.org/10.1016/j.jaap.2013.01.012.
5. Alper K, Tekin K, Karagöz S, Ragauskas AJ. Sustainable energy and fuels from biomass: a review focusing on hydrothermal biomass processing. Sustain Energy Fuels. 2020;4:4390–414. https://doi.org/10.1039/d0se00784f.
6. López Barreiro D, Prins W, Ronsse F, Brilman W. Hydrothermal liquefaction (HTL) of microalgae for biofuel production: state of the art review and future prospects. Biomass Bioenergy. 2013;53:113–27. https://doi.org/10.1016/j.biombioe.2012.12.029.
7. Ponnusamy VK, Nagappan S, Bhosale RR, Lay CH, Duc Nguyen D, Pugazhendhi A, Chang SW, Kumar G. Review on sustainable production of biochar through hydrothermal liquefaction: physico-chemical properties and applications. Bioresour Technol. 2020;310:123414. https://doi.org/10.1016/j.biortech.2020.123414.
8. Peterson AA, Vogel F, Lachance RP, Fröling M, Antal JMJ, Tester JW. Thermochemical biofuel production in hydrothermal media: a review of sub- and supercritical water technologies. Energy Environ Sci. 2008;1:32–65. https://doi.org/10.1039/b810100k.
9. Elliott DC, Biller P, Ross AB, Schmidt AJ, Jones SB. Hydrothermal liquefaction of biomass: developments from batch to continuous process. Bioresour Technol. 2015;178:147–56. https://doi.org/10.1016/j.biortech.2014.09.132.
10. Khandelwal K, Nanda S, Boahene P, Dalai AK. Conversion of biomass into hydrogen by supercritical water gasification: a review. Environ Chem Lett. 2023;21:2619–38. https://doi.org/10.1007/s10311-023-01624-z.
11. Demirbas A. Hydrogen-rich gas from fruit shells via supercritical water extraction. Int J Hydrog Energy. 2004;29:1237–43. https://doi.org/10.1016/j.ijhydene.2003.11.012.
12. Schmieder H, Boukis N, Dinjus E, Kruse A, Kluth M, Petrich G, Schacht M. Hydrothermal gasification of biomass and organic wastes. J Supercrit Fluids. 2000;17:145–53. https://doi.org/10.1016/S0896-8446(99)00051-0.
13. Couto E, Calijuri ML, Assemany P. Biomass production in high rate ponds and hydrothermal liquefaction: wastewater treatment and bioenergy integration. Sci Total Environ. 2020;724: 138104. https://doi.org/10.1016/j.scitotenv.2020.138104.
14. Ignatiev A, Chen X, Wu N, Lu Z, Smith L. Nanostructured thin solid oxide fuel cells with high power density. Dalton Trans. 2008;40:5501–6. https://doi.org/10.1039/b805658g.

15. Romano P, Stampone N, Di Giacomo G. Evolution and prospects of hydrothermal carbonization. Energies. 2023;16. https://www.mdpi.com/1996-1073/16/7/3125.
16. Anastas PT. Circularity what's the problem? ACS Sustain Chem Eng. 2020;8:13111. https://doi.org/10.1021/acssuschemeng.0c05714.
17. Li CS, Cai RR. Preparation of solid organic fertilizer by co-hydrothermal carbonization of peanut residue and corn cob: a study on nutrient conversion. Sci Total Environ. 2022;838: 155867. https://doi.org/10.1016/j.scitotenv.2022.155867.
18. Lin Z, Gao Q, Diao P. Promoting the electrocatalytic oxygen evolution reaction on NiCo(2)O (4) with infrared-thermal effect: a strategy to utilize the infrared solar energy to reduce activation energy during water splitting. J Colloid Interface Sci. 2023;638:54–62. https://doi.org/10.1016/j.jcis.2023.01.130.
19. Czapela FF, Kubeneck S, Preczeski KP, Dalastra C, Scapini T, Bonatto C, Stefanski FS, Camargo AF, Zanivan J, Mossi AJ, Fongaro G, Treichel H. Reactional ultrasonic systems and microwave irradiation for pretreatment of agro-industrial waste to increase enzymatic activity. Bioresour Bioprocess. 2020;7 https://doi.org/10.1186/s40643-020-00338-2.
20. Shao Y, Tan H, Shen D, Zhou Y, Jin Z, Zhou D, Lu W, Long Y. Synthesis of improved hydrochar by microwave hydrothermal carbonization of green waste. Fuel. 2020;266:117146. https://doi.org/10.1016/j.fuel.2020.117146.
21. Tirsoaga A, El Fergani M, Nuns N, Simon P, Granger P, Parvulescu VI, Coman SM. Multifunctional nanocomposites with non-precious metals and magnetic core for 5-HMF oxidation to FDCA. Appl Catal B Environ. 2020;278:119309. https://doi.org/10.1016/j.apcatb.2020.119309.
22. Dejong E, Dam R, Sipos L, Den Ouden D, Gruter GJ. Furandicarboxylic acid (FDCA) a versatile building block for a very interesting class of polyesters. Abstr Pap Am Chem Soc. 2011;241 https://doi.org/10.1021/bk-2012-1105.ch001.
23. Kim H, Choi J, Won W. Process synthesis and analysis of green plastic monomer production from cellulose. J Clean Prod. 2020;277:124072. https://doi.org/10.1016/j.jclepro.2020.124072.
24. Kim H, Lee S, Won W. Economical process for the co-production of renewable polymers and value-added chemicals from lignocellulosic biomass. J Clean Prod. 2020;276:124237. https://doi.org/10.1016/j.jclepro.2020.124237.
25. Motagamwala AH, Won W, Sener C, Alonso DM, Maravelias CT, Dumesic JA. Toward biomass-derived renewable plastics: production of 2,5-furandicarboxylic acid from fructose. Sci Adv. 2018;4:eaap9722. https://doi.org/10.1126/sciadv.aap9722.
26. Pandian M, Srinivasa RB, Maeum L, Jonathan HM, Gwang-Nam Y, Won HD. Integrated process towards sustainable renewable plastics: production of 2,5-furandicarboxylic acid from fructose in a base-free environment. Appl Catal A Gen. 2023;667 https://doi.org/10.1016/j.apcata.2023.119446.
27. Zhang L, Meng G, Zhang W, Li X, Zhang Z, Yang M, Wu Y, Wang D, Li Y. Oriented conversion of a LA/HMF mixture to GVL and FDCA in a biphasic solvent over a Ru single-atom/nanoparticle dual-site catalyst. ACS Catal. 2023;13:2268–76. https://doi.org/10.1021/acscatal.2c04726.
28. Manyepedza T, Makuve N. Catalysts for biomass upgrading: transforming 5-hydroxymethyl furfural into 2,5-furan dicarboxylic acid for fine chemicals, featuring mil-100 (Fe) Mof as a novel catalyst. ChemistrySelect. 2024;9 https://doi.org/10.1002/slct.202401412.
29. Aranha DJ, Gogate PR. A review on green and efficient synthesis of 5-hydroxymethylfurfural (HMF) and 2,5-furandicarboxylic acid (FDCA) from sustainable biomass. Ind Eng Chem Res. 2023;62:3053–78. https://doi.org/10.1021/acs.iecr.2c04047.
30. Zheng J, Li Z, Zhang D, Zhao X, Zhao Q, Wang Y. One-pot four-step direct synthesis of 2,5-furandicarboxylic acid from 2,5-diformylfuran under oxygen-free conditions. Appl Catal A Gen. 2024;685:119886. https://doi.org/10.1016/j.apcata.2024.119886.

31. Parate RD, Dharne MS, Rode CV. Integrated chemo and bio-catalyzed synthesis of 2,5-furandicarboxylic acid from fructose derived 5-hydroxymethylfurfural. Biomass Bioenergy. 2022;161:106474. https://doi.org/10.1016/j.biombioe.2022.106474.
32. Liu H, Tang X, Zeng X, Sun Y, Ke X, Li T, Zhang J, Lin L. Catalyst design strategy toward the efficient heterogeneously-catalyzed selective oxidation of 5-hydroxymethylfurfural. Green Energy Environ. 2022;7:900–32. https://doi.org/10.1016/j.gee.2021.10.004.
33. Zong P, Jiang Y, Tian Y, Li J, Yuan M, Ji Y, Chen M, Li D, Qiao Y. Pyrolysis behavior and product distributions of biomass six group components: starch, cellulose, hemicellulose, lignin, protein and oil. Energy Convers Manag. 2020;216:112777. https://doi.org/10.1016/j.enconman.2020.112777.
34. NaS R, NaS A. Thermo-kinetic assessment of glucose decomposition to 5-hydroxymethyl furfural and levulinic acid over acidic functionalized ionic liquid. Chem Eng J. 2018;335:221–30. https://doi.org/10.1016/j.cej.2017.10.112.
35. Binder JB, Raines RT. Simple chemical transformation of lignocellulosic biomass into furans for fuels and chemicals. J Am Chem Soc. 2009;131:1979–85. https://doi.org/10.1021/ja802172h.
36. Weingarten R, Rodriguez-Beuerman A, Cao F, Luterbacher JS, Alonso DM, Dumesic JA, Huber GW. Selective conversion of cellulose to hydroxymethylfurfural in polar aprotic solvents. ChemCatChem. 2014;6:2229–34. https://doi.org/10.1002/cctc.201402299.
37. Yang F, Liu Q, Bai X, Du Y. Conversion of biomass into 5-hydroxymethylfurfural using solid acid catalyst. Bioresour Technol. 2011;102:3424–9. https://doi.org/10.1016/j.biortech.2010.10.023.
38. Zhou D, Shen D, Lu W, Song T, Wang M, Feng H, Shentu J, Long Y. Production of 5-hydroxymethylfurfural from chitin biomass: a review. Molecules. 2020;25 https://doi.org/10.3390/molecules25030541.
39. Menegazzo F, Ghedini E, Signoretto M. 5-Hydroxymethylfurfural (HMF) production from real biomasses. Molecules. 2018;23 https://doi.org/10.3390/molecules23092201.
40. Olivito F, Algieri V, Tallarida MA, Jiritano A, Costanzo P, Maiuolo L, Nino AD. High-yield synthesis of HMF from glucose and fructose by selective catalysis with water-tolerant rare earth metal triflates assisted by choline chloride. Green Chem. 2023;25:1679–89. https://doi.org/10.1039/d2gc04046h.
41. He Q, Yu YX, Wang J, Suo XD, Liu YD. Kinetic study of the hydrothermal carbonization reaction of glucose and its product structures. Ind Eng Chem Res. 2021;60:4552–61. https://doi.org/10.1021/acs.iecr.0c06280.
42. Li K, Du M, Ji P. Multifunctional tin-based heterogeneous catalyst for catalytic conversion of glucose to 5-hydroxymethylfurfural. ACS Sustain Chem Eng. 2018;6:5636–44. https://doi.org/10.1021/acssuschemeng.8b00745.
43. Kabyemela BM, Adschiri T, Malaluan RM, Arai K. Glucose and fructose decomposition in subcritical and supercritical water: detailed reaction pathway, mechanisms, and kinetics. Ind Eng Chem Res. 1999;38:2888–95. https://doi.org/10.1021/ie9806390.
44. Yan Z-F, Lian J, Zhao Z, He C-Q, Yue X-P, Wang Y-P, Wu X, Lu J-J. Theoretical insight into the conversion of glucose to 5-hydroxymethylfurfural in subcritical water. J Fuel Chem Technol. 2021;49:1122–31. https://doi.org/10.1016/s1872-5813(21)60124-x.
45. Zhang R, Eronen A, Du X, Ma E, Guo M, Moslova K, Repo T. A catalytic approach via retro-aldol condensation of glucose to furanic compounds. Green Chem. 2021;23:5481–6. https://doi.org/10.1039/d1gc01429c.
46. Cao X, Peng X, Sun S, Zhong L, Chen W, Wang S, Sun RC. Hydrothermal conversion of xylose, glucose, and cellulose under the catalysis of transition metal sulfates. Carbohydr Polym. 2015;118:44–51. https://doi.org/10.1016/j.carbpol.2014.10.069.
47. Jaroslav H, Branimir K, Biserka M, Vitomir S. Mechanism of levulinic acid formation. Tetrahedron Lett. 1985;26:2111–4. https://doi.org/10.1016/s0040-4039(00)94793-2.

48. Yang Y, Qi H, Li H, Xu Z, Liu X, Yu S, Zhang ZC. Heterometallic PdII–Cl–CuI catalyst for efficient hydrolysis of β-1,4-glycosidic bonds in 1-butyl-3-methylimidazolium chloride. ACS Catal. 2021;11:11774–85. https://doi.org/10.1021/acscatal.1c03045.
49. Klemm D, Heublein B, Fink HP, Bohn A. Cellulose: fascinating biopolymer and sustainable raw material. Angew Chem Int Ed Engl. 2005;44:3358–93. https://doi.org/10.1002/anie.200460587.
50. Buendia-Kandia F, Mauviel G, Guedon E, Rondags E, Petitjean D, Dufour A. Decomposition of cellulose in hot-compressed water: detailed analysis of the products and effect of operating conditions. Energy Fuel. 2017;32:4127–38. https://doi.org/10.1021/acs.energyfuels.7b02994.
51. Barbier J, Charon N, Dupassieux N, Loppinet-Serani A, Mahé L, Ponthus J, Courtiade M, Ducrozet A, Quoineaud A-A, Cansell F. Hydrothermal conversion of lignin compounds. A detailed study of fragmentation and condensation reaction pathways. Biomass Bioenergy. 2012;46:479–91. https://doi.org/10.1016/j.biombioe.2012.07.011.
52. Wang K, Ma Q, Burns M, Sudibyo H, Sills DL, Goldfarb JL, Tester JW. Impact of feed injection and batch processing methods in hydrothermal liquefaction. J Supercrit Fluids. 2020;164:104887. https://doi.org/10.1016/j.supflu.2020.104887.
53. Guo K, Cheng Q, Jiang J, Xu J, Tan W. Qualitative analysis of liquid products generated from lignocellulosic biomass using post-target and nontarget analysis methods and liquefaction mechanism research. ACS Sustain Chem Eng. 2020;8:11099–113. https://doi.org/10.1021/acssuschemeng.0c00539.
54. Yang C, Wang S, Yang J, Xu D, Li Y, Li J, Zhang Y. Hydrothermal liquefaction and gasification of biomass and model compounds: a review. Green Chem. 2020;22:8210–32. https://doi.org/10.1039/d0gc02802a.
55. Yan Z, Lian J, Feng Y, Li M, Long F, Cheng R, Shi S, Guo H, Lu J. A mechanistic insight into glucose conversion in subcritical water: complex reaction network and the effects of acid-base catalysis. Fuel. 2021;289:119969. https://doi.org/10.1016/j.fuel.2020.119969.
56. Shen HS, Liu L. Cover feature: evolution process and controlled synthesis of humins with 5-hydroxymethylfurfural (HMF) as model molecule. ChemSusChem. 2020;13:436. https://doi.org/10.1002/cssc.201901489.
57. Ryu J, Suh Y-W, Suh DJ, Ahn DJ. Hydrothermal preparation of carbon microspheres from mono-saccharides and phenolic compounds. Carbon. 2010;48:1990–8. https://doi.org/10.1016/j.carbon.2010.02.006.
58. Zhang M, Yang H, Liu Y, Sun X, Zhang D, Xue D. Hydrophobic precipitation of carbonaceous spheres from fructose by a hydrothermal process. Carbon. 2012;50:2155–61. https://doi.org/10.1016/j.carbon.2012.01.024.
59. Kang X, Peng J, Ragauskas AJ, Ren X, Si C, Wang S, Song X. Competitive effects of glucan's main hydrolysates on biochar formation: a combined experiment and density functional theory analysis. Bioresour Technol. 2022;359:127427. https://doi.org/10.1016/j.biortech.2022.127427.
60. Qiu G, Wang X, Huang C, Li Y, Chen B. Niobium phosphotungstates: excellent solid acid catalysts for the dehydration of fructose to 5-hydroxymethylfurfural under mild conditions. RSC Adv. 2018;8:32423–33. https://doi.org/10.1039/c8ra05940c.
61. Kim Y, Mittal A, Robichaud DJ, Pilath HM, Etz BD, St. John PC, Johnson DK, Kim S. Prediction of hydroxymethylfurfural yield in glucose conversion through investigation of Lewis acid and organic solvent effects. ACS Catal. 2020;10:14707–21. https://doi.org/10.1021/acscatal.0c04245.
62. Román-Leshkov Y, Barrett CJ, Liu ZY, Dumesic JA. Production of dimethylfuran for liquid fuels from biomass-derived carbohydrates. Nature. 2007;447:982–5. https://doi.org/10.1038/nature05822.
63. Liu L, Chang H-M, Jameel H, Park J-Y, Park S. Catalytic conversion of biomass hydrolysate into 5-hydroxymethylfurfural. Ind Eng Chem Res. 2017;56:14447–53. https://doi.org/10.1021/acs.iecr.7b03635.

64. Huang Y-B, Fu Y. Hydrolysis of cellulose to glucose by solid acid catalysts. Green Chem. 2013;15:1095. https://doi.org/10.1039/c3gc40136g.
65. Bundhoo ZMA. Microwave-assisted conversion of biomass and waste materials to biofuels. Renew Sustain Energ Rev. 2018;82:1149–77. https://doi.org/10.1016/j.rser.2017.09.066.
66. Li B, Fan X, Yu S, Xia H, Nong Y, Bian J, Sun M, Zi W. Microwave heating of biomass waste residues for sustainable bioenergy and biomass materials preparation: a parametric simulation study. Energy. 2023;274:127347. https://doi.org/10.1016/j.energy.2023.127347.
67. Takagaki A, Ohara M, Nishimura S, Ebitani K. A one-pot reaction for biorefinery: combination of solid acid and base catalysts for direct production of 5-hydroxymethylfurfural from saccharides. Chem Commun (Camb). 2009:6276–8. https://doi.org/10.1039/b914087e.
68. Yang Y, Hu C-W, Abu-Omar MM. Conversion of carbohydrates and lignocellulosic biomass into 5-hydroxymethylfurfural using $AlCl_3{\cdot}6H_2O$ catalyst in a biphasic solvent system. Green Chem. 2012;14:509–13. https://doi.org/10.1039/c1gc15972k.
69. Kotadia DA, Soni SS. Symmetrical and unsymmetrical Brønsted acidic ionic liquids for the effective conversion of fructose to 5-hydroxymethyl furfural. Catal Sci Technol. 2013;3:469–74. https://doi.org/10.1039/c2cy20493b.
70. Stahlberg T, Fu W, Woodley JM, Riisager A. Synthesis of 5-(hydroxymethyl)furfural in ionic liquids: paving the way to renewable chemicals. ChemSusChem. 2011;4:451–8. https://doi.org/10.1002/cssc.201000374.
71. Lima S, Neves P, Antunes MM, Pillinger M, Ignatyev N, Valente AA. Conversion of mono/di/polysaccharides into furan compounds using 1-alkyl-3-methylimidazolium ionic liquids. Appl Catal A Gen. 2009;363:93–9. https://doi.org/10.1016/j.apcata.2009.04.049.
72. Walia M, Sharma U, Agnihotri VK, Singh B. Silica-supported boric acid assisted conversion of mono- and poly-saccharides to 5-hydroxymethylfurfural in ionic liquid. RSC Adv. 2014;4: 14414–8. https://doi.org/10.1039/c4ra01771d.
73. Tempelman C, Jacobs U, Herselman J, Van Driel R, Schraa F, Versijde J, Van Waversveld T, Yagci Y, Barg M, Smits F, Kuijpers F, Lamers K, Remijn T, Degirmenci V. Waste apple biomass conversion to 5-HMF over tin doped sulfonated activated carbon as a catalyst. Biomass Bioenergy. 2023;168:106661. https://doi.org/10.1016/j.biombioe.2022.106661.
74. Atanda L, Mukundan S, Shrotri A, Ma Q, Beltramini J. Catalytic conversion of glucose to 5-hydroxymethyl-furfural with a phosphated TiO_2 catalyst. ChemCatChem. 2015;7:781–90. https://doi.org/10.1002/cctc.201402794.
75. Souzanchi S, Nazari L, Venkateswara Rao KT, Yuan Z, Tan Z, Charles XC. Catalytic dehydration of glucose to 5-HMF using heterogeneous solid catalysts in a biphasic continuous-flow tubular reactor. J Ind Eng Chem. 2021;101:214–26. https://doi.org/10.1016/j.jiec.2021.06.010.
76. Kupiainen L, Ahola J, Tanskanen J. Kinetics of glucose decomposition in formic acid. Chem Eng Res Des. 2011;89:2706–13. https://doi.org/10.1016/j.cherd.2011.06.005.
77. Girisuta B, Janssen LPBM, Heeres HJ. A kinetic study on the decomposition of 5-hydroxymethylfurfural into levulinic acid. Green Chem. 2006;8:701. https://doi.org/10.1039/b518176c.
78. Yan P, Xia M, Chen S, Han W, Wang H, Zhu W. Unlocking biomass energy: continuous high-yield production of 5-hydroxymethylfurfural in water. Green Chem. 2020;22:5274–84. https://doi.org/10.1039/d0gc01446j.
79. Kubota SR, Choi KS. Electrochemical oxidation of 5-hydroxymethylfurfural to 2,5-furandicarboxylic acid (FDCA) in acidic media enabling spontaneous FDCA separation. ChemSusChem. 2018;11:2138–45. https://doi.org/10.1002/cssc.201800532.
80. Yang Y, Mu T. Electrochemical oxidation of biomass derived 5-hydroxymethylfurfural (HMF): pathway, mechanism, catalysts and coupling reactions. Green Chem. 2021;23: 4228–54. https://doi.org/10.1039/d1gc00914a.
81. Yu H, Kim K-A, Kang MJ, Hwang SY, Cha HG. Carbon support with tunable porosity prepared by carbonizing chitosan for catalytic oxidation of 5-hydroxymethylfurfural. ACS Sustain Chem Eng. 2018;7:3742–8. https://doi.org/10.1021/acssuschemeng.8b00105.

82. Hameed S, Liu W, Yu Z, Pang J, Luo W, Wang A. Base-free aerobic oxidation of 5-hydroxymethylfurfural to 2,5-furandicarboxylic acid over a Fe single-atom catalyst. Green Chem. 2024;26:7806–17. https://doi.org/10.1039/d4gc01777c.
83. Alvarez-Hernandez D, Megias-Sayago C, Penkova A, Centeno MA, Ivanova S. Highly effective non-noble MnO_2 catalysts for 5-hydroxymethylfurfural oxidation to 2,5-furandicarboxylic acid. ChemSusChem. 2024:e202400115. https://doi.org/10.1002/cssc.202400115.
84. Nie J, Xie J, Liu H. Activated carbon-supported ruthenium as an efficient catalyst for selective aerobic oxidation of 5-hydroxymethylfurfural to 2,5-diformylfuran. Chin J Catal. 2013;34:871–5. https://doi.org/10.1016/s1872-2067(12)60551-8.
85. Nie JF, Xie JH, Liu HC. Efficient aerobic oxidation of 5-hydroxymethylfurfural to 2,5-diformylfuran on supported Ru catalysts. J Catal. 2013;301:83–91. https://doi.org/10.1016/j.jcat.2013.01.007.
86. Zhang Z, Deng K. Recent advances in the catalytic synthesis of 2,5-furandicarboxylic acid and its derivatives. ACS Catal. 2015;5:6529–44. https://doi.org/10.1021/acscatal.5b01491.
87. Ferraz CP, Zieliński M, Pietrowski M, Heyte S, Dumeignil F, Rossi LM, Wojcieszak R. Influence of support basic sites in green oxidation of biobased substrates using Au-promoted catalysts. ACS Sustain Chem Eng. 2018;6:16332–40. https://doi.org/10.1021/acssuschemeng.8b03330.
88. Yuan Q, Hiemstra K, Meinds TG, Chaabane I, Tang Z, Rohrbach L, Vrijburg W, Verhoeven T, Hensen EJM, Van Der Veer S, Pescarmona PP, Heeres HJ, Deuss PJ. Bio-based chemicals: selective aerobic oxidation of tetrahydrofuran-2,5-dimethanol to tetrahydrofuran-2,5-dicarboxylic acid using hydrotalcite-supported gold catalysts. ACS Sustain Chem Eng. 2019;7:4647–56. https://doi.org/10.1021/acssuschemeng.8b03821.
89. Davis SE, Zope BN, Davis RJ. On the mechanism of selective oxidation of 5-hydroxymethylfurfural to 2,5-furandicarboxylic acid over supported Pt and Au catalysts. Green Chem. 2012;14:143–7. https://doi.org/10.1039/c1gc16074e.
90. Hayashi E, Yamaguchi Y, Kamata K, Tsunoda N, Kumagai Y, Oba F, Hara M. Effect of MnO_2 crystal structure on aerobic oxidation of 5-hydroxymethylfurfural to 2,5-furandicarboxylic acid. J Am Chem Soc. 2019;141:890–900. https://doi.org/10.1021/jacs.8b12108.
91. Donoeva B, Masoud N, De Jongh PE. Carbon support surface effects in the gold-catalyzed oxidation of 5-hydroxymethylfurfural. ACS Catal. 2017;7:4581–91. https://doi.org/10.1021/acscatal.7b00829.
92. Jin Y, Sarina S, Liu H, Martens W, Waclawik ER, Peiris E, Jia J, Shang J, Kou L, Guo C, Zhu H-Y. Aerobic oxidation of 5-hydroxymethyl-furfural to 2,5-furandicarboxylic acid at 20°C by optimizing adsorption on AgPd alloy nanoparticle catalysts. ACS Catal. 2022;12:11226–38. https://doi.org/10.1021/acscatal.2c03457.
93. Kumar R, Wang P, Xue W, Jia Y, Zhu Z, Luo L, Wong JW-C, Zhao J. Catalytic oxidation of 5-hydroxymethylfurfural to 2,5-furandicarboxylic acid using Co-N/C catalysts with stepwise base addition approach. Next Mater. 2024;5:100227. https://doi.org/10.1016/j.nxmate.2024.100227.
94. Jiang W, Wang D, Deng X, Gao Y, Wang W, Ge T, Zhao C, Sun Y. Oxygen vacancy boosted oxidation of 5-hydroxymethylfurfural to 2,5-furandicarboxylic acid over $CuCoO_x$. Mol Catal. 2024;556:113917. https://doi.org/10.1016/j.mcat.2024.113917.
95. Nguyen CV, Liao Y-T, Kang T-C, Chen JE, Yoshikawa T, Nakasaka Y, Masuda T, Wu KCW. A metal-free, high nitrogen-doped nanoporous graphitic carbon catalyst for an effective aerobic HMF-to-FDCA conversion. Green Chem. 2016;18:5957–61. https://doi.org/10.1039/c6gc02118b.
96. Verma S, Nadagouda MN, Varma RS. Porous nitrogen-enriched carbonaceous material from marine waste: chitosan-derived carbon nitride catalyst for aerial oxidation of 5-hydroxymethylfurfural (HMF) to 2,5-furandicarboxylic acid. Sci Rep. 2017;7:13596. https://doi.org/10.1038/s41598-017-13789-3.

97. Wei Y, Li C, Zhu C, Zhang Y, Zhu Z, Chen Y, Li X, Yan Y. Oxygen vacancy and support adsorption synergistic effect in aerobic oxidation of HMF to FDCA: a case study using nitrogen-doped porous carbon supported Bi-CeO_2. J Taiwan Inst Chem Eng. 2022;138: 104439. https://doi.org/10.1016/j.jtice.2022.104439.
98. Wu SK, Li Z. Whole-cell cascade biotransformations for one-pot multistep organic synthesis. ChemCatChem. 2018;10:2164–78. https://doi.org/10.1002/cctc.201701669.
99. Yang CF, Huang CR. Biotransformation of 5-hydroxymethylfurfural into 2,5-furandicarboxylic acid by bacterial isolate using thermal acid algal hydrolysate. Bioresour Technol. 2016;214:311–8. https://doi.org/10.1016/j.biortech.2016.04.122.
100. Yuan H, Liu H, Du J, Liu K, Wang T, Liu L. Biocatalytic production of 2,5-furandicarboxylic acid: recent advances and future perspectives. Appl Microbiol Biotechnol. 2020;104:527–43. https://doi.org/10.1007/s00253-019-10272-9.
101. Troiano D, Orsat V, Dumont M-J. Status of biocatalysis in the production of 2,5-furandicarboxylic acid. ACS Catal. 2020;10:9145–69. https://doi.org/10.1021/acscatal.0c02378.
102. Yang CF, Huang CR. Isolation of 5-hydroxymethylfurfural biotransforming bacteria to produce 2,5-furan dicarboxylic acid in algal acid hydrolysate. J Biosci Bioeng. 2018;125:407–12. https://doi.org/10.1016/j.jbiosc.2017.11.005.
103. Qian Y, Liu J, Song W, Chen X, Luo Q, Liu L. Production of β-alanine from fumaric acid using a dual-enzyme cascade. ChemCatChem. 2018;10:4984–91. https://doi.org/10.1002/cctc.201801050.
104. Crigler J, Eiteman MA, Altman E. Characterization of the furfural and 5-hydroxymethylfurfural (HMF) metabolic pathway in the novel isolate Pseudomonas putida ALS1267. Appl Biochem Biotechnol. 2020;190:918–30. https://doi.org/10.1007/s12010-019-03130-x.
105. Yuan H, Li J, Shin HD, Du G, Chen J, Shi Z, Liu L. Improved production of 2,5-furandicarboxylic acid by overexpression of 5-hydroxymethylfurfural oxidase and 5-hydroxymethylfurfural/furfural oxidoreductase in Raoultella ornithinolytica BF60. Bioresour Technol. 2018;247:1184–8. https://doi.org/10.1016/j.biortech.2017.08.166.
106. Yuan H, Liu Y, Lv X, Li J, Du G, Shi Z, Liu L. Enhanced 2,5-furandicarboxylic acid (FDCA) production in Raoultella ornithinolytica BF60 by manipulation of the key genes in FDCA biosynthesis pathway. J Microbiol Biotechnol. 2018;28:1999–2008. https://doi.org/10.4014/jmb.1808.08057.
107. Cunha JT, Romaní A, Domingues L. Whole cell biocatalysis of 5-hydroxymethylfurfural for sustainable biorefineries. Catalysts. 2022;12 https://doi.org/10.1016/j.jtice.2022.104439.
108. Cajnko MM, Novak U, Grilc M, Likozar B. Enzymatic conversion reactions of 5-hydroxymethylfurfural (HMF) to bio-based 2,5-diformylfuran (DFF) and 2,5-furandicarboxylic acid (FDCA) with air: mechanisms, pathways, and synthesis selectivity. Biotechnol Biofuels. 2020;13:66. https://doi.org/10.1186/s13068-020-01705-z.
109. Carro J, Ferreira P, Rodriguez L, Prieto A, Serrano A, Balcells B, Arda A, Jimenez-Barbero J, Gutierrez A, Ullrich R, Hofrichter M, Martinez AT. 5-hydroxymethylfurfural conversion by fungal aryl-alcohol oxidase and unspecific peroxygenase. FEBS J. 2015;282:3218–29. https://doi.org/10.1111/febs.13177.
110. Dijkman WP, Fraaije MW. Discovery and characterization of a 5-hydroxymethylfurfural oxidase from Methylovorus sp. strain MP688. Appl Environ Microbiol. 2014;80:1082–90. https://doi.org/10.1128/AEM.03740-13.
111. Dijkman WP, Groothuis DE, Fraaije MW. Enzyme-catalyzed oxidation of 5-hydroxymethylfurfural to furan-2,5-dicarboxylic acid. Angew Chem Int Ed Engl. 2014;53: 6515–8. https://doi.org/10.1002/anie.201402904.
112. Karich A, Kleeberg S, Ullrich R, Hofrichter M. Enzymatic preparation of 2,5-furandicarboxylic acid (FDCA)—a substitute of terephthalic acid—by the joined action of three fungal enzymes. Microorganisms. 2018;6:10. https://doi.org/10.3390/microorganisms6010005.

113. Cajnko MM, Grilc M, Likozar B. Mechanistic kinetic modelling of enzyme-catalysed oxidation reactions of 5-hydroxymethylfurfural (HMF) to 2,5-furandicarboxylic acid (FDCA). Chem Eng Sci. 2021;246:116982. https://doi.org/10.1016/j.ces.2021.116982.
114. Wang HF. Sum frequency generation vibrational spectroscopy (SFG-VS) for complex molecular surfaces and interfaces: spectral lineshape measurement and analysis plus some controversial issues. Prog Surf Sci. 2016;91:155–82. https://doi.org/10.1016/j.progsurf.2016.07.001.
115. Hu K, Zhang M, Liu B, Yang Z, Li R, Yan K. Efficient electrochemical oxidation of 5-hydroxymethylfurfural to 2,5-furandicarboxylic acid using the facilely synthesized 3D porous WO_3/Ni electrode. Mol Catal. 2021;504:111459. https://doi.org/10.1016/j.mcat.2021.111459.
116. Jiang X, Li W, Liu Y, Zhao L, Chen Z, Zhang L, Zhang Y, Yun S. Electrocatalytic oxidation of 5-hydroxymethylfurfural for sustainable 2,5-furandicarboxylic acid production—from mechanism to catalysts design. J Colloid Interface Sci. 2023;3:21. https://onlinelibrary.wiley.com/doi/full/10.1002/sus2.109
117. Sajid M, Zhao X, Liu D. Production of 2,5-furandicarboxylic acid (FDCA) from 5-hydroxymethylfurfural (HMF): recent progress focusing on the chemical-catalytic routes. Green Chem. 2018;20:5427–53. https://doi.org/10.1039/c8gc02539h.
118. Choi S, Balamurugan M, Lee KG, Cho KH, Park S, Seo H, Nam KT. Mechanistic investigation of biomass oxidation using nickel oxide nanoparticles in a CO2-saturated electrolyte for paired electrolysis. J Phys Chem Lett. 2020;11:2941–8. https://doi.org/10.1021/acs.jpclett.0c00425.
119. Cardiel AC, Taitt BJ, Choi KS. Stabilities, regeneration pathways, and electrocatalytic properties of nitroxyl radicals for the electrochemical oxidation of 5-hydroxymethylfurfural. ACS Sustain Chem Eng. 2017:11138–49. https://doi.org/10.1021/acssuschemeng.9b00203.
120. Song X, Liu X, Wang H, Guo Y, Wang Y. Improved performance of nickel boride by phosphorus doping as an efficient electrocatalyst for the oxidation of 5-hydroxymethylfurfural to 2,5-furandicarboxylic acid. Ind Eng Chem Res. 2020;59:17348–56. https://doi.org/10.1021/acs.iecr.0c01312.
121. Chen H, Wang J, Yao Y, Zhang Z, Yang Z, Li J, Chen K, Lu X, Ouyang P, Fu J. Cu–Ni bimetallic hydroxide catalyst for efficient electrochemical conversion of 5-hydroxymethylfurfural to 2,5-furandicarboxylic acid. ChemElectroChem. 2019;6:5797–801. https://doi.org/10.1002/celc.201901366.
122. Vandarkuzhali SA, Karthikeyan G, Pachamuthu M. Efficient oxidation of 5-hydroxymethylfurfural to 2,5-furandicarboxylic acid over FeNPs@NH2-SBA-15 catalyst in water. Mol Catal. 2021;516:111951. https://doi.org/10.1016/j.mcat.2021.111951.
123. Lee JS, Ko YS. Control of the molecular structure of ethylene-1-hexene copolymer by surface functionalization of SBA-15 with different compositions of amine groups. J Mol Catal A Chem. 2014;386:120–5. https://doi.org/10.1016/j.molcata.2014.01.015.
124. Costa JAS, De Jesus RA, Santos DO, Neris JB, Figueiredo RT, Paranhos CM. Synthesis, functionalization, and environmental application of silica-based mesoporous materials of the M41S and SBA-n families: a review. J Environ Chem Eng. 2021;9:105259. https://doi.org/10.1016/j.mcat.2021.111951.
125. Singh S, Kumar R, Setiabudi HD, Nanda S, Vo DVN. Advanced synthesis strategies of mesoporous SBA-15 supported catalysts for catalytic reforming applications: a state-of-the-art review. Appl Catal A Gen. 2018;559:57–74. https://doi.org/10.1016/j.apcata.2018.04.015.
126. Dubbink GHC, Geverink TRJ, Haar B, Koets HW, Kumar A, Van Den Berg H, Van Der Ham AGJ, Lange JP. Furfural to FDCA: systematic process design and techno-economic evaluation. Biofuels Bioprod Biorefin. 2021;15:1021–30. https://doi.org/10.1002/bbb.2204.
127. Kim H, Choi J, Park J, Won W. Production of a sustainable and renewable biomass-derived monomer: conceptual process design and techno-economic analysis. Green Chem. 2020;22:7070–9. https://doi.org/10.1039/D0GC01809A.

128. Mohanty AK, Wu F, Mincheva R, Hakkarainen M, Raquez JM, Mielewski DF, Narayan R, Netravali AN, Misra M. Sustainable polymers. Nat Rev Methods Primers. 2022;2:46. https://doi.org/10.1038/s44230-022-00046-z.
129. Ahn B, Park C, Liu JJ, Ok YS, Won W. Maximizing the utilization of lignocellulosic biomass: process development and analysis. Renew Energy. 2023;215:119004. https://doi.org/10.1016/j.renene.2023.119004.
130. Cristobal J, Caldeira C, Corrado S, Sala S. Techno-economic and profitability analysis of food waste biorefineries at European level. Bioresour Technol. 2018;259:244–52. https://doi.org/10.1016/j.biortech.2018.03.016.
131. Dessbesell L, Souzanchi S, Venkateswara Rao KT, Carrillo AA, Bekker D, Hall KA, Lawrence KM, Tait CLJ, Xu C. Production of 2,5-furandicarboxylic acid (FDCA) from starch, glucose, or high-fructose corn syrup: techno-economic analysis. Biofuels Bioprod Biorefin. 2019;13:1234–45. https://doi.org/10.1002/bbb.2014.
132. Li Q, Zhang Y, Hu G. Techno-economic analysis of advanced biofuel production based on bio-oil gasification. Bioresour Technol. 2015;191:88–96. https://doi.org/10.1016/j.biortech.2015.05.002.
133. Dessbesell L, Yuan Z, Hamilton S, Leitch M, Pulkki R, Xu C. Bio-based polymers production in a kraft lignin biorefinery: techno-economic assessment. Biofuels Bioprod Biorefin. 2017;12:239–50. https://doi.org/10.1002/bbb.1834.
134. Rajinipriya M, Nagalakshmaiah M, Robert M, Elkoun S. Importance of agricultural and industrial waste in the field of nanocellulose and recent industrial developments of wood-based nanocellulose: a review. ACS Sustain Chem Eng. 2018;6:2807–28. https://doi.org/10.1021/acssuschemeng.7b03437.
135. Liu H, Cao X, Wang T, Wei J, Tang X, Zeng X, Sun Y, Lei T, Liu S, Lin L. Efficient synthesis of bio-monomer 2,5-furandicarboxylic acid from concentrated 5-hydroxymethylfurfural or fructose in DMSO/H_2O mixed solvent. J Ind Eng Chem. 2019;77:209–14. https://doi.org/10.1016/j.jiec.2019.04.038.
136. Guo T, Li X, Liu X, Guo Y, Wang Y. Catalytic transformation of lignocellulosic biomass into arenes, 5-hydroxymethylfurfural, and furfural. ChemSusChem. 2018;11:2758–65. https://doi.org/10.1002/cssc.201800967.
137. Musolino M, Andraos J, Aricò F. An easy scalable approach to HMF employing DMC as reaction media: reaction optimization and comparative environmental assessment. ChemistrySelect. 2018;3:2359–65. https://doi.org/10.1002/slct.201800198.
138. Mika LT, Csefalvay E, Nemeth A. Catalytic conversion of carbohydrates to initial platform chemicals: chemistry and sustainability. Chem Rev. 2018;118:505–613. https://doi.org/10.1021/acs.chemrev.7b00395.
139. Zhu H, Zhang Y, Guo X, Cheng Y, Wang L, Li X. Efficient one-pot production of 5-hydroxymethylfurfural from glucose in an acetone–water solvent. Ind Eng Chem Res. 2022;61:5661–71. https://doi.org/10.1021/acs.iecr.2c00502.
140. Hernández B, Vlachos DG, Ierapetritou MG. Coupling process intensification and systems flowsheeting for economic and environmental analysis of 5-hydroxymethyl furfural modular microreactor plants. ACS Sustain Chem Eng. 2022;10:14955–71. https://doi.org/10.1021/acssuschemeng.2c05323.
141. Li S, Wang S, Wang Y, He J, Li K, Xu Y, Wang M, Zhao S, Li X, Zhong X, Wang J. Doped Mn enhanced NiS electrooxidation performance of HMF into FDCA at industrial-level current density. Adv Funct Mater. 2023;33 https://doi.org/10.1002/adfm.202214488.
142. Jun Y. Green process intensification using microreactor technology for the synthesis of biobased chemicals and fuels. Chem Eng Process. 2022;177 https://doi.org/10.1016/j.cep.2022.109400.
143. Zhitong W, Yansong Z, Peng Q, Chenfeng X, Wensheng F, Jian J, Lei H, Peilin D, Qiong SY, Rachel C, Xinlong T, Bo Y, Wei G, Tommaso DD, Yuanjie P, Shujiang D, Bao YX. Advanced catalyst design and reactor configuration upgrade in electrochemical carbon dioxide conversion. Adv Mater. 2023;35:e2303052. https://doi.org/10.1002/adma.202303052.

Chapter 8
Recent Advances in 2,5-Furandicarboxylic Acid (FDCA) Production from Biomass Derived Resources via Catalytic Processes

P. Díaz-Maizkurrena, A. Iriondo, A. Bueno, N. Viar, and J. Requies

Abstract Biomass is the only renewable resource capable of substituting petrochemical compounds derived from non-renewable raw materials such as crude oil and natural gas. Carbon-based compounds can be derived from biomass, utilizing non-edible raw materials from agriculture or forestry, presenting the most viable solution to replace fossil resources and, consequently, mitigate the associated environmental problems.

Among the monomers that can be obtained from agricultural residues and food wastes, 2,5-furandicarboxylic acid (FDCA) stands out due to its significant potential in the production of sustainable polymers and other valuable chemicals. FDCA is particularly important because it serves as a key building block for producing polyethylene furanoate (PEF), a promising biopolymer with superior barrier properties and mechanical strength compared to traditional plastics like polyethylene terephthalate (PET). This positions FDCA as a crucial component in advancing green chemistry and promoting the development of environmentally friendly materials.

This chapter explores the versatile potential of FDCA, considering its importance and the challenges associated with its production. A comprehensive overview of the different methods of producing FDCA from biomass-derived precursors is provided, with a particular focus on the conversion of 5-hydroxymethylfurfural (HMF), as it is the most extensively studied method in the literature. However, other production routes are also discussed in the chapter.

The chapter concludes with a discussion on the significance of FDCA in sustainable polymer synthesis, emphasizing its role in the production of PEF and other biopolymers. It highlights the applications and future perspectives of FDCA in promoting sustainable industrial practices and reducing reliance on fossil-based materials.

P. Díaz-Maizkurrena · A. Iriondo · A. Bueno · N. Viar · J. Requies (✉)
School of Engineering (UPV/EHU), Bilbao, Spain
e-mail: paula.diaz@ehu.eus; aitziber.iriondo@ehu.eus; ane.bueno@ehu.eus; nerea.viar@ehu.eus; jesus.requies@ehu.eus

Z. Fang et al. (eds.), *Production of Organic Acids and Alcohols from Agricultural Residues and Food Wastes*, Biofuels and Biorefineries 13,
https://doi.org/10.1007/978-981-96-1222-2_8

Keywords Renewable resources · 2,5-Furandicarboxylic acid (FDCA) · 5-Hydroxymethylfutfural (HMF) · Polyethylene furanoate (PEF) · Monomer synthesis

Abbreviations

AAO	Aryl-alcohol oxidase
ADH	Alcohol dehydrogenases
AMF	5-Alcoxymethylfurfural esthers
APIs	Active pharmaceutical ingredients
Au	Gold
BHMF	2,5-Bis(hydroxymethyl)furan
CF	Cobalt foam
Co	Cobalt
CoPz	Co thioporphyrazine
DFF	2,5-Diformylfuran
DMC	Dimethylcarbonate
DMF	2,5-Dimethylfuran
DMSO	Dimethyl sulfoxide
DMTHF	2,5-Dimethyltetrahydrofuran
EMIMBr	1-Ethyl-3-methylimifazolium bromide
FA	Furoic acid
FA-Cs	Cesium furoate
FDCA	2,5-Furandicarboxylic acid
Fe	Iron
FFCA	5-Formyl-2-furancarboxylic acid
$g\text{-}C_3N_4$	Graphitic carbon nitride
GO	Galactose oxidase
GVL	Gamma-valerolactone
H_2SO_4	Sulphuric acid
H_3PO_4	Phosphoric acid
HAc	Acetic acid
HAP	Hydroxyapatite
HCl	Hydrochloric acid
HCOOK	Potassium formate
HDO	1,6-Hexanediol
HF	Hydrofluoric acid
HMF	5-Hydroxymethylfurfural
HMFCA	5-Hydroxymethyl-2-furancarboxylic acid
HMFO	HMF oxidase
HRP	Horse radish peroxidase
HTFA	Trifluoroacetic acid
HTO	1,2,6-Hexanetriol
ILs	Ionic liquids

LA	Levulinic acid
LCB	Lignocellulosic biomass
MC	Amoco Mid-Century
MEG	Monoethylene glycol
MIBK	Methyl isobutyl ketone
MMF	5-Methoxymethylfurfural
MMFA	5-Methoxymethyl-furan-2-carboxylic acid or 5-methoxymethyl-2-furoic acid
Mn	Manganese
MOF	Metal organic framework
MOX	MeOH oxidase
Na_2CO_3	Sodium carbonate
NaOH	Sodium hydroxide
Ni	Nickel
NPs	Nanoparticles
OSC	O_2 storage capacity
PaoABC	Periplasmic aldehyde oxidase
Pd	Palladium
PEF	Polyethylene furanoate
PET	Polyethylene terephthalate
Pt	Platinum
PtHmfF	*P. thermopropionicum* HmfF
Ru	Ruthenium
t-BuOOH	tert-Butyl hydroperoxide
THF	Tetrahydrofuran
TPA	Terephtalic acid
TPPTS	Tris(3-sulfophenyl)phosphine trisodium salt
UPO	Unspecific peroxygenase

8.1 Introduction

The world energy system is mainly based on the use of non-renewable energy sources such as crude oil, coal and natural gas (see Fig. 8.1), being their contribution in the global primary energy consumption around 30%, 25% and 22%, respectively. This suggests that these fossil fuels present a total contribution close to 80% [1, 2].

In addition to serving as energy sources, the mentioned resources are also utilized as raw materials in the production of traditional fuels, chemicals and materials. According to Weerathunga et al. [3], the 86% and 96% of world's fuels and chemicals, respectively, are produced by fossil fuels and these markets are expected to grow in the near future. This observation highlights the increasing demand for fossil fuels to fulfil the requirements of the chemical commodities market. Nonetheless, owing to the heavy reliance of chemical production on the aforementioned

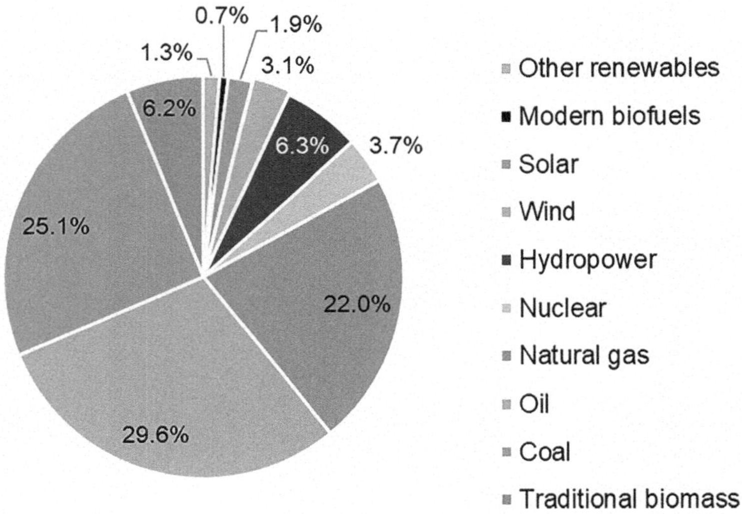

Fig. 8.1 Global primary energy consumption by source. (Data extracted from [2])

fossil fuels, particularly petroleum, various issues arise, including challenges related to supply stability, safety, price volatility, non-degradability, and environmental pollution [4–6].

Specifically, the chemical commodities sector contributes approximately 33% of greenhouse gas emissions [7]. Furthermore, our economies heavily rely on chemical products, particularly petrochemicals, making them integral to modern societies.

Considering these challenges and the prevailing global energy trends, it is evident that industries such as chemical and polymer are actively exploring renewable and sustainable feedstock to produce chemical commodities comparable to those derived from fossil resources. Among various renewable sources like solar, wind, geothermal, and hydropower, biomass stands out as the only option capable of addressing these issues through its conversion into chemicals. In fact, the European Commission in 2020 reported that biomass is a clear candidate as a feedstock to produce bio-based value-added chemicals [4, 8, 9].

Among various biomass resources, lignocellulosic biomass (LCB) has emerged as a promising alternative to petroleum [10]. It is the most abundant feedstock on the planet, estimated at around 200 billion tonnes [11], and is readily available in agricultural and forestry residues. The LCB has great potential as green raw material to produce bio-based products due to its structure composed by three different fractions constituted by biopolymers, cellulose (40–45%) and hemicellulose (25–35%), the main components, and lignin (10–15%), among other components such as inorganic ones [12, 13].

The cellulose structure is formed by cellulose layers with well-defined architectures, which are formed by D-glucose monomers, weakly bonded by Hydrogen bridges. This structure takes form of the so-called cellulose fibres. The hemicellulose, which is place around the cellulose fibres, allows this fibres and lignin adhesion, thanks to its more complex and heterogeneous structure than cellulose. In fact, the hemicellulose fraction is composed by different sugars, such as D-xylose, L-arabinose, D-glucose, L-galactose, among others, which are more branched. Lignin, an aromatic polymer that also presents complex structure, provides other properties to LCB such as impermeability, structural support and resistance to oxidation and microbial attack. Due to this structure, the LCB is also consider a natural composite material (cellulose fibres + hemicellulose-lignin matrix) [14].

Similar to petroleum, the conversion of LCB into value-added chemicals becomes viable through the separation of its polymeric components. This necessitates a pre-treatment step for LCB, facilitating the separation of cellulose, hemicellulose, and lignin. Numerous published studies focus on improving LCB pre-treatment to enhance cellulose accessibility and digestibility. This emphasis is due to cellulose being the primary component of LCB, characterized by a simpler structure compared to hemicellulose and lignin. Once cellulose is separated, it can be directly converted or undergo various steps to produce platform and/or intermediate chemicals. For the latter, cellulose undergoes a depolymerisation process through hydrolysis-type reactions to yield glucose and/or fructose (glucose isomer) [15–17].

8.2 HMF as a Versatile Precursor

HMF is often referred to as the "sleeping giant" among renewable intermediate chemicals, HMF has an immense market potential, making it a key compound in the sustainable future of chemical production [18].

8.2.1 Importance of HMF

Although there is a wide variety of biomass-derived platform chemicals that can be used to produce high value-added products, the HMF has gained a lot of attention. This fact is due to it presents multiple functionalities, providing to HMF a wide range of valorisation pathways [19–21]. Some of those reactions are presented in Fig. 8.2. In fact, this characteristic makes HMF a viable alternative to produce polymers and resins, solvents, pharmaceuticals, liquid fuels for transportation and fine chemicals [22].

HMF is derived from the glucose and/or fructose dehydration [23]. While the conversion of fructose into HMF has been extensively studied, considerable efforts have been directed towards HMF production using glucose as the primary raw material. Souzanchi et al. [24] attribute this emphasis to the higher cost of fructose compared to glucose, resulting in limited practical application.

One of the most investigated and promising bio-based monomer is the FDCA, which can be obtained from HMF, from 5-alkoxymethylfurfural ethers (AMFs) (such as 5-methoxymethylfurfural (MMF)) and/or directly from polysaccharides.

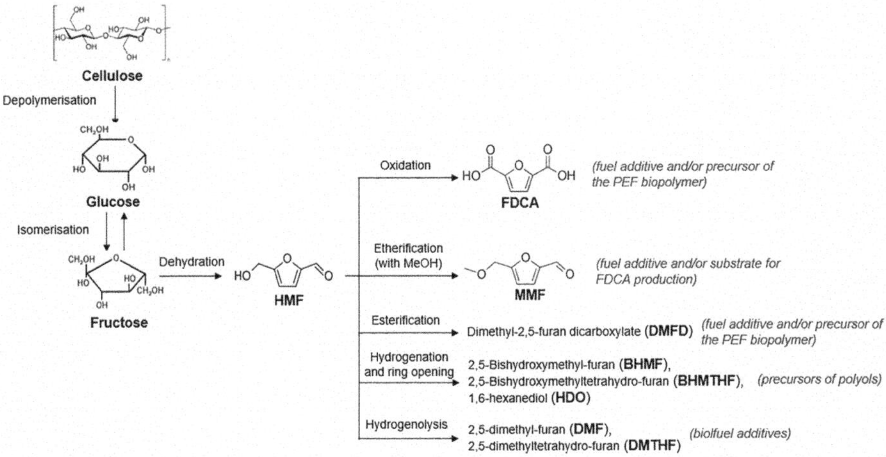

Fig. 8.2 HMF conversion reaction

The polymerization of FDCA, the real alternative of fossil-based terephthalic acid (TPA), with bio-based mono-ethylene glycol (MEG) produce PEF, which is the renewable substitute of PET. Among traditional thermoplastic polymers, PET presents a 22% contribution on global plastic packaging market indicating that PEF could also cover this market. PEF offers notable benefits when compared to PET. These benefits are related to higher biodegradability under composting conditions, renewability and recyclability, superior gas barrier properties, especially for CO_2 and O_2, higher mechanical, associated to higher tensile strength, and thermal properties, presenting greater resistance to high temperatures [25]. Moreover, it requires fewer additives to reach the desire characteristics.

The most investigated production process of bio-based FDCA is catalytic oxidation of HMF. This reaction usually takes place through catalysts, being those enzymatic, homogeneous and heterogeneous [4]. On the other hand, although most research studies are focused on the mentioned process, reaction through MMF or polysaccharides are also reported, being still very scarce. Some authors suggest that the conversion route of MMF into FDCA seems to be a promising prospect due to MMF exhibiting better stability during storage compared to HMF [9, 26].

8.2.2 HMF Production from Lignocellulosic Biomass

LCB, which includes materials such as agricultural residues and waste streams, has been recognized as the most promising feedstock for the production of valuable chemicals, fuels and power, due to its affordability and abundant availability [27].

Cellulose can be transformed into HMF through catalytic processes. Generally, this procedure involves three steps: (1) cellulose hydrolysis to glucose, (2) glucose isomerisation to fructose and, (3) fructose dehydration to HMF [28]. The ideal scenario would be to directly convert cellulose into HMF. However, when contemplating industrial-scale production, the technology for directly synthesizing HMF from cellulose still faces challenges related to poor selectivity and low yields, accompanied by complexities in separation and purification [29]. For this reason, the stages are often studied independently.

Based on catalyst type, processes are categorized as homogeneous (where the catalyst and the reaction media are in the same phase) and heterogeneous (the catalyst and the reaction media are in different state phase) [30]. While homogeneous inorganic acid and metal chloride catalysts exhibit considerable catalytic activity, their use present several drawbacks. The challenge lies in the difficulty of separating the catalyst from the products or the reaction system, as they are in the same phase. Additionally, the corrosive nature of inorganic acids leads to elevated maintenance costs due to material damage, and the resulting waste liquid requires further treatment to mitigate its environmental impact. To address these issues, solid heterogeneous catalysts emerge as a promising alternative.

Initial studies on cellulose hydrolysis to glucose focused on homogeneous mineral acid catalysts like hydrochloric acid (HCl), sulphuric acid (H_2SO_4), hydrofluoric acid (HF) and phosphoric acid (H_3PO_4) [31]. However, due to their drawbacks, research shifted to heterogeneous catalytic systems, investigating materials such as sulfonated carbon, zeolites, heteropoly acids, magnetic solid acids and carbonaceous solid acids [32]. Recently, ionic liquids have emerged as effective solvents for this process, being more environmentally friendly and stronger than traditional solvents [33]. Novel ionic liquids can yield up to 98% glucose with well-designed catalysts and conditions, due to their superior cellulose solubility [31].

For the direct production of HMF from glucose, systems combining solid catalysts with ionic liquids have been employed. For instance, the use of 1-ethyl-3-methylimidazolium bromide (EMIMBr) with the addition of a solid catalyst such as tin-phosphate or Al_2O_3 containing catalysts or even the combination of Sn and Al catalysts, have been reported to achieve HMF yields of approximately 60%, 50%, and 70%, respectively [34–36].

More recently, other studied systems include biphasic water-organic systems, as HMF is not stable in water and is extracted into the organic phase to optimize its production. Investigations have been focused on the study of biphasic reaction media, such as water-immiscible organic solvents such as methyl isobutyl ketone (MIBK) [37, 38], tetrahydrofuran (THF) [39], gamma-valerolactone (GVL) and dimethylcarbonate (DMC) [24], or combination of ionic liquids (ILs) and these organic solvents [40]. Regarding the catalytic material, bifunctional acid catalysts are required. On the one hand, Lewis acid sites are essential for the isomerisation of glucose to fructose, and on the other side, Brønsted acid sites are needed for the subsequent dehydration to HMF [41].

8.2.3 Conversion of HMF to Value Added Chemicals and Biofuels

As previously mentioned, HMF has versatile applications in various industries, playing a crucial role in the transition towards more sustainable and renewable alternatives in chemical and energy production. The attractiveness of HMF lies on its chemical structure, which exhibits various functionalities stemming from specific molecular features. It consists of an aldehyde group (-C=O) and a hydroxyl group (-OH) attached to a furan ring. With this diverse variety of functional groups, HMF serves as an intermediate for producing a range of value-added chemicals and biofuels [42].

On the one hand, HMF can be hydrogenated to produce compounds such as 2,5-bis(hydroxymethyl)furan (BHMF), which is a versatile molecule for polymer production. Additionally, the hydrogenolysis of HMF yields 2,5-dimethylfuran (DMF) and 2,5-dimethyltetrahydrofuran (DMTHF), which are promising biofuels or fuel additives [43]. Hydrogenation can be carried out under high-pressure hydrogen or by employing hydrogen donors such as alcohols, formic acid and silanes. Regarding the catalyst, noble metals such as platinum, palladium or ruthenium have been widely investigated. Noble metals catalysts exhibited exceptional performance in the hydrogenation of biomass-derived HMF. However, the industrial application of noble metal catalysts is restricted due to their high cost. For this reason, catalyst based on non-noble metals have been studied. Researchers have developed a series of transition metal catalysts, including copper, nickel and cobalt, for the hydrogenation of HMF [44].

Another pathway involves the opening of the furan ring. Through a Brønsted acid catalyst, HMF can be rehydrated into levulinic acid (LA), serving as a platform molecule to produce biofuels, lubricants, or plasticizers, as well as thermoplastics or polyols. Furthermore, the hydrogenation-ring-opening reaction of HMF produces 1,2,6-hexanetriol (HTO), which is employed in the production of cosmetics as humidity regulator and viscosity-controlling agent [28]. Lastly, 1,6-hexanediol (HDO) stands out as one of the most important diols obtained from lignocellulosic biomass, employed in the production of polyester and polyurethane, among others. Various reaction pathways have been reported in the literature, encompassing direct hydrogenation of HMF to HDO or stepwise hydrogenation through different intermediates [45].

Lastly, the partial and selective oxidation of HMF to furan-based products results in the production of 2,5-diformylfuran (DFF). This compound is used in the synthesis of fungicides and polymeric material. Subsequent oxidation of DFF leads to the production of FDCA. The following section is dedicated to the production of FDCA from HMF.

8.2.4 Challenges and Issues for Large Scale Production of HMF

Despite the considerable potential mentioned earlier, the current industrial-scale is restricted. This limitation arises because its current production is not inherently sustainable. The production process should prioritize the energy efficiency, minimize waste generation, utilize safe solvents, and employ catalysts that are as active and selective as possible [42].

Moreover, another challenge for the industrial production of HMF lies in the identification of low-cost feedstock. In the context of the bioeconomy, the production of platform chemicals from biomass can be conducted directly in decentralized biorefineries, employing agricultural residues as feedstock. Regrettably, the incorporation of biomass remains a challenge for industrial applications [46].

Therefore, achieving further optimization across all production stages holds the key to enabling the industrial-scale production of HMF.

8.3 Production of FDCA from HMF

8.3.1 Transformation of HMF to FDCA: A Brief Overview

The transformation of HMF to FDCA involves a series of key reactions that have been extensively studied in scientific literature (see Fig. 8.3).

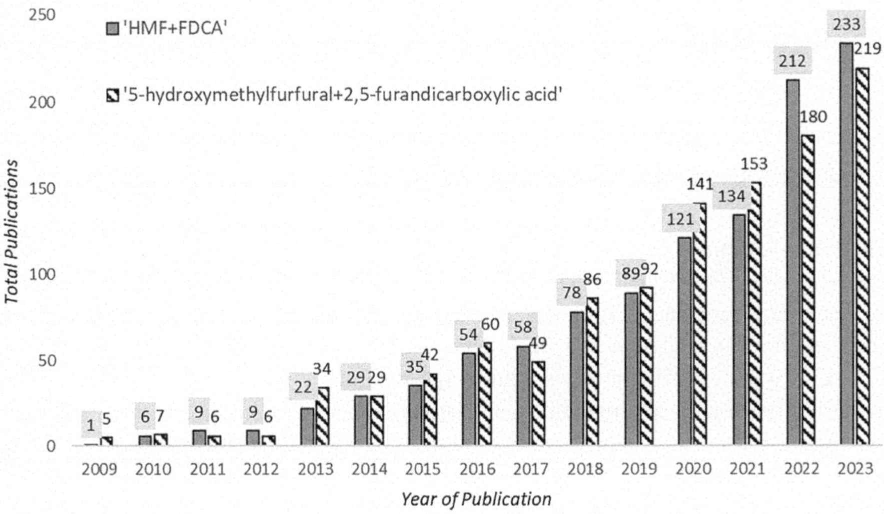

Fig. 8.3 Publications output distribution from 2009 to 2023 in ScienceDirect. Being search keywords 'HMF + FDCA' and '5-hydroxymethylfurfural + 2,5-furandicarboxylic acid'

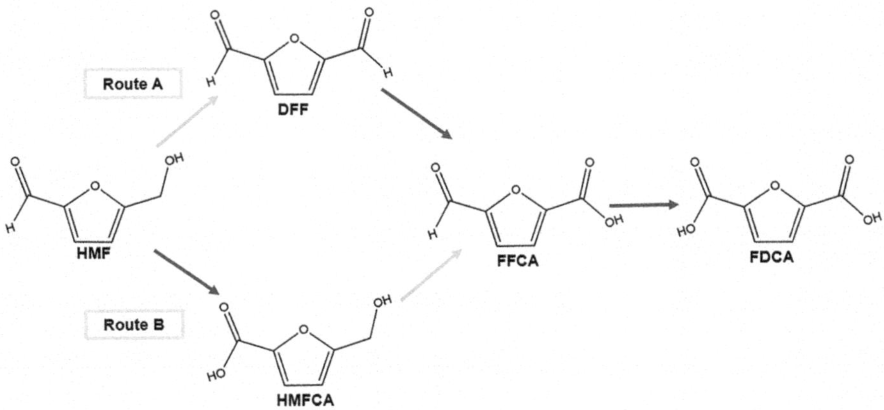

Fig. 8.4 Reaction mechanism from HMF to FDCA

The initial stage entails the oxidation of HMF to form DFF or to form 5-hydroxymethyl-2-furancarboxylic acid (HMFCA) (see Fig. 8.4). The next step always involves obtaining formylfurancarboxylic acid (FFCA). Finally, FFCA is converted to FDCA. The efficiency and selectivity of these reactions are critical for the economic viability and sustainability of this process.

Various catalysts have been explored to facilitate these transformations, including heterogeneous catalysts, enzymes, and homogeneous catalytic systems. The choice of catalyst and reaction conditions plays a crucial role in the effectiveness and yields of this process.

A brief exploration of different approaches and advancements in optimizing these stages will be addressed, emphasizing significant achievements and ongoing research areas. Noting that, as aforementioned, there are two pathways for this reaction to take place as the reaction system or the reaction conditions vary.

8.3.2 Homogeneous Catalytic Systems

The first mention of the transformation from HMF to FDCA using a homogeneous catalytic systems dates to 2001 [47]. In this investigation, the researchers reported the pioneering instance of aerobic oxidation reactions of HMF utilizing dioxygen and metal/bromide catalysts. In this approach, a 3 h batch reaction was conducted at 125 °C in water and 70 bar of air, achieving maximum yield of, approximately, 60% towards FDCA with optimized concentrations of Co, Mn, and Br. As elucidated by the authors, the furan ring tends to undergo more facile cleavage at elevated temperatures, resulting in higher yields of CO and CO_2. On the other hand, catalysts stabilities are not mentioned.

After this study, in the subsequent years, the utilization of homogeneous catalysts was scarcely explored. A study by Saha et al. [48] was conducted involving the addition of trifluoroacetic acid (HTFA) in the aerobic oxidation reaction of HMF with homogeneous metal salts, comprising transition-metal acetates and a bromide source, Co/Zn/Br, usually branded as the Amoco Mid-Century (MC). A 60% FDCA yield was achieved by introducing 1 wt% HTFA at temperatures as low as 90 °C and 10 bar of O_2. In contrast, in the absence of HTFA, DFF was identified as the exclusive oxidation product. The addition of HTFA and the use of milder reaction conditions provided the same HTFA yield as obtained one by Partenheimer et al. [47].

Zuo et al. [49] compared the oxidation of HMF with the oxidation of *p*-xylene. Employing Co/Mn/Br-based catalysts for the semi-continuous oxidation of HMF to FDCA, they aimed to reduce the temperatures typically used in the traditional Mid-Century process and to optimize the reaction conditions. Recognizing HMF's susceptibility to secondary reactions (such as over oxidation to CO and CO_2, esterification to produce 5-acetoxymethylfurfural and, acetic acid (HAc) and HMF burning), lowering temperatures and optimizing reaction conditions could prove highly beneficial in preventing undesired products. However, when compared to results obtained with *p*-xylene, the outcomes were suboptimal. Consequently, they opted to modify other reaction parameters, optimizing catalyst composition, water concentration in HAc solvent, and pressure. The results demonstrated that the FDCA yield could be enhanced to 90% at a 1/0.015/0.5 molar ratio of Co, Mn, and Br, 7% (v/v) in water, 30 bar ($CO_2/O_2 = 1/1$, mol/mol) and 180 °C.

Also in the breakthrough in the field of homogeneous catalysts, once again Zuo et al. [50] achieved optimal reaction conditions at 150 °C, 8 bar of air pressure, a molar ratio of 29:1:30 for Co/Mn/Br, and a HMF/HAc mass ratio of 1:9. Under these conditions, in mini-batch reactors, approximately 95% FDCA yield was attained in about 20 min, effectively reducing the conversion of the substrate into carbon oxides and mitigating other side reactions.

In a pioneering attempt using continuous flow micro reactors, the reaction was conducted in polytetrafluoroethylene capillary micro reactors, specifically, under gas-liquid slug flow. Employing Co/Mn/Br homogeneous catalyst with HAc as the solvent, and varying the medium temperature from 90 to 165 °C and pressure from 1 to 5 bar using either pure O_2 or air. After optimizing the reaction conditions to avoid the problems associated to the mass transfer in micro reactors, Hommes et al. [51] achieved FDCA yields close to 24%. Moreover, they detect continuous flow reactor clogging due to the FDCA precipitation in the HAc. This problem can be avoid using batch reactor or employing basic reaction media, as is described in the following sections.

Although homogeneous catalysis has encountered challenges in terms of processing and recycling, the majority of homogeneous catalysed oxidations of HMF occurred in substantial quantities of organic solvents or corrosive acids, in which the catalysts and reaction species are dissolve, and, in the particular case of HMF oxidation, quite high temperatures are require, generating the burning of HAc [49] and degradation of HMF. Taking into account this, homogeneous catalysts operate independently of molecules or ions, featuring uniform active centres, high activity, and high selectivity, but they commonly face issues concerning catalyst recovery and recyclability, low yield, elevated energy costs, by-product formation and environmental pollution [52]. For that reason, the majority of the reported studies are focused on the use of heterogeneous catalysts.

8.3.3 Heterogeneous Catalytic Systems

The utilization of heterogeneous catalysts is driven by their significant advantages compared to homogeneous or enzymatic catalysts. The use of a solid component eliminates the need for catalyst recovery from the product, as it can be easily separated due to its solid state. In contrast, homogeneous catalysts necessitate recovery and involve less conventional and, therefore, more cost-effective reactors than homogeneous or enzymatic reactors. Heterogeneous catalysts facilitate easy recycling and reuse.

Within heterogeneous catalysts, those based on noble metals are noteworthy. Various noble metals, including platinum (Pt), gold (Au), ruthenium (Ru), and palladium (Pd), have demonstrated effective catalytic activity, recyclability, and stability in the oxidation of HMF to FDCA [53]. In fact, as is reported [54, 55], the good performance of Pt and Pd catalysts are because they are able to oxidation of alcohol group of HMF, while Au catalysts are more active towards aldehyde group oxidation. However, apart from active sites, the catalysts could be presented structural characteristics (particle size, dispersion degree, interactions with other components of the catalyst) which play also an important role in this type of reactions [55].

The use of O_2 in combination with heterogeneous catalysts is closely aligned with the principles of "green and sustainable chemistry," representing a primary method for synthesizing FDCA from HMF. The activation of O_2, however, poses a significant challenge, which is the main reason for resorting to the use of noble metals [56].

In the following sections different catalytic systems and reaction conditions reported in the literature for HMF oxidation are included.

8.3.3.1 Platinum (Pt)

Pt was among the initial noble metals employed as a catalyst, alongside Ru, Pd, and Au. The application of Pt catalysts supported on carbon (Table 8.1) resulted in a significant conversion of HMF and a high yield of FDCA. Employing this catalyst occurred under mild temperature conditions (25 °C) using a 1.25 M sodium hydroxide (NaOH) solution. The reaction was carried out in a strongly alkaline aqueous medium and in the presence of oxygen and the FDCA yield reached 81% at 100% HMF conversion. The incorporation of Pb increased the FDCA yield from 81% to 99% under the same operating conditions. It appears that the presence of Pb inhibits Pt oxidation [57]. In fact, Rass et al. [58] suggested that activity loss of the tested Pt

Table 8.1 Some results of oxidation of HMF into FDCA over Pt-based catalysts

Catalyst	Base	Solvent	Oxidant	Temp (°C)	Time (h)	FDCA yield (%)	Reference
Pt/C	1.25 M NaOH	Alkaline aqueous medium	1 bar O_2	25	2	81	[57]
Pt-Pb/C	1.25 M NaOH	Alkaline aqueous medium	1 bar O_2	25	2	99	
Pt-Bi/C	2 equiv. Na_2CO_3	Basic aqueous solution	40 bar air	100	6	>99	[58]
Pt/TiO_2	2 equiv Na_2CO_3	Basic aqueous solution	40 bar air	100	6	84	
$Pt\text{-}Bi/TiO_2$	2 equiv Na_2CO_3	Basic aqueous solution	40 bar air	100	6	99	
$Pt/\gamma\text{-}Al_2O_3$	1 equiv Na_2CO_3	Aqueous solution	1 bar O_2	75	12	96	[60]
Pt/ZrO_2	1 equiv Na_2CO_3	Aqueous solution	1 bar O_2	75	12	94	
$Pt/Ce_{0.8}Bi_{0.2}O_{2-\delta}$	4 equiv NaOH	Aqueous solution	1 bar O_2	23	0.5	98	[61]
Pt/β-zeolite	6 equiv Na_2CO_3	Aqueous solution	1 bar O_2	90	24	90	[62]
Pt/nNiO	None	Aqueous solution	10 bar O_2	100	12	100	[63]

catalysts could have been due to catalyst covering by O_2 or over-oxidation of active sites during oxidation reaction when completed conversion of HMF were obtained.

Apart from the addition of Pt the incorporation of Bi (non-noble metal) has also been observed to positively influence the catalytic performance of the Pt/C catalyst. The Pt–Bi/C catalyst, featuring a Pt–Bi molar ratio of around 0.2, achieved a high FDCA yield of 99% after 6 h at 100 °C under 40 bar air, in a moderately basic aqueous solution using 2 equivalents of sodium carbonate (Na_2CO_3) [59]. The initial intermediates observed were HMFCA and DFF, undergoing rapid oxidation to FFCA, with the oxidation of FFCA identified as the rate-determining step. Furthermore, the Pt–Bi/C catalyst exhibited significantly greater stability compared to the Pt/C catalyst. The inclusion of Bi enhanced resistance to O_2 poisoning and prevented Pt leaching, resulting in avoiding catalyst deactivation during oxidation reaction and recycling operations [55]. The beneficial effect of adding Bi for catalyst stability can also be seen in Pt/TiO_2 catalysts. Through this addition, it can be observed that, under the same operating conditions, the yield increases from 84% to 99% [58]. The recyclability and durability of the Pt/TiO_2 catalyst was assessed by reusing it in successive catalytic tests. In the first run, HMF was fully oxidized within 2 h and nearly completely converted to FDCA after 12 h, but the catalyst's activity significantly decreased in subsequent runs due to over oxidation of the platinum sites, not

Pt leaching. Reactivating the catalyst with hydrogen fully restored its activity, demonstrating a recurring deactivation and reactivation pattern also observed with Pt on carbon. Apart from these benefits, the addition of non-noble metals, which are typically more inexpensive than noble ones, allows to reduce the noble metal catalyst costs and improve the catalytic properties because they act as co-catalysts or promoters.

Pt/γ-Al_2O_3 and Pt/ZrO_2 exhibited notable catalytic activity, yielding 96% and 94% FDCA, respectively. These findings indicate that Pt catalysts supported on non-reducible oxides (γ-Al_2O_3 and ZrO_2) display high catalytic efficiency. The author attributed this performance to the distinct O_2 storage capacity (OSC) of each support. Catalysts like Pt/γ-Al_2O_3, Pt/ZrO_2, and Pt/C are known for their low OSC, which maintains active sites in an uncovered state, as aforementioned, favouring HMF oxidation. Conversely, a high OSC, beneficial for reactions like CO oxidation, is disadvantageous in HMF oxidation [60]. The stability and reusability of the Pt/γAl_2O_3 catalyst was studied, showing consistent FDCA yields of around 90% and 100% HMF conversion across multiple runs. After each run, the catalyst was recovered, dried, and reactivated by hydrogen reduction, maintaining its performance. In another study, similar to the addition of Bi to Pt/C, it was discovered that incorporating Bi into the CeO_2 support significantly enhanced the catalytic performance of Pt/CeO_2 catalyst in water oxidized by O_2. Concretely, the Pt/$Ce_{0.8}Bi_{0.2}O_{2-\delta}$ catalyst achieved 100% HMF conversion and a 98% yield of FDCA [61]. In this study, it becomes clear that Bi-containing CeO_2 accelerates the O_2 reduction process due to the presence of a large number of O_2 vacancies and facilitates the cleavage of peroxide intermediates. They also concluded that the catalytic activity could be improved modulating the Pt particle size. In this sense, Ait Rass et al. [58] observed that poor activity of the Pt/TiO_2 and Pt/CeO_2 catalysts were ascribed to bigger Pt particle size and higher OSC than the γ-Al_2O_3 support. On the other hand, the catalyst was reused five times without much loss of FDCA selectivity.

In more recent research, the successful utilization of β-zeolite/encapsulated Pt nanoparticles has been reported, achieving FDCA production at 90 °C and 1 bar of O_2 with a yield of up to 90% using water as solvent. The encapsulation process improved the thermal stability of the nanoparticles, preventing Pt leaching, allowing the catalyst to be recycled for up to five uses [62]. In another very recent study, NiO was employed as a support for a Pt-based catalyst, achieving the highest reported values for FDCA productivity without the help of any base in the aqueous medium, under 10 bar O_2 at 100 °C [63]. This good performance seems to be related to higher number of metallic Pt active sites, the mesoporous character of the support which improves HMF transportation, and the electron transfer favored by Pt and NiO interaction. The composition, structure, and active sites of the used catalyst were nearly identical to those of its fresh state, enabling the Pt/NiO catalyst to be reusable.

8.3.3.2 Gold (Au)

Apart from Pt, Au exhibits excellent catalytic performance in the selective oxidation of HMF to produce FDCA. Casanova et al. [64] examined the impact of Au/TiO_2 and Au/CeO_2 and Cai et al. [65] examined the impact of, Au/C and Au/FeO_2 catalysts on the air oxidation of HMF. Operating at 65 °C and 10 bar of O_2, with NaOH as a base, they achieved a remarkable FDCA yield of 99% when TiO_2 and CeO_2 were used as supports. However, in terms of recyclability, although the catalyst successfully completed three cycles, deactivation was observed during the fourth cycle.

Alternative researchers utilized Au supported on CeO_2 with the addition of Bi as a dopant, aiming to prevent Au oxidation. The Bi-doped catalyst exhibited a higher yield at mild temperature conditions and a shorter reaction time compared to the catalyst supported solely on CeO_2 in alkaline aqueous solution [66]. The catalyst was tested for recyclability by filtering, washing, and drying it before reuse. ICP-OES analysis confirmed no Au leaching. The catalyst retained its activity for three cycles with minimal performance loss.

As observed in Table 8.2, all the Au catalytic systems, employed a basic medium through addition of NaOH. The addition of a soluble base is crucial to accelerate alkoxide formation and C–H activation in the catalytic cycle, as well as to help to prevent FDCA precipitation in the medium. However, the use of a base in the reaction presents a challenge, as it leads to the generation of an attendant aqueous waste stream during the subsequent acidification and purification of the reaction mixture. This has prompted some authors to focus on utilizing Au catalysts in the absence of any base [70].

Table 8.2 Some results of oxidation of HMF into FDCA over Au-based catalysts

Catalyst	Base	Solvent	Oxidant	Temp (°C)	Time (h)	FDCA yield (%)	Reference
Au/TiO_2	5 equiv NaOH	Aqueous solution	3 bar O_2	60	6	85	[65]
Au/CeO_2	4 equiv NaOH	Aqueous solution	10 bar air	130	5	96	[64]
$Au/Ce_{0.9}Bi_{0.1}O_2Pt/Ce_{0.8}Bi_{0.2}O_{2-\delta}$	4 equiv NaOH	Alkaline aqueous solution	1 bar O_2	65	2	>99	[66]
Au/Al_2O_3	4 equiv NaOH	Aqueous solution	10 bar O_2	70	4	99	[67]
Au/HY	5 equiv NaOH	Aqueous solution	3 bar O_2	60	6	>99	[65]
Au zeolite	NaOH	Aqueous solution	30 bar O_2	60	6	99	[68]
Au/HT	None	Aqueous solution	10 bar O_2	95	7	>99	[69]

Certain studies have established a base-free method for oxidizing HMF into FDCA utilizing a hydrotalcite-supported Au catalyst (Au/HT). Remarkably, FDCA was obtained with a high yield of 99% after 7 h at 95 °C under 1 bar O_2 in water [69]. The catalyst demonstrated reusability for at least three cycles without significant loss of activity and selectivity. Au deposited on neutral supports (Al_2O_3, C) exhibited minimal activity, while acidic SiO_2 showed no activity, indicating the significance of the basicity of HT in the oxidation of HMF into FDCA. As aforementioned, a basic reaction medium is crucial to avoid the FDCA precipitation, accelerate alkoxide formation and C–H activation and HT type materials can be precursors of these properties.

8.3.3.3 Palladium (Pd)

As well as aforementioned catalysts, the supported palladium ones exhibited also a good catalytic efficacy in the aerobic oxidation of HMF to FDCA. These supported Pd catalysts also demonstrated high selectivity, robust performance, and good recyclability, making them promising candidates for sustainable and efficient catalytic processes in the production of FDCA. Additionally, the choice of support materials played a crucial role in influencing the overall catalytic activity and stability.

In the first entry of the table (see Table 8.3), the result of the study conducted by Davis et al. [54] has been presented. In this study, carbon supported catalysts have been compared under identical conditions. The findings indicate that these two catalysts are comparable, showing very similar performance, achieving 100% HMF conversion with yields of 71% towards FDCA.

In the subsequent rows, the results obtained in a 2014 study are presented [71]. In summary, within this study, it can be stated that the $Pd/ZrO_2/La_2O_3$ catalyst exhibited the highest FDCA yield and a relatively stable catalytic performance. By-product formation was generally low in this case. For Ti_2O_3, characterized by a more robust interaction between the Pd nanoparticles and the support, catalytic performance remained stable, although lower yields of FDCA were achieved. This observation could be attributed to a more pronounced adsorption of HMF and

Table 8.3 Some results of oxidation of HMF into FDCA over Pd-based catalysts

Catalyst	Base	Solvent	Oxidant	Temp (°C)	Time (h)	FDCA yield (%)	Reference
Pd/C	2 equiv NaOH	Aqueous solution	7 bar O_2	23	6	71	[54]
Pd/Ti_2O_3	1.25 equiv NaOH	Aqueous solution	10 bar O_2	90	6	53	[71]
Pd/Al_2O_3	1.25 equiv NaOH	Aqueous solution	10 bar O_2	90	6	78	
$Pd/ZrO_2/La_2O_3$	1.25 equiv NaOH	Aqueous solution	10 bar O_2	90	6	90	

reaction intermediates on the catalyst surface, leading to a blockage of active surface sites during the initial stage of the reaction, as suggested by the lower carbon mass balance. They also detected changes in Pd structural properties, related to particle size and shape, and leaching phenomena, which intensifies when base concentration is high.

The catalytic stability of Pd-based materials (Pd/ZrO_2/La_2O_3, Pd/KF/Al_2O_3, and Pd/TiO_2) was assessed through reuse. The catalysts were separated, washed, and dried between runs. After each cycle, a slight loss in material (about 7%) was noted for Pd/ZrO_2/La_2O_3 and Pd/TiO_2. Despite complete conversion of HMF to FDCA within 1.5 h in all runs, a minor decline in FDCA yield was observed with each reuse: from 90% to 86% for Pd/ZrO_2/La_2O_3, and from 62% to 54% for Pd/TiO2 over three runs. Pd leaching decreased with each cycle for Pd/ZrO_2/La_2O_3.

The Pd/KF/Al_2O_3 catalyst showed a more significant drop in FDCA yield (91–55%) between the first and second run. Increased HMFCA yield was detected with repeated use. Similar trends in FDCA yield reduction were previously observed in other catalysts like Au/CeO_2 and Pt/Bi/C.

8.3.3.4 Ruthenium (Ru)

Ru has garnered increased attention due to its noteworthy activity achieved at reasonable prize when compared, for example, with Pt and Au. This affordability opens up promising avenues for the broader adoption of Ru-based catalysts, making it a cost-effective alternative for various catalytic applications. The lower cost of Ru enhances its accessibility and feasibility for researchers and industries seeking efficient and economical catalytic solutions.

Table 8.4 presents the results of a study by Zheng et al. [72]. However, it is worth noting that, in evaluating Ru catalysts in aqueous media, they have employed not only the commonly used bases but also other innovative ones, and in many instances, no base at all.

Considering the results shown in the table, it can be observed that the Ru/C catalyst exhibits the highest FDCA yield. Under the same optimum conditions used for Ru/C, TiO_2, Al_2O_3, and ZrO_2 led to a lower FDCA yield. Additionally, there was

Table 8.4 Some results of oxidation of HMF into FDCA over Ru-based catalysts

Catalyst	Base	Solvent	Oxidant	Temp (°C)	Time (h)	FDCA yield (%)	Reference
Ru/C	4.3 equiv Mg $(OH)_2$	Aqueous solution	10 bar O_2	108	8	97	[72]
Ru/Ti_2O_3	4.3 equiv Mg $(OH)_2$	Aqueous solution	10 bar O_2	108	8	82	
Ru/Al_2O_3	4.3 equiv Mg $(OH)_2$	Aqueous solution	10 bar O_2	108	8	80	
Ru/ZrO_2	4.3 equiv Mg $(OH)_2$	Aqueous solution	10 bar O_2	108	8	86	

a lower carbon balance than with active carbon (86–88% vs. 97%), potentially because of the advantageous generation of unidentified degradation or polymerization products of HMF as opposed to inert carbon. Moreover, the stability and reusability of the optimal carbon-supported catalyst was studied, concluding that the catalyst is stable after six uses.

Although the production of FDCA through noble metal catalytic oxidation of HMF presents advantages over homogenous process, it is important to mention that it seems to have some drawbacks that must be gained. These drawbacks are mainly ascribed to the cost, physicochemical properties and separation and reusability of the catalysts.

According to Chen et al. [62] the main goal of the biorefinery industry is to suppress the capital cost associated to use of noble catalysts. In this sense, the noble metal catalysts' leaching (influenced by the base concentration in the reaction medium [71]) makes cost drawback worse, because efficient recovery of the dissolved noble metal would be required. This drawback can be overcome developing catalysts based on transition metals (see Sect. 8.3.3.5) or adding a second non-noble metal, such as Bi. As aforementioned, the addition of Bi avoids the leaching phenomena, due to it provide stability to the noble metal, and therefore its recovery. Moreover, it could reduce the require amount of noble metal in the catalyst.

Regarding properties of the catalysts, related to their activity and stability, it seems important to do a good catalyst design taking into account: (1) a good dispersion of metal active sites, (2) a large mesoporous structure, that are crucial for substrate transportation, and (3) a low or moderate OSC to avoid catalyst deactivation by O_2 coverage and/or over-oxidation. The over-oxidation can also be prevented by modulating the pressure in the reaction system.

The catalyst separation from liquid phase, for later reuse, requires efficient filtration and centrifugation operations to avoid catalyst loss and significant energy consumption. This drawback could be gained using catalytic fixed bed reactor, which does not need separation process by filtration and centrifugation, or magnetic catalysts, which can be separated by magnetic process, and it could be transition metals or immobilised enzymes [55]. The reusability and stability of the catalysts are other aspects that they must be considering. For a catalyst to be reusable, it should not suffer leaching and/or metal active sites oxidation. In this last case, apart from adding of another metal, such as Bi, the catalyst can be reusable after H_2 treatment, employed to reduce the oxidised metal active sites.

In conclusion, the applicability of these catalysts in industrial scale requires overcoming many limitations related to the catalysts, even the reaction conditions.

8.3.3.5 Non-noble Metal-Based Catalysts

As above mentioned, the cost associated with noble metal catalysts has limited their practical application in the production of FDCA. Consequently, research has shifted towards non-noble metal catalysts for HMF oxidation, with a focus on economically

Table 8.5 Some results for HMF to FDCA oxidation with non-noble metal(s) catalysts [79

Catalyst	Base	Solvent	Oxidant	Temp (°C)	Time (h)	FDCA yield (%)	Reference
MnO_2	3 M $NaHCO_3$	Aqueous solution	10 bar O_2	100	24	91	[75]
MnO_x	2 M $NaHCO_3$	Aqueous solution	10 bar O_2	120	18	81	[78]
$MnCo_2O_4$	3 M $KHCO_3$	Aqueous solution	10 bar O_2	100	24	71	[79]
Co-Mn	2 M $NaHCO_3$	Aqueous solution	10 bar O_2	120	5	95	[78]
Fe^{III}-POP-1	None	Aqueous solution	10 bar air	100	10	79	[80]

viable options such as manganese (Mn), cobalt (Co), and iron (Fe) (a brief overview is listed in Table 8.5).

Among these, Mn-based catalysts have received extensive attention due to their varied oxidation states, ease of O_2 adsorption, reduction capabilities, and higher storage capacity [73]. Various Mn-based catalysts, including MnO_2, MnO_x–CeO_2 composite, $Mn_{0.75}/Fe_{0.25}$, and magnetic ferrites, have demonstrated efficient conversion of HMF to FDCA with promising yields [74–77]. According to Neațu et al. [74] synthetized a mixed Mn/Fe oxide for the oxidation of HMF under oxygen atmosphere in aqueous solution. The good performance of the $Mn_{0.75}/Fe_{0.25}$ catalysts can be ascribed to the presence of Mn^{3+} and Mn^{4+} with hematite phase. Moreover, no variations in the catalytic activity and selectivity were detected after three consecutive cycles. In the case of MnO_2 catalysts [75, 76], the β-MnO_2 catalyst provided good activity, yield and stability towards FDCA due to lower oxygen vacancy formation energy with oxygen as the sole oxidant. However, its catalytic behaviour could be improved increasing its surface area [55], because this textural property is important to improve the substrate transportation, as aforementioned.

Co-based heterogeneous catalysts have also shown promising results. In this context, Co supported on carbon-based catalyst, Co-SA_s/N@C, exhibited notable FDCA yield at 1 bar of oxygen pressure in an aqueous medium, exhibiting FDCA productivity (4.13 mol mol_{Co}^{-1} h^{-1}) comparable to some noble-metal catalysts [81].

Additionally, bimetallic catalyst such as nanoscale centre-hollowed $MnCo_2O_4$ spinel exhibited high activity for the oxidation of HMF, reaching almost complete conversion and 70.9% FDCA yield in aqueous media with pure oxygen as oxidant [79]. The Mn/Co ratio was optimized to favour the proportion of Mn^{3+}, which is stated as the active site of the catalyst. In this regard, bimetallic Co-Mn composite mixed oxide has also reported high activity towards HMF oxidation in aqueous solution pressurized with oxygen [78]. In this study, the ratio of Mn and Co was also optimized, along with the calcination temperature. The enhanced catalytic activity was ascribed to the high mobility of lattice oxygen and the presence of surface Mn^{4+} and Mn^{3+} species. Moreover, the recyclability of the catalyst was evaluated, demonstrating excellent stability after five cycles. Overall, the activity and stability of these catalysts make them viable options for industrial applications. The good performance of these catalysts are supported also on the fact that they present higher oxidation states, favouring electron transfer, and oxygen mobility [74, 78, 79, 81].

Fe, as a cost-effective and abundant element and magnetically separable, has been explored both in heterogeneous form and as an alloy with noble metals for HMF oxidation.

Fe-based catalysts have also been widely studied. Saha et al. [80] proposed an innovative porous organic polymer loaded with Fe. The resulting material exhibited highly cross-linked, a robust structured, and thermal stability. This catalyst underwent testing for HMF oxidation using molecular oxygen as the oxidant in an aqueous medium. The oxidation state of Fe remained unchanged after the reaction, allowing for the catalyst to be reused without experiencing significant loss in activity. Fe supported on conventional support such as SBA-15 has also been studied in aqueous media employing H_2O_2 as oxidant [82]. Although the catalyst demonstrated high activity in the oxidation of HMF to produce FDCA, it exhibited low stability under the operating conditions. Bimetallic nano Fe_3O_4-CoO_x catalysts were evaluated with dimethyl sulfoxide (DMSO) as solvent and tert-butyl hydroperoxide (t-BuOOH) as oxidizing agent [83]. The absence of a required base additive rendered this system considerably more economical and environmentally friendly. Moreover, it exhibited high stability throughout the reaction process.

Based on this background, it seems to be viable the use of non-noble metal catalysts on FDCA production from HMF oxidation, reaching similar conversion and yield to noble ones. However, the studies are scarcer, indicating that further research is needed in this area.

8.3.4 Whole Cells and Enzymes

The investigation into converting HMF to FDCA utilizing whole cells and enzymes has been prompted due to the potential represented by the fact that these reactions can be conducted under mild reaction conditions, being lower temperatures and pressure, with high selectivity. However, it is noteworthy that the use of these biochemical systems does not necessarily eliminate the need for harsh solvents, catalyst separation and reusability, among others [84].

Regarding whole-cell bio-catalytic processes, these are based on detoxification mechanisms developed by certain microorganisms, such as fungi and bacteria, to transform initially toxic molecules like HMF, which being an aldehyde and can cause damage to proteins, organelles and nucleic acids through formation of reactive O_2 species [84] into non-toxic ones. This mechanism can be extrapolated for converting HMF to FDCA.

An optimal whole-cell biocatalyst should be genetically manageable, exhibit rapid growth in uncomplicated media, generate substantial enzyme quantities, and be suitable for subsequent downstream processing procedures [85]. Hence, fine-tuning the biochemical characteristics of the whole-cell biocatalyst and reaction conditions (such as pH, temperature, substrate concentration, and biocatalyst dose) to enhance production is crucial in whole-cell biotransformation processes.

Up until 2015 only limited number of enzymes were identified as active toward HMF. Among these, only the ones investigated by Koopman et al. [86] and Carro et al. [56] were suitable for the production of FDCA.

Among the latest studies is the one conducted by Godan et al. [87], where, for the first time, a specific type of enzyme (*Acinetobacter oleivorans*) was employed, resulting in approximately a 65% yield of FDCA. The conditions for obtaining this were mild, as previously discussed, being these 30 °C, pH 7.0, and 200 rpm, with varying reaction times from 6 to 30 h. It is noteworthy that the desired product started to emerge between the 24th and 28th hour of the reaction.

Other types of bacteria that have been utilized in FDCA production include those from the *Pseudomonas clade*. For this reaction, different approaches have been employed, such as using an alternative alcohol oxidase, implementing a co-cultivation strategy, or adopting a hybrid process that combines whole-cell catalysis with purified enzymes [88, 89]. Additionally, several other species, such as *Raoultella ornithinolytica*, have been enlisted for FDCA production from HMF [90, 91]. Despite thorough exploration and examination of different enzymes, the difficulty persists in identifying the enzyme(s) accountable for aldehyde detoxification.

While whole-cell catalysis holds promise for future applications, certain aspects still require improvement, including the low product recovery, undesired side reactions, the necessity for specialized equipment and the continuous feeding of carbon sources [84, 92, 93].

Nowadays, two types of enzyme studies are distinguished: investigations involving isolated enzymes and the exploration of enzyme cascade systems.

The use of isolated enzymes always involves the use of mediators or electron acceptor substances. Among the enzymes utilized, those with the highest yields towards FDCA, in ascending order, are immobilized laccase, *HMF oxidase* (*HMFO*) and immobilized lipase. In experiments with laccase, where a 90% yield of FDCA was achieved, the reactions were conducted at 35 °C for 96 h [94]. For *HMFO*, under temperature conditions of 25 °C for a duration of 15 h, a FDCA yield of 95% was obtained [95]. With the use of lipase and additives in the medium, a 100% yield of FDCA was achieved at 40 °C for 24 h [96].

With isolated enzymes, higher productivities can often be reached, albeit at the expense of lower biocatalyst robustness under the process conditions. It is important to note that most of enzymes engaged in HMF conversion catalyse either alcohol or aldehyde groups but not both simultaneously and, for the conversion of HMF to FDCA, both alcohol and aldehyde must be oxidized simultaneously. Consequently, a cascade catalysis process is employed, utilizing multiple enzymes to biosynthesize FDCA from HMF [97].

Among the studies yielding the best results towards FDCA, using *Aryl-alcohol oxidase* (*AAO*) and *Unspecific peroxygenase* (*UPO*) a yield of a 90% was achieved for reaction time of 120 h [56]. With immobilized *Laccase* and *Lipase* yields around 94% in 20 h were obtained [98]. Employing the previously cited *HMFO* and lipase a yield of 94% in 60 h [99].

Using *Galactose oxidase* (*GO*), *Alcohol dehydrogenases* (*ADH*), and *Horse radish peroxidase* (*HRP*) 95% yield of FDCA was reached in 60 h [100]. With a cluster of *GO, HRP*, *Periplasmic aldehyde oxidase* (*PaoABC*), and *Catalase* a total conversion towards the desired product was achieved in 6 h [101]. Also, yields of 100% were attained using *AAO* and *Catalase*, maintaining the reaction for 288 h (100% in 288 h) [102].

8.3.5 Electrochemical Oxidation

This synthetic process is boosted by electron migration under mild conditions, enhancing the electrochemical potential and thereby eliminating the requirement for any external oxidant. The growing significance of HMF electrochemical oxidation lies in its reliance on non-noble metal catalysts at ambient temperatures, but nowadays, some non-precious metals, such as nickel (Ni), Co and Fe, have been identified for their notable electrochemical activity [103, 104]. Another noteworthy aspect of this process is the simultaneous production of H_2 at the cathode during the anodic oxidation of HMF [105].

The typical approach involves the use of heterogeneous catalysts to achieve high conversion rates and yields; however, homogeneous catalysis can also be employed with the aid of a mediator. In short, studies with some of the best results demonstrate FDCA yields above 90%.

One of these recent studies, is for example, the one of Wang et al. [106], where mesoporous δ-MnO_2 is employed as anode attaining FDCA yields nearby 98%. Also in 2022, Chen et al. [107] employed a Co-based Co_3O_4/CF (cobalt foam, CF) catalyst for electrochemical oxidation, achieving a FDCA yield of 93% along with the generation of H_2 gas.

Pang et al. [108], using Ni-MOF/Ag (metal organic framework, MOF) electrode stated that Ni^{3+}/Ag^+ is the main active species for HMF oxidation. Obtaining nearly 100% HMF conversion, with around 98% yield to FDCA.

The main drawback to the use of this method for FDCA production is the fact that recovering the desired product from the reaction media tends to be difficult and high current density would be necessary for industrial production with a high efficiency. Moreover, the performance of electrolytic cell and influence of its components on the reaction would requires more depth research and, resulting in design of new reactors which can reduce electrode space and improve flow path [109].

8.3.6 Photocatalytic Oxidation

The advantages of photocatalytic processes include mild reaction conditions, reduced need for harsh chemicals, and the potential for using sunlight as a renewable energy source. However, the key challenge for photocatalytic reactions lies in designing highly efficient and stable catalysts.

These photocatalytic studies are relatively recent, for instance, Xu et al. [110] demonstrated a photocatalyst composed of Co thioporphyrazine (CoPz) dispersed on graphitic carbon nitride (CoPz/g-C_3N_4), which shows remarkable catalytic activity in the selective oxidation of HMF to FDCA under simulated sunlight, using oxygen from air as a benign oxidant. They achieved an FDCA yield of 96% in an aqueous solution at pH = 9.18, under ambient temperature and air pressure.

In another study, Zhu et al. [111] synthesised Au/ZnO catalysts with varying oxygen vacancies were prepared and employed for the photocatalytic oxidation of HMF to FDCA under visible light irradiation. These catalysts exhibited excellent selectivity towards FDCA, exceeding 90%, under mild reaction conditions.

Obtaining poorer FDCA yields, Zhang et al. [112] employed plasmonic monometallic Ag or Au nanoparticles (NPs) and bimetallic Ag-Au single bond NPs supported on TiO_2 with a tuneable bandgap structure, achieving around 48% yield of FDCA. Interestingly, the monometallic Ag/TiO_2 catalyst only produced HMFCA, whereas Au/TiO_2 produced FDCA.

As in the case of other catalytic systems, the photocatalytic one must also meet a series of conditions for its practical applications, such as improving catalyst stability, yield and selectivity towards FDCA. In fact, most of the reported photocatalytic systems produced DFF as a main product [113, 114]. Moreover, the photocatalytic oxidation mechanisms are not well defined [113], and the study of mechanisms could provide information about how to enhance the production of FDCA.

8.3.7 *Comparative of Processes for FDCA Production by HMF Oxidation*

As seen in the previous sections, there are different methodologies for obtaining FDCA. Among all of them, homogeneous catalysis and heterogeneous catalysis stand out. Currently, the process closest to industrial FDCA production (Avantium) is based on homogeneous processes. However, these homogeneous processes present several challenges, such as the recovery of the used homogeneous catalyst, which is often hazardous and difficult to dispose of. Additionally, special equipment is required, which increases production costs.

Due to the characteristics of homogeneous processes, which present high FDCA yields, the use of heterogeneous catalysts is being studied as an alternative. These heterogeneous catalysts, being in a solid state, offer better characteristics for the process since the separation of the desired products is unnecessary, requiring no additional equipment. Moreover, they are not hazardous, thus eliminating the need for special equipment or neutralization treatments. This reduces energy requirements, improving the economic conditions and carbon footprint of the process; therefore, these are potentially more economical processes with less environmental impact.

The problem with these types of processes is that they do not achieve the yields of homogeneous systems. This is why research is being conducted to develop active and stable catalysts for FDCA production. As seen in Sect. 8.3.3, the catalysts being developed are based on noble metals such as Pt, Au, Ru, and Pd, which have shown the best FDCA performance. However, these noble catalysts are expensive, increasing the cost of the production system. Efforts are being made to find alternatives using metals such as Co, Mn, and Fe, though these catalysts currently offer lower performance. Reducing the amount of noble metal in the catalyst, improving metal dispersion, employing alloys, or using non-noble metals are all strategies being considered to enhance the economics of heterogeneous catalysis.

Other processes, such as whole cells and enzymes, and electrochemical oxidation, are presented as attractive new routes but are currently limited. The bioprocess offers several advantages, such as high FDCA production at lower temperatures and pressures, making the process cheaper and more sustainable from an environmental perspective. However, improving the performance of bacteria or enzymes capable of carrying out this reaction is necessary to make this process more viable. The main issue is the long time required to achieve high FDCA production, which prevents the process from becoming fully operational.

Finally, electrochemical oxidation is another alternative. In this process, FDCA is obtained through the oxidation of HMF in oxidation-reduction reactions, as mentioned in Sect. 8.3.5. Significant performances have been achieved using non-noble electrodes based on metals such as Ni, Mn, Ag, and Co with various supports for metal dispersion. This process allows operation at lower temperatures than homogeneous and heterogeneous catalytic processes but requires electricity, which could hinder the development of sustainable systems. The main challenge with this technology is separating the product from the reaction medium, which presents significant difficulties and prevents the process from being scaled up.

8.4 FDCA Production from MMF

As previously corroborated, the production of FDCA from HMF has been widely investigated. However, FDCA can also be obtained from MMF. Despite the growing demand for FDCA, research on its synthesis from MMF remains scarce. A review of the scientific literature revealed that few research articles have been published exploring this transformation, with a predominant focus on catalytic oxidation.

8.4.1 Transformation of MMF to FDCA: A Brief Overview

While research on the specific mechanisms of this reaction remains limited, the proposed pathway is based on findings from bibliography [115]. There are two potential routes for the transformation process. Beginning with MMF as the starting material, the initial step involves the formation of the intermediate 5-methoxymethyl-furan-2-carboxylic acid (also known as 5-methoxymethyl-2-furoic acid, MMFA, or MMFCA). This intermediate arises when the ether bond in MMF is cleaved, leaving the -CH_2OH (alcohol) group intact. From this point, the reaction can follow either Route A or Route B.

In Route A, MMFA first converts to HMFCA. This take places when the ether bond in MMF is broken and the -CH_2OH (alcohol) group undergoes further oxidation to form a -CHO (carbonyl) group. HMFCA is then oxidized to FFCA and subsequently to the desired product, FDCA. On the other hand, Route B involves a direct conversion of MMFA to FFCA, which is then further oxidized to yield FDCA. This proposed mechanism provides a look at the possible ways MMF can be oxidize to FDCA, explaining the steps and potential results of the process.

As previously mentioned, there are very few publications on the transformation of MMF to FDCA. However, the existing studies can be categorized into two main types: chemical catalytic systems and systems employing biocatalysts, such as whole cells and enzymes. These approaches offer different advantages and challenges, contributing valuable insights into the potential pathways and mechanisms for the conversion process.

8.4.2 Chemical Catalytic Systems

A pioneering study in this field was conducted by researchers who developed and patented the Avantium process. In batch reactors, using MMF as the substrate and Co/Mn/Br catalysts in an HAc solution, they achieved approximately a 70% yield of FDCA under optimal conditions: 180 °C, 60 bar air pressure, and a reaction time of 1 h. This study set a significant benchmark for the conversion of MMF to FDCA [116].

In another study [26], researchers explored the use of Ru/hydroxyapatite (Ru/HAP) catalysts in batch reactors. A solution of MMF and H_2O was fed into the reactor along using Ru/HAP as catalyst, and then, the system was pressurized at 10 bar of O_2. The best result obtained was a 65% yield of FDCA at 130 °C and 15 h of reaction time. Notably, this reaction predominantly followed Route A, highlighting the specific pathway favoured under these conditions.

A subsequent study by the same group [117] concluded that the addition of CeO_2 to the previously mentioned catalysts (RuOx/CeO_2-HAP) improved the FDCA yield obtained to 84% at the same temperature, but reducing pressure and reaction time to 5 bar and 12 h, respectively. This enhancement underscores the importance of catalyst composition in optimizing the conversion process.

The next catalyst studied was an O_2-vacancy-rich MnO_x supported RuO_x (RuO_x/MnO_x-VC) catalyst [118]. For performance of the mentioned catalysts batch reactor were used and, 140 °C, 15 bar of air, and a reaction time of 3 h were established. This last reaction process also appeared to follow Route A, suggesting that Route A might be the preferred pathway in reactions involving this type of catalytic system. This approach continues to build on the insights gained from previous studies, aiming to further optimize the yield and efficiency of FDCA production from MMF.

8.4.3 *Whole Cells and Enzymes*

The application of enzymatic and whole cell systems for the oxidation of MMF to FDCA remains a relatively unexplored area. Despite the limited research, a few notable studies have revealed potential methodologies and the effectiveness of these bio-catalytic processes.

One pioneering study [115] established a cascade reaction using three enzymes: *AAO*, *UPO*, and *MeOH oxidase* (*MOX*). In this process, MMF is first converted by *AAO* into MMFA. Subsequently, *UPO*, added in a second step, converts MMFA to FFCA by cleaving ether bonds in the presence of H_2O_2. *UPO* then oxidizes FFCA to FDCA. This process necessitates a significant amount of H_2O_2, as the quantity produced during the initial oxidation is insufficient. However, the hydrolysis reaction produces MeOH, which *MOX* can oxidize to formaldehyde while reducing O_2 to H_2O_2. This ensures an adequate supply of H_2O_2 for the *UPO*-catalysed reactions. Without the external addition of MeOH, a FDCA yield of 70% is achieved, while the addition of MeOH, the yield increases to 98%.

It is important to note that this enzymatic system tends to follow Route B of the proposed reaction pathways, highlighting a distinct mechanism compared to traditional chemical catalytic systems. The exploration of these bio-catalytic processes suggest that this unexplored approach holds promise for the sustainable production of FDCA.

8.5 Alternative Routes for FDCA Production

There are some alternatives for FDCA production, such as the use of furfural (derived from hemicellulose) which can serve as precursor in carboxylation reactions. Additionally, sugar acids, which can be obtained from glucose oxidation, offer a potentially sustainable and efficient route for FDCA synthesis.

8.5.1 *FDCA from Furfural*

For the production of FDCA from furfural, the first step is to convert furfural into furoic acid (FA). The subsequent reaction of FA to FDCA always involves C–H carboxylation with CO_2. The initial oxidation of furfural to FA is carried out in the presence of heterogeneous catalysts in alkaline or base-free solutions. The second step, converting FA to FDCA, is more complex and can be performed using various processes [119].

8.5.1.1 Homogeneous Catalysts for FA Carboxylation Through FDCA

Few studies have investigated the carboxylation of furan and its derivatives using homogeneous catalysis, typically involving Cu and Pd salts. This method primarily focuses on synthesizing FDCA through a multistep process that includes bromination, esterification, carboxylation, and hydrolysis [120].

For instance, one approach by Zhang et al. [121] used simple Pd salt such as Pd $(dba)_2$ with tris(3-sulfophenyl)phosphine trisodium salt (TPPTS), achieving a 93% yield of FDCA.

Yin et al. [122] studied the carbonylation of 5-bromo furoic acid with Pd-based catalysts ($PdCl_2$) in DMF, resulting in a 99% yield of FDCA. They examined the effects of various Pd precursors, phosphine ligands, and bases.

In another study, Nocito et al. [123] demonstrated that CO_2 indirectly aids the process by regenerating carbonate species in situ.

8.5.1.2 Heterogeneous Catalysts for FA Carboxylation Through FDCA

Kanan et al. [124] pioneered the use of CO_2 and a base for the carboxylation of FA, drawing from the principles of the Kolbe-Schmitt synthesis dating back to 1860 [125]. Their study achieved significant results, yielding 76% of $FDCA^{2-}$ after 12 h at 260 °C under a CO_2 flow of 40 mL min^{-1}, employing 10% excess Cs_2CO_3 in continuous reactor. In a pressurized batch reactor, they obtained an 89% yield of $FDCA^{2-}$ after 7 h at 200 °C with 8 bar of CO_2.

Recent studies have followed the work of Kanan et al., using Cs_2CO_3 as a catalyst. For example, Liu et al. [126] proposed a multistep process for the conversion of hemicellulose to FDCA, including hemicellulose-to-furfural conversion and sequential oxidation to FA, followed by the preparation of cesium furoate (FA-Cs) and its direct carboxylation with CO_2. The overall conversion yield of the targeted FDCA product in the hemicellulose-to-FDCA pathway reaches 29.6%. Greater success was achieved through the direct carboxylation reaction of FA-Cs with CO_2 using Cs_2CO_3 as a catalyst. Results show that an FDCA yield of 56.47% is achieved via the direct carboxylation of FA-Cs in continuous reactor at 290 °C for 12 h with a CO_2 flow rate of 40 mL min^{-1}. The study of this reaction and its improvement has been extensively investigated, including theoretical studies [127].

The high cost of Cs has led to the search for alternative catalytic systems, such as the use of K_2CO_3. Since the system with this catalyst is highly influenced by temperature, Han et al. [128] decided to add potassium formate (HCOOK), which allowed the reaction temperature to decrease by about 190 °C, achieving an FDCA yield of approximately 60%.

8.5.1.3 Whole Cells and Enzymes for FA Carboxylation Through FDCA

Regarding the use of whole cells and enzymes, most studies remain at the theoretical level or focus on detecting FDCA using these systems. In enzyme studies, various systems have been explored. For example, the results from Payne et al. [129] demonstrated the carboxylation ability of *HmfF* enzymes to produce FDCA. The yields were not very high (0.3% after 12 h), and the reaction required stringent control of reaction conditions. Kawanabe et al. [130] proposed a biocatalytic route to produce FDCA directly from furfural using bicarbonate as the carbon source and recombinant *Escherichia coli* cells co-expressing different enzymes.

Regarding whole cells, following the aforementioned enzymatic study, López-Lorenzo et al. [131] moved away from the use of isolated enzymes by supplementing a system with *E. coli* and the gene of *P. thermopropionicum HmfF* (*PtHmfF*), known to generate FDCA. The resulting whole-cell system allows for the production of FDCA, achieving yields up to 55% under mild conditions through the carboxylation of FA.

8.5.2 FDCA from Sugar Acids

The synthesis of FDCA was first reported in 1876 [132] using galactaric acid as the substrate, where heating with excess HBr produced FDCA. Subsequent methods also used excess H_2SO_4 or HBr, yielding around 50% FDCA [133, 134]. In the early 2000s, some approaches employed benzene sulfonic acid or *p*-toluene sulfonic acid for the dehydration of C6 aldaric acids, achieving similar yields [135, 136]. Zhao et al. [137] described a one-pot synthesis of mucic acid achieving FDCA yields around 53%. Taguchi et al. [138] demonstrated that heteropolyacids, like phosphotungstic acid, catalysed dehydration more efficiently than homogeneous acids, achieving 44% yield of FDCA ester 15 h. These findings suggest aldaric acids are a promising alternative for FDCA production, though significant enhancements are required for industrial viability. More recently, a study by Van Strien et al. [139] employed the dehydration of aldaric acids using solid acid catalysts in butanol, which results in furancarboxylates. They achieved over 90% yields of furancarboxylates using silica-supported acid catalysts, with selectivity towards FDCA and its esters reaching 80%.

As previously mentioned, other sugar acids have been employed for the production of FDCA, such as glucaric acid or gluconic acid. Several patents have been published transforming these sugar acids. For instance, using an HBr/H_2O/sulfone-catalysed glucaric acid dehydration system achieved yields close to 70% [140]. Other studies, like that of Kambourakis and Griffin [141], propose the conversion of glucose to glucaric acid and subsequently exploring various catalytic reaction media to obtain FFCA and FDCA. However, the use of these promising sugar acids is still in the developmental stage.

8.6 Significance in Sustainable Polymer Synthesis

The significance of sustainable polymer synthesis is paramount in addressing environmental challenges associated with conventional methods. In this context, polymers, crucial in various applications, are traditionally produced using non-renewable resources, energy-intensive processes, and generate substantial waste [142] here is when FDCA enters as a key building block for biodegradable and eco-friendly polymers. The global shift towards environmental sustainability necessitates a re-evaluation of polymer production methods. A key facet of this significance lies in reducing dependence on fossil fuels. By exploring alternative routes, such as utilizing plant-based feedstock, sustainable synthesis aims to minimize the environmental impact associated with resource depletion and greenhouse gas emissions. This approach aligns with the broader goal of promoting resource efficiency. Moreover, sustainable polymer synthesis strives to minimize environmental impact by adopting greener chemistry principles. This involves designing methodologies that are less harmful to ecosystems, reducing waste, energy consumption, and avoiding the use of toxic substances [143]. Embracing these principles contributes to a more sustainable materials life cycle.

Circular economy concepts further amplify the significance of sustainable polymer synthesis. Designing polymers for recyclability, biodegradability, or easy degradation into non-harmful by-products supports efficient resource use and addresses plastic pollution concerns [144]. This approach considers the entire life cycle of polymers, from production to disposal.

As stated before, the growing demand for greener alternatives, driven by stringent environmental regulations and consumer preferences, underscores the need for sustainable practices. The significance lies in manufacturers adapting to these changing dynamics, ensuring compliance with regulations, and meeting consumer expectations for environmentally friendly products.

Sustainable polymer synthesis not only encourages interdisciplinary collaboration and innovation but also holds the key to addressing multifaceted challenges associated with conventional polymer production. Researchers from diverse fields collaborate to develop novel methods and materials aligned with sustainability goals, showcasing the dynamic nature of the field and its potential to drive positive change in material science and industry practices.

The significance of sustainable polymer synthesis is further heightened with the emergence of materials like PEF, a bio-based alternative to traditional PET commonly used in beverage bottles. PEF's production involves renewable resources, reducing reliance on fossil fuels, and its potential to replace conventional plastics underscores the ongoing efforts to create environmentally friendly polymers aligned with circular economy principles. Integrating such innovations into the broader framework of sustainable polymer synthesis not only exemplifies ongoing endeavours but also offers a promising avenue for reducing the environmental impact of plastic materials.

8.6.1 PEF Synthesis from FDCA

It is imperative to provide a concise overview of the PEF synthesis system. As outlined earlier, the synthesis process of FDCA, an essential monomer for PEF production, entails the oxidation of HMF sourced from renewable feedstock like glucose or fructose. Polymerization of the PEF monomer is achieved through condensation-polymerization (see Fig. 8.5), catalysed by metal catalysts, with meticulous control over temperature, pressure, and catalyst concentration. Following polymerization, purification processes are implemented to eliminate residual monomers and catalysts. The refined PEF can be tailored into diverse forms, and the final product undergoes rigorous characterization and testing to ensure adherence to specific molecular, thermal, and mechanical properties, among other defined criteria.

The synthesis of PEF traces back to 1946, contemporaneous with the synthesis of PET. The Celanese Corporation of America patented the first reported synthesis, involving bulk polytransesterification at high temperatures (above 200 °C) and a high vacuum [145]. After a hiatus, research on PEF synthesis resumed in 2009 with Gandini et al., who employed polytransesterification of bis-(hydroxyethyl)-2,5-furandicarboxylate catalysed by antimony(III) oxide [146].

Various synthetic routes for PEF production have been explored, including polycondensation and polytransesterification. A two-stage melt polyesterification approach has been widely studied, but challenges such as yellow discoloration and difficulty in removing residual metals persist [147].

Efforts to make PEF synthesis sustainable and address these challenges have been reported. In 2018, a study addressed degradation and discoloration issues through a rapid ring-opening polymerization (ROP) technique, resulting in uncolored bottle-grade PEF within a short time [148]. One year later, Qu et al. focused on environmentally friendly processes, using metal-free ionic liquids as catalysts for the polycondensation of ethylene glycol with FDCA [149]. It is important to mention that some studies investigated PEF synthesis using an organic non-metal catalyst, achieving moderate molecular weight [150].

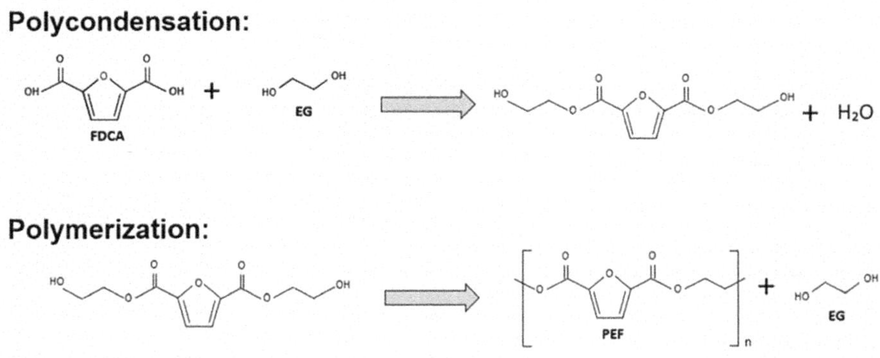

Fig. 8.5 Synthesis of PEF from its monomers, FDCA and EG

8.6.2 Role of FDCA in Sustainable Polymer Production

At the forefront of sustainable polymer production is FDCA, a game-changer in the shift toward environmentally friendly polymer synthesis [151]. Derived from biomass, particularly plant-based feedstock, FDCA stands as a renewable alternative to traditional petrochemical-based monomers.

FDCA's significance lies in its ability to utilize renewable feedstock, such as sugars found in crops like corn and sugarcane, or biomass residues [152]. This shift reduces the industry's reliance on finite and environmentally taxing fossil fuels. Moreover, polymers derived from FDCA offer improved biodegradability compared to conventional counterparts, contributing to the global effort to mitigate plastic pollution and environmental impact.

The synthesis of polymers using FDCA substantially reduces the carbon footprint associated with conventional methods. This shift towards renewable resources aligns with the industry's commitment to sustainability. Furthermore, FDCA-based polymers exhibit enhanced properties, including improved thermal stability, mechanical strength, and barrier properties, making them suitable for a wide range of applications.

In addition, it is worth noting that the adoption of FDCA in sustainable polymer production not only addresses environmental concerns but also aligns with the growing consumer demand for eco-friendly and responsibly sourced materials [153]. This further emphasizes the role of FDCA as a key player in shaping the future of polymer production towards a more sustainable and environmentally conscious industry.

8.6.3 Applications and Future Perspectives

The versatility of FDCA-based polymers spans various applications, promising a sustainable future for materials [145]. In the realm of packaging, these polymers emerge as sustainable alternatives, boasting enhanced barrier properties to meet the growing demand for eco-friendly packaging solutions.

Extending their reach into the textile industry, FDCA-based polymers are being explored for the creation of sustainable fibres with improved durability and comfort [154]. This aligns with the increasing consumer preference for environmentally conscious products. Likewise, in the healthcare sector, the biocompatibility and reduced toxicity of FDCA-based polymers position them as suitable materials for medical devices, offering a greener option across applications, from implants to drug delivery systems [155].

Looking ahead, ongoing research and development in the synthesis of FDCA-based polymers signal a transformative future. It is crucial for academia, industry, and policymakers to collaborate in order to accelerate the adoption of these innovations. As scalability, cost-effectiveness, and the diversification of FDCA-based polymers continue to evolve, the polymer industry is on the brink of a sustainable and environmentally conscious paradigm shift [156].

While PEF holds promise for the future, its economic viability is a crucial consideration. In 2015, PEF needed to be priced at approximately $1500 per ton to compete with established counterparts [157]. The economic analysis by the same authors suggests a production cost for PEF ranging from $150 to $1750 per ton when derived from wheat, contingent upon factors such as large-scale production, proper pricing for by-products, and maintaining wheat straw prices within a specific range. Notably, these figures have room for improvement, especially with the utilization of more efficient sugar sources, advancements in the conversion into FDCA, and progress in the PEF production process.

In conclusion, PEF shows the potential to be a competitive alternative to petro-based polymers. Although the cost of PEF remains a significant challenge for full market permeation, its exceptional properties position it for increased production and commercialization, particularly in specific niche markets.

8.7 Challenges and Opportunities

8.7.1 Current Challenges in FDCA Production

As indicated before, FDCA can be produced from various bio-based molecules, including LCB, carbohydrates, HMF and MMF. This is primarily achieved through oxidation reactions employing thermocatalytic, biological, electrochemical, and photocatalytic processes. Nonetheless, a few of these routes are in the early stages of development, and there is insufficient data available to evaluate their feasibility.

Efforts have been dedicated to mitigating environmental impact; however, the conventional methodology for oxidative transforming HMF to FDCA in the industry, still relies on substantial amounts of potent chemical oxidants. This traditional approach involves expensive catalysts and production methods, resulting in multiple oxidation products and side-products. Harsh conditions, including high O_2 pressure and high temperatures, are often employed. Developing an economical and efficient catalytic system for the mild oxidation of HMF remains a challenge pursued by both industry and researchers [53, 158–160].

Directly producing FDCA from carbohydrates derived from sugar, one-pot conversion, remains a challenging yet possible task. Multifunctional catalysts capable of conducting this conversion in a one-pot reaction with acidic, basic, and metal sites have been proposed [53, 161]. Consecutive carbohydrate oxidation poses a major challenge in the direct conversion of sugar to FDCA; however, this challenge can be partially addressed using polymeric membranes or molecular sieves in biphasic systems, as demonstrated by Kröger et al. [162], and Saha et al. [80]. Despite the limited FDCA yield and implementation challenges associated with this one-pot process, it opens up a new research field for FDCA production.

It is noteworthy that MMF, produced through the etherification of fructose or HMF in methanol, represents a thermally stable molecule suitable for various applications [163, 164]. However, side reactions, such as the formation of HMFCA, hindered the desired product yield and consumed a significant portion of the feedstock. Intriguingly, FDCA from MMF can also be produced through biological routes, although comprehensive details are limited, and further studies are needed [165].

Talking now about the thermocatalytic oxidation processes, they often occur at high temperatures and high O_2 pressures. Additionally, catalytic systems based on inexpensive first-row transition metal catalysts do not show as much promise as precious metal catalysts. Consequently, enzymatic routes have gained momentum due to their low-temperature operations. Nevertheless, enzymatic catalysts face challenges such as the high cost of genetic engineering for synthesis, lower activity, and lower stability and recyclability. Photocatalytic processes may present a promising route for FDCA production. However, detailed exploration of the photocatalytic production of FDCA is still pending and may become an intriguing area of research [113].

Nevertheless, one of the most important challenges, is that the economical production of FDCA must address several technological issues such as the instability of HMF caused by undesirable condensation reactions at moderate temperatures, especially in the presence of an acid and/or a base, the low solubility of FDCA in commonly used solvents, and the incomplete oxidation of HMF to FDCA, resulting in small quantities of partially oxidized intermediates, FFCA and HMFCA, which end the chain growth during the downstream polymerization process, leading to a poor-quality polymer [166]. These problems make the production and separation of FDCA, followed by the preparation of PEF, remain difficult.

8.7.2 Opportunities for Further Advancements

One of the major challenges, which also presents a potential opportunity for the future, is the synthesis of humins. The synthesis of humins poses a challenge in commercial HMF and FDCA production, leading to the formation of tar-like, insoluble by-products during bioprocessing. Humins, considered complex and poorly understood, result from condensation reactions between sugars and intermediates in the production of platform chemicals like HMF [167].

As mentioned earlier, the formation of humins poses a challenge in obtaining sugar-derived high-value molecules. The elemental composition of humins typically consists of approximately 50–60 wt% carbon, 4–5 wt% hydrogen, and 30–40 wt% oxygen [168, 169]. While optimizing reaction conditions or selecting appropriate solvents can mitigate humins' formation, it remains inevitable as a by-product in acid-catalysed valorisation steps. Researchers have proposed various formation mechanisms and molecular models for humins, yet their complex and intricate structures pose challenges to valorisation strategies.

Despite these challenges and the fact that humins have been usually discarded as waste, there are promising opportunities to devise innovative approaches for the valorization of humins by-products. These may involve creating bio-based composites for construction materials, enhancing the renewable carbon content and reducing costs. Also in construction industry, humins can also serve as an alternative for wood modification, showing similar properties to current resins used for this purpose. Another application for these 'residues' could be humins gasification, producing a graphitic carbon by-product and a volatile organic gas mixture suitable for syngas or hydrogen production. The humins could also be employed in the design of nanocomposites for diverse applications, production of fine chemicals (including active pharmaceutical ingredients (APIs)), and the development of opto-electronic materials. Additionally, humins have been explored in catalysis, where they contribute to the development of 'green' catalysts for selective oxidative hydrocarbon cleavage [170, 171].

These applications demonstrate the potential economic and environmental benefits of utilizing humin by-products in various industrial processes [167, 170, 172, 173].

One of the above-mentioned use for humins is their conversion into fine chemicals, which is where MMF falls into. Converting humins into fine chemicals is viable because a considerable portion of humins consists of soluble agglomerates of oligomeric species rather than being entirely composed of macromolecules [174]. One of those fine chemicals could be MMF, as reported by Pfab et al. [175]. In this study, leveraging the positive outcomes observed in literature studies on HMF hydrogenation, the researchers employed 5% Ru/C and 5% Pd/C catalysts in continuous flow. Both catalysts demonstrated exceptional conversion of MMF into hydrogenation products (100%), showcasing an interesting selectivity switch depending on reaction conditions. It was noted that Ru-based catalysts exhibited higher susceptibility to deactivation compared to Pd catalysts; nevertheless, the latter maintained remarkable stability after a few h under the investigated reaction conditions. This study demonstrates that, despite humins formation is generally a drawback in the formation of MMF, which is one of the molecule studied in this research, could also provide another pathway for its synthesis.

Other major challenge in the process, as mentioned before, is the necessary separation and purification of FDCA from the reaction medium to subsequently obtain PEF. Due to its poor solubility in standard solvents and high boiling point (420 °C), FDCA is notoriously difficult to purify using conventional techniques such as recrystallization and distillation [176].

The most studied and therefore most common method to obtain FDCA is through heterogeneous catalysts. As a result, the most common separation method involves phase separation, which allows for obtaining FDCA dissolved in the medium, by adjusting its pH to basic conditions through the addition of an alkaline salt [177] or another solvent such as DMSO [178]. After separating the liquid from the solid catalyst by centrifugation, FDCA is precipitated by adjusting the pH of the solution to acidic conditions, typically achieved by adding an acidic solution.

8.8 Conclusions and Future Outlook

HMF serves as an intriguing platform molecule due to its potential for producing a diverse array of chemicals and biofuels. HMF is synthesized from aqueous sugar solutions, predominantly hexoses, through isomerisation reactions followed by a triple dehydration process.

Various catalysts, both homogeneous and heterogeneous, are employed to obtain HMF from glucose. However, homogeneous catalysts come with inherent drawbacks, prompting the exploration of heterogeneous catalytic systems. These systems are preferred for their economic advantages and environmentally friendly processes.

Currently, HMF production occurs in heterogeneous systems and two-phase systems utilizing different solvents. To prevent the degradation of formed HMF, it is essential to transfer HMF from the aqueous to the organic phase. Various solvents like MIBK, THF, GVL, DMC, etc., are employed in this process. Additionally, the use of salts facilitates the transfer of HMF from the aqueous to the organic phase. The catalysts employed in this process are acidic and must encompass both Lewis and Brönsted acid centres. An alternative under study, as opposed to two-phase systems, is the utilization of ionic liquids. These liquids possess characteristics that can enhance the overall performance of the process, making them a focal point in research.

The oxidation of HMF to FDCA can be accomplished through homogeneous and heterogeneous catalysts, enzymes, and electrolytic oxidation. However, the use of homogeneous catalysts poses challenges due to the need for separation and recovery from the reaction medium. Often accompanied by inorganic acids, this requires specialized and costly systems, making the reaction setup more expensive and leading to environmental concerns. Notably, Co/Mn/Br salts have demonstrated yields exceeding 95% in FDCA.

Enzymatic catalysts face challenges related to the separation of the reaction medium, continuous feed of carbon sources, and low productivity. Nevertheless, lipase has achieved yields greater than 94% in 60 h.

Presently, attention is shifting towards employing redox reactions, specifically electrochemical oxidation, for FDCA production from HMF. This type of reaction allows for working with heterogeneous non-noble metals like those based on MnO_2 or Ni/MOF/Ag in electro-catalysis. Utilizing non-noble catalysts is advantageous, although a primary drawback lies in the recovery of the produced FDCA. In this system, H_2 is also generated as a by-product with high added value.

The most cost-effective and environmentally friendly method for producing FDCA from HMF involves heterogeneous systems. Noble catalysts, including Pt, Au, Pd, and Ru, have significant impact in this process. Pt catalysts, particularly when supported on C and combined with metals like Bi or Pb to prevent leaching and oxidation, exhibit high activity, achieving yields of 99% at room temperatures. Salts are commonly employed in reactions to enhance FDCA yield. Au and Pd catalysts are also active, albeit requiring higher temperatures, around 90 °C, to achieve comparable yields. Pd/C system achieved a yield of 90%, while Au catalysts supported on TiO_2 or CeO_2 achieved a yield of 99%. Ru, despite its lower cost compared to other metals, shows promising results but falls short of the performance achieved by previous catalysts, even at higher temperatures.

In the heterogeneous catalyst system for the oxidation of HMF to FDCA, addressing challenges involves utilizing non-noble catalysts, implementing continuous reaction systems, comprehending catalyst behaviour, and finding ways to exclude salts from the system. Promising catalytic systems based on Mn, Co-Mn, and Fe, as well as Pt/nNiO and Au/HT, have been identified. Some achieve yields of 100% and 99% without the need for salts. These results hold promise for the development of continuous systems.

In the context of the circular economy, the production of biopolymers is essential to address issues associated with conventional plastics. PEF, which surpasses PET in properties for use in beverages and foods, represents a promising field for exploration. Although the polymerization reactions of FDCA and EG to obtain PEF are relatively straightforward, optimization is necessary to enhance the characteristics of the resulting PEF. In this context, HMF plays an important role as a raw material for production of bio-based monomers and polymers.

References

1. Statistical review of world energy 2023. 2023.
2. Ritchie H, Rosado P, Roser M. Energy production and consumption. Our World Data. 2023; https://ourworldindata.org/energy-production-consumption. Accessed 30 Jan 2024.
3. Weerathunga H, Sarina S, Zhu HY, Waclawik ER. Oxidative esterification of 5-hydroxymethylfurfural into dimethyl 2,5-furandicarboxylate using gamma alumina-supported gold nanoparticles. ACS Omega. 2021;6:4740–8. https://doi.org/10.1021/acsomega.0c05541.
4. Sajid M, Zhao X, Liu D. Production of 2,5-furandicarboxylic acid (FDCA) from 5-hydroxymethylfurfural (HMF): recent progress focusing on the chemical-catalytic routes. Green Chem. 2018;20:5427–53. https://doi.org/10.1039/c8gc02680g.
5. Sardon H, Mecerreyes D, Basterretxea A, Avérous L, Jehanno C. From lab to market: current strategies for the production of biobased polyols. ACS Sustain Chem Eng. 2021;9:10664–77. https://doi.org/10.1021/acssuschemeng.1c02361.
6. Rao M, Wang M, Zheng X, Pan D, Hong C, Chen K, Lin L. One-pot preparation of CuOx-CoOy heterostructure decorated on graphene for highly efficient oxidation of 5-hydroxymethylfurfural (5-HMF) to 2,5-furandicarboxylic acid (FDCA). Chem Eng Sci. 2022;262:118004. https://doi.org/10.1016/j.ces.2022.118004.
7. Delparish A, Uslu A, Cao Y, De Groot T, Van Der Schaaf J, Noël T, d'Angelo MFN. Boosting the valorization of biomass and green electrons to chemical building blocks: a study on the kinetics and mass transfer during the electrochemical conversion of HMF to FDCA in a microreactor. Chem Eng J. 2022;438:135393. https://doi.org/10.1016/j.cej.2022.135393.
8. Yang H, Zhou F, Chen H, Li J, Ma H, Chen K, Lu X, Ouyang P, Fu J. Highly efficient production of 5-methoxymethylfurfural from fructose in dimethyl sulfoxide/amberlyst-15 catalytic system. Ind Eng Chem Res. 2020;59:4905–11. https://doi.org/10.1021/acs.iecr.9b06392.

9. Díaz-Maizkurrena P, Requies JM, Iriondo A, Arias PL, Mariscal R. 5-Methoxymethyl furfural production by acid heterogeneous catalytic etherification of 5-hydroxymethyl furfural. Catal Today. 2024;426:114374. https://doi.org/10.1016/j.cattod.2023.114374.
10. Mittal N, Kaur M, Singh V. Mild and selective catalytic oxidation of 5-HMF to 2,5-FDCA over metal loaded biomass derived graphene oxide using hydrogen peroxide in aqueous media. Mol Catal. 2023;546:113223. https://doi.org/10.1016/j.mcat.2023.113223.
11. Narisetty V, Cox R, Bommareddy R, Agrawal D, Ahmad E, Pant KK, Chandel AK, Bhatia SK, Kumar D, Binod P, Gupta VK, Kumar V. Valorisation of xylose to renewable fuels and chemicals, an essential step in augmenting the commercial viability of lignocellulosic biorefineries. Sustain Energy Fuels. 2022;6:29–65. https://doi.org/10.1039/d1se00927c.
12. Requies J, Agirre I, Iriondo A. Production of furanic biofuels with zeolite and metal oxide bifunctional catalysts for energy- and product-driven biorefineries. In: Production of biofuels and chemicals with bifunctional catalysts. Singapore: Springer; 2017. p. 239–71.
13. Oregui-Bengoechea M, Agirre I, Iriondo A, Lopez-Urionabarrenechea A, Requies JM, Agirrezabal-Telleria I, Bizkarra K, Barrio VL, Cambra JF. Heterogeneous catalyzed thermochemical conversion of lignin model compounds: an overview. Top Curr Chem. 2019;377:36. https://doi.org/10.1007/s41061-019-0260-5.
14. He ZJ, Chen K, Liu ZH, Li BZ, Yuan YJ. Valorizing renewable cellulose from lignocellulosic biomass toward functional products. J Clean Prod. 2023;414:137708. https://doi.org/10.1016/j.jclepro.2023.137708.
15. Zhou L, Yang X, Xu J, Shi M, Wang F, Chen C, Xu J. Depolymerization of cellulose to glucose by oxidation–hydrolysis. Green Chem. 2015;17:1519–24. https://doi.org/10.1039/C4GC02151G.
16. Paksung N, Pfersich J, Arauzo PJ, Jung D, Kruse A. Structural effects of cellulose on hydrolysis and carbonization behavior during hydrothermal treatment. ACS Omega. 2020;5:12210–23. https://doi.org/10.1021/acsomega.0c00737.
17. Deng W, Feng Y, Fu J, Guo H, Guo Y, Han B, Jiang Z, Kong L, Li C, Liu H, Nguyen PTT, Ren P, Wang F, Wang S, Wang Y, Wang Y, Wong SS, Yan K, Yan N, Yang X, Zhang Y, Zhang Z, Zeng X, Zhou H. Catalytic conversion of lignocellulosic biomass into chemicals and fuels. Green Energy Environ. 2023;8:10–114. https://doi.org/10.1016/j.gee.2022.07.003.
18. Kohli K, Prajapati R, Sharma BK. Bio-based chemicals from renewable biomass for integrated biorefineries. Energies. 2019;12:1–40. https://doi.org/10.3390/en12020233.
19. Eblagon KM, Malaika A, Ptaszyńska K, Pereira MFR, Kozłowski M, Figueiredo JL. Niobium oxide-phosphorylated carbon xerogel composites as solid acid catalysts for cascade conversion of glucose to 5-hydroxymethylfurfural (HMF) in pure water. Catal Today. 2023;418:114070. https://doi.org/10.1016/j.cattod.2023.114070.
20. Papanikolaou G, Chillè D, Perathoner S, Centi G, Migliori M, Giordano G, Lanzafame P. Use of zeolites in green chemicals and bio-fuel production via HMF valorisation. Microporous Mesoporous Mater. 2023;358:112330. https://doi.org/10.1016/j.micromeso.2022.112330.
21. Guo X, Si Y, Huang Y, Zhang J, Lyu X, Cheng Y, Li X. Base-free oxidative esterification of HMF using Mg-Al hydrotalcite-supported gold catalyst. Ind Eng Chem Res. 2023;62:16309–18. https://doi.org/10.1021/acs.iecr.3c02599.
22. Chen S, Wojcieszak R, Dumeignil F, Marceau E, Royer S. How catalysts and experimental conditions determine the selective hydroconversion of furfural and 5-hydroxymethylfurfural. Chem Rev. 2018;118:11023–117. https://doi.org/10.1021/acs.chemrev.8b00134.
23. Iriondo A, Agirre I, Viar N, Requies J. Value-added bio-chemicals commodities from catalytic conversion of biomass derived furan-compounds. Catalysts. 2020;10:1–25. https://doi.org/10.3390/catal10080895.
24. Souzanchi S, Nazari L, Venkateswara Rao KT, Yuan Z, Tan Z, Charles Xu C. Catalytic dehydration of glucose to 5-HMF using heterogeneous solid catalysts in a biphasic continuous-flow tubular reactor. J Ind Eng Chem. 2021;101:214–26. https://doi.org/10.1016/j.jiec.2021.06.010.

25. Stegmann P, Gerritse T, Shen L, Londo M, Puente Á, Junginger M. The global warming potential and the material utility of PET and bio-based PEF bottles over multiple recycling trips. J Clean Prod. 2023;395:136426. https://doi.org/10.1016/j.jclepro.2023.136426.
26. Lin G, Lin W, Wu J, Zhan Y, Okejiri F, Weng M, Fu J. Oxidation of 5-methoxymethylfurfural to 2,5-furandicarboxylic acid over Ru/hydroxyapatite catalyst in water. Chem Eng Sci. 2022;249:117343. https://doi.org/10.1016/j.ces.2021.117343.
27. Vaz S. Biomass and green chemistry building a renewable pathway. Springer; 2018. https://doi.org/10.1007/978-3-319-66736-2.
28. Li X, Xu R, Yang J, Nie S, Liu D, Liu Y, Si C. Production of 5-hydroxymethylfurfural and levulinic acid from lignocellulosic biomass and catalytic upgradation. Ind Crop Prod. 2019;130:184–97. https://doi.org/10.1016/j.indcrop.2018.12.082.
29. Liang J, Jiang J, Cai T, Liu C, Ye J, Zeng X, Wang K. Advances in selective conversion of carbohydrates into 5-hydroxymethylfurfural. Green Energy Environ. 2023;9:1384–406. https://doi.org/10.1016/j.gee.2023.11.005.
30. Hou Q, Qi X, Zhen M, Qian H, Nie Y, Bai C, Zhang S, Bai X, Ju M. Biorefinery roadmap based on catalytic production and upgrading 5-hydroxymethylfurfural. Green Chem. 2021;23:119–231. https://doi.org/10.1039/D0GC02770G.
31. Galadima A, Masudi A, Muraza O. Conversion of cellulose to glucose and further transformation into fuels over solid acid catalysts: a mini review. Microporous Mesoporous Mater. 2022;336:111846. https://doi.org/10.1016/j.micromeso.2022.111846.
32. Bai C, Hou Q, Bai X, Nie Y, Qian H, Zhen M, Ju M. Conversion of glucose to 5-hydroxymethylfurfural at high substrate loading: effect of catalyst and solvent on the stability of 5-hydroxymethylfurfural. Energy Fuel. 2020;34:16240–9. https://doi.org/10.1021/acs.energyfuels.0c03078.
33. Hu L, Lin L, Wu Z, Zhou S, Liu S. Chemocatalytic hydrolysis of cellulose into glucose over solid acid catalysts. Appl Catal B Environ. 2015;174–175:225–43. https://doi.org/10.1016/j.apcatb.2015.03.003.
34. Hou Q, Zhen M, Liu L, Chen Y, Huang F, Zhang S, Li W, Ju M. Tin phosphate as a heterogeneous catalyst for efficient dehydration of glucose into 5-hydroxymethylfurfural in ionic liquid. Appl Catal B Environ. 2018;224:183–93. https://doi.org/10.1016/j.apcatb.2017.09.049.
35. Hou Q, Zhen M, Li W, Liu L, Liu J, Zhang S, Nie Y, Bai C, Bai X, Ju M. Efficient catalytic conversion of glucose into 5-hydroxymethylfurfural by aluminum oxide in ionic liquid. Appl Catal B Environ. 2019;253:1–10. https://doi.org/10.1016/j.apcatb.2019.04.003.
36. Nie Y, Hou Q, Bai C, Qian H, Bai X, Ju M. Transformation of carbohydrates to 5-hydroxymethylfurfural with high efficiency by tandem catalysis. J Clean Prod. 2020;274:123023. https://doi.org/10.1016/j.jclepro.2020.123023.
37. Marianou AA, Michailof CM, Pineda A, Iliopoulou EF, Triantafyllidis KS, Lappas AA. Effect of Lewis and Brønsted acidity on glucose conversion to 5-HMF and lactic acid in aqueous and organic media. Appl Catal A Gen. 2018;555:75–87. https://doi.org/10.1016/j.apcata.2018.01.029.
38. Wang S, Zhang M, Guo D, Feng J, Pan H. Construction of Lewis-Brønsted bifunctional carbonaceous acidic catalyst for efficient transformation of glucose into 5-HMF. Fuel. 2024;363:130991. https://doi.org/10.1016/j.fuel.2024.130991.
39. Wang S, Eberhardt TL, Guo D, Feng J, Pan H. Efficient conversion of glucose into 5-HMF catalyzed by lignin-derived mesoporous carbon solid acid in a biphasic system. Renew Energy. 2022;190:1–10. https://doi.org/10.1016/j.renene.2022.03.021.
40. Hou Q, Li W, Zhen M, Liu L, Chen Y, Yang Q, Huang F, Zhang S, Ju M. An ionic liquid–organic solvent biphasic system for efficient production of 5-hydroxymethylfurfural from carbohydrates at high concentrations. RSC Adv. 2017;7:47288–96. https://doi.org/10.1039/C7RA10237B.

41. Shao Y, Ding Y, Dai J, Long Y, Hu ZT. Synthesis of 5-hydroxymethylfurfural from dehydration of biomass-derived glucose and fructose using supported metal catalysts. Green Synth Catal. 2021;2:187–97. https://doi.org/10.1016/j.gresc.2021.01.006.
42. Portillo-Pérez G, Mukherjee A, Dumont MJ. Insights into HMF catalysis. J Ind Eng Chem. 2019;70:1–34. https://doi.org/10.1016/j.jiec.2018.10.002.
43. Gao Z, Li C, Fan G, Yang L, Li F. Nitrogen-doped carbon-decorated copper catalyst for highly efficient transfer hydrogenolysis of 5-hydroxymethylfurfural to convertibly produce 2,5-dimethylfuran or 2,5-dimethyltetrahydrofuran. Appl Catal B Environ. 2018;226:523–33. https://doi.org/10.1016/j.apcatb.2018.01.006.
44. Zhang J, Li DN, Yuan HR, Wang SR, Chen Y. Advances on the catalytic hydrogenation of biomass-derived furfural and 5-hydroxymethylfurfural. Ranliao Huaxue Xuebao/J Fuel Chem Technol. 2021;49:1752–67. https://doi.org/10.1016/S1872-5813(21)60135-4.
45. Jing Y, Guo Y, Xia Q, Liu X, Wang Y. Catalytic production of value-added chemicals and liquid fuels from lignocellulosic biomass. Chem. 2019;5:2520–46. https://doi.org/10.1016/j.chempr.2019.05.022.
46. Świątek K, Olszewski MP, Kruse A. Continuous synthesis of 5-hydroxymethylfurfural from biomass in on-farm biorefinery. GCB Bioenergy. 2022;14:681–93. https://doi.org/10.1111/gcbb.12938.
47. Partenheimer W, Grushin VV. Synthesis of 2,5-diformylfuran and furan-2,5-dicarboxylic acid by catalytic air-oxidation of 5-hydroxymethylfurfural. Unexpectedly selective aerobic oxidation of benzyl alcohol to benzaldehyde with metal/bromide catalysts. Adv Synth Catal. 2001:102–11. https://doi.org/10.1002/1615-4169(20010129)343:1<102::AID-ADSC102>3.0.CO;2-Q.
48. Saha B, Dutta S, Abu-Omar MM. Aerobic oxidation of 5-hydroxylmethylfurfural with homogeneous and nanoparticulate catalysts. Catal Sci Technol. 2012;2:79–81. https://doi.org/10.1039/C1CY00321F.
49. Zuo X, Venkitasubramanian P, Busch DH, Subramaniam B. Optimization of Co/Mn/Br-catalyzed oxidation of 5-hydroxymethylfurfural to enhance 2,5-furandicarboxylic acid yield and minimize substrate burning. ACS Sustain Chem Eng. 2016;4:3659–68. https://doi.org/10.1021/acssuschemeng.6b00174.
50. Zuo X, Chaudhari AS, Snavely K, Niu F, Zhu H, Martin KJ, Subramaniam B. Kinetics of homogeneous 5-hydroxymethylfurfural oxidation to 2,5-furandicarboxylic acid with Co/Mn/Br catalyst. AICHE J. 2017;63:162–71. https://doi.org/10.1002/aic.15497.
51. Hommes A, Disselhorst B, Janssens HMM, Stevelink RJA, Heeres HJ, Yue J. Mass transfer and reaction characteristics of homogeneously catalyzed aerobic oxidation of 5-hydroxymethylfurfural in slug flow microreactors. Chem Eng J. 2021;413:127552. https://doi.org/10.1016/j.cej.2020.127552.
52. Cong H, Yuan H, Tao Z, Bao H, Zhang Z, Jiang Y, Huang D, Liu H, Wang T. Recent advances in catalytic conversion of biomass to 2,5-furandicarboxylic acid. Catalysts. 2021;11:1113. https://doi.org/10.3390/catal11091113.
53. Zhang Z, Deng K. Recent advances in the catalytic synthesis of 2,5-furandicarboxylic acid and its derivatives. ACS Catal. 2015;5:6529–44. https://doi.org/10.1021/acscatal.5b01491.
54. Davis SE, Houk LR, Tamargo EC, Datye AK, Davis RJ. Oxidation of 5-hydroxymethylfurfural over supported Pt, Pd and Au catalysts. Catal Today. 2011;160:55–60. https://doi.org/10.1016/j.cattod.2010.06.004.
55. Chen C, Wang L, Zhu B, Zhou Z, El-Hout SI, Yang J, Zhang J. 2,5-Furandicarboxylic acid production via catalytic oxidation of 5-hydroxymethylfurfural: catalysts, processes and reaction mechanism. J Energy Chem. 2021;54:528–54. https://doi.org/10.1016/j.jechem.2020.05.068.
56. Carro J, Ferreira P, Rodríguez L, Prieto A, Serrano A, Balcells B, Ardá A, Jiménez-Barbero J, Gutiérrez A, Ullrich R, Hofrichter M, Martínez AT. 5-Hydroxymethylfurfural conversion by fungal aryl-alcohol oxidase and unspecific peroxygenase. FEBS J. 2015;282:3218–29. https://doi.org/10.1111/febs.13177.

57. Verdeguer P, Merat N, Gaset A. Oxydation catalytique du HMF en acide 2,5-furane dicarboxylique. J Mol Catal. 1993;85:327–44. https://doi.org/10.1016/0304-5102(93)80059-4.
58. Ait Rass H, Essayem N, Besson M. Selective aerobic oxidation of 5-HMF into 2,5-furandicarboxylic acid with Pt catalysts supported on TiO_2- and ZrO_2-based supports. ChemSusChem. 2015;8:1206–17. https://doi.org/10.1002/cssc.201403390.
59. Ait Rass H, Essayem N, Besson M. Selective aqueous phase oxidation of 5-hydroxymethylfurfural to 2,5-furandicarboxylic acid over Pt/C catalysts: influence of the base and effect of bismuth promotion. Green Chem. 2013;15:2240. https://doi.org/10.1039/c3gc40727f.
60. Sahu R, Dhepe PL. Synthesis of 2,5-furandicarboxylic acid by the aerobic oxidation of 5-hydroxymethyl furfural over supported metal catalysts. React Kinet Mech Catal. 2014;112:173–87. https://doi.org/10.1007/s11144-014-0689-z.
61. Miao Z, Wu T, Li J, Yi T, Zhang Y, Yang X. Aerobic oxidation of 5-hydroxymethylfurfural (HMF) effectively catalyzed by a $Ce_{0.8}Bi_{0.2}O_{2-\delta}$ supported Pt catalyst at room temperature. RSC Adv. 2015;5:19823–9. https://doi.org/10.1039/c4ra16968a.
62. Liu X, Chen L, Xu H, Jiang S, Zhou Y, Wang J. Straightforward synthesis of beta zeolite encapsulated Pt nanoparticles for the transformation of 5-hydroxymethyl furfural into 2,5-furandicarboxylic acid. Chin J Catal. 2021;42:994–1003. https://doi.org/10.1016/S1872-2067(20)63720-2.
63. Zhang H, Gao T, Cao Q, Fang W. Tailoring the reactive oxygen species in mesoporous NiO for selectivity-controlled aerobic oxidation of 5-hydroxymethylfurfural on a loaded Pt catalyst. ACS Sustain Chem Eng. 2021;9:6056–67. https://doi.org/10.1021/acssuschemeng.1c01398.
64. Casanova O, Iborra S, Corma A. Biomass into chemicals: aerobic oxidation of 5-hydroxymethyl-2-furfural into 2,5-furandicarboxylic acid with gold nanoparticle catalysts. ChemSusChem. 2009;2:1138–44. https://doi.org/10.1002/cssc.200900137.
65. Cai J, Ma H, Zhang J, Song Q, Du Z, Huang Y, Xu J. Gold nanoclusters confined in a supercage of Y zeolite for aerobic oxidation of HMF under mild conditions. Chem Eur J. 2013;19:14215–23. https://doi.org/10.1002/chem.201301735.
66. Miao Z, Zhang Y, Pan X, Wu T, Zhang B, Li J, Yi T, Zhang Z, Yang X. Superior catalytic performance of $Ce_{1-x}Bi_xO_{2-\delta}$ solid solution and Au/$Ce_{1-x}Bi_xO_{2-\delta}$ for 5-hydroxymethylfurfural conversion in alkaline aqueous solution. Catal Sci Technol. 2015;5:1314–22. https://doi.org/10.1039/c4cy01060d.
67. Megías-Sayago C, Lolli A, Ivanova S, Albonetti S, Cavani F, Odriozola JA. Au/Al_2O_3—efficient catalyst for 5-hydroxymethylfurfural oxidation to 2,5-furandicarboxylic acid. Catal Today. 2019;333:169–75. https://doi.org/10.1016/j.cattod.2018.04.024.
68. Saxena S, Yadaw D, Tiwari K, Venkatesh V, Verma S, Pala RGS, Sivakumar S. Hollow zeolites encapsulating ultra-low noble metal nanoparticles for HMF oxidation. Catal Lett. 2021;151:921–31. https://doi.org/10.1007/s10562-020-03361-2.
69. Gupta NK, Nishimura S, Takagaki A, Ebitani K. Hydrotalcite-supported gold-nanoparticle-catalyzed highly efficient base-free aqueous oxidation of 5-hydroxymethylfurfural into 2,5-furandicarboxylic acid under atmospheric oxygen pressure. Green Chem. 2011;13:824–7. https://doi.org/10.1039/c0gc00911c.
70. Ardemani L, Cibin G, Dent AJ, Isaacs MA, Kyriakou G, Lee AF, Parlett CMA, Parry SA, Wilson K. Solid base catalysed 5-HMF oxidation to 2,5-FDCA over Au/hydrotalcites: fact or fiction? Chem Sci. 2015;6:4940–5. https://doi.org/10.1039/c5sc00854a.
71. Siyo B, Schneider M, Radnik J, Pohl MM, Langer P, Steinfeldt N. Influence of support on the aerobic oxidation of HMF into FDCA over preformed Pd nanoparticle based materials. Appl Catal A Gen. 2014;478:107–16. https://doi.org/10.1016/j.apcata.2014.03.020.
72. Zheng L, Zhao J, Du Z, Zong B, Liu H. Efficient aerobic oxidation of 5-hydroxymethylfurfural to 2,5-furandicarboxylic acid on Ru/C catalysts. Sci China Chem. 2017;60:950–7. https://doi.org/10.1007/s11426-016-0489-3.

73. Tang W, Wu X, Li D, Wang Z, Liu G, Liu H, Chen Y. Oxalate route for promoting activity of manganese oxide catalysts in total VOCs' oxidation: effect of calcination temperature and preparation method. J Mater Chem A. 2014;2:2544–54. https://doi.org/10.1039/c3ta13847j.
74. Neațu F, Marin RS, Florea M, Petrea N, Pavel OD, Pârvulescu VI. Selective oxidation of 5-hydroxymethyl furfural over non-precious metal heterogeneous catalysts. Appl Catal B Environ. 2016;180:751–7. https://doi.org/10.1016/j.apcatb.2015.07.043.
75. Hayashi E, Komanoya T, Kamata K, Hara M. Heterogeneously-catalyzed aerobic oxidation of 5-hydroxymethylfurfural to 2,5-furandicarboxylic acid with MnO_2. ChemSusChem. 2017;10: 654–8. https://doi.org/10.1002/cssc.201601443.
76. Hayashi E, Yamaguchi Y, Kamata K, Tsunoda N, Kumagai Y, Oba F, Hara M. Effect of MnO_2 crystal structure on aerobic oxidation of 5-hydroxymethylfurfural to 2,5-furandicarboxylic acid. J Am Chem Soc. 2019;141:899–900. https://doi.org/10.1021/jacs.8b09917.
77. Gawade AB, Nakhate AV, Yadav GD. Selective synthesis of 2,5-furandicarboxylic acid by oxidation of 5-hydroxymethylfurfural over $MnFe_2O_4$ catalyst. Catal Today. 2018;309:119–25. https://doi.org/10.1016/j.cattod.2017.08.061.
78. Rao KTV, Rogers JL, Souzanchi S, Dessbesell L, Ray MB, Charles Xu C. Inexpensive but highly efficient Co–Mn mixed-oxide catalysts for selective oxidation of 5-hydroxymethylfurfural to 2,5-furandicarboxylic acid. ChemSusChem. 2018;11:3323–34. https://doi.org/10.1002/cssc.201800989.
79. Zhang S, Sun X, Zheng Z, Zhang L. Nanoscale center-hollowed hexagon $MnCo_2O_4$ spinel catalyzed aerobic oxidation of 5-hydroxymethylfurfural to 2,5-furandicarboxylic acid. Catal Commun. 2018;113:19–22. https://doi.org/10.1016/j.catcom.2018.05.004.
80. Saha B, Gupta D, Abu-Omar MM, Modak A, Bhaumik A. Porphyrin-based porous organic polymer-supported iron(III) catalyst for efficient aerobic oxidation of 5-hydroxymethyl-furfural into 2,5-furandicarboxylic acid. J Catal. 2013;299:316–20. https://doi.org/10.1016/j.jcat.2012.12.024.
81. Zhao D, Su T, Wang Y, Varma RS, Len C. Recent advances in catalytic oxidation of 5-hydroxymethylfurfural. Mol Catal. 2020;495:111133. https://doi.org/10.1016/j.mcat.2020.111133.
82. Martínez-Vargas DX, Rivera De La Rosa J, Sandoval-Rangel L, Guzmán-Mar JL, Garza-Navarro MA, Lucio-Ortiz CJ, De Haro-Del Río DA. 5-Hydroxymethylfurfural catalytic oxidation under mild conditions by Co(II), Fe(III) and Cu(II) Salen complexes supported on SBA-15: synthesis, characterization and activity. Appl Catal A Gen. 2017;547:132–45. https://doi.org/10.1016/j.apcata.2017.08.035.
83. Wang S, Zhang Z, Liu B. Catalytic conversion of fructose and 5-hydroxymethylfurfural into 2,5-furandicarboxylic acid over a recyclable Fe_3O_4-CoO_x magnetite nanocatalyst. ACS Sustain Chem Eng. 2015;3:406–12. https://doi.org/10.1021/sc500702q.
84. Troiano D, Orsat V, Dumont M-J. Status of biocatalysis in the production of 2,5-furandicarboxylic acid. ACS Catal. 2020;10:9145–69. https://doi.org/10.1021/acscatal.0c02378.
85. Klatte S, Lorenz E, Wendisch VF. Whole cell biotransformation for reductive amination reactions. Bioengineered. 2013;5:56–62. https://doi.org/10.4161/bioe.27151.
86. Koopman F, Wierckx N, De Winde JH, Ruijssenaars HJ. Efficient whole-cell biotransformation of 5-(hydroxymethyl)furfural into FDCA, 2,5-furandicarboxylic acid. Bioresour Technol. 2010;101:6291–6. https://doi.org/10.1016/j.biortech.2010.03.050.
87. Godan TK, Rajesh RO, Loreni PC, Kumar Rai A, Sahoo D, Pandey A, Binod P. Biotransformation of 5-hydroxymethylfurfural by Acinetobacter oleivorans S27 for the synthesis of furan derivatives. Bioresour Technol. 2019;282:88–93. https://doi.org/10.1016/j.biortech.2019.02.125.
88. Hsu C-T, Kuo Y-C, Liu Y-C, Tsai S-L. Green conversion of 5-hydroxymethylfurfural to furan-2,5-dicarboxylic acid by heterogeneous expression of 5-hydroxymethylfurfural oxidase in Pseudomonas putida S12. Microb Biotechnol. 2020;13:1094–102. https://doi.org/10.1111/1751-7915.13564.

89. Tan H, Zhou F, Liao D, Ouyang J, Zheng Z. Improved biosynthesis of 2,5-Furandicarboxylic acid through coupling of heterologous pathways in Escherichia coli and native pathways in Pseudomonas putida. Biochem Eng J. 2020;161:107657. https://doi.org/10.1016/j.bej.2020.107657.
90. Sheng Y, Tan X, Zhou X, Xu Y. Bioconversion of 5-hydroxymethylfurfural (HMF) to 2,5-furandicarboxylic acid (FDCA) by a native obligate aerobic bacterium, Acinetobacter calcoaceticus NL14. Appl Biochem Biotechnol. 2020;192:455–65. https://doi.org/10.1007/s12010-020-03325-7.
91. Sayed M, Gaber Y, Junghus F, Valdés Martín E, Pyo S-H, Hatti-Kaul R. Oxidation of 5-hydroxymethylfurfural with a novel aryl alcohol oxidase from Mycobacterium sp. MS1601. Microb Biotechnol. 2022;15:2176–90. https://doi.org/10.1111/1751-7915.14052.
92. Kroutil W, Mang H, Edegger K, Faber K. Biocatalytic oxidation of primary and secondary alcohols. Adv Synth Catal. 2004;346:125–42. https://doi.org/10.1002/adsc.200303177.
93. Dong J, Fernández-Fueyo E, Hollmann F, Paul CE, Pesic M, Schmidt S, Wang Y, Younes S, Zhang W. Biocatalytic oxidation reactions: a chemist's perspective. Angew Chem Int Ed. 2018;57:9238–61. https://doi.org/10.1002/anie.201800343.
94. Wang K-F, Liu C, Sui K, Guo C, Liu C-Z. Efficient catalytic oxidation of 5-hydroxymethylfurfural to 2,5-furandicarboxylic acid by magnetic laccase catalyst. Chembiochem. 2018;19:654–9. https://doi.org/10.1002/cbic.201800008.
95. Dijkman WP, Groothuis DE, Fraaije MW. Enzyme-catalyzed oxidation of 5-hydroxymethylfurfural to furan-2,5-dicarboxylic acid. Angew Chem. 2014;53:6515–8. https://doi.org/10.1002/anie.201402904.
96. Krystof M, Pérez-Sánchez M, Domínguez de María P. Lipase-mediated selective oxidation of furfural and 5-hydroxymethylfurfural. ChemSusChem. 2013;6:826–30. https://doi.org/10.1002/cssc.201200954.
97. Rudroff F, Mihovilovic MD, Gröger H, Snajdrova R, Iding H, Bornscheuer UT. Opportunities and challenges for combining chemo- and biocatalysis. Nat Catal. 2018;1:12–22. https://doi.org/10.1038/s41929-017-0010-4.
98. Chang X, Zhang C, Gao L, Liu X, You S, Qi W, Wang K, Guo X, Su R, Lu H, He Z. Tandem biocatalysis by CotA-TJ102@UIO-66-NH2 and Novozym 435 for highly selective transformation of HMF into FDCA. Trans Tianjin Univ. 2019;25:488–96. https://doi.org/10.1007/s12209-019-00215-w.
99. Wu S, Liu Q, Tan H, Zhang F, Yin H. A novel 2,5-furandicarboxylic acid biosynthesis route from biomass-derived 5-hydroxymethylfurfural based on the consecutive enzyme reactions. Appl Biochem Biotechnol. 2020;191:1470–82. https://doi.org/10.1007/s12010-020-03290-1.
100. Jia H-Y, Zong M-H, Zheng G-W, Li N. One-pot enzyme cascade for controlled synthesis of furancarboxylic acids from 5-hydroxymethylfurfural by H_2O_2 internal recycling. ChemSusChem. 2019;12:4764–8. https://doi.org/10.1002/cssc.201902199.
101. McKenna SM, Mines P, Law P, Kovacs-Schreiner K, Birmingham WR, Turner NJ, Leimkühler S, Carnell AJ. The continuous oxidation of HMF to FDCA and the immobilisation and stabilisation of periplasmic aldehyde oxidase (PaoABC). Green Chem. 2017;19:4660–5. https://doi.org/10.1039/C7GC01696D.
102. Serrano A, Calviño E, Carro J, Sánchez-Ruiz MI, Cañada FJ, Martínez AT. Complete oxidation of hydroxymethylfurfural to furandicarboxylic acid by aryl-alcohol oxidase. Biotechnol Biofuels. 2019;12:217. https://doi.org/10.1186/s13068-019-1555-z.
103. Taitt BJ, Nam D-H, Choi K-S. A comparative study of nickel, cobalt, and iron oxyhydroxide anodes for the electrochemical oxidation of 5-hydroxymethylfurfural to 2,5-furandicarboxylic acid. ACS Catal. 2019;9:660–70. https://doi.org/10.1021/acscatal.8b04003.
104. Zhao Y, Cai M, Xian J, Sun Y, Li G. Recent advances in the electrocatalytic synthesis of 2,5-furandicarboxylic acid from 5-(hydroxymethyl)furfural. J Mater Chem A. 2021;9:20164–83. https://doi.org/10.1039/D1TA04981J.

105. Yang Y, Mu T. Electrochemical oxidation of biomass derived 5-hydroxymethylfurfural (HMF): pathway, mechanism, catalysts and coupling reactions. Green Chem. 2021;23: 4228–54. https://doi.org/10.1039/D1GC00914A.
106. Wang C, Bongard HJ, Weidenthaler C, Wu Y, Schüth F. Design and application of a high-surface-area mesoporous δ-MnO_2 electrocatalyst for biomass oxidative valorization. Chem Mater. 2022;34:3123–32. https://doi.org/10.1021/acs.chemmater.1c04223.
107. Chen C, Zhou Z, Liu J, Zhu B, Hu H, Yang Y, Chen G, Gao M, Zhang J. Sustainable biomass upgrading coupled with H_2 generation over in-situ oxidized Co_3O_4 electrocatalysts. Appl Catal B Environ. 2022;307:121209. https://doi.org/10.1016/j.apcatb.2022.121209.
108. Pang X, Zhao H, Huang Y, Luo B, Bai H, Fan W. Electrochemically induced NiOOH/Ag^+ active species for efficient oxidation of 5-hydroxymethylfurfural. Appl Surf Sci. 2023;608: 155152. https://doi.org/10.1016/j.apsusc.2022.155152.
109. Guo L, Zhang X, Gan L, Pan L, Shi C, Huang Z-F, Zhang X, Zou J-J. Advances in selective electrochemical oxidation of 5-hydroxymethylfurfural to produce high-value chemicals. Adv Sci. 2023;10:e2205540. https://doi.org/10.1002/advs.202205540.
110. Xu S, Zhou P, Zhang Z, Yang C, Zhang B, Deng K, Bottle S, Zhu H. Selective oxidation of 5-hydroxymethylfurfural to 2,5-furandicarboxylic acid using O_2 and a photocatalyst of co-thioporphyrazine bonded to g-C_3N_4. J Am Chem Soc. 2017;139:14775–82. https://doi.org/10.1021/jacs.7b08861.
111. Zhu P, Zhang W, Li Q, Xia H. Visible-light-driven photocatalytic oxidation of 5-hydroxymethylfurfural to 2,5-furandicarboxylic acid over plasmonic Au/ZnO catalyst. ACS Sustain Chem Eng. 2022;10:8778–87. https://doi.org/10.1021/acssuschemeng.2c01143.
112. Zhang W, Li Q, Xia H. Photocatalytic oxidation of 5-hydroxymethylfurfural to furandicarboxylic acid over the Au-Ag/TiO_2 catalysts under visible light irradiation. Appl Surf Sci. 2023;613:156036. https://doi.org/10.1016/j.apsusc.2022.156036.
113. Li C, Na Y. Recent advances in photocatalytic oxidation of 5-hydroxymethylfurfural. ChemPhotoChem. 2021;5:502–11. https://doi.org/10.1002/cptc.202000261.
114. Meng Y, Yang S, Li H. Electro- and photocatalytic oxidative upgrading of bio-based 5-hydroxymethylfurfural. ChemSusChem. 2022;15:e202102581. https://doi.org/10.1002/cssc.202102581.
115. Carro J, Fernández-Fueyo E, Fernández-Alonso C, Cañada J, Ullrich R, Hofrichter M, Alcalde M, Ferreira P, Martínez AT. Self-sustained enzymatic cascade for the production of 2,5-furandicarboxylic acid from 5-methoxymethylfurfural. Biotechnol Biofuels. 2018;11:86. https://doi.org/10.1186/s13068-018-1091-2.
116. Muñoz de Diego C, Schammel WP, Dam MA, Gruter GJM, Method for the preparation of 2,5-furandicarboxylic acid and esters thereof. WO 2011/043660 A2. 2011. https://patentscope.wipo.int/search/en/detail.jsf?docId=WO2011043660.
117. Lin G, Fan H, Zhan Y, Chen C, Zhao X, Lin W, Li B, Wang J, Li J, Song W, Fu J. Strongly electron-interacting Ru–Ce pair sites in RuO_x/CeO_2–HAP for efficient oxidation of MMF to FDCA. ACS Catal. 2024;14:1862–73. https://doi.org/10.1021/acscatal.3c04329.
118. Wu J, Xie W, Zhang Y, Ke X, Li T, Fang H, Sun Y, Zeng X, Lin L, Tang X. Oxygen-vacancy-rich MnO supported RuO for efficient base-free oxidation of 5-hydroxymethylfurfural and 5-methoxymethylfurfural to 2,5-furandicarboxylic acid. J Energy Chem. 2024;95:670–83. https://doi.org/10.1016/j.jechem.2024.04.019.
119. Drault F, Snoussi Y, Paul S, Itabaiana I, Wojcieszak R. Recent advances in carboxylation of furoic acid into 2,5-furandicarboxylic acid: pathways towards bio-based polymers. ChemSusChem. 2020;13:5164–72. https://doi.org/10.1002/cssc.202001393.
120. Faba L, Rapado P, Ordóñez S. Carboxylation reactions for integrating CO_2 capture with the production of renewable monomers. Greenhouse Gases. 2023;13:227–44. https://doi.org/10.1002/ghg.2175.
121. Zhang S, Shen G, Deng Y, Lei Y, Xue J-W, Chen Z, Yin G. Efficient synthesis of 2,5-furandicarboxylic acid from furfural based platform through aqueous-phase carbonylation. ACS Sustain Chem Eng. 2018;6:13192–8. https://doi.org/10.1021/acssuschemeng.8b02780.

122. Yin G, Shen G, Shi J, Lei Y, Fu C, Chen Z, Andrioletti B. Aqueous carbonylation of furfural-derived 5-bromofuroic acid to 2,5-furandicarboxylic acid with supported palladium catalyst. Ind Eng Chem Res. 2019;58:22951–7. https://doi.org/10.1021/acs.iecr.9b05134.
123. Nocito F, Ditaranto N, Dibenedetto A. Valorization of C5 polyols by direct carboxylation to FDCA: synthesis and characterization of a key intermediate and role of carbon dioxide. J CO2 Utiliz. 2019;32:170–7. https://doi.org/10.1016/j.jcou.2019.04.013.
124. Kanan MW, Banerjee A, Dick GR, Yoshino T. Carbon dioxide utilization via carbonate-promoted C–H carboxylation. Nature. 2016;531:215–9. https://doi.org/10.1038/nature17185.
125. Kolbe H, Lautemann E. Ueber die constitution und basicität der salicylsäure. Ann Chem Pharm. 1860;115:157–206.
126. Liu Y, Wang X, Hu M, Yu Y. Catalytic conversion of 2,5-furandicarboxylic acid production from hemicellulose. Biomass Convers Bioref. 2023;14:20275–85. https://doi.org/10.1007/s13399-023-04162-4.
127. Wang Y-G, Guo C-Y, Shen J, Sun Y-Q, Niu Y-X, Li P, Liu G, Wei X-Y. A sustainable and green route to furan-2,5-dicarboxylic acid by direct carboxylation of 2-furoic acid and CO_2. J CO2 Utiliz. 2021;48:101524. https://doi.org/10.1016/j.jcou.2021.101524.
128. Han X, Wang Y, Liu G, Wang M, Guo C, Shen J. A sustainable and low-cost route to 2,5-furandicarboxylic acid by carboxylation of biomass-based furoic acid and CO_2. J CO2 Utiliz. 2023;75:102572. https://doi.org/10.1016/j.jcou.2023.102572.
129. Payne KAP, Marshall SA, Fisher K, Cliff MJ, Cannas DM, Yan C, Heyes DJ, Parker DA, Larrosa I, Leys D. Enzymatic carboxylation of 2-furoic acid yields 2,5-furandicarboxylic acid (FDCA). ACS Catal. 2019;9:2854–65. https://doi.org/10.1021/acscatal.8b04862.
130. Kawanabe K, Aono R, Kino K. 2,5-Furandicarboxylic acid production from furfural by sequential biocatalytic reactions. J Biosci Bioeng. 2021;132:18–24. https://doi.org/10.1016/j.jbiosc.2021.03.001.
131. López-Lorenzo X, Asem H, Stamm A, Subramaniyan S, Hakkarainen M, Syrén P. Whole-cell mediated carboxylation of 2-furoic acid towards the production of renewable platform chemicals and biomaterials. ChemCatChem. 2023;15:e202201483. https://doi.org/10.1002/cctc.202201483.
132. Fittig R. Mittheilungen aus dem chemischen Institut der Universität Strassburg. Ber Dtsch Chem Ges. 1876;9:1189–99. https://doi.org/10.1002/cber.18760090250.
133. Haworth WN, Jones WGM, Wiggins LF. The conversion of sucrose into furan compounds. J Chem Soc. 1945;1:1–4.
134. Kuhn R, Dury K. Ringschlüsse mit α,α′-Dioxymuconsäure-estern. Justus Liebigs Ann Chem. 1951;571:44–68. https://doi.org/10.1002/jlac.19515710107.
135. Brătulescu G. Cyclisation under the action of the microwaves of the D-saccaric acid to 2,5-furandicarboxylic acid. Revue Roumaine de Chimie. 2000;45:883–5.
136. Lewkowski J. Synthesis, chemistry and applications of 5-hydroxymethyl-furfural and its derivatives. ARKIVOC. 2001;2001:17–54. https://doi.org/10.3998/ark.5550190.0002.102.
137. Zhao D, Delbecq F, Len C. One-pot FDCA diester synthesis from mucic acid and their solvent-free regioselective polytransesterification for production of glycerol-based furanic polyesters. Molecules. 2019;24:1030. https://doi.org/10.3390/molecules24061030.
138. Taguchi Y, Oishi A, Iida H. One-step synthesis of dibutyl furandicarboxylates from galactaric acid. Chem Lett. 2008;37:50–1. https://doi.org/10.1246/cl.2008.50.
139. Van Strien N, Rautiainen S, Asikainen M, Thomas DA, Linnekoski J, Niemelä K, Harlin A. A unique pathway to platform chemicals: aldaric acids as stable intermediates for the synthesis of furandicarboxylic acid esters. Green Chem. 2020;22:8271–7. https://doi.org/10.1039/D0GC02293D.
140. Guo N. Acids-catalyzed dehydration of glucaric acid to 2,5-furandicarboxylic acid (FDCA). WO2017083297. 2017.
141. Kambourakis S, Griffin BM. Synthesis of FDCA and FDCA precursors from gluconic acid derivatives. WO2017030668. 2017.

142. Rosenboom J-G, Langer R, Traverso G. Bioplastics for a circular economy. Nat Rev Mater. 2022;7:117–37. https://doi.org/10.1038/s41578-021-00407-8.
143. Ali SS, Abdelkarim EA, Elsamahy T, Al-Tohamy R, Li F, Kornaros M, Zuorro A, Zhu D, Sun J. Bioplastic production in terms of life cycle assessment: a state-of-the-art review. Environ Sci Ecotechnol. 2023;15:100254. https://doi.org/10.1016/j.ese.2023.100254.
144. Iroegbu AOC, Ray SS, Mbarane V, Bordado JC, Sardinha JP. Plastic pollution: a perspective on matters arising: challenges and opportunities. ACS Omega. 2021;6:19343–55. https://doi.org/10.1021/acsomega.1c02760.
145. Loos K, Zhang R, Pereira I, Agostinho B, Hu H, Maniar D, Sbirrazzuoli N, Silvestre AJD, Guigo N, Sousa AF. A perspective on PEF synthesis, properties, and end-life. Front Chem. 2020;8:585. https://doi.org/10.3389/fchem.2020.00585.
146. Boyer C, Liu J, Wong L, Tippett M, Bulmus V, Davis TP. Stability and utility of pyridyl disulfide functionality in RAFT and conventional radical polymerizations. J Polym Sci A Polym Chem. 2008;46:7207–24. https://doi.org/10.1002/pola.23028.
147. Banella MB, Bonucci J, Vannini M, Marchese P, Lorenzetti C, Celli A. Insights into the synthesis of poly(ethylene 2,5-furandicarboxylate) from 2,5-furandicarboxylic acid: steps toward environmental and food safety excellence in packaging applications. Ind Eng Chem Res. 2019;58:8955–62. https://doi.org/10.1021/acs.iecr.9b00661.
148. Rosenboom J-G, Hohl DK, Fleckenstein P, Storti G, Morbidelli M. Bottle-grade polyethylene furanoate from ring-opening polymerisation of cyclic oligomers. Nat Commun. 2018;9:2701. https://doi.org/10.1038/s41467-018-05147-y.
149. Qu X, Jiang M, Wang B, Deng J, Wang R, Zhang Q, Zhou G, Tang J. A Brønsted acidic ionic liquid as an efficient and selective catalyst system for bioderived high molecular weight poly (ethylene 2,5-furandicarboxylate). ChemSusChem. 2019;12:4927–35. https://doi.org/10.1002/cssc.201902958.
150. Wu J, Xie H, Wu L, Li BG, Dubois P. DBU-catalyzed biobased poly(ethylene 2,5-furandicarboxylate) polyester with rapid melt crystallization: synthesis, crystallization kinetics and melting behavior. RSC Adv. 2016;6:101578–86. https://doi.org/10.1039/c6ra21135f.
151. Bueno A, Barredo A, Viar N, Requies J. Main routes of production of high-value-added 2,5-furandincarboxylic acid using heterogeneous catalytic systems. Catalysts. 2023;13:880. https://doi.org/10.3390/catal13050880.
152. Al Ghatta A, Wilton-Ely JDET, Hallett JP. From sugars to FDCA: a techno-economic assessment using a design concept based on solvent selection and carbon dioxide emissions. Green Chem. 2021;23:1716–33. https://doi.org/10.1039/D0GC03991H.
153. Heo JB, Lee Y-S, Chung C-H. Conversion of inulin-rich raw plant biomass to 2,5-furandicarboxylic acid (FDCA): progress and challenge towards biorenewable plastics. Biotechnol Adv. 2021;53:107838. https://doi.org/10.1016/j.biotechadv.2021.107838.
154. Egan J, Salmon S. Strategies and progress in synthetic textile fiber biodegradability. SN Appl Sci. 2021;4:22. https://doi.org/10.1007/s42452-021-04851-7.
155. Hwang K-R, Jeon W, Lee SY, Kim M-S, Park Y-K. Sustainable bioplastics: recent progress in the production of bio-building blocks for the bio-based next-generation polymer PEF. Chem Eng J. 2020;390:124636. https://doi.org/10.1016/j.cej.2020.124636.
156. Fei X, Wang J, Zhang X, Jia Z, Jiang Y, Liu X. Recent progress on bio-based polyesters derived from 2,5-furandicarbonxylic acid (FDCA). Polymers. 2022;14:625. https://doi.org/10.3390/polym14030625.
157. Eerhart AJJE, Faaij APC, Patel MK. Replacing fossil based PET with biobased PEF; process analysis, energy and GHG balance. Energy Environ Sci. 2012;5:6407–22. https://doi.org/10.1039/C2EE02480B.
158. Neukum D, Baumgarten L, Wüst D, Sarma BB, Saraçi E, Kruse A, Grunwaldt J-D. Challenges of green production of 2,5-furandicarboxylic acid from bio-derived 5-hydroxymethylfurfural: overcoming deactivation by concomitant amino acids. ChemSusChem. 2022;15:e202200418. https://doi.org/10.1002/cssc.2022004.

159. Gidi L, Amalraj J, Tenreiro C, Ramírez G. Recent progress, trends, and new challenges in the electrochemical production of green hydrogen coupled to selective electrooxidation of 5-hydroxymethylfurfural (HMF). RSC Adv. 2023;13:28307–36. https://doi.org/10.1039/D3RA05623F.
160. Prasad S, Khalid AJ, Narishetty V, Kumar V, Dutta S, Ahmad E. Recent advances in the production of 2,5-furandicarboxylic acid from biorenewable resources. Mater Sci Energy Technol. 2023;6:502–21. https://doi.org/10.1016/j.mset.2023.04.005.
161. Deshan ADK, Atanda L, Moghaddam L, Rackemann DW, Beltramini J, Doherty WOS. Heterogeneous catalytic conversion of sugars into 2,5-furandicarboxylic acid. Front Chem. 2020;8:659. https://doi.org/10.3389/fchem.2020.00659.
162. Kroger M, Pruße U, Vorlop K-D. A new approach for the production of 2,5-furandicarboxylic acid by in situ oxidation of 5-hydroxymethylfurfural starting from fructose. Top Catal. 2000;13:237–42. https://doi.org/10.1023/A:1009017929727.
163. Zhu H, Cao Q, Li C, Mu X. Acidic resin-catalysed conversion of fructose into furan derivatives in low boiling point solvents. Carbohydr Res. 2011;346:2016–8. https://doi.org/10.1016/j.carres.2011.05.026.
164. Douša M, Gibala P, Břicháč J, Havlíček J. The formation of furfural compounds in selected saccharide- and polysaccharide-based pharmaceutical excipients. J Pharm Sci. 2012;101:1811–20. https://doi.org/10.1002/jps.23072.
165. Yuan H, Liu H, Du J, Liu K, Wang T, Liu L. Biocatalytic production of 2,5-furandicarboxylic acid: recent advances and future perspectives. Appl Microbiol Biotechnol. 2020;104:527–43. https://doi.org/10.1007/s00253-019-10272-9.
166. Motagamwala AH, Won W, Sener C, Alonso DM, Maravelias CT, Dumesic JA. Toward biomass-derived renewable plastics: production of 2,5-furandicarboxylic acid from fructose. Sci Adv. 2018;4:eaap9722. https://doi.org/10.1126/sciadv.aap9722.
167. Marshall A, Jiang B, Gauvin RM, Thomas CM. 2,5-Furandicarboxylic acid: an intriguing precursor for monomer and polymer synthesis. Molecules. 2022;27:4071. https://doi.org/10.3390/molecules27134071.
168. Girisuta B, Janssen LPBM, Heeres HJ. A kinetic study on the decomposition of 5-hydroxymethylfurfural into levulinic acid. Green Chem. 2006;8:701. https://doi.org/10.1039/b518176c.
169. Hu X, Lievens C, Larcher A, Li C-Z. Reaction pathways of glucose during esterification: effects of reaction parameters on the formation of humin type polymers. Bioresour Technol. 2011;102:10104–13. https://doi.org/10.1016/j.biortech.2011.08.040.
170. Sangregorio A, Muralidhara A, Guigo N, Thygesen LG, Marlair G, Angelici C, De Jong E, Sbirrazzuoli N. Humin based resin for wood modification and property improvement. Green Chem. 2020;22:2786–98. https://doi.org/10.1039/C9GC03620B.
171. Ouyang W, Balu AM, ALOthman ZA, Wang X, Wang W, Luque R. Valorisation of humins to high value-added products: integrating biorefinery process towards a more sustainable future. Curr Opin Green Sustain Chem. 2023;39:100717. https://doi.org/10.1016/j.cogsc.2022.100717.
172. Hoang TMC, Van Eck ERH, Bula WP, Gardeniers JGE, Lefferts L, Seshan K. Humin based by-products from biomass processing as a potential carbonaceous source for synthesis gas production. Green Chem. 2015;17:959–72. https://doi.org/10.1039/C4GC01324G.
173. Mija A, Van Der Waal JC, Pin J-M, Guigo N, De Jong E. Humins as promising material for producing sustainable carbohydrate-derived building materials. Constr Build Mater. 2017;139:594–601. https://doi.org/10.1016/j.conbuildmat.2016.11.019.
174. Tsilomelekis G, Orella MJ, Lin Z, Cheng Z, Zheng W, Nikolakis V, Vlachos DG. Molecular structure, morphology and growth mechanisms and rates of 5-hydroxymethyl furfural (HMF) derived humins. Green Chem. 2016;18:1983–93. https://doi.org/10.1039/C5GC01938A.
175. Pfab E, Filiciotto L, Romero AA, Luque R. Valorization of humins-extracted 5-methoxymethylfurfural: toward high added value furanics via continuous flow catalytic

hydrogenation. Ind Eng Chem Res. 2019;58:16065–70. https://doi.org/10.1021/acs.iecr.9b00893.

176. Engel CAR, van der Meer J, Geers LFG, Crockatt M. Method for purifying and removing color of FDCA. WO2019132663A1. 2019. https://patents.google.com/patent/WO2019132663A1/en.

177. Han X, Geng L, Guo Y, Jia R, Liu X, Zhang Y, Wang Y. Base-free aerobic oxidation of 5-hydroxymethylfurfural to 2,5-furandicarboxylic acid over a Pt/C–O–Mg catalyst. Green Chem. 2016;18:1597–604. https://doi.org/10.1039/C5GC02114F.

178. Huang Y-T, Wong J-J, Chen J-H. Method for purifying crude of 2,5-furandicarboxylic acid by crystallization. US10344010B2. 2019.

Chapter 9
Fermentative Production of Amino Acids from Agricultural Sidestreams and Food Wastes

Nora Junker, Berna Sariyar Akbulut, and Volker F. Wendisch

Abstract Fermentation industry relies on glucose from starch hydrolysis or sucrose- and fructose-containing molasses as carbon sources for the million-ton-scale production of amino acids. Extensive metabolic engineering of the amino acid producing bacteria *Escherichia coli* and *Corynebacterium glutamicum* to access alternative carbon and nitrogen substrates has provided improved production strains that have the potential to be used industrially for amino acid production from sustainable feedstocks that do not have competing uses as food or feed. About 14% of all food is thought to be lost between harvest and retail giving rise to wastes or sidestreams, e.g., straws, shells, peels, seeds from food processing. This chapter discusses the current state and prospects of amino acid production from agricultural sidestreams and food wastes. Exemplarily, we will summarize hydrolysis options for agricultural sidestreams and food wastes, utilization of monomeric and polymeric compounds as well as amino acid production using the complex mixtures of substrates present in hydrolysates.

Keywords Amino acids · Metabolic engineering · Synthetic biology · Microbial bioprocesses · Renewable substrates · *Escherichia coli* · *Corynebacterium glutamicum*

N. Junker · V. F. Wendisch (✉)
Genetics of Prokaryotes, Faculty of Biology & Center for Biotechnology (CeBiTec), Bielefeld University, Bielefeld, Germany
e-mail: nora.junker@uni-bielefeld.de; volker.wendisch@uni-bielefeld.de

B. Sariyar Akbulut
Department of Bioengineering, Marmara University, Istanbul, Turkey
e-mail: berna.akbulut@marmara.edu.tr

Z. Fang et al. (eds.), *Production of Organic Acids and Alcohols from Agricultural Residues and Food Wastes*, Biofuels and Biorefineries 13,
https://doi.org/10.1007/978-981-96-1222-2_9

9.1 Bacteria Used for Amino Acid Production

Amino acids find direct applications in various industries including food and beverage, pharmaceuticals, and animal feed. Additionally, they serve as precursors for the synthesis of biodegradable plastics, surfactants, and specialty chemicals [1]. The demand for amino acid production is high and is steadily increasing. While advances in the chemical industry have made it possible to synthesize even the most intricate molecules, serious limitations may arise during their production. In the case of amino acid production, chiral purity and selectivity can be regarded among the major advantages offered by bacterial production routes as opposed to chemical synthesis. Due to stereochemical specificity of enzymatic reactions, only one enantiomer of the amino acid is produced, leading to the desired chiral purity. This circumvents the need for selective separation of the L-form, which is an economically infeasible process [2]. Extraction using protein hydrolysates derived from plant or animal sources yields the L-forms of amino acids, but is costly [3]. With growing concerns about sustainability issues, bacterial production is now receiving increasing attention. Production through bacterial fermentation processes not only eliminates the need for harsh chemicals or solvents commonly used in chemical synthesis methods but also utilizes renewable feedstocks reducing reliance on finite fossil sources [4, 5].

Early attempts to find suitable production hosts for amino acids relied on traditional screening and random mutagenesis [6–8]. Advances in metabolic engineering have shifted economically competitive amino acid synthesis to the use of rationally designed cells [9–11]. Now through metabolic engineering approaches, microbial systems can be engineered to overproduce specific amino acids, addressing their growing demand in various industries. Based on their natural metabolic capabilities and ability to undergo genetic manipulations, several bacteria stand out as attractive hosts for amino acid production.

9.1.1 Escherichia coli

Gram-negative *E. coli* has been traditionally studied as the model organism in molecular biology and biotechnology. Due to its rapid growth rate, well-understood genetics and metabolism, and ease of genetic manipulation on top of its robust fermentation capabilities [12], it has been recognized as a production host for a wide range of different molecules that also include amino acids. It is among the hosts that can be used for the production of molecules that hold GRAS status [13, 14]; however, product purification is demanding due to the presence of endotoxins. One of the pioneering studies for amino acid production with *E. coli* was on the industrially important amino acid, L-lysine. Reverend et al. [15] was able to demonstrate that recombinant DNA techniques can be used to manipulate its metabolism and achieve L-lysine overproduction. Later, *E. coli* cells have been engineered to overproduce other amino acids with significant market volumes in food, feed, beverage, pharmaceutical, and cosmetics industries such as L-arginine [16], L-valine [17], and aromatic amino acids [18, 19], along with non-proteinogenic amino acids L-homoserine [20], γ-aminobutyric acid (GABA), δ-aminolevulinic acid (ALA) [21]. Engineering efforts were also extended to fine chemicals derived from amino acids such as the precursor of the neurotransmitters dopamine 4-dihydroxy-L-phenylalanine (L-DOPA) [22] and the pigment melanin [23].

Considering the utilization of lignocellulosic hydrolysates, the versatility of *E. coli* to utilize a wide range of carbohydrates makes it an attractive production host, but the hierarchy in sugar utilization complicates the development of production strains [24]. On the other hand, to mitigate the inhibitory effects of toxic molecules derived from pre-treatment operations of lignocellulosic biomass, strain engineering for their assimilation or conversion to less toxic molecules is essential [25–27]. While simplicity of genetic manipulations in *E. coli* is an advantage, vulnerability to phage contamination is a major concern, which is difficult and costly to overcome on industrial scale production [28, 29]. Moreover, contamination with endotoxins is a serious issue for human use.

9.1.2 Corynebacterium glutamicum

The discovery of the Gram-positive *C. glutamicum* in 1957 has been a milestone for fermentative production of amino acids through biotechnology. Following its isolation by the Japanese company, Kyowa Hakko Kogyo (now Kyowa Kirin) [8] as an L-glutamate producer, with the enormous global demand for L-glutamate and L-lysine, studies focused on obtaining stable *C. glutamicum* strains for scalable production [30]. It not only grows fast and reaches high cell densities in cost-effective minimal media, but it is also genetically stable and free of endotoxins. More importantly, it can also be used for the production of generally recognized as safe (GRAS) status food ingredients that extends to amino acids [31–33]. With advances in metabolic engineering, various *C. glutamicum* strains have also been engineered to produce other amino acids such as L-valine [34], L-glutamine [35], L-threonine [36], L-tyrosine [10], and L-phenylalanine [2], found in building blocks of proteins, as well as L-ornithine, L-citrulline [37], GABA [38], and L-DOPA [39].

In order to enhance the utilization of lignocellulosic hydrolysates, the metabolism of *C. glutamicum* has already been extended to assimilate a wide range of different sugars derived from biomass [1]. Major advantages in its use as a production host, are its natural ability to assimilate lignin-derived aromatics [40], as well as acetic acid [41], and its relatively high tolerance to inhibitors that are formed during pre-treatment [42–44]. Its tolerance has further been improved by adaptive evolution [45] and studies indicate that there is still room for improvement [46, 47]. Currently, *C. glutamicum* constitutes the major industrial workhorse of amino acid production and can be used to valorize the main carbon components of biomass hydrolysates. However, when compared to *E. coli*, the complexity of genetic manipulations can be one disadvantage associated with strain construction of *C. glutamicum*.

9.1.3 Emerging Hosts

While bacterial fermentation for amino acid production has been realized mainly by using conventional *E. coli* and *C. glutamicum* strains, there is continuing search for alternative hosts desirable for industrial production. In this respect, the Gram-positive *Bacillus methanolicus* MGA3 is an emerging industrial workhorse for amino acid production [48]. It is a natural L-glutamate producer [49, 50] but with the development of genetic manipulation tools, its product spectrum has been expanded to L-lysine [11, 51], and the derivatives of these two amino acids, cadaverine and GABA [52]. One of its major advantages is its capability to utilize methanol as the sole carbon and energy source [53, 54]. Furthermore, its thermotolerance which necessitates growth at 50 °C, reduces the cost of cooling water during production. Growing concerns on sustainability and environmental impacts indicate that this bacterium possesses great potential as a production host. The fact that only a few genetic tools have been developed [55, 56] could be regarded as the major bottleneck which hampers production in this host.

In the search for robust strains for amino acid production, the Gram-negative acidophilic bacterium of the *Enterobacteriaceae* family, *Pantoea ananatis* AJ13355 has been isolated as a potential host [57]. Its capacity has already been modified for the overproduction of L-glutamate [58] and L-cysteine [59]. Since it is capable of growing at acidic and neutral pH, it is also resistant to high concentrations of L-glutamic acid, making it an ideal host for the production of molecules that drop pH during fermentation. Although *P. ananatis* has been recognized as a phytopathogenic host, it has passed all tests required for industrialization and general genetic tools are available to construct engineered strains [58].

Overall, amino acid production will be dominated by *E. coli* and *C. glutamicum* for the future. This is irrespective of the carbon source used, be it sidestreams or pure sugars. However, it is possible that an emerging host will become dominant, if it outperforms the traditional hosts on sidestreams and if regulations do not stand in the way.

9.2 Sidestreams Relevant for Amino Acid Production

The agricultural and food processing industries generate vast quantities of organic by-products [60]. These by-products include inedible components, such as peels, husks, stems, seeds, plant juices and other residues, as well as sidestreams from processing. While some of these by-products either remain on the field during harvesting or are spread as organic fertilizer, other by-products can find a usage as animal feed. In some regions, farmers continue to resort to open field burning, and some food processing factories employ incineration of wastes or use landfilling as disposal strategies [60–62]. These practices not only contribute to significant environmental challenges but also represent a lost opportunity for resource recovery and sustainable utilization. Agricultural sidestreams are valuable sources of carbohydrates, proteins, lipids, and other nutrients of interest, but the high potential of the residues is rarely utilized. Microbial biotechnology, however, enables the conversion of biomass waste into valuable products, thereby following the principles of a circular bioeconomy and sustainable resource management.

Over the past two decades, with the intensification of agricultural activities, there has been an overall increase in global agricultural production of more than 50% in primary crops, meat and milk production, and aquaculture, thereby even outpacing population growth [63, 64]. The global production of primary crop commodities reached 9.5 billion metric tons, with cereals representing the primary contributor (32% of total primary crop production) [63]. The sheer magnitude of cereal crop production translates into substantial volumes of by-products, especially for the three main cereals produced: maize (corn), rice, and wheat. Their husks and straws serve as prime candidates for biotechnological valorization. For wheat straw, as an example for these lignocellulosic by-products, the amount of produced straw worldwide can be estimated to be about 650 million metric tons, if the proportion of wheat straw [60] and the global wheat production value of 771 million metric tons per year (in 2021) [63] is considered. A composition of 32–52 wt% cellulose, 18–34 wt% hemicellulose and 14–19 wt% lignin is reported for this residue [65–68]. Cellulose is a linear polysaccharide of β(1–4)-linked D-glucopyranose ring units [69], while hemicellulose represents a more heterogeneous group of branched polysaccharides. Besides glucose, it includes further hexoses (such as mannose and galactose) as well as sugar acids (e.g. glucuronic acid), however, in most cases the sugar with the highest amount is the pentose sugar xylose, followed by arabinose [70]. In contrast, lignin is a diverse group of aromatic polymers consistent of phenylpropane units [71]. Due to high cellulose and hemicellulose contents, the main sugars obtained after hydrolysis of cereal husk and straws are glucose, xylose and arabinose. Hydrolysates of wheat bran, rice straw, maize stover and maize husk were implemented as feedstocks for fermentative amino acid production [72–74] (Table 9.1). In addition to inedible plant residues, sidestreams from processing industries can also be utilized. One such example is the wheat sidestream from the starch and paper industry. Its acidic hydrolysate contains a significantly higher concentration of glucose (about 30 g L^{-1}), in comparison to the pentose sugars xylose and arabinose (less than 6 g L^{-1} each) and was used for the microbial synthesis of L-lysine and its derivatives [75].

Fermentative production of L-lysine was demonstrated with the liquid by-products generated from maize and grass silages [76] (Table 9.1). During ensiling, lactic acid bacteria convert a significant amount of water-soluble carbohydrates into lactate. Producer strains, such as *C. glutamicum*, can subsequently utilize the lactate as a carbon source for amino acid synthesis. The same process is employed for the pre-treatment of the sidestream produced by the green crop-drying industry. The production of green pellets begins with the pressing of green plants (e.g., rye grass, clover, and alfalfa), whereby plant juices are obtained as a by-product. Depending on whether heat is applied, this plant juice is referred to as green or brown juice. A pre-treatment method utilizing lactic acid bacteria has been

Table 9.1 Relevant agricultural sidestreams and their utilization for fermentative amino acid production

Sector	Characteristic components	Sidestream	Main sources in hydrolysate [g L^{-1}]	Use	Amino acid product	Strain	Reference
Cereals	Cellulose and hemicellulose	Maize husk	215 Xylitol[a] 77 Sorbitol[a] 45 Xylose[a] 37 Glucose[a] 13 Acetic acid[a]	Supplement	L-Tyrosine (0.44 g L^{-1})	*E. coli* XT02	[74]
		Rice straw	7.2 Glucose 24.9 Xylose 9.9 Arabinose	Sole C source	L-Glutamate (13.7 ± 0.6 g L^{-1})	*C. glutamicum* WT (pEKEx3-*xylA*)(pVWEx1 *araBAD*)	[72]
					L-Lysine (6.1 ± 0.3 g L^{-1})	*C. glutamicum* DM1729 (pEKEx3-*xylA*)(pVWEx1 *araBAD*)	
		Wheat bran	22.5 Glucose 9.3 Xylose 9.6 Arabinose	Sole C source	L-Glutamate (14.1 ± 0.6 g L^{-1})	*C. glutamicum* WT (pEKEx3-*xylA*)(pVWEx1 *araBAD*)	
					L-Lysine (6.1 ± 0.4 g L^{-1})	*C. glutamicum* DM1729 (pEKEx3-*xylA*)(pVWEx1 *araBAD*)	
		Maize stover	124.2 Glucose 16.9 Xylose 2.1 Acetic acid	Supplement	L-Lysine (33.8 g L^{-1})	*C. glutamicum* SIIM B253	[73]
Starch industry		Wheat sidestream	33.2 Glucose 5.7 Xylose/Fructose	Complete medium, accept	L-Lysine (11.8 ± 0.9 g L^{-1})	*C. glutamicum* LYS-XA	[75]

			3.1 Arabinose	N-source and pH buffer			
		Potato juice	(n.d.)		L-Lysine (35 g L^{-1})	*C. glutamicum* V5 ATCC 700239	[77]
Silages/ fermented sidestreams	Lactic acid	Corn silage	30.0 ± 2.0 Lactic acid 2.0 ± 1.5 Glucose 1.5 ± 0.5 Maltose	Complete medium	L-Lysine (4.9 g L^{-1})	*C. glutamicum* $lysC^{fbr}$ dld_{Psod} pyc_{Psod} $malE_{Psod}$ fbp_{Psod} $gapX_{Psod}$	[76]
		Grass silage	40.0 ± 7.5 Lactic acid 4.0 ± 2.0 Glucose 20.0 ± 4.0 Fructose	Complete medium	L-Lysine (4.9 g L^{-1})		
		Supernatant from lactic acid-fermented green juice + fermented, evaporated brown juice	(n.d.)	Supplement	L-Lysine (65 g L^{-1})	*C. glutamicum* B031	[77]
Fruits and vegetables	Pectin	Orange peel	17.8 ± 2.1 Glucose 11.7 ± 1.2 Xylose 7.5 ± 0.5 Arabinose 4.2 ± 0.5 Galacturonic acid	Complete medium, accept N-source and pH buffer	L-Valine (0.28 ± 0.19 g L^{-1})	*C. glutamicum* VAL1	[46]
					L-Tyrosine (0.67 ± 0.03 g L^{-1})	*C. glutamicum* AROM3	
					L-Serine (0.52	*C. glutamicum* SER1	

(continued)

Table 9.1 (continued)

Sector	Characteristic components	Sidestream	Main sources in hydrolysate [g L^{-1}]	Use	Amino acid product	Strain	Reference
					± 0.03 g L^{-1})		
					L-Lysine (2.9 ± 0.8 g L^{-1})	*C. glutamicum* GRLys1 Δ*sugR* Δ*ldhA*	
					L-Arginine (1.7 ± 0.6 g L^{-1})	*C. glutamicum* ARG5	
Aquaculture	Chitin, protein	Shrimp waste	2.8 GlcNAc 2.3 $(GlcNAc)_2$ 1.6 $(GlcNAc)_3$	Sole C and N source	L-Tyrosine (0.91 g L^{-1})	*E. coli* MG1655 Δ*recA*_Δ*endA*_Δ*pheA*_Δ*tyrR*_*DE3* (NP19)	[92]
					L-DOPA (0.41 g L^{-1})	*E. coli* MG1655 Δ*recA*_Δ*endA*_Δ*pheA*_Δ*tyrR*_*DE3* (LDOPA5)	
	Mannitol	Brown seaweed (*Laminaria digitata*)	53–75 Mannitol 16–22 Glucose 0.5–3.5 Fructose	Sole C source	L-Lysine (2.6 g L^{-1})	*C. glutamicum* SEA-7	[96]

[a] Concentration is given in [g kg^{-1}]

established for the purpose of imitating the ensiling process. The resulting sidestream based medium with lactate as the main carbon source was successfully used for amino acid production [77] (Table 9.1). As indicated by Thomsen et al. [78], the brown juice sidestream of a single green crop drying factory (with a production capacity of 45,000 metric tons of fodder pellets per year) is sufficient to provide the fermentation medium for a 11,000-metric ton L-lysine factory.

Compared to green and brown juice, the utilization spectrum is reverse for fruits: juice factories are interested in extracting the liquid whereas they discard the solid remains, including peels, pulps, and seeds. About 910 million metric tons of fruits are produced globally per year [63]. Oranges are among the top five produced fruits globally together with bananas, watermelons, apples, and grapes. From orange juice production alone, a sidestream of 0.7–0.9 million metric tons per year is estimated [79]. Besides cellulose, hemicellulose, and lignin (14–23, 6–14, and 1–7 wt%, respectively), orange peels contain high amounts of pectin (18–29 wt%) [80–83]. This family of complex structured polysaccharides consists of approximately 70 wt% covalently linked galacturonic acid monomers, which can be released by hydrolysis. An orange peel hydrolysate-based medium was established for cultivation of *C. glutamicum*, enabling the production of different amino acids, as demonstrated for L-arginine, L-lysine, L-serine, L-valine and L-tyrosine [46] (Table 9.1).

Non-lignocellulosic sidestreams are derived from livestock farming. With a worldwide milk production of about 966 million metric tons [84], the by-products of milk processing, including dairy sludge and whey, are abundant. Milk-derived by-products are characterized by high lactose concentrations (about 45–60 g L^{-1} lactose in whey [85]), which can be utilized by lactic acid bacteria, as demonstrated for the production of L-glutamate [86].

The fastest growing food production sector globally is aquaculture, including the processing of shrimps (6.3 million metric tons [64]). The shrimp's head and body shell (carapace), both of which are discarded as waste, make up approximately 48–56 wt% of the shrimp [87]. The dry weight of shrimp contains about 10–19 wt % of the aminopolysaccharide chitin, which is comprised of β(1–4)-linked *N*-acetylglucosamine (GlcNAc) and glucosamine monomers [88–90]. These amino sugars can support growth of *E. coli* not only as a sole carbon source, but also as a sole nitrogen source [91]. The high protein content of shrimps (41–49 wt%) [88–90] further contributes to the supply of carbon and nitrogen. Utilizing hydrolysate of shrimp by-products as feedstock has been demonstrated for the synthesis of L-tyrosine and its derivative L-DOPA [92] (Table 9.1).

With nearly 36 million tons (wet weight), cultivated algae, particularly seaweeds (i.e. marine macroalgae), account for 30 wt% of global aquaculture production [93]. In contrast to the previously introduced sidestreams, which can be classified as second-generation feedstocks, seaweed belongs to the third-generation. In addition to fixing CO_2 from the environment, seaweed does not require fresh water, chemical fertilizers, or arable land for cultivation [94]. In brown algae, the main cell wall polysaccharides are alginate, which is a composed of D-mannuronic acid and L-guluronic acid, and a diverse group of fucose-containing sulfated polysaccharides. Together, these two groups account for up to 45% of the brown algal dry weight, while up to 20–30% can be represented by the storage carbohydrate mannitol, depending upon the harvest period [95]. These carbohydrates are the most prominent, although many more can be found in the algal biomass. Fermentative production of L-lysine was realized using hydrolysates of brown seaweed as carbon source [96] (Table 9.1).

Besides carbon sources, a sufficient nitrogen supply is of importance for amino acid production and should be considered for sidestream selection [97]. While brown seaweeds have a low protein content of 4–24% of dry weight, green or red seaweeds clearly outperform them in this regard with 8 up to 47% [98]. The protein content of some algae species thus reaches the high protein levels found in soybeans. With a protein content of about 40–49 wt% [99], soybean meal, a by-product from oil extraction, is a valuable nitrogen source. Other mentioned by-products have lower protein contents, such as wheat bran with about 17 wt% [100], wheat straw with about 2 wt% [65], or orange peels with about 7 wt% [101]. When designing fermentation processes, it is important to consider additional nutrients for metabolism and growth. Previous studies have demonstrated that hydrolysates derived from wheat sidestream or orange peels can replace other essential nutrients required for growth and production, including phosphorus and sulfur sources, as well as trace elements. For the production of amino acids with the biotin-auxotrophic bacterium *C. glutamicum*, the agricultural by-products can also provide the necessary biotin [46, 75]. In addition, sidestreams and their hydrolysates can serve special functions, e.g. as antioxidant. This can be important to stabilize products such as L-DOPA, which are prone to auto-oxidation. Sidestream hydrolysates could also potentially be implemented as antifoam agents, as foam reduction has been reported upon their addition [102].

The production of different variants of amino acids may have different requirements. However, it is not possible to provide a definitive answer as to which type of amino acid can be produced most efficiently on a given sidestream without conducting experimental tests. As demonstrated for the production of amino acids on orange peel hydrolysate, the synthesis of L-tyrosine was more efficient than the production of other amino acids, such as L-arginine, relative to its respective production on minimal medium [46]. The metabolic pathways and their regulations are very complex, as is the composition of the sidestreams and their hydrolysates. Importantly, the prevalence of a sidestream used for biotechnological applications will depend not only on the presence of nutrients, which facilitate microbial growth, but also on antimicrobial components that may be present in agricultural byproducts. Such components are naturally present in certain sidestreams, i.e., D-limonene, which is found in citrus peel [103] or occur as a result of pre-treatment methods. Therefore, the choice of a sidestream will also depend on required pre-treatment methods and their influence on the contents of nutrients and inhibitory compounds, as will be discussed in the next section (Sect. 9.3). In addition, the selection of a sidestream should consider its availability, depending on alternative usages. Not every sidestream equals a waste product. Cereal straw, for example, often either remains or returns to the field after use for animal bedding, and thus plays a role in maintenance of soil organic matter and circular nutrients recovery on the harvested fields [104, 105].

In essence, agricultural sidestreams have the potential to provide all the necessary medium components required for fermentative amino acid production, starting with the carbon and nitrogen source, as well as sources required for growth and metabolism, but also for supporting the process itself as antioxidant or antifoam agent. In the future, the combination of different agricultural sidestreams may allow for precise coordination of the required sources to optimize process efficiency. For instance, Ghazanfari et al. [86] demonstrated the use of dairy sludge as a carbon source in combination with soybean meal as a nitrogen source.

9.3 Pre-treatment Options

Biomass originating from agricultural sidestreams typically requires various pre-treatment operations to release the fermentable sugars to be utilized in a bioprocess. Based on the different forms of carbohydrates present in such biomass, different pre-treatment options could be preferred. For instance, lignocellulosic sidestreams are predominantly characterized by recalcitrant structures and necessitate rigorous pre-treatment operations. Conversely, milder conditions, even simple liquid hot water treatment (a hydrothermal treatment) with water under high temperature [106], maybe sufficient for seaweed, which contain carbohydrate forms devoid of lignin. For a pre-treatment process to be effective, it must adhere to specific criteria. Primarily, it should help to maximize the free sugar concentration obtained from the biomass. Additionally, it should minimize the formation of lignin derived degradation products (phenolic compounds, catechol, etc.), the degradation of sugars or polysaccharides to any by-products (furfural, hydroxymethylfurfural or their derivatives) and the release of organic acids (levulinic acid, formic acid, etc.) that will reduce sugar yield or have inhibitory/toxic effects in consequent processes [107]. Finally, it should be economically feasible and adaptable to large scale applications [108]. Although different classifications are available, the major pre-treatment operations are usually categorized into four main groups: physical, chemical, physico-chemical, and biological [109]. Usually, combinatorial approaches are more effective when dealing with complex structures. Specifically for amino acid production, in the few number of studies where rice straw, wheat bran [72], corn stover [74] or orange peel [46] were used, mechanical treatment such as milling or grinding was followed by acid treatment and sometimes further by enzymatic hydrolysis. Here, pre-treatment options within these four groups are briefly explained and then summarized with respect to their mechanisms in Table 9.2 as possible future options.

Table 9.2 Pre-treatment methods for agricultural wastes

Pre-treatment	Mechanism/properties	Reference
Physical		
Electron beam	Exposure to a beam of high-energy electrons to induce physical and chemical changes	[124]
Extrusion (thermomechanical)	Generation of high shear forces along the extruder by the rotating extruder screws to change the physical (shortening of fiber and defibrillation) as well as chemical structure	[125]
Gamma ray	Exposure to gamma rays emitted from radioactive isotopes (cobalt-60 or cesium-137) to induce physical and chemical changes	[126]
Milling, grinding, crushing	Application of impact, friction or compression forces reduce size	[127]
Microwave	Penetration of microwaves (non-ionizing electromagnetic radiations) into the material and rapid heating facilitates the disruption of recalcitrant structures	[128]
Pulsed electric field	Exposure to high voltage (0.1–100 kV/cm) for very short pulse durations (ns–ms) to disrupt structure	[112]
Soaking/maceration	Soaking in a liquid for a certain period to disrupt structure	[129]
Ultrasonication	Propagation of high-frequency sound waves (20 kHz to several MHz) associated with cavitation, heat, and mechanical actions through the biomass induce cell wall breakage, particle size reduction, matrix degradation	[130, 131]
Chemical		
Acid hydrolysis	Solubilization in dilute/concentrated organic or inorganic acids to break down glycosidic bonds and solubilize various components	[132]
Base/alkali hydrolysis	Solubilization in different alkali solutions to delignify and to cleavage glycosidic and intermolecular bonds involving polysaccharides	
Deep eutectic solvents (DESs)	Exposure (soaking, extraction, etc.) to deep eutectic solvents (ionic liquid-like systems formed by the formation of hydrogen bonds between two or more components) to disrupt hydrogen bonding, reduce crystallinity, solubilize hemicellulose, delignify	[115, 133, 134]
Ionic liquids (ILs)	Treatment in liquids composed of anions and cations to break down the hydrogen bonds leading to the disruption of structure	[135]
Organosolv process	Treatment in a mixture of organic solvents and water under elevated temperature and pressure where solvent penetrates the biomass to swell and then cleave linkages of lignin and glycosidic bonds	[113]
Ozonolysis	Exposure to ozone in a controlled environment for a predetermined time where ozone decomposes itself	[114]

(continued)

Table 9.2 (continued)

Pre-treatment	Mechanism/properties	Reference
	to radicals and reacts both directly and indirectly with various organic compounds in the biomass	
Physicochemical		
Ammonia fiber explosion (AFEX) (thermochemical)	Exposure to anhydrous ammonia under moderate temperatures and high pressures for a short time followed by sudden drop of pressure where ammonia vaporizes and leads to an explosive decompression and considerable fiber disruption	[119, 136]
CO_2 explosion	Exposure to supercritical CO_2 followed by an explosive release of the CO_2 pressure to facilitate structural transformations	[137]
Liquid hot water treatment (hydrothermal)	Penetration of water under high pressure and temperature to disrupt structure	[120]
Steam explosion	Exposure to sudden high-pressure saturated steam for few minutes followed by abrupt release of pressure, which leads to chemical autohydrolysis to cleave glycosidic bonds cleavage and mechanical disruption of the lignocellulosic matrix	[117]
Biological		
Enzymatic	Treatment with enzymes (cellulases, hemicellulases, ligninases, lyases, etc.) to disrupt various structural units	[138, 139]
Whole-cell	Incubated with bacteria and fungi that synthesize extracellular cellulases, hemicellulases, ligninases, and lyases that disrupt various structural units	[138, 139]

9.3.1 *Physical*

Physical pre-treatments generally reduce the size of the biomass particles, which in turn decreases the degree of polymerization and increases the surface area. An increase in the surface area does not necessarily increase the obtained sugars by itself, however it makes the biomass easily accessible to subsequent processes. These treatments can also increase the porosity of the material and may help to disrupt crystalline structures. Most physical pre-treatments use temperature, pressure, and mechanical force to obtain particles of smaller sized and altered structures without modifying the chemical compositions. Among the different types, milling, grinding, crushing, chipping, and extrusion are known as mechanical pre-treatments. In these processes, energy consumption for size reduction increases with moisture content [108]. Physical pre-treatments that are assisted by waves include microwave, ultrasound, electron beam, or gamma ray irradiation [110, 111]. These are short treatments but are not economical. Furthermore, they are difficult to scale-up due to challenges in maintaining uniform heat distribution within the biomass. The emerging technology, pulsed electric field pre-treatment, is based on exposure to high voltage. Physical pre-treatments are considered eco-friendly with rare production of toxic byproducts but with high energy demands [112]. For effective sugar recovery, physical pre-treatments are usually coupled to chemical, physicochemical or biological treatments.

9.3.2 Chemical

Chemical pre-treatments are characterized by the application of chemical agents, such as acids, bases, ozone, ionic liquids or organic solvents to biomass feedstocks under controlled conditions of temperature, pressure, and duration to facilitate break down of the biomass structure. Here, the chemical interacts with biomass components to disrupt the structure by hydrolysis, swelling or fragmentation. They are fast and versatile, meaning they can be adopted to a wide range of biomass feedstock from woody biomass and energy crops to seaweed. The effectiveness and scalability of acid/base hydrolysis make them attractive for frequent use. Unfortunately, most chemicals utilized with these pre-treatments are usually toxic or are coupled with the production of toxic or inhibitory by-products for subsequent fermentation processes. On the other hand, the organosolv method, which uses a mixture of organic solvents and water [113], not only produces fewer inhibitory by-products but also offers higher selectivity for lignin removal. Ozonolysis produces virtually no inhibitory by-products, but the high cost renders the process economically infeasible [114]. Compared to conventional solvents, pre-treatment with ionic liquids emerges as an environmentally safer method. Similar to ionic liquids, deep eutectic solvents [115] are considered as environmentally benign media in pre-treatment processes [116].

9.3.3 Physicochemical

Physicochemical pre-treatments combine physical and chemical methods. Steam explosion (catalyzed and uncatalyzed), CO_2 explosion, ammonia fiber explosion, and liquid hot water pre-treatment are the most widely known physicochemical pre-treatments. Steam explosion and CO_2 explosion disrupt biomass structure through exposure to high-pressure saturated steam or high-pressure CO_2, respectively, followed by rapid reduction of pressure [117, 118]. Ammonia fiber explosion is quite similar to steam explosion, where liquid ammonia replaces steam [119]. In liquid hot water pre-treatment, biomass is subjected to water at elevated temperatures and pressures but pressure is not dropped abruptly [120]. Its moderate operational conditions without chemical additions facilitate its scalability. All these methods lead to swelling, which promote subsequent disintegration and separation of the biomass.

9.3.4 *Biological*

Biological pre-treatment methods exploit the potential of hydrolytic and lignolytic enzymes or whole-cells that synthesize these enzymes to release the fermentable sugars of biomass. The effectiveness is greatly dependent on the microorganism or selected enzymes, the type of biomass, temperature, and pH of the process. Based on low energy consumption, high product yield, and absence of toxic inhibitors, biological pre-treatments are attracting significant attention. Unfortunately, the slow and long incubation nature and microbial consumption of the end products (released sugars) when whole-cells are used are the major disadvantages of these ecofriendly treatment processes [121].

Taken together, the choice and optimization of the pre-treatment procedure have to be made according to the sidestream used (e.g., hardwood, macroalgae, shrimp wastes) focusing on maximizing nutrient content, while minimizing the presence of inhibitory compounds. The latter may be present in the sidestream already or arise during preprocessing. For example, the hydrolysis of polymer structures may result in the release of toxic components, such as phenolic compounds from lignin. Additionally, harsh hydrolysis conditions may lead to the formation of toxic compounds, e.g. by degrading pentose and hexose sugars into furfural and 5-hydroxymethylfurfural, respectively [122]. Since these compounds affect the microbial biocatalyst to different extent, the choice of the microbial host is important. In the future, it is foreseeable that microbial hosts will be engineered to degrade inhibitory compounds during conversion of the nutrients to the desired products [123].

9.4 Exemplary Amino Acid Productions from Agricultural Sidestreams

This section is dedicated to the production of proteinogenic amino acids such as L-glutamate, L-lysine, and L-tyrosine as well as examples of their derived amino acids such as the ω-amino acid L-GABA, the dicarboxylate dipicolinic acid, and L-DOPA from alternative carbon sources with a focus on utilization of agricultural sidestreams as fermentation substrates.

9.4.1 *Glutamate Production Using Rice Straw Hydrolysate*

The umami taste-eliciting amino acid L-glutamate typically is produced from molasses by wild-type strains of *C. glutamicum*, an industrial show-case since more than six decades [1, 8]. To establish L-glutamate production from lignocellulosic hydrolysates, such as from rice straw, a number of genetic modifications were introduced.

Glucose is a native carbon and energy source, but the hemicellulosic pentose sugars xylose and arabinose are not. Metabolic engineering of *C. glutamicum* enabled the conversion of both pentoses to the innate central carbon metabolism, i.e., to the pentose phosphate pathway intermediate xylulose-5-phosphate [140, 141]. Access to xylose was possible by plasmid-borne overexpression of the heterologous xylulose isomerase gene *xylA* from *Xanthomonas campestris* and the endogenous xylulokinase gene *xylB* [72, 140, 142, 143]. This strain produced L-glutamate from both the xylose and glucose fractions of rice stray hydrolysate, thus, a higher titer was achieved as compared to the wild type that produced L-glutamate from the glucose fraction only [72]. An arabinose utilizing strain that was obtained by heterologous plasmid-borne expression of the *araBAD* operon from *E. coli,* co-utilized arabinose with glucose, and thus produced more L-glutamate from rice straw hydrolysate than *C. glutamicum* wild type. The combined overexpression of *xylA-xylB* and *araBAD* using two plasmids entailed co-consumption of glucose with both arabinose and xylose which increased the L-glutamate titer further. This concept was transferred for the production of many target compounds from hydrolysates [142, 144–147].

9.4.2 Production of Dipicolinic Acid from Hemicellulosic Pentoses

Dipicolinic acid (DPA) is found in heat-resistant bacterial endospores. It complexes transition metals, lanthanides, actinides, and also finds application as a precursor in bioplastics and pharmaceutical industries. As biosynthesis of dipicolinic acid initiates from an intermediate of L-lysine biosynthesis, namely, hydroxy-2,3,4,5-tetrahydrodipicolinic acid (HTPA), *C. glutamicum* strains overproducing L-lysine were chosen as the starting point for production of dipicolinic acid from xylose and arabinose [148]. The chosen base strain had genomic modifications improving flux into the initial phase of L-lysine biosynthesis: pyc^{P458S} (improved anaplerosis and precursor supply), hom^{V59A} (decreased flux to homoserine and derived side products), $lysC^{T311I}$ (desensitizing of feedback inhibition), Δ*pck* (abolishing decarboxylating flux counteracting anaplerosis) (Fig. 9.1).

Dehydration and oxidation of HTPA in two sequential reactions to dipicolinic acid resulted from introduction of the respective enzymes by plasmid-borne expression of the *dpaAB* operon from the endospore-forming *Paenibacillus sonchi* genomovar *riograndensis* SBR5 [148]. Upon transformation with either a plasmid for xylose utilization or a plasmid for arabinose utilization, the strains produced 0.77 ± 0.03 g L^{-1} and 0.85 ± 0.02 g L^{-1} dipicolinic acid, respectively [148].

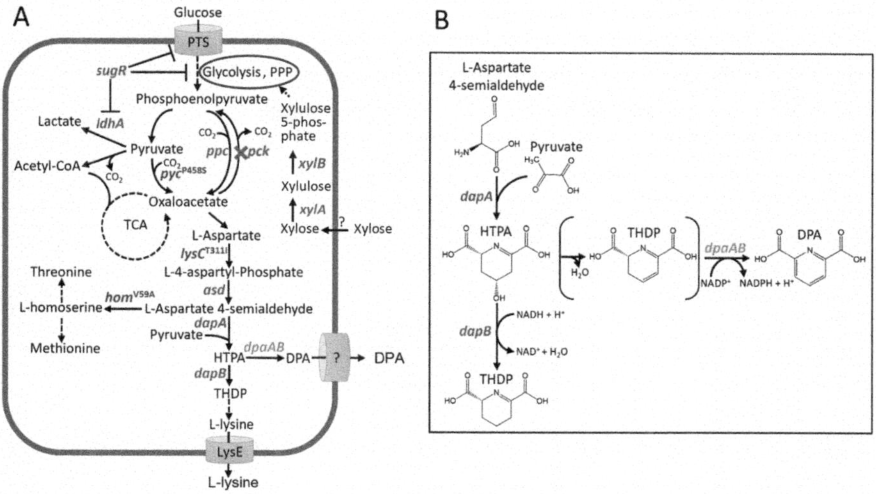

Fig. 9.1 Scheme of the metabolic pathway for DPA biosynthesis in engineered *C. glutamicum* Dpa1 (**a**) and of the reactions at the branchpoint intermediate HTPA (**b**) (Reproduced from Schwardmann et al. [148]). Single catalytic steps are depicted by solid lines and multiple catalytic steps by dashed lines. It includes endogenous genes (blue) and heterologous dipicolinate synthase genes *dpaAB* from *P. sonchi* SBR5 (green). Variants of endogenous genes (purple), namely feedback-resistant aspartokinase *lysC*^{T311I}, pyruvate carboxylase *pyc*^{P458S} and homoserine dehydrogenase *hom*^{V59A}, increase L-aspartate conversion and anaplerosis via pyruvate carboxylase, while decreasing drain of aspartate-4-semialdehyde towards threonine and methionine biosynthesis. DPA shares the first three steps of L-lysine biosynthesis via aspartokinase (*lysC*), aspartate-semialdehyde dehydrogenase (*asd*) and HTPA synthase (*dapA*). A red cross indicates the deletion of carboxykinase gene *pck* (red) to prevent decarboxylation of oxaloacetate. Glucose uptake is PTS (phosphotransferase system) mediated and L-lysine is exported via LysE. Brackets depict putative spontaneous dehydration of HTPA. Xylose is taken up by an unknown transporter, its catabolism in the pentose phosphate pathway (PPP) requires heterologous xylose isomerase and endogenous xylulose kinase

9.4.3 Utilization of Straw Hydrolysates for L-Lysine Production

The feed amino acid L-lysine could be produced from the agricultural residues rice straw and wheat straw as well as from a sidestream of the wheat starch industry. Plasmid-borne expression of pentose utilization genes (Sect. 9.4.1) enabled growth and production based on hydrolysates of dried oat spelts, rice straw and a sidestream of industrial starch production [72, 75, 149]. A genome-based mode of gene expression to enable xylose utilization was established and employed in production based on wheat straw [102]. Upon replacing the gene *actA* coding for CoA transferase [150] by a synthetic operon consisting of *xylA* encoding the xylose isomerase from *X. campestris* SCC1758 and *xylB* encoding the native xylulose kinase, slow growth on xylose was observed. Repeated re-inoculation yielded a strain by adaptive laboratory evolution that grew three times faster and genome sequencing of this strain revealed a single genetic change: a duplicated 53 bp region with parts of the untranslated region and the initial coding sequence of *xylA* that enhanced transcription and translation of *xylA* [102]. Transfer of this region into the genomes of strains overproducing L-lysine, L-glutamate, dipicolinic acid, *N*-methyl-phenylalanine, and fatty alcohols was sufficient to enable production of these compounds from xylose. In batch cultures, a mixture of 15 g L^{-1} glucose and 5 g L^{-1} xylose ("mock" wheat hydrolysate), as present in the wheat straw hydrolysate, led to 8.9 ± 0.6 g L^{-1} L-lysine, 3.7 ± 0.2 L-glutamate, 0.7 ± 0.0 g L^{-1} dipicolinic acid, and 1.2 ± 0.0 g L^{-1} *N*-methyl-phenylalanine, respectively, while a "real-world" wheat straw hydrolysate (i.e., from a cellulosic ethanol plant) supported production of 2.45 ± 0.09 g L^{-1} fatty alcohols in fed-batch bioreactor cultivation [102].

9.4.4 Aromatic Amino Acid Production on Corn Husks via Polyols

Enzymatic treatment of agricultural sidestreams with cellulases and xylanases to depolymerize their cellulose and hemicellulose content into fermentable biomolecules, such as D-glucose and D-xylose, but often is not cost-effective [151–153]. As an alternative, abiotic hydrogenolysis methods are proposed to rapidly break down cellulose and hemicellulose into sorbitol and xylitol at elevated temperatures [154]. However, high cost associated with the extraction of these valuable polyol products from the aqueous medium limits their use. To circumvent this challenge, Fung et al. [74] have developed a one-pot reaction to convert corn husk into polyols by hydrolysis in dilute sulfuric acid with instantaneous hydrogenation of sugars using a commercial Ru/C (ruthenium on carbon catalyst). Then, this crude hydrolysate was directly utilized without any purification for L-tyrosine overproduction by an engineered *E. coli* strain. The cells were able to produce L-tyrosine with all the major hydrogenolysis products (D-mannitol, D-glucose, and D-sorbitol and D-xylose) except xylitol. This is because the wild-type *E. coli* is missing the enzyme that enables the oxidation of xylitol into D-xylulose required for xylitol assimilation. By plasmid-based expression of the xylitol dehydrogenase (encoded by *gutB*) from *Bacillus subtilis* [74, 155] to oxidize xylitol into D-xylulose in the overproducer cells, the highest L-tyrosine titer was obtained with xylitol compared to the other tested polyols. Therefore, hydrogenolysis process was optimized to maximize xylitol concentration from raw biomass by adjusting the catalyst loading. Resulting strain was able to accumulate 0.44 g L^{-1} L-tyrosine from 10 g L^{-1} crude corn husk hydrolysate during fermentation under optimized conditions [74].

9.4.5 Production of L-Tyrosine and L-DOPA from Orange Peel Hydrolysate

Orange peel, a by-product especially abundant in juice-producing factories, has found application as a valuable feedstock. Following acidic hydrolysis and subsequent pH neutralization using ammonia and potassium hydroxide, the resulting liquid proved to be usable as a complete medium supporting growth of *C. glutamicum*. Among various amino acid productions, the synthesis of 0.67 ± 0.03 g L^{-1} L-tyrosine on orange peel hydrolysate stands out as particularly successful compared to production on minimal medium [46] (Fig. 9.2).

The employed tyrosine producing *C. glutamicum* strain harbors a feedback-resistant variant of 3-deoxy-D-arabino-heptulosonate-7-phosphate (DAHP) synthase from *E. coli* (encoded by $aroG^{D146N}$). Furthermore, the translational start codons ATG in *pheA* and *trpE* have been replaced by the less preferred codon TTG in the chromosome, thereby reducing the flux into the competing synthesis pathways for L-phenylalanine and L-tryptophan [10]. In addition to the aforementioned gene modifications for L-tyrosine overproduction, gene deletions in this strain have

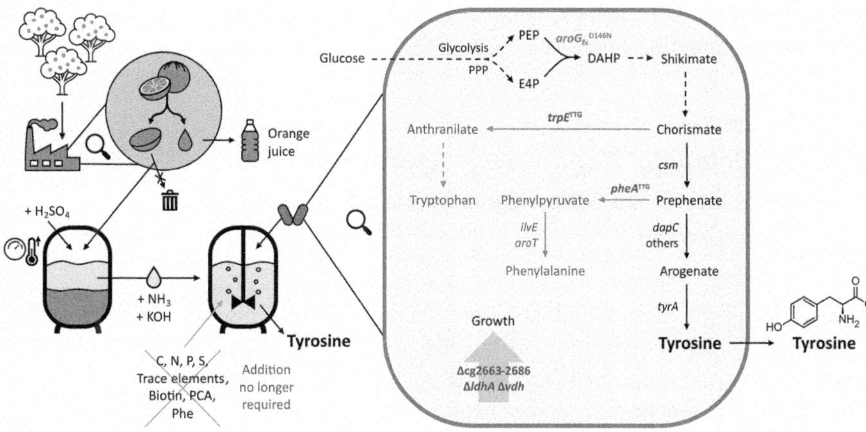

Fig. 9.2 Production of L-tyrosine on orange peel hydrolysate by engineered *C. glutamicum* (according to Junker et al. [46]). Orange peels as a by-product from juice producing factories were hydrolyzed by addition of 2% (v/v) H_2SO_4 at 121 °C for 50 min. After pH neutralization by addition of NH_3 and KOH, the derived liquid was used as a fermentation medium for L-tyrosine production by the engineered *C. glutamicum* strain AROM3. AROM3 is based on genome reduced chassis strain C1* with chromosomal integration of $aroG_{Ec}^{D146N}$ (depicted in green) encoding a feedback-resistant version of 3-deoxy-7-phosphoheptulonate synthase from *E. coli* (Δvdh::P_{ilvC}-$aroG_{Ec}^{D146N}$) and translational start codon exchanges from ATG to TTG in the genes *trpE* and *pheA* (depicted in red), encoding anthranilate synthase and prephenate dehydratase respectively. Single catalytic steps are depicted by solid lines and multiple catalytic steps by dashed lines. Knockouts (shown in red) which released growth inhibition on orange peel hydrolysate derived medium are cg2663–2686, *ldh* and *vdh* (encoding lactate dehydrogenase and vanillin dehydrogenase, respectively). Native enzymes are encoded by: *csm* chorismate mutase, *ilvE* branched-chain amino acid aminotransferase, *aroT* phenylalanine aminotransferase, *dapC* *N*-succinyl-aminooxopimelate aminotransferase, *tyrA* L-arogenate decarboxylase. *PPP* pentose phosphate pathway, *PEP* phosphoenolpyruvate, *E4P* erythrose 4-phosphate, *DAHP* 3-Deoxy-D-arabinoheptulosonate 7-phosphate, *PCA* protocatechuate, *Phe* L-phenylalanine.

contributed to a better growth performance on orange peel hydrolysate compared to the parental WT strain. Among the 414 gene deletions for genome reduction purposes and to avoid by-product formation, one gene deletion contributing to a higher growth rate was of *vdh*, encoding vanillin dehydrogenase, which resulted in reduced formation of 5-hydroxymethyl-2-furancarboxylic acid (HMFCA). HMFCA is one of two detoxification products formed from 5-(hydroxymethyl)furfural (HMF), a growth inhibitor typically present in lignocellulosic derived hydrolysates. Further modifications leading to enhanced growth performance included the deletions of the gene cluster cg2663–2686 and of *ldh*, encoding lactate dehydrogenase. However, further research is required to understand their precise roles in releasing growth inhibition [46].

Tyrosine serves as the precursor to L-DOPA, which is the most commonly used drug for the treatment of Parkinson's disease. The expression of the tyrosinase gene of *Ralstonia solanacearum* in *C. glutamicum* wild type allowed for the oxidation of tyrosine yielding L-DOPA [39]. When this strain was used in a whole-cell biotransformation after growth with hazelnut husk acid hydrolysate as the sole carbon source it converted added L-tyrosine to 20 ± 0.4 mg L^{-1} L-DOPA [156]. The production of L-DOPA required the addition of an antioxidant to prevent the formation of the pigment melanin. The use of orange peel hydrolysate as a medium, not only provided a nutritious feedstock, but also served as antioxidant and antifoam reagent during the fermentation process.

9.4.6 A Metabolic Engineering Approach to Chitin-Based L-Lysine Production by a Synthetic Microbial Consortium

Chitin is one of the most abundant biopolymers on Earth. Chitin is readily available as sidestream from the shrimp and mushroom industries as it is a polysaccharide component of crustacean shells and fungal mycelia. It is a linear long-chain polymer made up of the monomeric components *N*-acetyl-glucosamine and glucosamine linked by β-(1 → 4)-glycosidic bonds. The breakdown of chitin to glucosamine involves the enzymatic reactions of a chitinase yielding chitin oligomers and *N*-acetyl-glucosamine disaccharides. Subsequent liberation of acetate from *N*-acetyl-glucosamine disaccharides by a chitin deacetylase and a glucosaminidase cleavage of the resulting glucosamine disaccharide yields glucosamine (Fig. 9.3). In a bottom-up approach, a chitin deacetylase-expressing L-lysine auxotrophic *E. coli* strain in co-culture with a L-lysine producing *C. glutamicum* strain enabled growth of the consortium with *N*-acetyl-glucosamine as sole carbon source [157]. Since metabolic engineering of the *E. coli* strain abolished its utilization of *N*-acetyl-glucosamine, it could only grow with the acetate fraction of the chitin breakdown products. The *C. glutamicum* strain was engineered to be able to utilize glucosamine, but neither *N*-acetyl-glucosamine nor acetate. Thus, the carbon source *N*-acetyl-glucosamine was

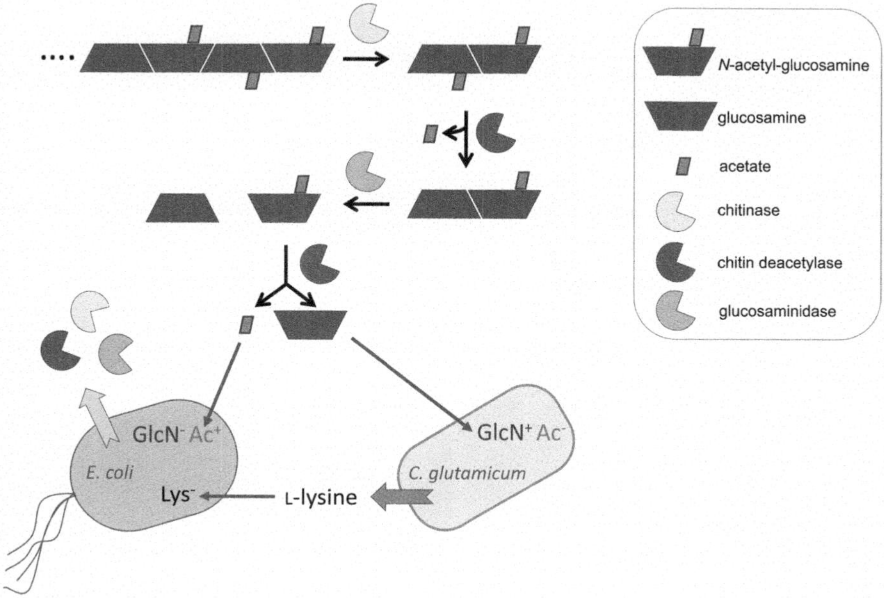

Fig. 9.3 Design of a mutualistic synthetic microbial consortium for chitin-based L-lysine production with a chitin degrading *E. coli* strain and a L-lysine producing *C. glutamicum* strain (modified from Vortmann et al. [157]). Chitin is a polymer made up of the monomeric compounds *N*-acetyl-glucosamine and glucosamine (GlcN). The recombinant chitin-degrading *E. coli* strain secretes chitinase, chitin deacetylase and glucosaminidase, while it is lysine auxotrophic and unable to utilize glucosamine. The recombinant L-lysine producing *C. glutamicum* strain is able to utilize glucosamine, but cannot grow with acetate (Ac)

divided between both bacteria such that *E. coli* grew with acetate and *C. glutamicum* with glucosamine. The *C. glutamicum* strain overproduced L-lysine with more L-lysine accumulating than was required to supplement the L-lysine auxotrophic *E. coli* strain [157].

Full implementation of this concept for conversion of colloidal chitin to L-lysine was not achieved, likely due to insufficient activities of chitinase and chitin deacetylase in the medium supernatant. The bottleneck might be low enzyme activities and/or insufficient secretion, while disaccharide cleavage may not be the bottleneck since *C. glutamicum* possesses β-*N*-acetylglucosaminidase [158]. Future research is required to enable utilization of colloidal chitin for growth and amino acid production.

9.4.7 GABA from Seaweed

The non-proteogenic amino acid, GABA, is a value-added product with significant medical uses. It is synthesized from L-glutamate, catalyzed by the enzyme glutamate

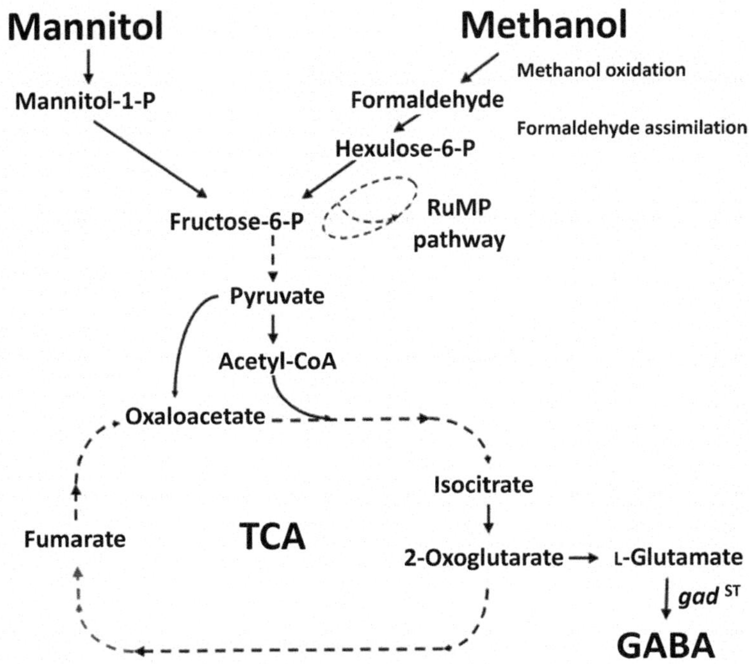

Fig. 9.4 Production of GABA by engineered *B. methanolicus* (Modified from Hakvåg et al. [52]). Single catalytic steps are depicted by solid lines and multiple catalytic steps by dashed lines. *B. methanolicus* can utilize methanol, mannitol, and extract of the brown seaweed *S. latissimi* to produce GABA by expression of glutamate decarboxylase *gad*ST or *gadB*. To prepare seaweed extract, hot water (60–70 °C) was added to the milled *S. latissima* (biomass:water, 1:1, weight basis). After adjusting pH to 3.5 with 1 M H_2SO_4, extraction was carried out by incubation at 70 °C for 1 h. After centrifugation (7000 rpm, 12,220 × G, 15 min) the supernatant was sieved and used for cultivations

decarboxylase. The anticipated increase in the market size of GABA [159] has made its microbial production an attractive route. In this respect, brown seaweed, rich in structural (alginate and cellulose) and storage (laminarin and mannitol) carbohydrates [160, 161] has been exploited for sustainable GABA production [52, 162].

Since microbial cultivations in seaweed-based media with high salt content might be challenging, strains displaying high salt tolerance are advantageous. Based on this fact, Hakvåg et al. [52] have evaluated the brown seaweed *Saccharina latissima* with significant mannitol and laminarin content as a carbon source for amino acid production by the salt tolerant *B. methanolicus* MGA3 (Fig. 9.4). The fact that *B. methanolicus* MGA3 can utilize mannitol as the sole carbon and energy source has made it an attractive bacterium to valorize the extract of *S. latissima* for growth and production [163]. The product range of this L-glutamate producing methylotrophic bacterium was already extended by Irla et al. [164] to achieve methanol-based production of GABA, a molecule with applications as a food additive, as a drug, and as a constituent in biopolymers. GABA was made from

L-glutamate by heterologous overexpression of the glutamate decarboxylase genes from *Sulfobacillus thermosulfidooxidans* (*gad*St) or *E. coli* (*gadB*) and a titer of 9 g L^{-1} was obtained in fed-batch fermentations. The GABA producer strain with *gad*St was able to grow on extracts of the brown seaweed *S. latissima*, and biomass formation was higher with 50% (v/v) seaweed extract medium than with 10 g L^{-1} mannitol medium. Interestingly, while the GABA titer on this mannitol-based medium with 152 ± 4 mg L^{-1} was lower than the titer obtained on methanol-based media (260 mg L^{-1}), titer on the 50% (v/v) seaweed extract medium was very low (13 ± 1 mg L^{-1}). In conclusion, *B. methanolicus* could produce GABA upon growth in seaweed-based media but levels were low compared to methylotrophic growth.

Later in another study, Lee et al. [162] have evaluated the potential of *Levilactobacillus brevis* to synthesize GABA using the *Saccharina japonica* extract. When the brown seaweed extract was used as a supplement after thermal acid hydrolysis and enzymatic saccharification, 2.63 g L^{-1} of GABA was synthesized by *L. brevis*.

The examples chosen to exemplify production of amino acids from agricultural sidestreams as fermentation substrates highlight the potential of this approach also for amino acid derived compounds. Of course, there is no limitation of using sidestreams for the production of non-nitrogenous compounds. Thus, the strategies discussed here may also serve as guidance for developing sidestream-based production processes for alcohols, carboxy acids, etc., provided that the nitrogen content of the sidestream is factored into such processes, e.g., to support growth of the microbial hosts in the process itself or in the seed train.

9.5 Conclusions and Future Outlook

Fermentative production of amino acids from agricultural sidestreams and food wastes has been developed but has achieved various technical readiness levels. We can expect further developments for improving upstream pre-treatment, strain engineering, process intensification and downstream processing. In our eyes, the biocatalyst is key and, therefore its optimization has priority. Moreover, metabolic engineering strategies are steadily improving, thus, a technology push will drive biocatalyst development.

Regarding the enzymes used to utilize agricultural sidestreams and food wastes, the natural variation has not been fully exploited since biology lacks many of the reactions used by synthetic chemists [165]. On the one hand, more gene and amino acid sequences as well as three-dimensional structure prediction (e.g., AlphaFold2) allow us to access natural diversity [166], while on the other hand, enzyme evolution [167] enables us to even select for new-to-nature catalytic potential as shown for new-to-nature (bio)synthesis [165, 168]. In the future, we will likely see many applications of enzyme evolution regarding catabolic pathways for improving access to nutrients in agricultural sidestreams and food wastes and for degrading inhibitors present in hydrolysates. These enzymes may also be scaffolded to orchestrate their action as inspired by nature [169], e.g., using synthetic cellulosomes [170].

Adaptive laboratory evolution (ALE) operates on the cell level with cell growth as readout. Thus, it can easily be adapted to select for strains that grow faster with a non-native substrate such as xylose [102] or in the presence of an inhibitor such as methanol [171]. The ALE approach has been extended to include evolution of consortia [172] and product traits after flux enforcement [173, 174] or using biosensors [175–177].

CRISPR technologies will also shape strain engineering [178]. Genome editing by gene deletion [179] or base editing [180] can be facilitated by various CRISPR systems. Gene repression by CRISPR interference can be used for metabolic engineering [181] and for parallel testing of gene targets in screening applications [182, 183]. Gene activation by CRISPRa, likewise, accelerates gene target identification [184].

Notwithstanding this bright future of metabolic engineering for strain optimization, other factors will govern how successful fermentative production of amino acids from food wastes and agricultural sidestreams will be. Amongst others, there are and will be economical requirements (e.g., petrol oil and sugar pricing) and regulatory constraints (e.g., competing uses of substrates for human nutrition). Overall, we believe that we are facing a bright future for biotechnology based on sidestreams.

Acknowledgements This work was supported in part by the Scientific and Technological Research Council of Turkey (TUBITAK) grant 120N728 and by the Bundesministerium für Bildung und Forschung (BMBF) grant 01DL21001. The funders had no role in the design of the study, in the collection, analyses, or interpretation of data, in the writing of the manuscript, or in the decision to publish the results.

We wish to thank Eldin Kurpejović, Emre Abdullahoglu, and Sara-Sophie Poethe for fruitful discussions.

References

1. Wendisch VF. Metabolic engineering advances and prospects for amino acid production. Metab Eng. 2020;58:17–34. https://doi.org/10.1016/j.ymben.2019.03.008.
2. Nakayama K, Araki K, Kase H. Microbial production of essential amino acid with *Corynebacterium glutamicum* mutants. Adv Exp Med Biol. 1978;105:649–61. https://doi.org/10.1007/978-1-4684-3366-1_31.
3. Sandeaux J, Sandeaux R, Gavach C, Grib H, Sadat T, Belhocine D, Mameri N. Extraction of amino acids from protein hydrolysates by electrodialysis. J Chem Technol Biotechnol. 1998;71:267–73.
4. Gaur S, Kaur M, Kalra R, Rene ER, Goel M. Application of microbial resources in biorefineries: current trend and future prospects. Heliyon. 2024;10:e28615. https://doi.org/10.1016/j.heliyon.2024.e28615.
5. Pérez-García F, Brito LF, Bakken TI, Brautaset T. Riboflavin overproduction from diverse feedstocks with engineered *Corynebacterium glutamicum*. Biofabrication. 2024;16:045012. https://doi.org/10.1088/1758-5090/ad628e.

6. Ikeda S, Fujita I, Yoshinaga F. Screening of L-isoleucine producers among ethionine resistant mutants of L-threonine producing bacteria. Agric Biol Chem. 1976;40:511–6. https://doi.org/10.1080/00021369.1976.10862090.
7. Crook N, Alper HS. Classical strain improvement. In: Patnaik R, editor. Engineering complex phenotypes in industrial strains. Wiley; 2012. p. 1–33.
8. Udaka S. Screening method for microorganisms accumulating metabolites and its use in the isolation of *Micrococcus glutamicus*. J Bacteriol. 1960;79:754–5. https://doi.org/10.1128/jb.79.5.754-755.1960.
9. Bailey JE. Toward a science of metabolic engineering. Science. 1991;252:1668–75. https://doi.org/10.1126/science.2047876.
10. Kurpejović E, Burgardt A, Bastem GM, Junker N, Wendisch VF, Sariyar Akbulut B. Metabolic engineering of *Corynebacterium glutamicum* for L-tyrosine production from glucose and xylose. J Biotechnol. 2023;363:8–16. https://doi.org/10.1016/j.jbiotec.2022.12.005.
11. Nærdal I, Netzer R, Irla M, Krog A, Heggeset TMB, Wendisch VF, Brautaset T. L-Lysine production by *Bacillus methanolicus*: genome-based mutational analysis and L-lysine secretion engineering. J Biotechnol. 2017;244:25–33. https://doi.org/10.1016/j.jbiotec.2017.02.001.
12. Lee SY. High cell-density culture of *Escherichia coli*. Trends Biotechnol. 1996;14:98–105. https://doi.org/10.1016/0167-7799(96)80930-9.
13. U.S. Food and Drug Administration. GRAS Notice 659. 2016. https://www.fda.gov/media/100020/download.
14. U.S. Food and Drug Administration. GRAS Notice 897. 2019. https://www.fda.gov/media/136755/download.
15. Reverend BD-L, Boitel M, Deschamps AM, Lebeault J-M, Sano K, Takinami K, Patte J-C. Improvement of *Escherichia coli* strains overproducing lysine using recombinant DNA techniques. Eur J Appl Microbiol Biotechnol. 1982;15:227–31. https://doi.org/10.1007/BF00499961.
16. Ginesy M, Belotserkovsky J, Enman J, Isaksson L, Rova U. Metabolic engineering of *Escherichia coli* for enhanced arginine biosynthesis. Microb Cell Factories. 2015;14:29. https://doi.org/10.1186/s12934-015-0211-y.
17. Park JH, Lee KH, Kim TY, Lee SY. Metabolic engineering of *Escherichia coli* for the production of L-valine based on transcriptome analysis and in silico gene knockout simulation. Proc Natl Acad Sci USA. 2007;104:7797–802. https://doi.org/10.1073/pnas.0702609104.
18. Rodriguez A, Martínez JA, Flores N, Escalante A, Gosset G, Bolivar F. Engineering *Escherichia coli* to overproduce aromatic amino acids and derived compounds. Microb Cell Factories. 2014;13:126. https://doi.org/10.1186/s12934-014-0126-z.
19. Liu Y, Xu Y, Ding D, Wen J, Zhu B, Zhang D. Genetic engineering of *Escherichia coli* to improve L-phenylalanine production. BMC Biotechnol. 2018;18:5. https://doi.org/10.1186/s12896-018-0418-1.
20. Sun Y, Wu J, Xu J, Yang L. Metabolic engineering of *Escherichia coli* for the production of L-homoserine. Chem Bio Eng. 2024;1:223–30. https://doi.org/10.1021/cbe.3c00077.
21. Yang Y, Zou Y, Chen X, Sun H, Hua X, Johnston L, Zeng X, Qiao S, Ye C. Metabolic engineering of *Escherichia coli* for the production of 5-aminolevulinic acid based on combined metabolic pathway modification and reporter-guided mutant selection (RGMS). Biotechnol Biofuels Bioprod. 2024;17:82. https://doi.org/10.1186/s13068-024-02530-4.
22. Fordjour E, Adipah FK, Zhou S, Du G, Zhou J. Metabolic engineering of *Escherichia coli* BL21 (DE3) for de novo production of L-DOPA from D-glucose. Microb Cell Factories. 2019;18:74. https://doi.org/10.1186/s12934-019-1122-0.
23. Chávez-Béjar MI, Balderas-Hernandez VE, Gutiérrez-Alejandre A, Martinez A, Bolívar F, Gosset G. Metabolic engineering of *Escherichia coli* to optimize melanin synthesis from glucose. Microb Cell Factories. 2013;12:108. https://doi.org/10.1186/1475-2859-12-108.

24. Bañares AB, Nisola GM, Valdehuesa KNG, Lee W-K, Chung W-J. Engineering of xylose metabolism in *Escherichia coli* for the production of valuable compounds. Crit Rev Biotechnol. 2021;41:649–68. https://doi.org/10.1080/07388551.2021.1873243.
25. Willson BJ, Herman R, Langer S, Thomas GH. Improved furfural tolerance in *Escherichia coli* mediated by heterologous NADH-dependent benzyl alcohol dehydrogenases. Biochem J. 2022;479:1045–58. https://doi.org/10.1042/BCJ20210811.
26. Wu W, Liu F, Singh S. Toward engineering *E. coli* with an autoregulatory system for lignin valorization. Proc Natl Acad Sci USA. 2018;115:2970–5. https://doi.org/10.1073/pnas.1720129115.
27. Clarkson SM, Giannone RJ, Kridelbaugh DM, Elkins JG, Guss AM, Michener JK. Construction and optimization of a heterologous pathway for protocatechuate catabolism in *Escherichia coli* enables bioconversion of model aromatic compounds. Appl Environ Microbiol. 2017;83:e01313–7. https://doi.org/10.1128/AEM.01313-17.
28. Gabig-Ciminska M, Los M, Holmgren A, Albers J, Czyz A, Hintsche R, Wegrzyn G, Enfors SO. Detection of bacteriophage infection and prophage induction in bacterial cultures by means of electric DNA chips. Anal Biochem. 2004;324:84–91. https://doi.org/10.1016/j.ab.2003.09.020.
29. Theisen M, Liao JC. Industrial biotechnology: *Escherichia coli* as a host. In: Wittmann C, Liao JC, editors. Industrial biotechnology—microorganisms. 1st ed. Weinheim: Wiley-VCH Verlag GmbH & Co. KGaA; 2017. p. 149–81.
30. Wendisch VF. Microbial production of amino acids and derived chemicals: synthetic biology approaches to strain development. Curr Opin Biotechnol. 2014;30:51–8. https://doi.org/10.1016/j.copbio.2014.05.004.
31. Lee J-Y, Na Y-A, Kim E, Lee H-S, Kim P. The actinobacterium *Corynebacterium glutamicum*, an industrial workhorse. J Microbiol Biotechnol. 2016;26:807–22. https://doi.org/10.4014/jmb.1601.01053.
32. U.S. Food and Drug Administration. Generally recognized as safe (GRAS). 2023. https://www.fda.gov/food/food-ingredients-packaging/generally-recognized-safe-gras.
33. U.S. Food and Drug Administration. AGRN 64—GRAS—Response Letter. 2024. https://www.fda.gov/media/180346/download.
34. Hasegawa S, Suda M, Uematsu K, Natsuma Y, Hiraga K, Jojima T, Inui M, Yukawa H. Engineering of *Corynebacterium glutamicum* for high-yield L-valine production under oxygen deprivation conditions. Appl Environ Microbiol. 2013;79:1250–7. https://doi.org/10.1128/AEM.02806-12.
35. Lv Q, Hu M, Tian L, Liu F, Wang Q, Xu M, Rao Z. Enhancing L-glutamine production in *Corynebacterium glutamicum* by rational metabolic engineering combined with a two-stage pH control strategy. Bioresour Technol. 2021;341:125799. https://doi.org/10.1016/j.biortech.2021.125799.
36. Diesveld R, Tietze N, Fürst O, Reth A, Bathe B, Sahm H, Eggeling L. Activity of exporters of *Escherichia coli* in *Corynebacterium glutamicum*, and their use to increase L-threonine production. J Mol Microbiol Biotechnol. 2009;16:198–207. https://doi.org/10.1159/000142530.
37. Jensen JVK, Eberhardt D, Wendisch VF. Modular pathway engineering of *Corynebacterium glutamicum* for production of the glutamate-derived compounds ornithine, proline, putrescine, citrulline, and arginine. J Biotechnol. 2015;214:85–94. https://doi.org/10.1016/j.jbiotec.2015.09.017.
38. Shi F, Li Y. Synthesis of γ-aminobutyric acid by expressing *Lactobacillus brevis*-derived glutamate decarboxylase in the *Corynebacterium glutamicum* strain ATCC 13032. Biotechnol Lett. 2011;33:2469–74. https://doi.org/10.1007/s10529-011-0723-4.
39. Kurpejović E, Wendisch VF, Sariyar Akbulut B. Tyrosinase-based production of L-DOPA by *Corynebacterium glutamicum*. Appl Microbiol Biotechnol. 2021;105:9103–11. https://doi.org/10.1007/s00253-021-11681-5.

40. Shen X-H, Zhou N-Y, Liu S-J. Degradation and assimilation of aromatic compounds by *Corynebacterium glutamicum*: another potential for applications for this bacterium? Eur J Appl Microbiol Biotechnol. 2012;95:77–89. https://doi.org/10.1007/s00253-012-4139-4.
41. Gerstmeir R, Wendisch VF, Schnicke S, Ruan H, Farwick M, Reinscheid D, Eikmanns BJ. Acetate metabolism and its regulation in *Corynebacterium glutamicum*. J Biotechnol. 2003;104:99–122. https://doi.org/10.1016/S0168-1656(03)00167-6.
42. Zhou P, Khushk I, Gao Q, Bao J. Tolerance and transcriptional analysis of *Corynebacterium glutamicum* on biotransformation of toxic furaldehyde and benzaldehyde inhibitory compounds. J Ind Microbiol Biotechnol. 2019;46:951–63. https://doi.org/10.1007/s10295-019-02171-9.
43. Mhatre A, Shinde S, Jha AK, Rodriguez A, Wardak Z, Jansen A, Gladden JM, George A, Davis RW, Varman AM. *Corynebacterium glutamicum* as an efficient omnivorous microbial host for the bioconversion of lignocellulosic biomass. Front Bioeng Biotechnol. 2022;10: 827386. https://doi.org/10.3389/fbioe.2022.827386.
44. Becker J, Kuhl M, Kohlstedt M, Starck S, Wittmann C. Metabolic engineering of *Corynebacterium glutamicum* for the production of cis, cis-muconic acid from lignin. Microb Cell Factories. 2018;17:115. https://doi.org/10.1186/s12934-018-0963-2.
45. Wang X, Khushk I, Xiao Y, Gao Q, Bao J. Tolerance improvement of *Corynebacterium glutamicum* on lignocellulose derived inhibitors by adaptive evolution. Eur J Appl Microbiol Biotechnol. 2018;102:377–88. https://doi.org/10.1007/s00253-017-8627-4.
46. Junker N, Sariyar Akbulut B, Wendisch VF. Utilization of orange peel waste for sustainable amino acid production by *Corynebacterium glutamicum*. Front Bioeng Biotechnol. 2024;12: 1419444. https://doi.org/10.3389/fbioe.2024.1419444.
47. Gu H, Hao X, Liu R, Shi Z, Zhao Z, Chen F, Wang W, Wang Y, Shen X. Small protein Cgl2215 enhances phenolic tolerance by promoting MytA activity in *Corynebacterium glutamicum*. Stress Biol. 2022;2:49. https://doi.org/10.1007/s44154-022-00071-0.
48. Brautaset T, Jakobsen ØM, Josefsen KD, Flickinger MC, Ellingsen TE. *Bacillus methanolicus*: a candidate for industrial production of amino acids from methanol at 50 degrees C. Appl Microbiol Biotechnol. 2007;74:22–34. https://doi.org/10.1007/s00253-006-0757-z.
49. Brautaset T, Williams MD, Dillingham RD, Kaufmann C, Bennaars A, Crabbe E, Flickinger MC. Role of the *Bacillus methanolicus* citrate synthase II gene, *citY*, in regulating the secretion of glutamate in L-lysine-secreting mutants. Appl Environ Microbiol. 2003;69:3986–95. https://doi.org/10.1128/AEM.69.7.3986-3995.2003.
50. Krog A, Heggeset TMB, Ellingsen TE, Brautaset T. Functional characterization of key enzymes involved in L-glutamate synthesis and degradation in the thermotolerant and methylotrophic bacterium *Bacillus methanolicus*. Appl Environ Microbiol. 2013;79:5321–8. https://doi.org/10.1128/AEM.01382-13.
51. Jakobsen OM, Brautaset T, Degnes KF, Heggeset TMB, Balzer S, Flickinger MC, Valla S, Ellingsen TE. Overexpression of wild-type aspartokinase increases L-lysine production in the thermotolerant methylotrophic bacterium *Bacillus methanolicus*. Appl Environ Microbiol. 2009;75:652–61. https://doi.org/10.1128/AEM.01176-08.
52. Hakvåg S, Nærdal I, Heggeset TMB, Kristiansen KA, Aasen IM, Brautaset T. Production of value-added chemicals by *Bacillus methanolicus* strains cultivated on mannitol and extracts of seaweed *Saccharina latissima* at 50°C. Front Microbiol. 2020;11:680. https://doi.org/10.3389/fmicb.2020.00680.
53. Irla M, Drejer EB, Brautaset T, Hakvåg S. Establishment of a functional system for recombinant production of secreted proteins at 50 °C in the thermophilic *Bacillus methanolicus*. Microb Cell Factories. 2020;19:151. https://doi.org/10.1186/s12934-020-01409-x.
54. Müller JEN, Heggeset TMB, Wendisch VF, Vorholt JA, Brautaset T. Methylotrophy in the thermophilic *Bacillus methanolicus*, basic insights and application for commodity production from methanol. Eur J Appl Microbiol Biotechnol. 2015;99:535–51. https://doi.org/10.1007/s00253-014-6224-3.

55. Irla M, Heggeset TMB, Nærdal I, Paul L, Haugen T, Le SB, Brautaset T, Wendisch VF. Genome-based genetic tool development for *Bacillus methanolicus*: theta- and rolling circle-replicating plasmids for inducible gene expression and application to methanol-based cadaverine production. Front Microbiol. 2016;7:1481. https://doi.org/10.3389/fmicb.2016.01481.
56. Schultenkämper K, Brito LF, López MG, Brautaset T, Wendisch VF. Establishment and application of CRISPR interference to affect sporulation, hydrogen peroxide detoxification, and mannitol catabolism in the methylotrophic thermophile *Bacillus methanolicus*. Appl Microbiol Biotechnol. 2019;103:5879–89. https://doi.org/10.1007/s00253-019-09907-8.
57. Hara Y, Kadotani N, Izui H, Katashkina JI, Kuvaeva TM, Andreeva IG, Golubeva LI, Malko DB, Makeev VJ, Mashko SV, Kozlov YI. The complete genome sequence of *Pantoea ananatis* AJ13355, an organism with great biotechnological potential. Appl Microbiol Biotechnol. 2012;93:331–41. https://doi.org/10.1007/s00253-011-3713-5.
58. Usuda Y, Nishio Y, Nonaka G, Hara Y. Microbial production potential of *Pantoea ananatis*: from amino acids to secondary metabolites. Microorganisms. 2022;10:1133. https://doi.org/10.3390/microorganisms10061133.
59. Takumi K, Ziyatdinov MK, Samsonov V, Nonaka G. Fermentative production of cysteine by *Pantoea ananatis*. Appl Environ Microbiol. 2017;83:e02502–16. https://doi.org/10.1128/AEM.02502-16.
60. Millati R, Cahyono RB, Ariyanto T, Azzahrani IN, Putri RU, Taherzadeh MJ. Agricultural, industrial, municipal, and forest wastes. In: Sustainable resource recovery and zero waste approaches. Elsevier; 2019. p. 1–22.
61. Gupta GK, Mondal MK. Bioenergy generation from agricultural wastes and enrichment of end products. In: Refining biomass residues for sustainable energy and bioproducts. Elsevier; 2020. p. 337–56.
62. Amin T, Bashir O, Bhat SA, et al., editors. Integrated waste management approaches for food and agricultural byproducts. 1st ed. Palm Bay, FL: Apple Academic Press; 2023.
63. FAO. Agricultural production statistics 2000–2021. FAOSTAT analytical briefs, No. 60. Rome: Food and Agriculture Organization of the United Nations; 2022.
64. FAO. Fishery and aquaculture statistics—yearbook 2021. Rome: Food and Agriculture Organization of the United Nations; 2024.
65. Adapa P, Tabil L, Schoenau G. Compaction characteristics of barley, canola, oat and wheat straw. Biosyst Eng. 2009;104:335–44. https://doi.org/10.1016/j.biosystemseng.2009.06.022.
66. Del Río JC, Rencoret J, Prinsen P, Martínez ÁT, Ralph J, Gutiérrez A. Structural characterization of wheat straw lignin as revealed by analytical pyrolysis, 2D-NMR, and reductive cleavage methods. J Agric Food Chem. 2012;60:5922–35. https://doi.org/10.1021/jf301002n.
67. Yu B, Fan G, Zhao S, Lu Y, He Q, Cheng Q, Yan J, Chai B, Song G. Simultaneous isolation of cellulose and lignin from wheat straw and catalytic conversion to valuable chemical products. Appl Biol Chem. 2021;64(1):15. https://doi.org/10.1186/s13765-020-00579-x.
68. Harper SHT, Lynch JM. The chemical components and decomposition of wheat straw leaves, internodes and nodes. J Sci Food Agric. 1981;32:1057–62. https://doi.org/10.1002/jsfa.2740321103.
69. Heinze T. Cellulose: structure and properties. In: Rojas OJ, editor. Cellulose chemistry and properties: fibers, nanocelluloses and advanced materials, vol. 271. Cham: Springer International Publishing; 2016. p. 1–52.
70. Schädel C, Blöchl A, Richter A, Hoch G. Quantification and monosaccharide composition of hemicelluloses from different plant functional types. Plant Physiol Biochem. 2010;48:1–8. https://doi.org/10.1016/j.plaphy.2009.09.008.
71. Del Río JC, Rencoret J, Gutiérrez A, Elder T, Kim H, Ralph J. Lignin monomers from beyond the canonical monolignol biosynthetic pathway: another brick in the wall. ACS Sustain Chem Eng. 2020;8:4997–5012. https://doi.org/10.1021/acssuschemeng.0c01109.
72. Gopinath V, Meiswinkel TM, Wendisch VF, Nampoothiri KM. Amino acid production from rice straw and wheat bran hydrolysates by recombinant pentose-utilizing *Corynebacterium*

glutamicum. Appl Microbiol Biotechnol. 2011;92:985–96. https://doi.org/10.1007/s00253-011-3478-x.

73. Chen Z, Liu G, Zhang J, Bao J. A preliminary study on L-lysine fermentation from lignocellulose feedstock and techno-economic evaluation. Bioresour Technol. 2019;271:196–201. https://doi.org/10.1016/j.biortech.2018.09.098.
74. Fung V, Xiao Y, Tan ZJD, Ma X, Zhou JFJ, Panda S, Yan N, Zhou K. Producing aromatic amino acid from corn husk by using polyols as intermediates. Biomaterials. 2022;287:121661. https://doi.org/10.1016/j.biomaterials.2022.121661.
75. Burgardt A, Prell C, Wendisch VF. Utilization of a wheat sidestream for 5-aminovalerate production in *Corynebacterium glutamicum*. Front Bioeng Biotechnol. 2021;9:732271. https://doi.org/10.3389/fbioe.2021.732271.
76. Neuner A, Wagner I, Sieker T, Ulber R, Schneider K, Peifer S, Heinzle E. Production of L-lysine on different silage juices using genetically engineered *Corynebacterium glutamicum*. J Biotechnol. 2013;163:217–24. https://doi.org/10.1016/j.jbiotec.2012.07.190.
77. Andersen M, Kiel P. Method for treating organic waste products. Eur. Pat. Appl. (WO 00/56912); 1999.
78. Thomsen MH, Bech D, Kiel P. Manufacturing of stabilised brown juice for L-lysine production—from university lab scale over pilot scale to industrial production. Chem Biochem Eng Q. 2004;18.
79. de Castro LA, Lizi JM, Chagas EGL, de Carvalho RA, Vanin FM. From orange juice by-product in the food industry to a functional ingredient: application in the circular economy. Food Secur. 2020;9:593. https://doi.org/10.3390/foods9050593.
80. Bicu I, Mustata F. Cellulose extraction from orange peel using sulfite digestion reagents. Bioresour Technol. 2011;102:10013–9. https://doi.org/10.1016/j.biortech.2011.08.041.
81. de Castro SC, Stanisic D, Tasic L. Sequential extraction of hesperidin, pectin, lignin, and cellulose from orange peels: towards valorization of agro-waste. Biofuels Bioprod Biorefin. 2024; https://doi.org/10.1002/bbb.2606.
82. Cypriano DZ, Da Silva LL, Tasic L. High value-added products from the orange juice industry waste. Waste Manag. 2018;79:71–8. https://doi.org/10.1016/j.wasman.2018.07.028.
83. de La Torre I, Ravelo M, Segarra S, Tortajada M, Santos VE, Ladero M. Study on the effects of several operational variables on the enzymatic batch saccharification of orange solid waste. Bioresour Technol. 2017;245:906–15. https://doi.org/10.1016/j.biortech.2017.08.094.
84. FAO. Dairy market review: overview of global dairy market in 2023. Rome: Food and Agriculture Organization of the United Nations; 2024.
85. Tsermoula P, Khakimov B, Nielsen JH, Engelsen SB. WHEY—the waste-stream that became more valuable than the food product. Trends Food Sci Technol. 2021;118:230–41. https://doi.org/10.1016/j.tifs.2021.08.025.
86. Ghazanfari N, Fallah S, Vasiee A, Tabatabaei Yazdi F. Optimization of fermentation culture medium containing food waste for L-glutamate production using native lactic acid bacteria and comparison with industrial strain. LWT. 2023;184:114871. https://doi.org/10.1016/j.lwt.2023.114871.
87. Sachindra NM, Bhaskar N, Mahendrakar NS. Carotenoids in different body components of Indian shrimps. J Sci Food Agric. 2005;85:167–72. https://doi.org/10.1002/jsfa.1977.
88. Deng J-J, Mao H-H, Fang W, Li Z-Q, Shi D, Li Z-W, Zhou T, Luo X-C. Enzymatic conversion and recovery of protein, chitin, and astaxanthin from shrimp shell waste. J Clean Prod. 2020;271:122655. https://doi.org/10.1016/j.jclepro.2020.122655.
89. Synowiecki J, Al-Khateeb NAAQ. The recovery of protein hydrolysate during enzymatic isolation of chitin from shrimp *Crangon crangon* processing discards. Food Chem. 2000;68:147–52. https://doi.org/10.1016/S0308-8146(99)00165-X.
90. Tokatlı K, Demirdöven A. Optimization of chitin and chitosan production from shrimp wastes and characterization. J Food Process Preserv. 2018;42:e13494. https://doi.org/10.1111/jfpp.13494.

91. Alvarez-Añorve LI, Calcagno ML, Plumbridge J. Why does *Escherichia coli* grow more slowly on glucosamine than on *N*-acetylglucosamine? Effects of enzyme levels and allosteric activation of GlcN6P deaminase (NagB) on growth rates. J Bacteriol. 2005;187:2974–82. https://doi.org/10.1128/jb.187.9.2974-2982.2005.
92. Ma X, Gözaydın G, Yang H, Ning W, Han X, Poon NY, Liang H, Yan N, Zhou K. Upcycling chitin-containing waste into organonitrogen chemicals via an integrated process. Proc Natl Acad Sci USA. 2020;117:7719–28. https://doi.org/10.1073/pnas.1919862117.
93. Cai J, Lovatelli A, Aguilar-Manjarrez J, Cornish L, Dabbadie L, Desrochers A, Diffey S, Garrido Gamarro E, Geehan J, Hurtado A, Lucente D, Mair G, Miao W, Potin P, Przybyla C, Reantaso M, Roubach R, Tauati M, Yuan X. Seaweeds and microalgae: an overview for unlocking their potential in global aquaculture development. FAO fisheries and aquaculture circular, no. 1229. Rome: Food and Agriculture Organization of the United Nations; 2021.
94. Torres MD, Kraan S, Domínguez H. Seaweed biorefinery. Rev Environ Sci Biotechnol. 2019;18:335–88. https://doi.org/10.1007/s11157-019-09496-y.
95. Rioux L-E, Turgeon SL. Seaweed carbohydrates. In: Seaweed sustainability. Elsevier; 2015. p. 141–92.
96. Hoffmann SL, Kohlstedt M, Jungmann L, Hutter M, Poblete-Castro I, Becker J, Wittmann C. Cascaded valorization of brown seaweed to produce L-lysine and value-added products using *Corynebacterium glutamicum* streamlined by systems metabolic engineering. Metab Eng. 2021;67:293–307. https://doi.org/10.1016/j.ymben.2021.07.010.
97. Wendisch VF, Nampoothiri KM, Lee J-H. Metabolic engineering for valorization of agri- and aqua-culture sidestreams for production of nitrogenous compounds by *Corynebacterium glutamicum*. Front Microbiol. 2022;13:835131. https://doi.org/10.3389/fmicb.2022.835131.
98. Thiviya P, Gamage A, Gama-Arachchige NS, Merah O, Madhujith T. Seaweeds as a source of functional proteins. Phycology. 2022;2:216–43. https://doi.org/10.3390/phycology2020012.
99. Banaszkiewicz T. Nutritional value of soybean meal. In: El-Shemy H, editor. Soybean and nutrition. InTech; 2011.
100. Dornez E, Gebruers K, Wiame S, Delcour JA, Courtin CM. Insight into the distribution of arabinoxylans, endoxylanases, and endoxylanase inhibitors in industrial wheat roller mill streams. J Agric Food Chem. 2006;54:8521–9. https://doi.org/10.1021/jf061728n.
101. Rivas B, Torrado A, Torre P, Converti A, Domínguez JM. Submerged citric acid fermentation on orange peel autohydrolysate. J Agric Food Chem. 2008;56:2380–7. https://doi.org/10.1021/jf073388r.
102. Werner F, Schwardmann LS, Siebert D, Rückert-Reed C, Kalinowski J, Wirth M-T, Hofer K, Takors R, Wendisch VF, Blombach B. Metabolic engineering of *Corynebacterium glutamicum* for fatty alcohol production from glucose and wheat straw hydrolysate. Biotechnol Biofuels Bioprod. 2023;16:116. https://doi.org/10.1186/s13068-023-02367-3.
103. Pourbafrani M, Forgács G, Horváth IS, Niklasson C, Taherzadeh MJ. Production of biofuels, limonene and pectin from citrus wastes. Bioresour Technol. 2010;101:4246–50. https://doi.org/10.1016/j.biortech.2010.01.077.
104. Blair N, Faulkner RD, Till AR, Poulton PR. Long-term management impacts on soil C, N and physical fertility. Soil Tillage Res. 2006;91:30–8. https://doi.org/10.1016/j.still.2005.11.002.
105. Powlson DS, Riche AB, Coleman K, Glendining MJ, Whitmore AP. Carbon sequestration in European soils through straw incorporation: limitations and alternatives. Waste Manag. 2008;28:741–6. https://doi.org/10.1016/j.wasman.2007.09.024.
106. Kooren R, Sumithra TG, Jaseera KV, Sunithakumari K, Hasan S, Sayooj P, Kaladharan P. A comparative study on pre-treatment methods for enhanced saccharification from tropical seaweeds to aid in bioethanol production. Aquat Bot. 2023;184:103594. https://doi.org/10.1016/j.aquabot.2022.103594.
107. Jönsson LJ, Martín C. Pretreatment of lignocellulose: formation of inhibitory by-products and strategies for minimizing their effects. Bioresour Technol. 2016;199:103–12. https://doi.org/10.1016/j.biortech.2015.10.009.

108. Alawad I, Ibrahim H. Pretreatment of agricultural lignocellulosic biomass for fermentable sugar: opportunities, challenges, and future trends. Biomass Conv Bioref. 2024;14:6155–83. https://doi.org/10.1007/s13399-022-02981-5.
109. Chaudhary G, Chaudhary N, Saini S, Gupta Y, Vivekanand V, Panghal A. Assessment of pretreatment strategies for valorization of lignocellulosic biomass: path forwarding towards lignocellulosic biorefinery. Waste Biomass Valor. 2024;15:1–36. https://doi.org/10.1007/s12649-023-02219-z.
110. Zhang YQ, Fu EH, Liang JH. Effect of ultrasonic waves on the saccharification processes of lignocellulose. Chem Eng Technol. 2008;31:1510–5. https://doi.org/10.1002/ceat.200700407.
111. Jablonowski ND, Pauly M, Dama M. Microwave assisted pretreatment of Szarvasi (*Agropyron elongatum*) biomass to enhance enzymatic saccharification and direct glucose production. Front Plant Sci. 2021;12:767254. https://doi.org/10.3389/fpls.2021.767254.
112. Vorobiev E, Lebovka N. Application of pulsed electric energy for lignocellulosic biorefinery. In: Miklavčič D, editor. Handbook of electroporation. Cham: Springer International Publishing; 2017. p. 2843–61.
113. Nair LG, Agrawal K, Verma P. Organosolv pretreatment: an in-depth purview of mechanics of the system. Bioresour Bioprocess. 2023;10:50. https://doi.org/10.1186/s40643-023-00673-0.
114. Travaini R, Martín-Juárez J, Lorenzo-Hernando A, Bolado-Rodríguez S. Ozonolysis: an advantageous pretreatment for lignocellulosic biomass revisited. Bioresour Technol. 2016;199:2–12. https://doi.org/10.1016/j.biortech.2015.08.143.
115. Abbott AP, Capper G, Davies DL, Rasheed RK, Tambyrajah V. Novel solvent properties of choline chloride/urea mixtures. Chem Commun (Camb). 2003:70–1. https://doi.org/10.1039/B210714G.
116. Procentese A, Johnson E, Orr V, Garruto Campanile A, Wood JA, Marzocchella A, Rehmann L. Deep eutectic solvent pretreatment and subsequent saccharification of corncob. Bioresour Technol. 2015;192:31–6. https://doi.org/10.1016/j.biortech.2015.05.053.
117. Duque A, Manzanares P, Ballesteros I, Ballesteros M. Steam explosion as lignocellulosic biomass pretreatment. In: Biomass fractionation technologies for a lignocellulosic feedstock based biorefinery. Elsevier; 2016. p. 349–68.
118. Gu T. Pretreatment of lignocellulosic biomass using supercritical carbon dioxide as a green solvent. In: Gu T, editor. Green biomass pretreatment for biofuels production. Dordrecht, Netherlands: Springer; 2013. p. 107–25.
119. Chundawat SPS, Pal RK, Zhao C, Campbell T, Teymouri F, Videto J, Nielson C, Wieferich B, Sousa L, Dale BE, Balan V, Chipkar S, Aguado J, Burke E, Ong RG. Ammonia fiber expansion (AFEX) pretreatment of lignocellulosic biomass. J Vis Exp. 2020; https://doi.org/10.3791/57488.
120. Jimenez-Gutierrez JM, Verlinden RAJ, van der Meer PC, van der Wielen LAM, Straathof AJJ. Liquid hot water pretreatment of lignocellulosic biomass at lab and pilot scale. Processes. 2021;9:1518. https://doi.org/10.3390/pr9091518.
121. Kuhad RC, Rapoport A, Kumar V, Singh D, Kumar V, Tiwari SK, Ahlawat S, Singh B. Biological pretreatment of lignocellulosic biomass: an environment-benign and sustainable approach for conversion of solid waste into value-added products. Crit Rev Environ Sci Technol. 2024;54:771–96. https://doi.org/10.1080/10643389.2023.2277670.
122. Jönsson LJ, Alriksson B, Nilvebrant N-O. Bioconversion of lignocellulose: inhibitors and detoxification. Biotechnol Biofuels. 2013;6:16. https://doi.org/10.1186/1754-6834-6-16.
123. Ujor VC, Okonkwo CC. Microbial detoxification of lignocellulosic biomass hydrolysates: biochemical and molecular aspects, challenges, exploits and future perspectives. Front Bioeng Biotechnol. 2022;10:1061667. https://doi.org/10.3389/fbioe.2022.1061667.
124. Sarosi O, Sulaeva I, Fitz E, Sumerskii I, Bacher M, Potthast A. Lignin resists high-intensity electron beam irradiation. Biomacromolecules. 2021;22:4365–72. https://doi.org/10.1021/acs.biomac.1c00926.

125. Konan D, Koffi E, Ndao A, Peterson EC, Rodrigue D, Adjallé K. An overview of extrusion as a pretreatment method of lignocellulosic biomass. Energies. 2022;15:3002. https://doi.org/10.3390/en15093002.
126. Hamza H, Jiménez-Araujo A, Ben Miloud N, Guillén-Bejarano R, Ghorbal A, Rodríguez-Arcos R, Benabderrahim MA, Elfalleh W. Impact of gamma irradiation pretreatment on the characteristics of native and defatted date palm seed flour and oil. Food Biosci. 2023;56:103288. https://doi.org/10.1016/j.fbio.2023.103288.
127. Arce C, Kratky L. Mechanical pretreatment of lignocellulosic biomass toward enzymatic/fermentative valorization. iScience. 2022;25:104610. https://doi.org/10.1016/j.isci.2022.104610.
128. Hoang AT, Nižetić S, Ong HC, Mofijur M, Ahmed SF, Ashok B, van Bui TV, Chau MQ. Insight into the recent advances of microwave pretreatment technologies for the conversion of lignocellulosic biomass into sustainable biofuel. Chemosphere. 2021;281:130878. https://doi.org/10.1016/j.chemosphere.2021.130878.
129. Kim TH, Taylor F, Hicks KB. Bioethanol production from barley hull using SAA (soaking in aqueous ammonia) pretreatment. Bioresour Technol. 2008;99:5694–702. https://doi.org/10.1016/j.biortech.2007.10.055.
130. Abolore RS, Jaiswal S, Jaiswal AK. Green and sustainable pretreatment methods for cellulose extraction from lignocellulosic biomass and its applications: a review. Carbohydr Polym Technol Appl. 2024;7:100396. https://doi.org/10.1016/j.carpta.2023.100396.
131. Calcio Gaudino E, Grillo G, Tabasso S, Stevanato L, Cravotto G, Marjamaa K, Pihlajaniemi V, Koivula A, Aro N, Uusitalo J, Ropponen J, Kuutti L, Kivinen P, Kanerva H, Arshanitsa A, Jashina L, Jurkjane V, Andersone A, Dreyer T, Schories G. Optimization of ultrasound pretreatment and enzymatic hydrolysis of wheat straw: from lab to semi-industrial scale. J Clean Prod. 2022;380:134897. https://doi.org/10.1016/j.jclepro.2022.134897.
132. Zhang J, Bao J. Lignocellulose pretreatment using acid as catalyst. In: Bisaria V, editor. Handbook of biorefinery research and technology. Dordrecht, Netherlands: Springer; 2020. p. 1–14.
133. Huet G, Hadad C, González-Domínguez JM, Courty M, Jamali A, Cailleu D, van Nhien AN. IL versus DES: impact on chitin pretreatment to afford high quality and highly functionalizable chitosan. Carbohydr Polym. 2021;269:118332. https://doi.org/10.1016/j.carbpol.2021.118332.
134. Li P, Yang C, Jiang Z, Jin Y, Wu W. Lignocellulose pretreatment by deep eutectic solvents and related technologies: a review. J Bioresour Bioprod. 2023;8:33–44. https://doi.org/10.1016/j.jobab.2022.11.004.
135. Mora-Pale M, Meli L, Doherty TV, Linhardt RJ, Dordick JS. Room temperature ionic liquids as emerging solvents for the pretreatment of lignocellulosic biomass. Biotechnol Bioeng. 2011;108:1229–45. https://doi.org/10.1002/bit.23108;10.1021/ja025790m.
136. Balan V, Bals B, Chundawat SPS, Marshall D, Dale BE. Lignocellulosic biomass pretreatment using AFEX. Methods Mol Biol. 2009;581:61–77. https://doi.org/10.1007/978-1-60761-214-8_5.
137. Zheng Y, Lin H, Tsao GT. Pretreatment for cellulose hydrolysis by carbon dioxide explosion. Biotechnol Prog. 1998;14:890–6. https://doi.org/10.1021/bp980087g.
138. Andlar M, Rezić T, Marđetko N, Kracher D, Ludwig R, Šantek B. Lignocellulose degradation: an overview of fungi and fungal enzymes involved in lignocellulose degradation. Eng Life Sci. 2018;18:768–78. https://doi.org/10.1002/elsc.201800039.
139. Mathews SL, Pawlak J, Grunden AM. Bacterial biodegradation and bioconversion of industrial lignocellulosic streams. Eur J Appl Microbiol Biotechnol. 2015;99:2939–54. https://doi.org/10.1007/s00253-015-6471-y.
140. Kawaguchi H, Vertès AA, Okino S, Inui M, Yukawa H. Engineering of a xylose metabolic pathway in *Corynebacterium glutamicum*. Appl Environ Microbiol. 2006;72:3418–28. https://doi.org/10.1128/AEM.72.5.3418-3428.2006.

141. Kawaguchi H, Sasaki M, Vertès AA, Inui M, Yukawa H. Engineering of an L-arabinose metabolic pathway in *Corynebacterium glutamicum*. Appl Microbiol Biotechnol. 2008;77: 1053–62. https://doi.org/10.1007/s00253-007-1244-x.
142. Gopinath V, Murali A, Dhar KS, Nampoothiri KM. *Corynebacterium glutamicum* as a potent biocatalyst for the bioconversion of pentose sugars to value-added products. Appl Microbiol Biotechnol. 2012;93:95–106. https://doi.org/10.1007/s00253-011-3686-4.
143. Meiswinkel TM, Gopinath V, Lindner SN, Nampoothiri KM, Wendisch VF. Accelerated pentose utilization by *Corynebacterium glutamicum* for accelerated production of lysine, glutamate, ornithine and putrescine. Microb Biotechnol. 2013;6:131–40. https://doi.org/10.1111/1751-7915.12001.
144. Pérez-García F, Brito LF, Wendisch VF. Microbial production of amine chemicals from sustainable substrates. In: Fang Z, Smith Jr RL, Xu L, editors. Production of *N*-containing chemicals and materials from biomass, vol. 12. Singapore: Springer Nature Singapore; 2023. p. 189–248.
145. Choi JW, Jeon EJ, Jeong KJ. Recent advances in engineering *Corynebacterium glutamicum* for utilization of hemicellulosic biomass. Curr Opin Biotechnol. 2019;57:17–24. https://doi.org/10.1016/j.copbio.2018.11.004.
146. Wendisch VF, Brito LF, Gil Lopez M, Hennig G, Pfeifenschneider J, Sgobba E, Veldmann KH. The flexible feedstock concept in industrial biotechnology: metabolic engineering of *Escherichia coli, Corynebacterium glutamicum, Pseudomonas, Bacillus* and yeast strains for access to alternative carbon sources. J Biotechnol. 2016;234:139–57. https://doi.org/10.1016/j.jbiotec.2016.07.022.
147. Zha J, Zhao Z, Xiao Z, Eng T, Mukhopadhyay A, Koffas MA, Tang YJ. Biosystem design of *Corynebacterium glutamicum* for bioproduction. Curr Opin Biotechnol. 2023;79:102870. https://doi.org/10.1016/j.copbio.2022.102870.
148. Schwardmann LS, Dransfeld AK, Schäffer T, Wendisch VF. Metabolic engineering of *Corynebacterium glutamicum* for sustainable production of the aromatic dicarboxylic acid dipicolinic acid. Microorganisms. 2022;10:730. https://doi.org/10.3390/microorganisms10040730.
149. Buschke N, Schröder H, Wittmann C. Metabolic engineering of *Corynebacterium glutamicum* for production of 1,5-diaminopentane from hemicellulose. Biotechnol J. 2011;6:306–17. https://doi.org/10.1002/biot.201000304.
150. Veit A, Rittmann D, Georgi T, Youn J-W, Eikmanns BJ, Wendisch VF. Pathway identification combining metabolic flux and functional genomics analyses: acetate and propionate activation by *Corynebacterium glutamicum*. J Biotechnol. 2009;140:75–83. https://doi.org/10.1016/j.jbiotec.2008.12.014.
151. Chukwuma OB, Rafatullah M, Tajarudin HA, Ismail N. Lignocellulolytic enzymes in biotechnological and industrial processes: a review. Sustainability. 2020;12:7282. https://doi.org/10.3390/su12187282.
152. Kuthiala T, Thakur K, Sharma D, Singh G, Khatri M, Arya SK. The eco-friendly approach of cocktail enzyme in agricultural waste treatment: a comprehensive review. Int J Biol Macromol. 2022;209:1956–74. https://doi.org/10.1016/j.ijbiomac.2022.04.173.
153. Prajapati AS, Panchal KJ, Pawar VA, Noronha MJ, Patel DH, Subramanian RB. Review on cellulase and xylanase engineering for biofuel production. Ind Biotechnol. 2018;14:38–44. https://doi.org/10.1089/ind.2017.0027.
154. Ribeiro LS, Órfão JJM, Pereira MFR. Simultaneous catalytic conversion of cellulose and corncob xylan under temperature programming for enhanced sorbitol and xylitol production. Bioresour Technol. 2017;244:1173–7. https://doi.org/10.1016/j.biortech.2017.08.015.
155. Ng K, Ye R, Wu XC, Wong SL. Sorbitol dehydrogenase from *Bacillus subtilis*. Purification, characterization, and gene cloning. J Biol Chem. 1992;267:24989–94. https://doi.org/10.1016/S0021-9258(19)73995-7.

156. Pakalın B, Kurpejović E, Bastem GM, Sayar NA, Wendisch VF, Akbulut BS. Valorization of hazelnut husk as a carbon source for L-DOPA production with *Corynebacterium glutamicum*. Biochem Eng J. 2023;190:108768. https://doi.org/10.1016/j.bej.2022.108768.
157. Vortmann M, Stumpf AK, Sgobba E, Dirks-Hofmeister ME, Krehenbrink M, Wendisch VF, Philipp B, Moerschbacher BM. A bottom-up approach towards a bacterial consortium for the biotechnological conversion of chitin to L-lysine. Appl Microbiol Biotechnol. 2021;105:1547–61. https://doi.org/10.1007/s00253-021-11112-5.
158. Matano C, Kolkenbrock S, Hamer SN, Sgobba E, Moerschbacher BM, Wendisch VF. *Corynebacterium glutamicum* possesses β-*N*-acetylglucosaminidase. BMC Microbiol. 2016;16:177. https://doi.org/10.1186/s12866-016-0795-3.
159. QYResearch Global GABA Market Insights, Forecast to 2030. https://www.qyresearch.com/reports/2708051/gaba.
160. Schiener P, Black KD, Stanley MS, Green DH. The seasonal variation in the chemical composition of the kelp species *Laminaria digitata, Laminaria hyperborea, Saccharina latissima* and *Alaria esculenta*. J Appl Phycol. 2015;27:363–73. https://doi.org/10.1007/s10811-014-0327-1.
161. Pérez-García F, Klein VJ, Brito LF, Brautaset T. From brown seaweed to a sustainable microbial feedstock for the production of riboflavin. Front Bioeng Biotechnol. 2022;10:863690. https://doi.org/10.3389/fbioe.2022.863690.
162. Lee SY, Ra CH. Evaluation of pretreatment and GABA production using *Levilactobacillus brevis* fermentation of the seaweed *Saccharina japonica*. Biotechnol Bioproc E. 2023;28:568–76. https://doi.org/10.1007/s12257-023-0073-9.
163. Jakobsen ØM, Benichou A, Flickinger MC, Valla S, Ellingsen TE, Brautaset T. Upregulated transcription of plasmid and chromosomal ribulose monophosphate pathway genes is critical for methanol assimilation rate and methanol tolerance in the methylotrophic bacterium *Bacillus methanolicus*. J Bacteriol. 2006;188:3063–72. https://doi.org/10.1128/jb.188.8.3063-3072.2006.
164. Irla M, Nærdal I, Brautaset T, Wendisch VF. Methanol-based γ-aminobutyric acid (GABA) production by genetically engineered *Bacillus methanolicus* strains. Ind Crop Prod. 2017;106:12–20. https://doi.org/10.1016/j.indcrop.2016.11.050.
165. Huang J, Quest A, Cruz-Morales P, Deng K, Pereira JH, van Cura D, Kakumanu R, Baidoo EEK, Dan Q, Chen Y, Petzold CJ, Northen TR, Adams PD, Clark DS, Balskus EP, Hartwig JF, Mukhopadhyay A, Keasling JD. Complete integration of carbene-transfer chemistry into biosynthesis. Nature. 2023;617:403–8. https://doi.org/10.1038/s41586-023-06027-2.
166. Assessing the laboratory performance of AI-generated enzymes. Nat Biotechnol. 2024. https://doi.org/10.1038/s41587-024-02239-7.
167. Yang J, Li F-Z, Arnold FH. Opportunities and challenges for machine learning-assisted enzyme engineering. ACS Cent Sci. 2024;10:226–41. https://doi.org/10.1021/acscentsci.3c01275.
168. Gergel S, Soler J, Klein A, Schülke KH, Hauer B, Garcia-Borràs M, Hammer SC. Engineered cytochrome P450 for direct arylalkene-to-ketone oxidation via highly reactive carbocation intermediates. Nat Catal. 2023;6:606–17. https://doi.org/10.1038/s41929-023-00979-4.
169. Dueber JE, Wu GC, Malmirchegini GR, Moon TS, Petzold CJ, Ullal AV, Prather KLJ, Keasling JD. Synthetic protein scaffolds provide modular control over metabolic flux. Nat Biotechnol. 2009;27:753–9. https://doi.org/10.1038/nbt.1557.
170. Barba-Cedillo V, Montanier CY. Effect of multimodularity and spatial organization of glycoside hydrolases on catalysis. Essays Biochem. 2023;67:629–38. https://doi.org/10.1042/EBC20220167.
171. Leßmeier L, Wendisch VF. Identification of two mutations increasing the methanol tolerance of *Corynebacterium glutamicum*. BMC Microbiol. 2015;15:216. https://doi.org/10.1186/s12866-015-0558-6.
172. Zuchowski R, Schito S, Neuheuser F, Menke P, Berger D, Hollmann N, Gujar S, Sundermeyer L, Mack C, Wirtz A, Weiergräber OH, Polen T, Bott M, Noack S, Baumgart

M. Discovery of novel amino acid production traits by evolution of synthetic co-cultures. Microb Cell Factories. 2023;22:71. https://doi.org/10.1186/s12934-023-02078-2.

173. Prell C, Busche T, Rückert C, Nolte L, Brandenbusch C, Wendisch VF. Adaptive laboratory evolution accelerated glutarate production by *Corynebacterium glutamicum*. Microb Cell Factories. 2021;20:97. https://doi.org/10.1186/s12934-021-01586-3.
174. Haupka C, Delépine B, Irla M, Heux S, Wendisch VF. Flux enforcement for fermentative production of 5-aminovalerate and glutarate by *Corynebacterium glutamicum*. Catalysts. 2020;10:1065. https://doi.org/10.3390/catal10091065.
175. Velasquez-Guzman JC, Huttanus HM, Morales DP, Werner TS, Carroll AL, Guss AM, Yeager CM, Dale T, Jha RK. Biosensors for the detection of chorismate and cis,cis-muconic acid in *Corynebacterium glutamicum*. J Ind Microbiol Biotechnol. 2024;51:kuae024. https://doi.org/10.1093/jimb/kuae024.
176. Stella RG, Gertzen CGW, Smits SHJ, Gätgens C, Polen T, Noack S, Frunzke J. Biosensor-based growth-coupling and spatial separation as an evolution strategy to improve small molecule production of *Corynebacterium glutamicum*. Metab Eng. 2021;68:162–73. https://doi.org/10.1016/j.ymben.2021.10.003.
177. Mahr R, Gätgens C, Gätgens J, Polen T, Kalinowski J, Frunzke J. Biosensor-driven adaptive laboratory evolution of L-valine production in *Corynebacterium glutamicum*. Metab Eng. 2015;32:184–94. https://doi.org/10.1016/j.ymben.2015.09.017.
178. Schultenkämper K, Brito LF, Wendisch VF. Impact of CRISPR interference on strain development in biotechnology. Biotechnol Appl Biochem. 2020;67:7–21. https://doi.org/10.1002/bab.1901.
179. Jinek M, Chylinski K, Fonfara I, Hauer M, Doudna JA, Charpentier E. A programmable dual-RNA-guided DNA endonuclease in adaptive bacterial immunity. Science. 2012;337:816–21. https://doi.org/10.1126/science.1225829.
180. Nishida K, Arazoe T, Yachie N, Banno S, Kakimoto M, Tabata M, Mochizuki M, Miyabe A, Araki M, Hara KY, Shimatani Z, Kondo A. Targeted nucleotide editing using hybrid prokaryotic and vertebrate adaptive immune systems. Science. 2016;353:aaf8729. https://doi.org/10.1126/science.aaf8729.
181. Cleto S, Jensen JV, Wendisch VF, Lu TK. *Corynebacterium glutamicum* metabolic engineering with CRISPR interference (CRISPRi). ACS Synth Biol. 2016;5:375–85. https://doi.org/10.1021/acssynbio.5b00216.
182. Göttl VL, Schmitt I, Braun K, Peters-Wendisch P, Wendisch VF, Henke NA. CRISPRi-library-guided target identification for engineering carotenoid production by *Corynebacterium glutamicum*. Microorganisms. 2021;9:670. https://doi.org/10.3390/microorganisms9040670.
183. Sun L, Zheng P, Sun J, Wendisch VF, Wang Y. Genome-scale CRISPRi screening: a powerful tool in engineering microbiology. Eng Microbiol. 2023;3:100089. https://doi.org/10.1016/j.engmic.2023.100089.
184. Fontana J, Sparkman-Yager D, Faulkner I, Cardiff R, Kiattisewee C, Walls A, Primo TG, Kinnunen PC, Garcia Martin H, Zalatan JG, Carothers JM. Guide RNA structure design enables combinatorial CRISPRa programs for biosynthetic profiling. Nat Commun. 2024;15:6341. https://doi.org/10.1038/s41467-024-50528-1.

Part II
Production of Biofuels from Agricultural Residues and Food Wastes

Chapter 10
Catalytic Production of Organic Alcohols from Agricultural Residues and Food Wastes

Mengying Liu, Zhe-Hui Zhang, Zhuohua Sun, and Tong-Qi Yuan

Abstract The catalytic conversion of lignocellulosic biomass, primarily sourced from agricultural residues, yields an extensive range of valuable products including biofuels (such as ethanol, biodiesel, and biogas) as well as platform chemicals. This chapter mainly introduces the existing research strategies on the catalytic conversion of cellulose in lignocellulosic components to produce aliphatic alcohol, sugar alcohol, ethanol and other organic alcohol value-added chemicals, the catalytic conversion of hemicellulose to produce furfuryl alcohol, sugar alcohol and other organic alcohols, the catalytic conversion of lignin to produce cyclohexanol and other organic alcohols, and the catalytic conversion of glycerol to produce propanol and propylene glycol and other organic alcohol chemicals. Future development and prospects for organic alcohol chemicals from agricultural waste are summarized and highlighted.

Keywords Agricultural residues · Food wastes · Catalysis · Alcohols · Glycerol

10.1 Introduction

Agricultural residue encompass the byproducts and wastes generated throughout diverse agricultural activities, encompassing organic materials like crop residues, animal manure, food processing waste, and other agricultural byproducts. These waste materials are derived from farming operations, livestock rearing, and food processing industries [1]. Agricultural residues can exist in both solid and liquid forms, posing environmental challenges if not effectively managed. However, it also presents opportunities for recycling, reuse, and conversion into valuable resources

M. Liu · Z.-H. Zhang · Z. Sun (✉) · T.-Q. Yuan
State Key Laboratory of Efficient Production of Forest Resources, Beijing Forestry University, Beijing, China

Beijing Key Laboratory of Lignocellulosic Chemistry, Beijing Forestry University, Beijing, China
e-mail: sunzhuohua@bjfu.edu.cn

Z. Fang et al. (eds.), *Production of Organic Acids and Alcohols from Agricultural Residues and Food Wastes*, Biofuels and Biorefineries 13,
https://doi.org/10.1007/978-981-96-1222-2_10

[2]. For instance, crop residues can serve as animal feed or be composted to enrich soil quality or even converted into biofuels. Animal manure can be utilized as fertilizer or processed into biogas for energy production purposes. Efficient management of agricultural residues is pivotal for sustainable agriculture practices and environmental conservation efforts [3]. Proper disposal methods along with treatment techniques can significantly minimize pollution levels while reducing greenhouse gas emissions simultaneously promoting circular economy principles within this sector [4]. Furthermore, innovative technologies alongside practices are being developed to optimize the utilization of agricultural residues while extracting its potential value in a cost-effective manner that aligns with environmentally friendly approaches.

Approximately 70–80% of global agricultural residues comprises crop residues, including straw, husk, and leaves. In 2019, the worldwide production of cereal straw alone reached approximately 2.14 billion tons. Crop residues can be utilized as animal feed, biomass energy sources, or soil amendments [2, 5]. Animal waste is generated from livestock farming, encompassing manure and slurry [6]. In the United States, approximately 3.55 billion tons of animal manure are produced annually. Animal waste constitutes a substantial contributor to global greenhouse gas emissions, particularly methane release [7].

Globally, around 1.3 billion tons of food waste is generated each year. Approximately 40% of the food produced in the United States goes to waste, resulting in an estimated annual amount of 133 billion pounds discarded [8]. Food waste can be repurposed for composting, animal feed utilization, or biogas production purposes. Furthermore, alongside advancements in modern agriculture practices emerge various novel forms of waste that accumulate significantly on an annual basis; examples include agricultural plastic and chemical residues. In China alone during 2019, roughly 4.6 million tons of agricultural plastic waste were generated independently as a result of agricultural activities conducted within the country's borders exclusively [9]. The environmental consequences associated with agricultural plastic residue can be severe due to its potential contamination effects on soil and water resources alike. Additionally, an approximate quantity of 1.3 million tons pesticide residues are annually produced globally as chemical byproducts originating from diverse sources such as agriculture-related applications among others [10]; these chemical wastes possess inherent risks towards soil quality degradation along with groundwater pollution tendencies while simultaneously posing threats to both human health and overall ecological well-being.

In recent years, there has been growing awareness of the environmental and economic impacts of agricultural residues. Governments, researchers, and farmers are exploring sustainable waste management practices, such as waste reduction, recycling, and converting waste into valuable resources [11, 12]. These efforts can help minimize the environmental impact of agricultural residues while promoting resource efficiency and supporting sustainable agricultural practices.

Agricultural residues can also be transformed into various chemicals through innovative processes and technologies. These chemicals find applications in industries such as plastics, fuels, and fertilizers. The commonly employed methods encompass bioconversion, thermochemical conversion, and chemical conversion processes [13, 14]. Bioconversion entails the utilization of microorganisms, enzymes, or whole cells to convert agricultural residues into valuable chemicals. Thermochemical conversion involves subjecting agricultural residues to high temperatures in an oxygen-free environment to generate chemicals. Chemical conversion employs chemical reactions and processes to convert agricultural residues into useful chemicals. Among these approaches, the catalytic process for converting agricultural residues has garnered significant attention due to its ability to transform these materials into valuable chemicals.

Catalytic conversion of lignocellulosic biomass, primarily derived from agricultural residues, typically yields a diverse range of valuable products including biofuels (such as ethanol, biodiesel, and biogas) and platform chemicals [15, 16]. While extensive reviews have been conducted on most of these processes in recent years, the literature lacks comprehensive evaluations specifically focusing on catalytic systems for the production of valuable organic alcohols. Alcohols play a pivotal role across various industries; they serve as essential ingredients in food products like sauces, marinades, and desserts. Moreover, alcohols are utilized as solvents, preservatives, and excipients in pharmaceutical manufacturing encompassing tablets, capsules, and ointments. Furthermore, they constitute crucial raw materials for the synthesis of diverse chemicals such as plastics, synthetic fibers, and lubricants [17]. In this chapter's context, the aim is to provide an overview of catalytic processes involved in transforming agricultural and food waste into valuable alcohols.

10.2 Organic Alcohols from Lignocellulosic Agricultural Residues

Lignocellulosic agricultural residues refer to the non-edible plant components, primarily comprising cellulose, hemicellulose, and lignin. These constituents are predominantly present in the stems, branches, leaves, and husks of crops such as corn, sugarcane, and bamboo. Lignocellulosic agricultural residues are widely recognized as a renewable resource with immense potential for applications in biofuel production, bioplastics synthesis, and other value-added products [12, 18]. As shown in Table 10.1, lignocellulosic agricultural residues could be converted to different alcohols depend on the catalytic systems used.

The annual generation of lignocellulosic agricultural residues varies depending on the region, agricultural practices, and crop types cultivated, making it challenging to accurately estimate. However, it is widely believed that a substantial global quantity of millions of tons of lignocellulosic agricultural residues is generated each year [19, 20]. In the United States, for instance, approximately 130 million tons of corn stover (the residual plant material following corn harvest) are generated annually. In Brazil, sugarcane bagasse (the residue left behind after sugarcane

Table 10.1 Literature review on the preparation of alcohol compounds from agricultural waste and its cellulose, hemicellulose and lignin

Feedstock	Catalyst	Reaction condition			Yield	Reference
		Temperature/°C	Time/h	Solvent		
Maple sawdust	$Cu_{20}PMO$	300–320	0.5–8	Supercritical methanol	C2–C6 aliphatic alcohols (5.9 wt%)	[21]
Cellulose	CuMgAlOx	300	4	Methanol	C2–C6 mono-alcohols (26.6%)	[22]
Cellulose	Pt/Al_2O_3	190	24	H_2O	Sorbitol (25%) and mannitol (6%)	[23]
Cellulose	Ru (0.2)/H–USY	190	3	H_2O	Sugar alcohol (66%)	[24]
Rice straw	$Pt/\gamma\text{-}Al_2O_3$	230	24	Ethanol	Mannitol (5.5%), sorbitol (0.8%)	[25]
Wheat straw	$CrPO_4$	180	1.5	H_2O/THF	Furfural (67%), HMF (32%)	[26]
Cellulose	WCx/MC	245	0.5	H_2O	Polyols (82.1%)	[27]
Corn straw	Ni-W/MCM	250	2	H_2O	Ethylene glycol (63.6%)	[28]
Cellulose	$H_2WO_4 + Pt/ZrO_2$	250	5	H_2O	Ethanol (32%), ethylene glycol (24%)	[29]
Hemicellulose	$Ir\text{-}ReOx/SiO_2$	140	12	n-dodecane + H_2O	Xylitol (79%)	[30]
Corncob	Ru/C + 0.1% H_2SO_4	140	3	Water-isopropanol	Xylitol (19%)	[31]
Phenol	Raney Ni	80–120	3	Isopropanol	Cyclohexanol (78–95%)	[32]
Poplar	Raney Ni	180	3	2-PrOH/H_2O	Cyclohexanol mixture	[33]
4-Propylguaiacol	$Ni/SiO_2\text{-}Al_2O_3$	250	–	Hexadecane	4-n-propylcyclohexanol (85%)	[34]
4-Propylguaiacol	Raney NiCu	160	6	2-propanol	4-propylcyclohexanol (87.8%)	[35]

harvest) amounts to around 150 million tons per year. Being the largest traditional agricultural country globally, China possesses a substantial quantity of diverse crop straw with a yield of 1.04 billion tonnes in 2015, accounting for nearly one-third of global production.

10.2.1 Organic Alcohols from the Carbohydrate Part of Lignocellulosic Agricultural Residues

The carbohydrate part of lignocellulosic agricultural residues contains cellulose and hemicellulose. Cellulose and hemicellulose are two important structural polysaccharides found in the cell walls of plants and microorganisms. They play a crucial role in providing structural support and protection for these organisms. Cellulose is a linear polymer of β-D-glucose units linked by β-1,4-glycosidic bonds. It is the most abundant organic compound on Earth, making up about 30% of the world's plant biomass [36]. In plants, cellulose forms microfibrils that aggregate to form cellulose fibers, which provide strength and rigidity to the cell walls. Cellulose is also an important component of the cell walls of bacteria and fungi. Hemicellulose is a complex mixture of polysaccharides, including xylose, galactose, and mannose, linked by various glycosidic bonds. The structure of hemicellulose varies among different plant species, but it generally exists as a network of microfibrils that contribute to the overall strength and flexibility of the cell wall [37]. Hemicellulose plays a vital role in plant growth and development, as it allows the cell wall to expand and contract during cell division and growth. Both cellulose and hemicellulose have significant industrial applications. They are primary sources of raw material for the production of paper, cardboard, and other cellulosic products. The breakdown of cellulose and hemicellulose into simpler sugars is also an area of intense research, as it has the potential to provide sustainable energy and reduce our reliance on fossil fuels [38, 39]. Enzymes produced by microorganisms, such as bacteria and fungi, are key to breaking down these complex polymers and converting them into useful compounds. As shown in Fig. 10.1, cellulose can be converted to alcohols with different structures depends on the catalytic systems.

10.2.1.1 From Cellulose to Mixed Aliphatic Alcohols

Efficient methodologies for converting biomass solids into liquid fuels have the potential to reduce dependence on imported petroleum and mitigate atmospheric carbon dioxide emissions. In this context, Ford et al. reported a quantitative catalytic system utilizing copper-doped porous metal oxide (Cu-PMO), composed of earth-abundant materials, for the conversion of cellulosic solids to a mixture of C2–C6 aliphatic alcohols and their methylated derivatives in supercritical methanol [21]. This smart system soon received great attentions of the field and several

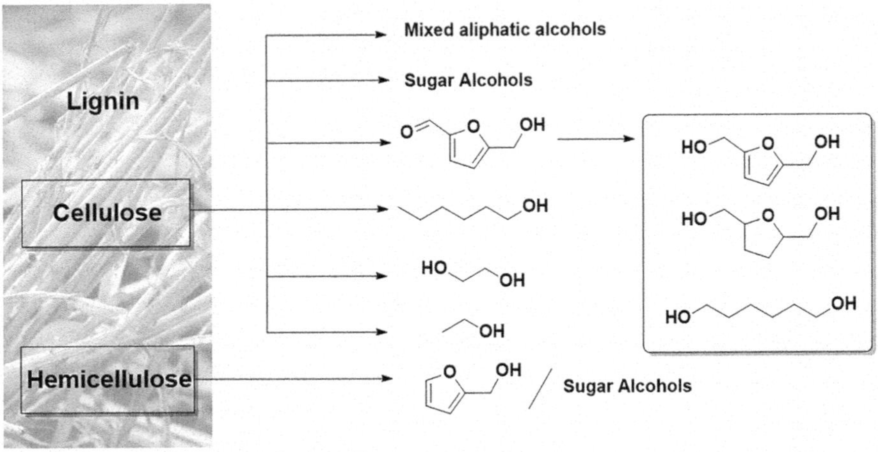

Fig. 10.1 Summary of organic alcohols obtained from the carbohydrate part of lignocellulosic agricultural residues via catalytic processes

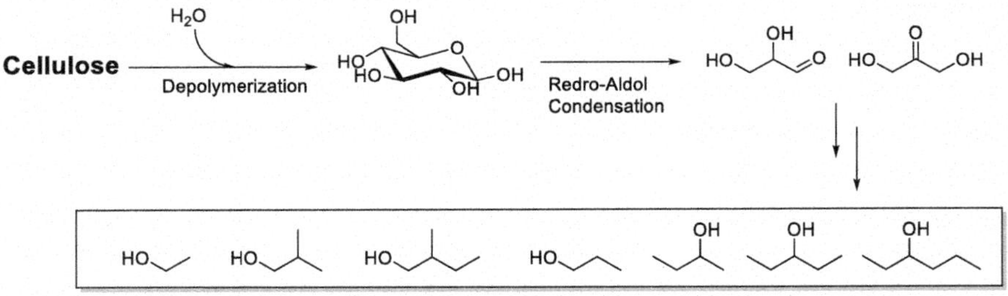

Fig. 10.2 Proposed reaction pathway for alcohol production from cellulose in supercritical methanol with CuMgAl mixed metal-oxide catalyst [22]

follow-up studies were reported. For example, a series of CuO–MO/Al_2O_3 (M = Ce, Mg, Mn, Ni, Zn) catalysts were prepared and investigated in the same system [40]. It achieved a remarkable cellulose conversion rate of 88% and an exceptional content of C4–C7 alcohols above 95% in the liquid products. Furthermore, this catalyst demonstrated excellent stability under the reaction conditions and could be reused for up to four runs.

The reaction pathway and products of cellulose supercritical methanol depolymerization and hydrodeoxygenation (SCM-DHDO) were further investigated by Huber et al. (Fig. 10.2) [22]. It was discovered that monoalcohols, diols, alcohol ethers, and methyl esters were generated from cellulose at 300 °C using a CuMgAl mixed metal-oxide catalyst. Time-course experiments revealed that cellulose rapidly solubilized and depolymerized within 1 h, with C2–C4 diols serving as intermediates. Furthermore, experiments conducted with ^{13}C labeled glucose demonstrated that methanol was incorporated into all liquid products, accounting for approximately 30–40% of the carbon content in total products. Model compound experiments using dihydroxyacetone, isosorbide, and 5-hydroxymethylfurfural indicated that the primary reaction pathway for cellulose involved retro-aldol condensation of solubilized cellulose followed by recondensation with methanol. In conclusion, it can be inferred that the reaction pathway involves rapid solubilization and depolymerization of cellulose followed by retro-aldol condensation to form C2–C4 oxygenates. These retro-aldol condensation products undergo hydrodeoxygenation reactions along with extensive carbon-carbon coupling to produce C2–C7 alcohols or other oxygenates.

10.2.1.2 From Cellulose to Sugar Alcohols

Sugar alcohols, also referred to as polyols, are a class of organic compounds containing multiple alcohol functional groups. These substances occur naturally in certain fruits and vegetables and are industrially synthesized for diverse applications. The significance of sugar alcohols lies in their role as low-calorie sweeteners, impact on oral health, and extensive utilization within the pharmaceutical and food industries [41, 42].

The conversion of cellulose to sugar alcohols typically involves hydrolysis using an acid catalyst, followed by rapid metal-catalyzed hydrogenation of glucose into hexitols. Hexitols, including sorbitol, mannitol and sorbitan, are crucial platform molecules and renewable chemical precursors. Fukuoka and Dhepe observed that catalysts supported with Pt or Ru exhibit remarkable catalytic activity in the conversion of cellulose into sugar alcohols [41]. Additionally, they demonstrated for the first time that supported metal catalysts can environmentally convert cellulose into sugar alcohols. The Pt/Al_2O_3 catalyst exhibited a 31% yield of sugar alcohols, with sorbitol (25%) and mannitol (6%) as the major products. Notably, sorbitol was the predominant product, exhibiting a molar ratio of sorbitol to mannitol of at least 4:1. Building upon this research, Sels et al. further enhanced the hexitol yield by employing a novel catalytic system comprising minute quantities of mineral acid and a solid bifunctional Ru/H–USY or Ru/H–MOR catalyst [24]. The system demonstrates exceptional activity and selectivity, leading to rapid cellulose conversion and outstanding yields (>90%) for hexitols.

The similar system could also be applied for the treatment of agricultural residues. For instance, Tsubaki et al. proposed a novel environmentally friendly and feasible approach for the direct conversion of rice straw into sugar alcohol, along with other aqueous sugars, through a two-step process in a single reactor [25]. Firstly, rice straw was hydrolyzed to glucose and other monosaccharides using solid acid; subsequently, glucose was hydrogenated using in situ hydrogen generated from ethanol steam reforming. The conversion efficiency of rice straw over $Pt/\gamma\text{-}Al_2O_3$ at 503 K reached 73.1%, while the yields of mannitol and sorbitol were 5.5% and 0.8%, respectively.

10.2.1.3 From Cellulose to HMF and the Derived Alcohols

Due to the recalcitrance and highly functionalized nature of biomass, the conversion process for obtaining desired products necessitates a series of intricate procedures involving deconstruction, catalytic conversion, as well as separation and purification techniques. Within the biorefinery roadmap, 5-hydroxymethylfurfural (HMF) emerges as a pivotal link connecting biomass feedstocks with alternative fuels, chemicals, and materials that have the potential to replace petroleum-derived products [43–45].

One of the most prevalent techniques for obtaining HMF (5-hydroxymethylfurfural) from cellulose involves acid-catalyzed hydrolysis [46, 47]. In this process, cellulose is subjected to treatment with a potent acid, such as sulfuric acid, in a liquid medium at elevated temperatures (180–200 °C) and pressures (up to 100 bar partial pressure). The hydrolysis reaction disassembles the cellulose polymer into smaller molecules, including HMF. Alternatively, cellulose can be hydrolyzed using a basic catalyst like sodium hydroxide or potassium hydroxide in an aqueous solution [48, 49]. This method typically operates at lower temperatures (around 100 °C) and pressures (up to 10 bar partial pressure). The base catalyst facilitates the cleavage of cellulose chains, leading to the formation of HMF and other furan compounds. Enzymatic hydrolysis represents an alternative approach for the production of HMF from cellulose [50, 51]. In this process, cellulolytic enzymes, such as cellulase and beta-glucosidase, are employed to degrade cellulose into smaller molecular fragments. The reaction occurs under mild temperatures (around 50–60 °C) and in aqueous buffer solutions. These enzymes act as catalysts for the hydrolysis of cellulose, yielding HMF that can be further transformed into furan derivatives. Recently, there has been a growing interest in exploring MOF catalysts for the conversion of cellulose to HMF [52, 53]. These porous materials possess high surface areas which facilitate the efficient conversion of cellulose to HMF. Notably, MOF catalysts incorporating metal centers (e.g. zinc, iron or copper) and organic ligands have exhibited promising activity and selectivity towards the hydrolysis of cellulose to produce HMF. Reactions typically occur at moderate temperatures (100–150 °C) and pressures (up to 10 bar partial pressure). Another innovative approach for obtaining HMF from cellulose involves the utilization of photocatalysts [54, 55]. In this process, cellulose is subjected to UV or visible light irradiation under aqueous conditions, while a photocatalyst (such as titanium dioxide or zinc oxide) facilitates the hydrolysis of cellulose to yield HMF. This method offers the advantage of harnessing solar energy to drive the reaction, rendering it more sustainable and environmentally benign. The reaction proceeds at ambient temperatures and pressures.

The direct utilization of agricultural residues has also been investigated. For example, Fang et al. examined the impact of Brønsted acids (H_2SO_4, HCl, and HNO_3) and alkali metal cations (Li^+, Na^+, and K^+) on the conversion of corn stover in an H_2O/THF medium system [56]. Among the catalysts tested, H_2SO_4/Na_2SO_4 achieved a 76% yield at 190 °C for 1 h. The $CrPO_4$ catalyst demonstrated remarkable efficiency in facilitating the production of furfural and HMF from wheat straw, yielding 67% furfural and 32% HMF at a temperature of 180 °C for a duration of 90 min [26]. The $SnCl_2$-PTA/H β zeolite, synthesized by incorporating $SnCl_2$ and phosphotungstic acid (PTA) onto H β zeolite, exhibited a 30% yield of HMF and a 71% yield of furfural from wheat straw [57].

As one of the most pivotal platform chemicals, HMF has been extensively investigated for its downstream applications. The conversion of HMF into high-value fuels, fine chemicals, and materials encompasses a series of intricate reaction processes involving selective oxidation, hydrogenation, etherification, coupling, functionalization, condensation reactions as well as the integration of diverse reaction networks. Achieving selective modification of C=O, C–O and furan ring functional groups in HMF towards desired products necessitates the development of catalysts with tailored compositions, structures and active sites that exhibit both activity and selectivity. Furthermore, it requires the exploration of suitable reaction media and meticulous regulation of reaction conditions.

In order to produce value-added alcohols, highly selective hydrogenation catalysts are necessary. For instance, the selective hydrogenation of HMF to BHMF (2,5-bishydroxymethylfuran) is crucial for the production of ethers, ketones, and polymers [58]. This reaction can be achieved using the commercially available Ru/C catalyst in an aqueous solution. At a mild condition (50 °C, 6 h, 30 bar H_2 partial pressure), this system enabled the selective hydrogenation of HMF to BHMF with a yield of 93.0% [59]. Furthermore, when diphenylsilane (Ph_2SiH_2) is added as a hydrogen donor under mild conditions (25 °C, 2 h), selective reduction of HMF into BHMF over K_2CO_3 can be achieved with a high yield of 94% [60]. Kuai et al. reported a TiO_2 supported single-atom palladium (Pd1) and Pd nanoparticles (PdNPs) catalyst which enables the selective hydrogenation of HMF to BHMF at room temperature. The synergistical effect was discovered where PdNPs mainly catalyzed the dissociation of H_2 molecules to H atoms, while Pd1 was responsible for the activation of the C=O group [61]. In addition of using high pressure hydrogen, this reaction can also be achieved via the selective catalytic transfer hydrogenation (CTH) reaction, where alcohols such as 2-butanol or isopropanol function as both the hydrogen donor and solvent [62].

Similar to BHMF, BHMTHF (2,5-bishydroxymethyltetrahydrofuran) represents a promising platform chemical for bio-based polyesters and serves as a potential feedstock for the synthesis of 1,6-hexanediol and 2,5-dimethyltetrahydrofuran [63]. Fulignati et al. demonstrated that employing a commercial Ru/C catalyst under mild conditions (50 °C, 6 h, 30 bar H_2 partial pressure) enabled selective hydrogenation of HMF to BHMF. By increasing the temperature and pressure (100 ° C, 6 h, 50 bar H_2 partial pressure), BHMTHF could be obtained in aqueous solution with an impressive yield of 95.3% [59]. Furthermore, through ring-opening and hydrogenolysis processes over the Pt–WOx/TiO_2 catalyst, BHMTHF could be further converted into 1,6-hexanediol with a remarkable yield of up to 70% [64].

10.2.1.4 From Cellulose to Ethylene Glycol

Ethylene glycol (EG) is a colorless and odorless chemical compound that exhibits high versatility, finding extensive applications across various sectors of the chemical industry. With an annual consumption of 20 million metric tons, it holds a significant market share primarily attributed to its utilization in the production of poly(ethylene teraphthalate) (PET) fibers and bottles. However, the current predominant method for EG synthesis involves multiple steps including cracking, epoxidation, and hydration from petroleum-derived ethylene [65, 66]. In contrast to the petroleum-dependent multistep process, the biomass route offers significant advantages as a one-pot process utilizing renewable feedstock.

In this field, the research group led by Zhang has made significant contributions to this system [67]. Their initial publication demonstrated the potential of tungsten carbide as a replacement for precious metals in catalyzing cellulose degradation under mild conditions [68]. Notably, tungsten carbide exhibits a remarkable advantage over platinum and ruthenium catalysts due to its high yield of ethylene glycol (EG) compared to other polyols. Moreover, through the synergistic effect between nickel and W_2C, the addition of a small amount of nickel significantly increased the EG yield to 61%. Subsequently, they employed a nanocasting approach to prepare mesoporous carbon (MC) with three-dimensional interconnected mesopores, enhancing dispersion and accessibility of tungsten carbides [27]. The resulting catalyst exhibited high selectivity towards EG (up to 72.9%) in an aqueous solution. Furthermore, their ongoing work involved synthesizing a series of W-based bimetallic catalysts including Ni-W, Pd-W, Pt-W, Ru-W, and Ir-W on SBA-15 support material; achieving maximum EG yield (75.4%) on Ni-W/SBA-15 catalyst [69].

The reaction pathway for cellulose conversion to EG is illustrated in Fig. 10.3. In the initial step, cellulose undergoes hydrolysis into oligosaccharides and glucose under the catalysis of H^+. This catalytic activity is facilitated by both hot water and dissolved H_xWO_3. Subsequently, these sugar intermediates rapidly undergo C–C cleavage through a retro-aldol reaction, resulting in the formation of glycolaldehyde as a key intermediate. The homogeneous catalysis of dissolved H_xWO_3 plays a

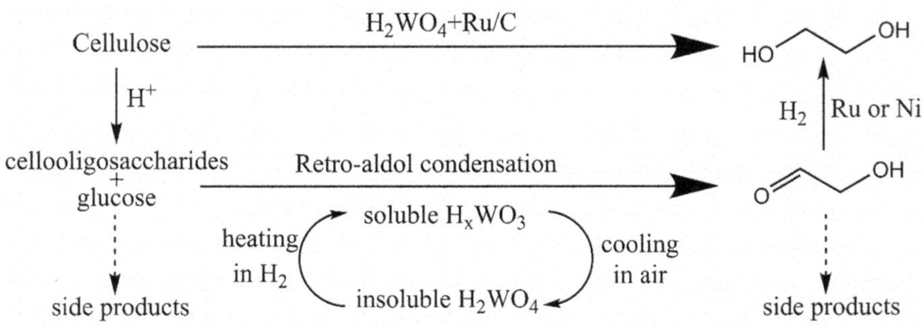

Fig. 10.3 Illustration of the reaction pathway for cellulose conversion to EG over tungsten-based catalysts [67]. (Reprinted with permission from [67], Copyright © 2013, American Chemical Society)

crucial role in this process. Finally, glycolaldehyde is efficiently hydrogenated into EG with the aid of heterogeneous catalysts such as Ru or Ni transition metals. Additionally, alongside the three consecutive reactions leading to EG formation, various side reactions occur due to metastable intermediate sugars and glycolaldehyde.

Compared to cellulose substrates, the conversion of agricultural residues into high-value products is significantly more complex due to the diverse chemical structures and properties that dictate the composition of cellulose, hemicellulose, lignin, and other contaminants. Recently, numerous tests have also been conducted on bimetallic Ni-W catalysts with different supports for the conversion of raw biomass. Ribeiro et al. investigated the efficacy of Ni-W catalyst supported by CNT for direct conversion of forestry, agricultural, and urban wastes [70]. Cotton wool exhibited the highest feedstock conversion and EG yield among all tested materials. The remarkable 43.1% yield can be attributed to the naturally low lignin content present in cotton. By employing an alternative catalyst prepared through the impregnation method, the catalytic conversion of alkali-pretreated corn stalk into polyol was investigated. The utilization of Ni-W/MCM catalysts proves to be a highly advantageous pursuit, resulting in an impressive 63.6% yield of ethylene glycol from cornstalk pretreated with 2% NaOH [28]. In contrast, Li et al.'s attempt at one-pot catalytic conversion of cornstalk into liquid alkylcyclohexanes and polyols over RuO/C catalyst was less successful, yielding only up to 23.4% EG [71].

10.2.1.5 From Cellulose to Ethanol

Bioethanol is a form of alcohol generated through the fermentation process of organic matter, primarily sourced from plants. It serves as a sustainable fuel option, particularly as an alternative to gasoline in transportation applications. Typically derived from crops like corn or sugarcane, bioethanol production involves fermenting the carbohydrates present in these crops into ethanol [72]. In the realm of biofuels, bioethanol is classified as a "first-generation" biofuel due to its direct derivation from plant materials. Second-generation biofuels, such as cellulosic ethanol, are produced using non-food cellulosic materials like wood chips, grasses, or agricultural residues [73].

While the sugar industry has traditionally utilized fermentation for centuries to produce ethanol, alternative routes involving direct carbohydrate transformation remain exceedingly rare. Recently, a groundbreaking aqueous-phase catalytic method was developed for synthesizing cellulosic ethanol, representing a significant advancement as it offers the potential to utilize all carbon atoms in sugars with enhanced kinetics [74]. In this approach, the utilization of subcritical aqueous environments is imperative, thus necessitating experiments to be conducted within a hermetically sealed reactor as the reference scenario. These experiments encompass cascade reactions including hydrolysis, retro-aldol condensation, and hydrogenolysis.

In 2017, Zhang's group reported the first chemo-catalytic route for producing bioethanol from cellulose. The process involved converting cellulose to methyl glycolate through alcoholysis using a Ni-W catalyst, followed by hydrogenation with a Cu/SiO_2 catalyst to produce ethanol with a yield of 29% [75]. Later, the yield was improved to nearly 44% using a bifunctional Pt-Cu/SiO_2 catalyst; however, this modification resulted in economic inferiority due to additional separation and retreatment processes. To address these issues, the group developed a Mo/Pt/WOx catalyst that could couple multiple reactions in a one-pot approach [76, 77].

Following the investigation conducted by Zhang et al., Wang and colleagues achieved a carbon yield of 32.0% for cellulosic ethanol through the physical combination of H_2WO_4 and Pt/ZrO_2 [29]. In another study, Fu's group identified a Ru-WOx/HZSM-5 catalyst that yielded ethanol at 42.3% from 4 wt% cellulose [78, 79]. Subsequently, Ma et al. synthesized a Ni@C catalyst and employed it for the direct conversion of cellulose to ethanol in the presence of phosphoric acid, achieving a remarkable maximum yield of 69.1% [80]. Since then, numerous transition metal-based catalysts have been systematically developed to enable efficient production of cellulosic ethanol [81]. For example, Mauriello et al. synthesized a Pd/Fe_3O_4 catalyst that exhibited a remarkable selectivity of 51.0% for ethanol, comparable to the Pt-based catalysts prepared by Ding et al. [82, 83] This finding provides compelling evidence for the current focus on efficient production of cellulosic ethanol through chemo-catalytic methods. It highlights the feasibility of direct conversion of cellulose into bio-ethanol via a series of tandem reactions, emphasizing the crucial role played by processing conditions and multifunctional catalysts in orchestrating the intricate network.

In order to facilitate further industrial applications, it is imperative to consider the entire lignocellulosic biomass as a potential feedstock. However, this approach also presents unexpected challenges due to the presence of co-existing components, such as lignin and hemicellulose. For instance, Li et al. demonstrated the utilization of a Ru-WOx/HZSM-5 catalyst for the chemical conversion of cellulose, bagasse, and corn stalk under identical conditions, resulting in ethanol yields of 45.2%, 34.9%, and 26.8% respectively [78]. To address this issue, a continuous process was subsequently developed for the hydrogenolysis of cellulose extracted from corn straw [84]. This process demonstrated the capability to yield ethanol at 6.1 wt% using concentrated feedstock (up to 30 wt%), while no detectable by-products were observed. As shown in Fig. 10.4, the initial stage in hydrothermal reactors focused on the conversion of accumulated cellulose into polyols through hydrogenolysis, followed by the cleavage of C–O bonds to produce cellulosic ethanol with relatively high conversion and selectivity rates of 75% and 84%, respectively. This multifunctional approach offers a viable pathway for utilizing raw biomass with economically competitive advantages.

Based on the aforementioned discussions, it is indisputable that cellulosic ethanol serves as a renewable and environmentally friendly energy carrier, exhibiting immense potential in revolutionizing the current utilization of fossil fuels within biorefining industries, thereby facilitating the establishment of a carbon-neutral and circular economy. The contributions made by chemo-catalytical approaches

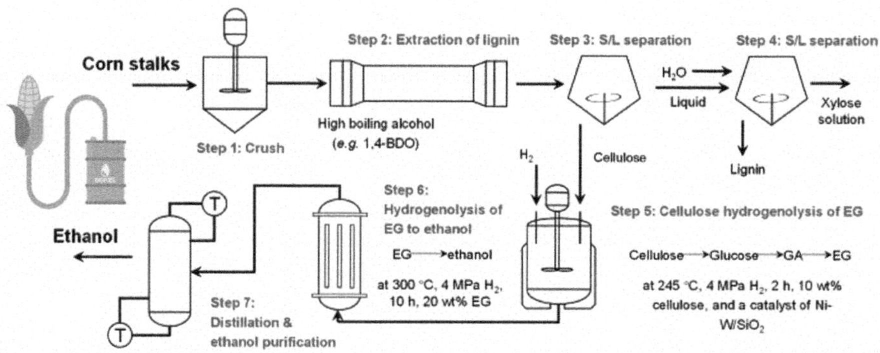

Fig. 10.4 Schematic diagram of the continuous hydrogenolysis of biomass-derived cellulose to bio-ethanol [84]. (Reprinted with permission from [84], Copyright © 2021, Elsevier)

efficiently address several drawbacks associated with traditional cellulose-to-ethanol technologies, such as prolonged reaction times and imbalanced carbon atom ratios. Despite its promising prospects, this pathway still faces various challenges at present (e.g. low selectivity, unstable systems, and limited industrial data).

10.2.1.6 From Hemicellulose to Furfuryl Alcohol and Sugar Alcohols

Hemicellulose, a complex carbohydrate polymer presents in the cell walls of plants such as trees and crops, is composed of various sugar molecules including xylose, glucose, and mannose that are linked together in a long branching structure [37]. This versatile biomaterial can be utilized for the production of valuable platform chemicals like sugar alcohols, furfural, and furfuryl alcohols.

Traditionally, hemicellulose is hydrolyzed using enzymes and mineral acids to produce sugar monomers [85, 86]. However, due to their low efficiency and harsh conditions, efforts have been focused on developing greener routes based on the application of solid acid and supported metal catalysts. The production of xylitol via hydrogenation of xylose using a RANEY® nickel catalyst in a three-phase slurry reactor is a well-established method; however, the deactivation of catalytically active sites poses a significant challenge [87]. In addition to this catalyst, supported Ru and Pt metal catalysts have been employed for the conversion of xylose into xylitol [88]. Latterly, there has been an exploration of combining a supported metal catalyst (Ru/C) with a mineral acid (H_2SO_4) for achieving high-yield production of xylitol (83%) from hemicellulose (xylan) [89]. In 2014, Tathod and Dhepe reported an efficient one-pot method for the conversion of hemicellulose (xylan) and xylose into sugar alcohols (xylitol + arabitol) using a PtSn/AL bimetallic catalyst [90]. By incorporating a small amount of Sn (0.43 wt%) into the Pt (3.5)/AL catalyst (33%), the yields of sugar alcohols in xylose reactions can be increased up to 79%. Moreover, when applied to hemicellulose (xylan) reactions, the addition of 0.43 wt% Sn to the Pt (3.5)/AL catalyst resulted in a remarkable enhancement in

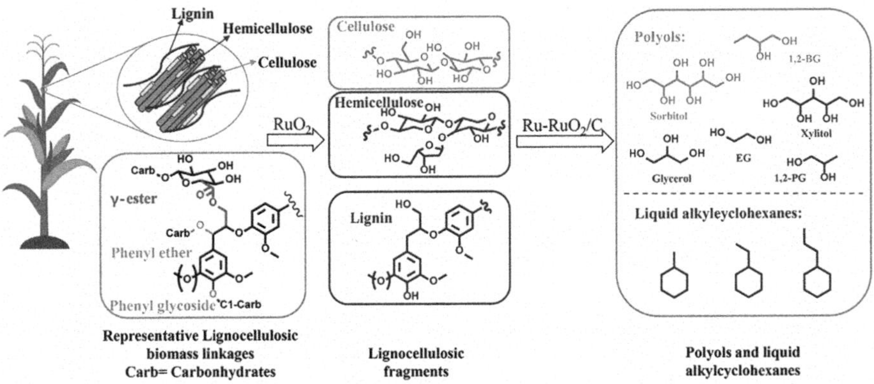

Fig. 10.5 Schematic representation of the direct catalytic transformation of lignocellulosic biomass into liquid alkylcyclohexanes and polyols over partially reduced Ru/C catalysts [71]. (Reprinted with permission from [71], Copyright © 2018, American Chemical Society)

sugar alcohol yields from 18% to 50%, representing a significant improvement by approximately 2.8 times. Latterly, the research group led by Tomishige discovered that the combination of Ir–ReO_x/SiO_2 with acids exhibited remarkable efficacy in converting hemicellulose (xylan) into three distinct products, namely n-pentane, pentanols, and xylitol [30]. Notably, a high yield of 79% xylitol was achieved through the utilization of Ir–ReOx/SiO_2 combined with H_2SO_4 in an aqueous phase at 413 K for a duration of 12 h.

Direct conversion of agricultural residues poses a greater challenge compared to xylose transformation. Wang et al. present a one-pot catalytic process using Ru/C catalysts for the conversion of cornstalk into liquid alkylcyclohexanes (from lignin fraction) and polyols (from cellulose and hemicellulose components) [71]. The study found that the best performance was achieved with Ru/C catalyst reduced at 300 °C, resulting in a mole yield of 97.2% for liquid alkylcyclohexanes and 52.7% for polyols (including sorbitol, xylitol, and C2–C4 polyols) at 200 °C (Fig. 10.5). The hydrolysis-hydrogenation of hemicellulose for direct xylitol production from corncob was investigated by Delgado-Arcaño et al. [31] They discovered that simultaneous hydrogenation of the formed xylose to produce xylitol could be achieved through hydrolyzing the lignocellulosic material, but the addition of acid was necessary to enhance xylan hydrolysis into xylose, which was subsequently reduced to xylitol, thereby eliminating intermediate purification steps. In the presence of H_2 pressure, a xylitol yield of 19% was obtained, while when isopropanol was utilized, a yield of 7% was achieved. Sugar beet fiber is an agricultural by-product in the sugar manufacturing and an available biomass source with a rich hemicellulose component. In the study conducted by Fukuoka and colleagues, the hydrogenolysis of beet fiber was investigated using supported metal catalysts under pressurized hydrogen conditions [91]. It was observed that activated carbon-supported Ru exhibited superior catalytic activity, resulting in arabitol as the major product with a remarkable yield of 83% during the hydrogenolysis of the hemicellulose component within this material.

As a bio-based product derived from hemicellulose, furfural is considered a highly value-added commodity with a global production capacity of approximately 300 kt/year. It has great potential for use in the production of herbicides, stabilizers, resins, oil refining products, plastics, agrochemicals and pharmaceuticals [92]. Approximately 60% of the global furfural production is dedicated to the synthesis of furfuryl alcohol, a crucial monomer in the polymer industry for manufacturing poly(furfuryl alcohol), a chemically resistant resin with significant implications [93].

At the initial stage, the industrial reduction of furfural to furfuryl alcohol is typically conducted through catalytic hydrogenation in either liquid or vapor phase, employing a heterogeneous chromium-based catalyst. However, due to its carcinogenic potential, extensive efforts have been devoted to replacing these catalysts with non-toxic alternatives. Among the major systems investigated are RANEY®-type catalysts and noble metals supported on oxides [94–96]. Instead of employing a two-step strategy, an innovative approach was reported for the synthesis of furfuryl alcohol from xylose using a dual heterogeneous catalyst system, enabling the reaction to proceed in a single step with exceptional selectivity [97]. Additionally, a chemoenzymatic approach was employed to synthesize furfuryl alcohol from various biomass sources such as corncob, bamboo shoot shell, and rice straw [98]. This method involved the utilization of tandem catalysis with Lewis acid ($SnCl_4$ or solid acid SO_4^{2-}/SnO^{2-} bentonite) and biocatalyst in a one-pot manner.

Since cellulose itself is a macromolecular polysaccharide composed of a series of glucose, hydrolysis and further conversion of cellulose into various sugar alcohols is a common way of cellulose value-added utilization. The main types of sugar alcohols obtained by different processes are also different, which can better obtain the desired target products. If the $Pt/\gamma\text{-}Al_2O_3$ catalyst is used to react at 230 °C for 24 h, the sugar alcohol products dominated by sorbitol can be obtained. With the increase of people's demand for different chemicals and the deepening of research on cellulose, cellulose conversion to produce ethanol, ethylene glycol, hexanediol and other alcohols, the main reactions involved are hydrolysis and hydrolysis, and methyl glycolate can be obtained by alcoholysis and then catalytic hydrogenation to produce ethanol.

10.2.2 *Organic Alcohols from the Lignin Part of Lignocellulosic Agricultural Residues*

Lignin, a complex and abundant polymer present in plant cell walls, imparts structural support and plays a pivotal role in water transport. Composed of phenolic compounds, lignin exhibits exceptional resistance to degradation owing to its

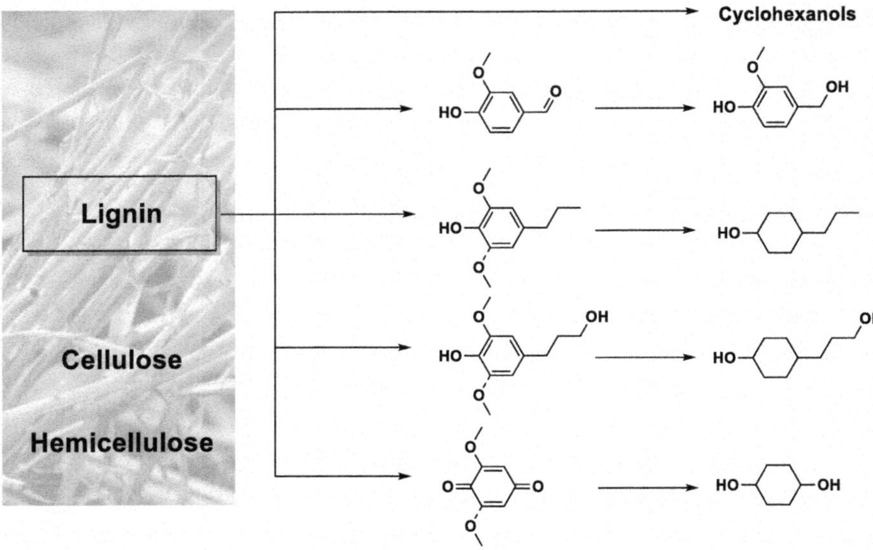

Fig. 10.6 Summary of organic alcohols obtained from the lignin part of lignocellulosic agricultural residues via catalytic processes

intricate and irregular structure. Given its renewable nature and abundance, lignin has garnered significant attention as a promising feedstock for chemical production [99].

The depolymerization processes can effectively convert lignin into smaller aromatic compounds, which can be achieved through various methods including thermochemical, biochemical, or enzymatic processes [100–102]. Consequently, the resulting aromatic compounds possess significant potential for conversion into valuable chemicals such as phenols, vanillin, syringaldehyde, and guaiacol. These chemicals find wide-ranging applications in industries such as pharmaceuticals, flavors, fragrances, and polymers. Lignin can be transformed into biofuels through processes such as pyrolysis or hydrothermal liquefaction, enabling the breakdown of lignin into smaller molecules that can be further refined into bio-oil or biochar. Moreover, lignin serves as a valuable precursor for the synthesis of carbon materials including activated carbon, carbon fibers, and carbon nanomaterials, which find diverse applications in energy storage, water treatment, and electronics. Lignin can be modified and harnessed as a natural adhesive or resin in various applications, showcasing its potential for synergistic integration with other polymers to augment their mechanical properties while mitigating their environmental impact. Notably, the valorization of lignin represents an active field of investigation, with ongoing efforts focused on developing efficient and cost-effective conversion processes. However, substantial technical and economic challenges must be addressed before achieving large-scale commercialization of lignin utilization in chemical production.

Among all the chemicals derived from lignin, alcohols constitute a substantial proportion. As depicted in Fig. 10.6, lignin can be directly valorized into a blend of cyclohexanols or depolymerized to yield diverse platform chemicals that can subsequently undergo upgrading processes to obtain organic alcohols. In this section, we will discuss all these related catalytic systems.

In 2012, Wang and Rinaldi pioneered the utilization of isopropanol as both a hydrogen donor and solvent in conjunction with Raney Ni for the defunctionalization of lignin monomers. Remarkable yields (78–95%) of cyclohexanols were achieved under mild reaction conditions (80–120 °C) [32]. Following this strategy, the research group then successfully developed an integrated catalytic biorefining process capable of efficiently converting lignin into a non-pyrolytic bio-oil. Subsequently, under remarkably mild conditions, the obtained bio-oil was further transformed into a mixture of cyclohexanols through hydrogen transfer reactions in the presence of Raney Ni and 2-PrOH [33].

Vanillin alcohol is a synthetic compound by hydrogenation of vanillin, which is a natural compound found in vanilla beans or produced from oxidative depolymerization of lignin [103]. It could find extensive application as a flavor and fragrance constituent in diverse products, including baked goods, ice cream, confectionery, and perfumery. Moreover, vanillin alcohol is employed in the manufacturing of certain pharmaceutical preparations such as cough syrups, mouthwashes, and throat lozenges. The catalytic hydrogenation of vanillin to vanillin alcohol could be achieved by various supported metal catalysts [104, 105]. For instance, Guo et al. discovered that Pd/C, Pt/C, and Au/C catalysts exhibit potential for the aqueous-phase hydrogenation of vanillin to vanillyl alcohol under mild reaction conditions [106]. The catalytic performance followed the sequence of Pd/C > Pt/C > Au/C. Notably, the Pd/C catalyst demonstrated exceptional catalytic activity (>99% vanillin conversion and >99% vanillyl alcohol selectivity) at optimized reaction conditions (i.e., 30 °C and 0.7 MPa H_2 for 90 min), along with remarkable reusability and stability (up to four consecutive runs). Furthermore, the support material could be synthesized through thermo-chemical activation of a fermentation residue derived from a mixture of food waste and sludge to produce a nitrogen-enriched advanced carbon material.

In the past decade, various approaches have emerged for the separation of high-quality lignin under mild conditions or through elegant catalytic systems for transforming lignin into chemicals, fuels, and materials [99, 107]. Among these promising strategies, the lignin-first biorefinery stands out as one of the most brilliant and extensively studied methods due to its ability to achieve high yields and selectivity of several mono-phenolic derivatives [108, 109].

The most widely reported platform chemicals produced from this strategy are 4-propylmethoxyphenols, which can be converted to 4-propylcyclohexanol. This compound holds great potential as an important building block for the synthesis of polymers and flavors. Sels and co-workers conducted a comprehensive investigation on various home-made supported Ni catalysts [110]. Among them, the best performing catalyst was found to be Ni/CeO_2, achieving over 80% yields of 4-alkylcyclohexanol. Notably, this catalyst demonstrated excellent performance in converting guaiacol mixtures derived from real softwood sources and could be reused in subsequent experiments. Additionally, commercial catalysts including

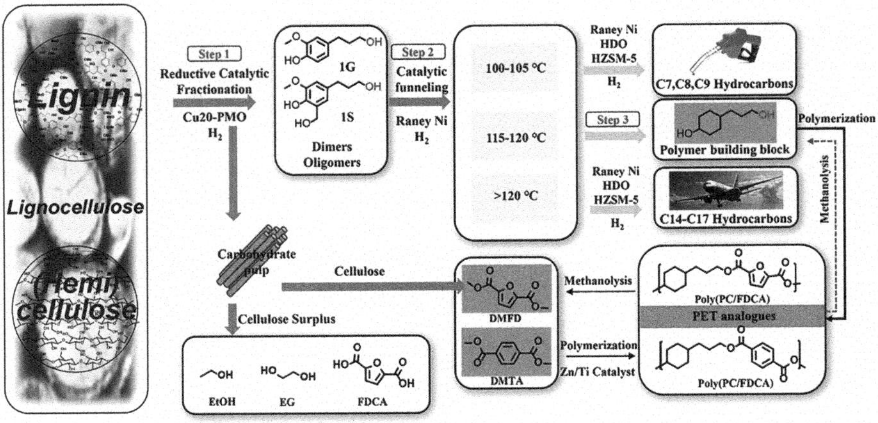

Fig. 10.7 Biorefinery strategy to produce fully lignocellulose-based, recyclable PET analogs [111]. (Reprinted with permission from [111], Copyright © 2022, Springer Nature)

Ru/C, Pd/C, and Ni/SiO_2–Al_2O_3 were evaluated and resulted in a satisfactory yield of 85% for producing 4-propylcyclohexanol [34]. A Raney NiCu catalyst was prepared by employing a ball milling and dealloying process, as reported by Sun et al. The resulting catalysts exhibited enhanced activity compared to the commonly used Raney Ni catalyst, with a remarkable yield of 87.8% for 4-propylcyclohexanol achieved at 160 °C when utilizing 2-propanol as the hydrogen donor [35].

Another platform chemical widely reported in the lignin-first strategy is 4-propanolmethoxyphenols. Exploiting its distinctive structure, which includes a hydroxy group in the side chain, Sun et al. proposed an intelligent approach to convert them into diol monomers for the synthesis of PET analogues [111]. As depicted in Fig. 10.7, the core of this concept lies in a two-step catalytic process devoid of noble metals (Cu-PMO-catalyzed reductive catalytic fractionation and Raney Ni-mediated catalytic funneling), enabling the synthesis of a single aliphatic diol, namely 4-(3-hydroxypropyl)-cyclohexanol, with remarkable isolated yield (11.7 wt%). This diol is co-polymerized with methyl esters derived from cellulose residues, specifically terephthalic acid and furan dicarboxylic acid, yielding polyesters exhibiting competitive molecular weight and thermal properties. Moreover, these polymers exhibit exceptional chemical recyclability in methanol, making them promising contenders for the circular economy.

Leveraging the advantages of the RCF process, in 2021, Samec's group discovered that selective conversion of the high molecular weight fraction obtained after depolymerization of native lignin to 2,6-dimethoxybenzoquinone (DMBQ) could be achieved through an elegant oxidation process employing Bobitt's salt [112]. Building upon this groundbreaking work, Barta et al. subsequently identified DMBQ as a viable precursor for synthesizing 1,4-cyclohexanediol (CHDO) via a reductive demethoxylation and hydrogenation pathway utilizing commercial Raney Ni catalyst [113]. CHDO has demonstrated its utility as an intermediate chemical for manufacturing diverse polymer materials such as scratch-resistant coatings and supramolecular associations [114–116]. Subsequently, Sun et al. proposed a hydrogen-free process that is more sustainable [117]. The focal point of this study lies in the preparation of environmentally friendly ball-milled Mn-modified Raney Ni catalysts (Raney NiMn), resulting in a comparable yield as high as 86.1%.

According to the time sequence, the process of preparing cyclohexanol, cyclohexanol, vanillin and other organic alcohols from lignin in agricultural waste was discussed. Lignin can be converted by pyrolysis, biological enzyme method, oxidation catalysis and other methods, and the more popular transformation method is the lignin priority strategy. It can achieve high yield and high selectivity of several monophenol derivatives.

10.3 Organic Alcohols from Glycerol

Glycerol is a naturally occurring byproduct obtained during the processing of fats and oils, derived from both plant sources (e.g. palm oil, soybean oil) and animal sources (e.g. lard, tallow) [118]. As depicted in Fig. 10.8, glycerol is a significant byproduct of biodiesel production processes, constituting approximately 10 wt% of the total biodiesel yield. With the increasing production and utilization of biomass,

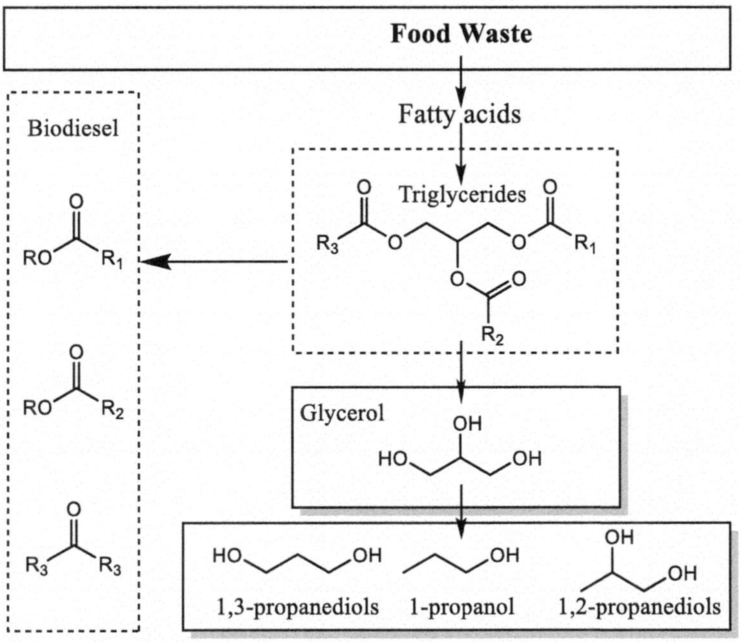

Fig. 10.8 General process of the production of glycerol and its transformation to propanol and propanediols

glycerol has become more readily available as an abundant renewable resource. Consequently, there has been considerable interest in developing innovative technologies for converting glycerol into value-added chemicals such as propanol and propanediols [119, 120].

Propanediols are significant commodity chemicals. 1,2-PD serves as a less toxic alternative to other compounds in the domains of paint, liquid detergent, cosmetics, food and tobacco [121]. Additionally, it finds utility as an antifreeze coolant and deicing reagent. On the other hand, 1,3-PD primarily functions as a monomer in the production of polymers like polytrimethylene terephthalate (PTT), which is a biodegradable polyester with immense potential for application in carpet and textile manufacturing alongside polyethers and polyurethanes [122].

Glycerol can be converted into propanediols through catalytic hydrogenolysis, offering a promising novel synthesis pathway. The catalytic hydrogenolysis reactions and processes for glycerol to propanediols have been extensively investigated over the past few decades. Hydrogenolysis of glycerol involves the breaking (lysis) of C–C or C–O bonds induced by hydrogen molecules. This process initiates with dehydration, encompassing cleavage of C–O and C–H bonds, which is facilitated by catalysts containing acidic sites. Dehydration at primary hydroxyl groups is promoted by Lewis acids, while Bronsted acids induce dehydration at secondary hydroxyl groups. Subsequently, a hydrogenation step occurs where C–H and O–H bonds are formed and/or another dehydration step takes place.

Table 10.2 gives examples of various hydrogenolysis reactions on glycerol, and classifies the catalysts used in these reactions as precious metal catalysts and non-precious metal catalysts. The first is the research on the use of non-precious metal catalysts. In the study conducted by Shan et al., Cu catalysts supported on SBA-15 were employed for liquid-phase hydrogenolysis of glycerol, resulting in a remarkable selectivity of approximately 97% towards 1,2-propanediol [123]. The hydrogenolysis reaction of glycerol in an alkaline environment was investigated by De Andrade et al., utilizing NaY-zeolite supports loaded with 20 wt.% metal catalysts including Cu, Ni, and Cu-Ni [124]. The reaction was conducted in a continuous flow mode for 30 h at a temperature of 260 °C and pressure of 4.6 MPa. Cu-Ni catalyst exhibited superior effectiveness in the production of 1,2-propanediol. The addition of NaOH to the reaction significantly enhanced the yield for 1,2-propanediol. A glycerol to NaOH ratio of 5 (on a molar basis) resulted in an impressive glycerol conversion rate of 96.4% and a substantial yield of 31.8% for 1,2-propanediol. The 6Cu–4Ru/MgO catalyst exhibited nearly complete conversion and a propylene glycol selectivity of approximately 75% at an H_2/glycerol ratio of 66.8 (mole basis) under a reaction pressure of 0.8 MPa. Sharma et al. employed a novel catalyst comprising 3Cu/2Zn/1Cr/3Zr, operating at 240 °C and 4 MPa, for the identical reaction. As a result, propylene glycol selectivity was significantly enhanced to 97%, while complete conversion of glycerol was achieved within a duration of 10 h [125].

The utilization of noble metal-based catalysts, such as Pt-Re and Pt-WOx, has demonstrated remarkable efficacy in the synthesis of 1,3-propanediol [126]. The catalysts employed in this study consisted of Pt, WO_x and Al_2O_3 for the glycerol

Table 10.2 Performance of different catalysts used in hydrogenolysis of glycerol

Catalyst		Temperature (°C)	Pressure (MPa)	Time (h)	Conversion of glycerol (%)	Selectivity			Reference
						1,2-PDO	1,3-PDO	1-Propanol	
Non-noble metal	Ni/γ-Al_2O_3	220	–	5	100	87.0	–	0.1	[130]
	Ni/CS-P*	260	–	5	100	0.0	–	71.0	[130]
	Ni/CS-P*	260	–	2	100	0.0	–	91.0	[130]
	Ni/Y-Zeolite	260	4.5	23–30	95.6	42.9	–	–	[124]
	Cu/Y-Zeolite	260	4.5	23–30	71.1	12.6	–	–	[124]
	Ni-Cu/Y-Zeolite	260	4.5	23–30	96.4	43.9	–	–	[124]
	Cu/SBA-15	230	4	1.5	90.3	97.3	–	–	[123]
	Cu/ZnO	200	2	16	37.0	92	–	–	[131]
	Cu/Al_2O_3	220	5	6	61	93.3	–	–	[132]
	Cu/MgO	180	3	20	72	97.6	–	–	[133]
	Cu/MgAlO	180	3	20	80	98.2	–	–	[134]
Noble metal	Ag/Al_2O_3	220	1.5	10	46.0	96	–	–	[135]
	Pt/SiO2-Al_2O_3	220	4.5	24	19.8	32	–	–	[136]
	Ru/CeO_2	180	5	10	85.2	62.7	–	–	[137]
	Cu-Pd/MgAlO	180	2	10	77	97.2	–	–	[138]
	Pt (0.5)-Ir-ReOx/SiO_2	190	2	17	30	19	–	4.1	[139]
	Ir-ReOx/SiO_2	120	8	36	81	–	46	–	[140]
	Pt-WO_3/ZrO_2	170	8	18	86	–	28	–	[141]
	Pt/WO_3-TiO_2/SiO_2	180	5.6	12	15.3	–	50	–	[142]
	Pt/WOx-TiOx	180	5.6	12	18.4	–	40.3	–	[143]

hydrogenolysis reaction. It was observed that the Pt and W content exerted significant influences on the yield of 1,3-propanediol. Notably, the highest yield achieved was 28.4 wt.% at a Pt/W ratio (on a molar basis) of 6/12.9 [127]. The Cu-Ru/MgO catalyst was employed by Pandey and Biswas for the production of propylene glycol (1,2-propanediol), revealing its capability to operate under high reaction pressures. Conversely, atmospheric pressure proved more suitable for hydroxyacetone (acetol) production [128]. The Ir-ReOx catalyst was utilized in conjunction with various supports, including C, CeO_2, TiO_2, Al_2O_3, ZrO_2 and MgO, for the production of 1,3-propanediol via glycerol hydrogenolysis. It was observed that the productivity of 1,3-propanediol depended on the metal loading. Notably, a catalyst with an Ir loading of 6 wt% supported on TiO_2 (with a surface area of 6 m^2/g) exhibited the highest level of catalytic activity [129].

10.4 Discussion and Future Outlook

Biomass residues can be divided into primary, secondary and tertiary. The initial residue is usually produced by the grain harvest and forest products (such as corn straw and wheat straw) in an area. When the harvest is fully ripe and processed in the final form of the product, secondary residues are extracted. Bagasse, palm kernel cake and wood chips are sources of agricultural and dietary waste. Tertiary residues can be obtained by humans and/or animals after consumption of biomass-derived products [144]. In fact, these waste streams exist in the form of municipal solid waste (MSW) and/or soluble organic distillate wastewater [144]. Previous studies of further value-added use of agricultural waste have yielded many high-value products, but their use has led to serious associated environmental problems: soil and water pollution, increased greenhouse gas emissions (grown outside traditional agricultural Settings), reduced biodiversity, and the spread of crop diseases and pests.

Taking the conversion of cellulose to ethanol as an example, the production of ethanol from cellulose is mainly through a cascade catalytic reaction, including cellulose hydrolysis, acetal reaction and hydrogenation reaction. However, chemical catalytic processes also involve some side reactions, such as dehydration and aldol condensation. In order to obtain high-yield ethanol, it is necessary to design a multifunctional catalyst that can balance the main reaction and the side reaction. Multifunctional and bimetallic catalysts are promising catalysts for cellulose conversion to ethanol, but the poor hydrothermal stability of some catalysts or the use of toxic organic solvents limit their large-scale application. This also indicates that there is more room for improvement in catalyst preparation in the future.

From an economic point of view, raw material costs account for an increasing proportion of overall operating costs. The difference in energy consumption between different processes is mainly due to their pre-treatment, where raw material processing is one of the most energy-intensive processes, producing grey water and so on [145]. This prompts us to continue to study more processes that directly transform agricultural waste to produce alcohol chemicals without pretreatment.

10.5 Conclusions and Future Outlook

In this chapter, we have comprehensively summarized the catalytic processes involved in the production of organic alcohols from agricultural residues and food wastes. The extensive research conducted on agricultural residues, particularly lignocellulosic biomass components, has resulted in the sustainable synthesis of various aliphatic alcohols. Glycerol, a significant waste product from the food industry, serves as a crucial building block in current industrial applications. The hydrolysis processes enable the transformation of glycerol into diverse value-added alcohols such as propanol and propanediols. We firmly believe that continuous research efforts in the near future will yield even more valuable products.

Acknowledgements This work was supported by the Fundamental Research Funds for the Central Universities (QNTD202302), National Natural Science Foundation of China (22378024).

References

1. Maina S, Kachrimanidou V, Koutinas A. A roadmap towards a circular and sustainable bioeconomy through waste valorization. Curr Opin Green Sustain. 2017;8:18–23. https://doi.org/10.1016/j.cogsc.2017.07.007.
2. Koul B, Yakoob M, Shah MP. Agricultural waste management strategies for environmental sustainability. Environ Res. 2022;206:112285. https://doi.org/10.1016/j.envres.2021.112285.
3. Scarlat N, Dallemand J-F, Monforti-Ferrario F, Nita V. The role of biomass and bioenergy in a future bioeconomy: policies and facts. Environ Dev. 2015;15:3–34. https://doi.org/10.1016/j.envdev.2015.03.006.
4. Zabed H, Sahu JN, Boyce AN, Faruq G. Fuel ethanol production from lignocellulosic biomass: an overview on feedstocks and technological approaches. Renew Sust Energ Rev. 2016;66:751–74. https://doi.org/10.1016/j.rser.2016.08.038.
5. Peng X, Jiang Y, Chen Z, Osman AI, Farghali M, Rooney DW, Yap P-S. Recycling municipal, agricultural and industrial waste into energy, fertilizers, food and construction materials, and economic feasibility: a review. Environ Chem Lett. 2023;21:765–801. https://doi.org/10.1007/s10311-022-01551-5.
6. Banković-Ilić IB, Stojković IJ, Stamenković OS, Veljkovic VB, Hung Y-T. Waste animal fats as feedstocks for biodiesel production. Renew Sust Energ Rev. 2014;32:238–54. https://doi.org/10.1016/j.rser.2014.01.038.
7. Khalil M, Berawi MA, Heryanto R, Rizalie A. Waste to energy technology: the potential of sustainable biogas production from animal waste in Indonesia. Renew Sust Energ Rev. 2019;105:323–31. https://doi.org/10.1016/j.rser.2019.02.011.
8. Knorr D, Augustin MA. From food to gods to food to waste. Crit Rev Food Sci Nutr. 2024;64(16):5379–97. https://doi.org/10.1080/10408398.2022.2153795.
9. Ren S, Wang K, Zhang J, Li J, Zhang H, Qi R, Xu W, Yan C, Liu X, Zhang F, Jones DL, Chadwick DR. Potential sources and occurrence of macro-plastics and microplastics pollution

in farmland soils: a typical case of China. Crit Rev Environ Sci Technol. 2024; https://doi.org/10.1080/10643389.2023.2259275.
10. Aidoo OF, Osei-Owusu J, Chia SY, Dofuor AK, Antwi-Agyakwa AK, Okyere H, Gyan M, Edusei G, Ninsin KD, Duker RQ, Siddiqui SA, Borgemeister C. Remediation of pesticide residues using ozone: a comprehensive overview. Sci Total Environ. 2023;894:164933. https://doi.org/10.1016/j.scitotenv.2023.164933.
11. Zhang G-H, Zhu J-F, Okuwaki A. Prospect and current status of recycling waste plastics and technology for converting them into oil in China. Resour Conserv Recycl. 2007;50:231–9. https://doi.org/10.1016/j.resconrec.2006.11.007.
12. Periyasamy S, Karthik V, Senthil Kumar P, Isabel JB, Temesgen T, Hunegnaw BM, Melese BB, Mohamed BA, Vo D-VN. Chemical, physical and biological methods to convert lignocellulosic waste into value-added products. A review. Environ Chem Lett. 2022;20:1129–52. https://doi.org/10.1007/s10311-021-01374-w.
13. Huang S, Zheng X, Luo L, Ni Y, Yao L, Ni W. Biostimulants in bioconversion compost of organic waste: a novel booster in sustainable agriculture. J Clean Prod. 2021;319:128704. https://doi.org/10.1016/j.jclepro.2021.128704.
14. Lee D, Nam H, Won Seo M, Hoon Lee S, Tokmurzin D, Wang S, Park Y-K. Recent progress in the catalytic thermochemical conversion process of biomass for biofuels. Chem Eng J. 2022;447:137501. https://doi.org/10.1016/j.cej.2022.137501.
15. Gómez Millán G, Hellsten S, Llorca J, Luque R, Sixta H, Balu AM. Recent advances in the catalytic production of platform chemicals from holocellulosic biomass. ChemCatChem. 2019;11:2022–42. https://doi.org/10.1002/cctc.201801843.
16. Afreen G, Upadhyayula S. Alkylation of phenol and substituted phenols with C1–C4 alcohols/olefins as an upgrading route for bio-oil oxygenates: a review. Renew Sust Energ Rev. 2021;147:111189. https://doi.org/10.1016/j.rser.2021.111189.
17. Wu X, De Bruyn M, Trimmel G, Zangger K, Barta K. High-performance thermoplastics from a unique bicyclic lignin-derived diol. ACS Sustain Chem Eng. 2023;11:2819–29. https://doi.org/10.1021/acssuschemeng.2c05998.
18. Chandra R, Takeuchi H, Hasegawa T. Methane production from lignocellulosic agricultural crop wastes: a review in context to second generation of biofuel production. Renew Sust Energ Rev. 2012;16:1462–76. https://doi.org/10.1016/j.rser.2011.11.035.
19. Das H, Singh SK. Useful byproducts from cellulosic wastes of agriculture and food industry—a critical appraisal. Crit Rev Food Sci Nutr. 2004;44:77–89. https://doi.org/10.1080/10408690490424630.
20. Bian H, Gao Y, Luo J, Jiao L, Wu W, Fang G, Dai H. Lignocellulosic nanofibrils produced using wheat straw and their pulping solid residue: from agricultural waste to cellulose nanomaterials. Waste Manag. 2019;91:1–8. https://doi.org/10.1016/j.wasman.2019.04.052.
21. Lindman B, Karlström G, Stigsson L. On the mechanism of dissolution of cellulose. J Mol Liq. 2010;156:76–81. https://doi.org/10.1016/j.molliq.2010.04.016.
22. Mäki-Arvela P, Salmi T, Holmbom B, Willför S, Murzin DY. Synthesis of sugars by hydrolysis of hemicelluloses—a review. Chem Rev. 2011;111:5638–66. https://doi.org/10.1021/cr2000042.
23. Geboers J, Van De Vyver S, Carpentier K, De Blochouse K, Jacobs P, Sels B. Efficient catalytic conversion of concentrated cellulose feeds to hexitols with heteropoly acids and Ru on carbon. Chem Commum. 2010;46:3577–9. https://doi.org/10.1039/c001096k.
24. Negoi A, Trotus IT, Mamula Steiner O, Tudorache M, Kuncser V, Macovei D, Parvulescu VI, Coman SM. Direct synthesis of sorbitol and glycerol from cellulose over ionic Ru/magnetite nanoparticles in the absence of external hydrogen. ChemSusChem. 2013;6:2090–4. https://doi.org/10.1002/cssc.201300554.
25. Matson TD, Barta K, Iretskii AV, Ford PC. One-pot catalytic conversion of cellulose and of woody biomass solids to liquid fuels. J Am Chem Soc. 2011;133:14090–7. https://doi.org/10.1021/ja205436c.

26. Galebach PH, Mcclelland DJ, Eagan NM, Wittrig AM, Buchanan JS, Dumesic JA, Huber GW. Production of alcohols from cellulose by supercritical methanol depolymerization and hydrodeoxygenation. ACS Sustain Chem Eng. 2018;6:4330–44. https://doi.org/10.1021/acssuschemeng.7b04820.
27. Liu X, Bouxin FP, Fan J, Budarin VL, Hu C, Clark JH. Microwave-assisted catalytic depolymerization of lignin from birch sawdust to produce phenolic monomers utilizing a hydrogen-free strategy. J Hazard Mater. 2021;402:123490. https://doi.org/10.1016/j.jhazmat.2020.123490.
28. Geboers J, Van De Vyver S, Carpentier K, Jacobs P, Sels B. Efficient hydrolytic hydrogenation of cellulose in the presence of Ru-loaded zeolites and trace amounts of mineral acid. Chem Commum. 2011;47:5590–2. https://doi.org/10.1039/c1cc10422e.
29. Zhang X, Zhao T, Hara N, Jin Y, Zeng C, Yoneyama Y, Tsubaki N. Direct conversion of rice straw catalyzed by solid acid supported-Pt catalyst using in situ H_2 by ethanol steam reforming. Fuel. 2014;116:34–8. https://doi.org/10.1016/j.fuel.2013.07.111.
30. Xu S, Pan D, Wu Y, Song X, Gao L, Li W, Das L, Xiao G. Efficient production of furfural from xylose and wheat straw by bifunctional chromium phosphate catalyst in biphasic systems. Fuel Process Technol. 2018;175:90–6. https://doi.org/10.1016/j.fuproc.2018.04.005.
31. Zhang Y, Wang A, Zhang T. A new 3D mesoporous carbon replicated from commercial silica as a catalyst support for direct conversion of cellulose into ethylene glycol. Chem Commun. 2010;46:862–4. https://doi.org/10.1039/b919182h.
32. Pan G-Y, Ma Y-L, Ma X-X, Sun Y-G, Lv J-M, Zhang J-L. Catalytic hydrogenation of corn stalk into polyol over Ni–W/MCM-41 catalyst. Chem Eng J. 2016;299:386–92. https://doi.org/10.1016/j.cej.2016.04.074.
33. Song H, Wang P, Li S, Deng W, Li Y, Zhang Q, Wang Y. Direct conversion of cellulose into ethanol catalysed by a combination of tungstic acid and zirconia-supported Pt nanoparticles. Chem Commum. 2019;55:4303–6. https://doi.org/10.1039/c9cc00619b.
34. Liu S, Okuyama Y, Tamura M, Nakagawa Y, Imai A, Tomishige K. Selective transformation of hemicellulose (xylan) into n-pentane, pentanols or xylitol over a rhenium-modified iridium catalyst combined with acids. Green Chem. 2016;18:165–75. https://doi.org/10.1039/c5gc02183a.
35. Delgado-Arcaño Y, Mandelli D, Carvalho WA, Pontes LAM. Valorization of corncob by hydrolysis-hydrogenation to obtain xylitol under mild conditions. Waste Biomass Valoriz. 2021;12:5109–20. https://doi.org/10.1007/s12649-021-01348-7.
36. Wang X, Rinaldi R. Exploiting H-transfer reactions with RANEY® Ni for upgrade of phenolic and aromatic biorefinery feeds under unusual, low-severity conditions. Energy Environ Sci. 2012;5:8244. https://doi.org/10.1039/c2ee21855k.
37. Ferrini P, Rinaldi R. Catalytic biorefining of plant biomass to non-pyrolytic lignin bio-oil and carbohydrates through hydrogen transfer reactions. Angew Chem Int Ed Engl. 2014;53:8634–9. https://doi.org/10.1002/anie.201403747.
38. Schutyser W, Van Den Bossche G, Raaffels A, Van Den Bosch S, Koelewijn S-F, Renders T, Sels BF. Selective conversion of lignin-derivable 4-alkylguaiacols to 4-alkylcyclohexanols over noble and non-noble-metal catalysts. ACS Sustain Chem Eng. 2016;4:5336–46. https://doi.org/10.1021/acssuschemeng.6b01580.
39. Wang X-Q, Zhang Z-H, Ren X, Rong Z, Sun Z, Yuan T-Q. A sustainable and environmental benign catalytic process for the production of valuable flavors and fragrances from lignin platform chemicals. Ind Crop Prod. 2022;187:115460. https://doi.org/10.1016/j.indcrop.2022.115460.
40. Zhao G, Rui K, Dou SX, Sun W. Heterostructures for electrochemical hydrogen evolution reaction: a review. Adv Funct Mater. 2018;28:1803291. https://doi.org/10.1002/adfm.201803291.
41. Fukuoka A, Dhepe PL. Catalytic conversion of cellulose into sugar alcohols. Angew Chem Int Ed Engl. 2006;45:5161–3. https://doi.org/10.1002/anie.200601921.

42. Huber GW, Cortright RD, Dumesic JA. Renewable alkanes by aqueous-phase reforming of biomass-derived oxygenates. Angew Chem Int Ed Engl. 2004;43:1549–51. https://doi.org/10.1002/anie.200353050.
43. Kumar Vaidyanathan V, Saikia K, Senthil Kumar P, Karanam Rathankumar A, Rangasamy G, Dattatraya Saratale G. Advances in enzymatic conversion of biomass derived furfural and 5-hydroxymethylfurfural to value-added chemicals and solvents. Bioresour Technol. 2023;378:128975. https://doi.org/10.1016/j.biortech.2023.128975.
44. Wang Y, Wang H, Kong X, Zhu Y. Catalytic conversion of 5-hydroxymethylfurfural to high-value derivatives by selective activation of C–O, C=O, and C=C bonds. ChemSusChem. 2022;15:e202200421. https://doi.org/10.1002/cssc.202200421.
45. Martins FCOL, Alcantara GMRN, Silva AFS, Melchert WR, Rocha FRP. The role of 5-hydroxymethylfurfural in food and recent advances in analytical methods. Food Chem. 2022;395:133539. https://doi.org/10.1016/j.foodchem.2022.133539.
46. Kang S, Fu J, Zhang G. From lignocellulosic biomass to levulinic acid: a review on acid-catalyzed hydrolysis. Renew Sust Energ Rev. 2018;94:340–62. https://doi.org/10.1016/j.rser.2018.06.016.
47. Yemiş O, Mazza G. Optimization of furfural and 5-hydroxymethylfurfural production from wheat straw by a microwave-assisted process. Bioresour Technol. 2012;109:215–23. https://doi.org/10.1016/j.biortech.2012.01.031.
48. Rajendran DS, Venkataraman S, Senthil Kumar P, Bhattacharya T, Ramachandran K, Vaidyanathan VK. Understanding the impact of different pretreatment methods on the conversion of Casuarina equisetifolia biomass to 5-hydroxymethylfurfural and their energy cost assessment. Ind Crops Prod. 2022;186:115275. https://doi.org/10.1016/j.indcrop.2022.115275.
49. Ge WTZ, Zhang K, Chen M, Qin S, Chen X, Liu W. Preparation of cellulose and synthesisof 5-hydroxymethyl furfural from agricultural waste cottonseed hull. Appl Eng Agric. 2016;32:661–7. https://doi.org/10.13031/aea.32.11013.
50. Wu H, Huang T, Cao F, Zou Q, Wei P, Ouyang P. Co-production of HMF and gluconic acid from sucrose by chemo-enzymatic method. Chem Eng J. 2017;327:228–34. https://doi.org/10.1016/j.cej.2017.06.107.
51. Neves LNDO, De Oliveira MAL. Effects of enzymatic lactose hydrolysis on thermal markers in lactose-free UHT milk. J Food Sci Technol. 2020;57:3518–24. https://doi.org/10.1007/s13197-020-04561-9.
52. Zhao J, Zhang Y, Wang K, Yan C, Da Z, Li C, Yan Y. Development of hierarchical porous MOF-based catalyst of UiO-66(Hf) and its application for 5-hydroxymethylfurfural production from cellulose. ChemistrySelect. 2018;3:11476–85. https://doi.org/10.1002/slct.201802423.
53. Jin P, Zhang Y, Chen Y, Pan J, Dai X, Liu M, Yan Y, Li C. Facile synthesis of hierarchical porous catalysts for enhanced conversion of fructose to 5-hydroxymethylfurfural. J Taiwan Inst Chem Eng. 2017;75:59–69. https://doi.org/10.1016/j.jtice.2017.03.038.
54. Ye X, Zhong M, Cao Z, Han C, Wang A, Yao C, Li X. Plasmon resonance enhanced palygorskite-based composite toward the photocatalytic reformation of cellulose biomass under full spectrum. Appl Clay Sci. 2023;231:106755. https://doi.org/10.1016/j.clay.2022.106755.
55. Nayebi M, Faraji A, Bahadoran A, Othman ZJ, Arghavani S, Kargar PG, Sajjadinezhad SM, Varma RS. TiO_2/g-C_3N_4/SO_3H(IL): unique usage of ionic liquid-based sulfonic acid as an efficient photocatalyst for visible-light-driven preparation of 5-HMF from cellulose and glucose. ACS Appl Mater Interfaces. 2023;15:8054–65. https://doi.org/10.1021/acsami.2c20480.
56. Fang X, Wang Z, Song W, Li S, Lin W. Preparation of furans from catalytic conversion of corn stover in H_2O–THF co-solvent system—the effects of acids combined with alkali metal cations. J Taiwan Inst Chem Eng. 2019;97:105–11. https://doi.org/10.1016/j.jtice.2019.02.024.

57. Xu S, Pan D, Wu Y, Xu N, Yang H, Gao L, Li W, Xiao G. Direct conversion of wheat straw components into furan compounds using a highly efficient and reusable SnCl2-PTA/β zeolite catalyst. Ind Eng Chem Res. 2019;58:9276–85. https://doi.org/10.1021/acs.iecr.9b00984.
58. Hu L, Xu J, Zhou S, He A, Tang X, Lin L, Xu J, Zhao Y. Catalytic advances in the production and application of biomass-derived 2,5-dihydroxymethylfuran. ACS Catal. 2018;8:2959–80. https://doi.org/10.1021/acscatal.7b03530.
59. Fulignati S, Antonetti C, Licursi D, Pieraccioni M, Wilbers E, Heeres HJ, Raspolli Galletti AM. Insight into the hydrogenation of pure and crude HMF to furan diols using Ru/C as catalyst. Appl Catal A Gen. 2019;578:122–33. https://doi.org/10.1016/j.apcata.2019.04.007.
60. Hou Q, Qi X, Zhen M, Qian H, Nie Y, Bai C, Zhang S, Bai X, Ju M. Biorefinery roadmap based on catalytic production and upgrading 5-hydroxymethylfurfural. Green Chem. 2021;23: 119–231. https://doi.org/10.1039/d0gc02770g.
61. Kuai L, Chen Z, Liu S, Kan E, Yu N, Ren Y, Fang C, Li X, Li Y, Geng B. Titania supported synergistic palladium single atoms and nanoparticles for room temperature ketone and aldehydes hydrogenation. Nat Commun. 2020;11:48. https://doi.org/10.1038/s41467-019-13941-5.
62. Hu L, Dai X, Li N, Tang X, Jiang Y. Highly selective hydrogenation of biomass-derived 5-hydroxymethylfurfural into 2,5-bis(hydroxymethyl)furan over an acid–base bifunctional hafnium-based coordination polymer catalyst. Sustain Energy Fuels. 2019;3:1033–41. https://doi.org/10.1039/C8SE00545A.
63. Tang X, Wei J, Ding N, Sun Y, Zeng X, Hu L, Liu S, Lei T, Lin L. Chemoselective hydrogenation of biomass derived 5-hydroxymethylfurfural to diols: key intermediates for sustainable chemicals, materials and fuels. Renew Sust Energ Rev. 2017;77:287–96. https://doi.org/10.1016/j.rser.2017.04.013.
64. He J, Burt SP, Ball M, Zhao D, Hermans I, Dumesic JA, Huber GW. Synthesis of 1,6-hexanediol from cellulose derived tetrahydrofuran-dimethanol with Pt-WO_x/TiO_2 catalysts. ACS Catal. 2018;8:1427–39. https://doi.org/10.1021/acscatal.7b03593.
65. Song H, Jin R, Kang M, Chen J. Progress in synthesis of ethylene glycol through C1 chemical industry routes. Chinese J Catal. 2013;34:1035–50. https://doi.org/10.1016/s1872-2067(12)60529-4.
66. Yue H, Zhao Y, Ma X, Gong J. Ethylene glycol: properties, synthesis, and applications. Chem Soc Rev. 2012;41:4218. https://doi.org/10.1039/c2cs15359a.
67. Wang A, Zhang T. One-pot conversion of cellulose to ethylene glycol with multifunctional tungsten-based catalysts. Acc Chem Res. 2013;46:1377–86. https://doi.org/10.1021/ar3002156.
68. Ji N, Zhang T, Zheng M, Wang A, Wang H, Wang X, Chen JG. Direct catalytic conversion of cellulose into ethylene glycol using nickel-promoted tungsten carbide catalysts. Angew Chem Int Ed Engl. 2008;47:8510–3. https://doi.org/10.1002/anie.200803233.
69. Zheng M-Y, Wang A-Q, Ji N, Pang J-F, Wang X-D, Zhang T. Transition metal–tungsten bimetallic catalysts for the conversion of cellulose into ethylene glycol. ChemSusChem. 2010;3:63–6. https://doi.org/10.1002/cssc.200900197.
70. Ribeiro LS, Pires ALF, Órfão JJDM, Pereira MFR. Paving the way towards an eco- and budget-friendly one-pot catalytic conversion of cellulose and lignocellulosic residues into ethylene glycol over Ni–W/CNT catalysts. Renew Energy. 2022;200:1008–22. https://doi.org/10.1016/j.renene.2022.10.026.
71. Li X, Guo T, Xia Q, Liu X, Wang Y. One-pot catalytic transformation of lignocellulosic biomass into alkylcyclohexanes and polyols. ACS Sustain Chem Eng. 2018;6:4390–9. https://doi.org/10.1021/acssuschemeng.8b00012.
72. Balat M, Balat H, Öz C. Progress in bioethanol processing. Prog Energy Combust Sci. 2008;34:551–73. https://doi.org/10.1016/j.pecs.2007.11.001.
73. Aditiya HB, Mahlia TMI, Chong WT, Nur H, Sebayang AH. Second generation bioethanol production: a critical review. Renew Sust Energ Rev. 2016;66:631–53. https://doi.org/10.1016/j.rser.2016.07.015.

74. Zhuang X, Wang H, Jiang S, Hu X, Su T, Zhang X, Ma L. A review on the chemo-catalytic conversion of cellulose to bio-ethanol. Green Chem Eng. 2023;5:276–89. https://doi.org/10.1016/j.gce.2023.08.002.
75. Xu G, Wang A, Pang J, Zhao X, Xu J, Lei N, Wang J, Zheng M, Yin J, Zhang T. Chemocatalytic conversion of cellulosic biomass to methyl glycolate, ethylene glycol, and ethanol. ChemSusChem. 2017;10:1390–4. https://doi.org/10.1002/cssc.201601714.
76. Yang M, Qi H, Liu F, Ren Y, Pan X, Zhang L, Liu X, Wang H, Pang J, Zheng M, Wang A, Zhang T. One-pot production of cellulosic ethanol via tandem catalysis over a multifunctional Mo/Pt/WO_x catalyst. Joule. 2019;3:1937–48. https://doi.org/10.1016/j.joule.2019.05.020.
77. Yang C, Miao Z, Zhang F, Li L, Liu Y, Wang A, Zhang T. Hydrogenolysis of methyl glycolate to ethanol over a Pt–Cu/SiO_2 single-atom alloy catalyst: a further step from cellulose to ethanol. Green Chem. 2018;20:2142–50. https://doi.org/10.1039/c8gc00309b.
78. Li C, Xu G, Wang C, Ma L, Qiao Y, Zhang Y, Fu Y. One-pot chemocatalytic transformation of cellulose to ethanol over Ru-WO_x/HZSM-5. Green Chem. 2019;21:2234–9. https://doi.org/10.1039/c9gc00719a.
79. Li C, Xu G, Li K, Wang C, Zhang Y, Fu Y. A weakly basic Co/CeO_x catalytic system for one-pot conversion of cellulose to diols: Kungfu on eggs. Chem Commun. 2019;55:7663–6. https://doi.org/10.1039/c9cc04020j.
80. Liu Q, Wang H, Xin H, Wang C, Yan L, Wang Y, Zhang Q, Zhang X, Xu Y, Huber GW, Ma L. Selective cellulose hydrogenolysis to ethanol using Ni@C combined with phosphoric acid catalysts. ChemSusChem. 2019;12:3977–87. https://doi.org/10.1002/cssc.201901110.
81. Chu D, Luo Z, Xin Y, Jiang C, Gao S, Wang Z, Zhao C. One-pot hydrogenolysis of cellulose to bioethanol over Pd-Cu-WO_x/SiO_2 catalysts. Fuel. 2021;292:120311. https://doi.org/10.1016/j.fuel.2021.120311.
82. Gumina B, Espro C, Galvagno S, Pietropaolo R, Mauriello F. Bioethanol production from unpretreated cellulose under neutral selfsustainable hydrolysis/hydrogenolysis conditions promoted by the heterogeneous Pd/Fe_3O_4 catalyst. ACS Omega. 2019;4:352–7. https://doi.org/10.1021/acsomega.8b03088.
83. Wu Y, Dong C, Wang H, Peng J, Li Y, Samart C, Ding M. One-pot ethanol production from cellulose transformation over multifunctional Pt/WO_x and hollow Pt@HZSM-5 catalysts. ACS Sustain Chem Eng. 2022;10:2802–10. https://doi.org/10.1021/acssuschemeng.1c08204.
84. Chu D, Xin Y, Zhao C. Production of bio-ethanol by consecutive hydrogenolysis of corn-stalk cellulose. Chinese J Catal. 2021;42:844–54. https://doi.org/10.1016/s1872-2067(20)63709-3.
85. Saha BC. Hemicellulose bioconversion. J Ind Microbiol Biotechnol. 2003;30:279–91. https://doi.org/10.1007/s10295-003-0049-x.
86. Lu Y, Mosier NS. Kinetic modeling analysis of maleic acid-catalyzed hemicellulose hydrolysis in corn stover. Biotechnol Bioeng. 2008;101:1170–81. https://doi.org/10.1002/bit.22008.
87. Gallezot P, Nicolaus N, Flèche G, Fuertes P, Perrard A. Glucose hydrogenation on ruthenium catalysts in a trickle-bed reactor. J Catal. 1998;180:51–5. https://doi.org/10.1006/jcat.1998.2261.
88. Li N, Tompsett GA, Zhang T, Shi J, Wyman CE, Huber GW. Renewable gasoline from aqueous phase hydrodeoxygenation of aqueous sugar solutions prepared by hydrolysis of maple wood. Green Chem. 2011;13:91–101. https://doi.org/10.1039/c0gc00501k.
89. Yi G, Zhang Y. One-pot selective conversion of hemicellulose (xylan) to xylitol under mild conditions. ChemSusChem. 2012;5:1383–7. https://doi.org/10.1002/cssc.201200290.
90. Tathod AP, Dhepe PL. Towards efficient synthesis of sugar alcohols from mono- and polysaccharides: role of metals, supports & promoters. Green Chem. 2014;16:4944–54. https://doi.org/10.1039/c4gc01264j.
91. Guha SK, Kobayashi H, Hara K, Kikuchi H, Aritsuka T, Fukuoka A. Hydrogenolysis of sugar beet fiber by supported metal catalyst. Catal Commun. 2011;12:980–3. https://doi.org/10.1016/j.catcom.2011.02.017.

92. Zhang T, Li W, Xiao H, Jin Y, Wu S. Recent progress in direct production of furfural from lignocellulosic residues and hemicellulose. Bioresour Technol. 2022;354:127126. https://doi.org/10.1016/j.biortech.2022.127126.
93. Corma A, Iborra S, Velty A. Chemical routes for the transformation of biomass into chemicals. Chem Rev. 2007;107:2411–502. https://doi.org/10.1021/cr050989d.
94. Kijeński J, Winiarek P, Paryjczak T, Lewicki A, Mikołajska A. Platinum deposited on monolayer supports in selective hydrogenation of furfural to furfuryl alcohol. Appl Catal A Gen. 2002;233:171–82. https://doi.org/10.1016/S0926-860X(02)00140-0.
95. Nagaraja BM, Padmasri AH, David Raju B, Rama Rao KS. Vapor phase selective hydrogenation of furfural to furfuryl alcohol over Cu–MgO coprecipitated catalysts. J Mol Catal A Chem. 2007;265:90–7. https://doi.org/10.1016/j.molcata.2006.09.037.
96. Merlo AB, Vetere V, Ruggera JF, Casella ML. Bimetallic PtSn catalyst for the selective hydrogenation of furfural to furfuryl alcohol in liquid-phase. Catal Commun. 2009;10:1665–9. https://doi.org/10.1016/j.catcom.2009.05.005.
97. Perez RF, Fraga MA. Hemicellulose-derived chemicals: one-step production of furfuryl alcohol from xylose. Green Chem. 2014;16:3942. https://doi.org/10.1039/c4gc00398e.
98. Qin L-Z, He Y-C. Chemoenzymatic synthesis of furfuryl alcohol from biomass in tandem reaction system. Appl Biochem Biotechnol. 2020;190:1289–303. https://doi.org/10.1007/s12010-019-03154-3.
99. Sun Z, Fridrich B, De Santi A, Elangovan S, Barta K. Bright side of lignin depolymerization: toward new platform chemicals. Chem Rev. 2018;118:614–78. https://doi.org/10.1021/acs.chemrev.7b00588.
100. Liu W-J, Jiang H, Yu H-Q. Thermochemical conversion of lignin to functional materials: a review and future directions. Green Chem. 2015;17:4888–907. https://doi.org/10.1039/c5gc01054c.
101. Gall DL, Ralph J, Donohue TJ, Noguera DR. Biochemical transformation of lignin for deriving valued commodities from lignocellulose. Curr Opin Biotechnol. 2017;45:120–6. https://doi.org/10.1016/j.copbio.2017.02.015.
102. Sheng Y, Lam SS, Wu Y, Ge S, Wu J, Cai L, Huang Z, Le QV, Sonne C, Xia C. Enzymatic conversion of pretreated lignocellulosic biomass: a review on influence of structural changes of lignin. Bioresour Technol. 2021;324:124631. https://doi.org/10.1016/j.biortech.2020.124631.
103. Fache M, Boutevin B, Caillol S. Vanillin production from lignin and its use as a renewable chemical. ACS Sustain Chem Eng. 2016;4:35–46. https://doi.org/10.1021/acssuschemeng.5b01344.
104. Pongthawornsakun B, Kaewsuanjik P, Kittipreechakun P, Ratova M, Kelly P, Mekasuwandumrong O, Praserthdam P, Panpranot J. Deposition of Pt nanoparticles on TiO_2 by pulsed direct current magnetron sputtering for selective hydrogenation of vanillin to vanillyl alcohol. Catal Today. 2020;358:51–9. https://doi.org/10.1016/j.cattod.2019.08.045.
105. Qiu Z, He X, Ma S, Li Z, Xiong Y, Cao Y. Ni nanoparticles embedded in nitrogen doped carbon derived from metal–organic frameworks for the efficient hydrogenation of vanillin to vanillyl alcohol. New J Chem. 2022;46:10347–56. https://doi.org/10.1039/d2nj01110g.
106. Guo X, Gao G, Remón J, Ma Y, Jiang Z, Shi B, Tsang DCW. Selective hydrogenation of vanillin to vanillyl alcohol over Pd, Pt, and Au catalysts supported on an advanced nitrogen-containing carbon material produced from food waste. Chem Eng J. 2022;440:135885. https://doi.org/10.1016/j.cej.2022.135885.
107. Schutyser W, Renders T, Van Den Bosch S, Koelewijn SF, Beckham GT, Sels BF. Chemicals from lignin: an interplay of lignocellulose fractionation, depolymerisation, and upgrading. Chem Soc Rev. 2018;47:852–908. https://doi.org/10.1039/c7cs00566k.
108. Sun Z, Cheng J, Wang D, Yuan T-Q, Song G, Barta K. Downstream processing strategies for lignin-first biorefinery. ChemSusChem. 2020;13:5199–212. https://doi.org/10.1002/cssc.202001085.

109. Abu-Omar MM, Barta K, Beckham GT, Luterbacher JS, Ralph J, Rinaldi R, Román-Leshkov Y, Samec JSM, Sels BF, Wang F. Guidelines for performing lignin-first biorefining. Energy Environ Sci. 2021;14:262–92. https://doi.org/10.1039/d0ee02870c.
110. Schutyser W, Van Den Bosch S, Dijkmans J, Turner S, Meledina M, Van Tendeloo G, Debecker DP, Sels BF. Selective nickel-catalyzed conversion of model and lignin-derived phenolic compounds to cyclohexanone-based polymer building blocks. ChemSusChem. 2015;8:1805–18. https://doi.org/10.1002/cssc.201403375.
111. Wu X, Galkin MV, Stern T, Sun Z, Barta K. Fully lignocellulose-based PET analogues for the circular economy. Nat Commun. 2022;13:3376. https://doi.org/10.1038/s41467-022-30735-4.
112. Subbotina E, Rukkijakan T, Marquez-Medina MD, Yu X, Johnsson M, Samec JSM. Oxidative cleavage of C–C bonds in lignin. Nat Chem. 2021;13:1118–25. https://doi.org/10.1038/s41557-021-00783-2.
113. Wu S, Chang L, Tian L. Identification and characterization of two regiospecific tricetin UDP-dependent glycosyltransferases from pomegranate (Punica granatum L.). Plants. 2022;11:810. https://doi.org/10.3390/plants11060810.
114. Moon J-I, Lee Y-H, Kim H-J. Synthesis and characterization of elastomeric polyester coatings for automotive pre-coated metal. Prog Org Coat. 2012;74:125–33. https://doi.org/10.1016/j.porgcoat.2011.11.025.
115. Asaad JN, Abd-El-Messieh SL. Effect of glycol and filler types on some polymeric composite properties. J Appl Polym. 2008;109:1916–24. https://doi.org/10.1002/app.27264.
116. Bian S, Pagan C, Andrianova "Artemyeva" AA, Du G. Synthesis of polycarbonates and poly (ether carbonate)s directly from carbon dioxide and diols promoted by a Cs_2CO_3/CH_2Cl_2 System. ACS Omega. 2016;1:1049–57. https://doi.org/10.1021/acsomega.6b00278.
117. Zhang Z-H, Wu X, Ren X, Rong Z, Sun Z, Barta K, Yuan T-Q. High yield production of 1,4-cyclohexanediol from lignin derived 2,6-dimethoxybenzoquinone via Raney NiMn catalyst in hydrogen free conditions. J Energy Chem. 2023;83:275–86. https://doi.org/10.1016/j.jechem.2023.04.023.
118. Pagliaro M, Ciriminna R, Kimura H, Rossi M, Della Pina C. From glycerol to value-added products. Angew Chem Int Ed Engl. 2007;46:4434–40. https://doi.org/10.1002/anie.200604694.
119. Gholami Z, Abdullah AZ, Lee K-T. Dealing with the surplus of glycerol production from biodiesel industry through catalytic upgrading to polyglycerols and other value-added products. Renew Sust Energ Rev. 2014;39:327–41. https://doi.org/10.1016/j.rser.2014.07.092.
120. Tan HW, Abdul Aziz AR, Aroua MK. Glycerol production and its applications as a raw material: a review. Renew Sust Energ Rev. 2013;27:118–27. https://doi.org/10.1016/j.rser.2013.06.035.
121. Martin A, Armbruster U, Gandarias I, Arias PL. Glycerol hydrogenolysis into propanediols using in situ generated hydrogen—a critical review. Eur J Lipid Sci Technol. 2013;115:9–27. https://doi.org/10.1002/ejlt.201200207.
122. Schlaf M. Selective deoxygenation of sugar polyols to α,ω-diols and other oxygen content reduced materials—a new challenge to homogeneous ionic hydrogenation and hydrogenolysis catalysis. Dalton Trans. 2006;(39):4645–53. https://doi.org/10.1039/B608007C.
123. Shan J, Liu H, Lu K, Zhu S, Li J, Wang J, Fan W. Identification of the dehydration active sites in glycerol hydrogenolysis to 1,2-propanediol over Cu/SiO_2 catalysts. J Catal. 2020;383:13–23. https://doi.org/10.1016/j.jcat.2019.12.032.
124. De Andrade TS, Souza MMVM, Manfro RL. Hydrogenolysis of glycerol to 1,2-propanediol without external H_2 addition in alkaline medium using Ni-Cu catalysts supported on Y zeolite. Renew Energy. 2020;160:919–30. https://doi.org/10.1016/j.renene.2020.06.060.
125. Sharma RV, Kumar P, Dalai AK. Selective hydrogenolysis of glycerol to propylene glycol by using Cu:Zn:Cr:Zr mixed metal oxides catalyst. Appl Catal A Gen. 2014;477:147–56. https://doi.org/10.1016/j.apcata.2014.03.007.
126. Wang Y, Zhou J, Guo X. Catalytic hydrogenolysis of glycerol to propanediols: a review. RSC Adv. 2015;5:74611–28. https://doi.org/10.1039/C5RA11957J.

127. Lei N, Zhao X, Hou B, Yang M, Zhou M, Liu F, Wang A, Zhang T. Effective hydrogenolysis of glycerol to 1,3-propanediol over metal-acid concerted $Pt/WO_x/Al_2O_3$ catalysts. ChemCatChem. 2019;11:3903–12. https://doi.org/10.1002/cctc.201900689.
128. Pandey DK, Biswas P. Continuous production of propylene glycol (1,2-propanediol) by the hydrogenolysis of glycerol over a bi-functional Cu–Ru/MgO catalyst. React Chem Eng. 2020;5:2221–35. https://doi.org/10.1039/D0RE00285B.
129. Liu L, Asano T, Nakagawa Y, Tamura M, Okumura K, Tomishige K. Selective hydrogenolysis of glycerol to 1,3-propanediol over rhenium-oxide-modified iridium nanoparticles coating rutile titania support. ACS Catal. 2019;9:10913–30. https://doi.org/10.1021/acscatal.9b03824.
130. Gatti MN, Cerioni JL, Pompeo F, Santori GF, Nichio NN. High yield to 1-propanol from crude glycerol using two reaction steps with Ni catalysts. Catalysts. 2020;10:615. https://doi.org/10.3390/catal10060615.
131. Balaraju M, Rekha V, Sai Prasad PS, Prasad RBN, Lingaiah N. Selective hydrogenolysis of glycerol to 1,2 propanediol over Cu–ZnO catalysts. Catal Lett. 2008;126:119–24. https://doi.org/10.1007/s10562-008-9590-6.
132. Hirunsit P, Luadthong C, Faungnawakij K. Effect of alumina hydroxylation on glycerol hydrogenolysis to 1,2-propanediol over Cu/Al_2O_3: combined experiment and DFT investigation. RSC Adv. 2015;5:11188–97. https://doi.org/10.1039/C4RA14698K.
133. Yuan Z, Wang J, Wang L, Xie W, Chen P, Hou Z, Zheng X. Biodiesel derived glycerol hydrogenolysis to 1,2-propanediol on Cu/MgO catalysts. Bioresour Technol. 2010;101:7088–92. https://doi.org/10.1016/j.biortech.2010.04.016.
134. Yuan Z, Wang L, Wang J, Xia S, Chen P, Hou Z, Zheng X. Hydrogenolysis of glycerol over homogenously dispersed copper on solid base catalysts. Appl Catal B. 2011;101:431–40. https://doi.org/10.1016/j.apcatb.2010.10.013.
135. Zhou J, Zhang J, Guo X, Mao J, Zhang S. Ag/Al_2O_3 for glycerol hydrogenolysis to 1,2-propanediol: activity, selectivity and deactivation. Green Chem. 2012;14:156–63. https://doi.org/10.1039/C1GC15918F.
136. Gandarias I, Arias PL, Requies J, Güemez MB, Fierro JLG. Hydrogenolysis of glycerol to propanediols over a Pt/ASA catalyst: the role of acid and metal sites on product selectivity and the reaction mechanism. Appl Catal B. 2010;97:248–56. https://doi.org/10.1016/j.apcatb.2010.04.008.
137. Feng J, Xiong W, Xu B, Jiang W, Wang J, Chen H. Basic oxide-supported Ru catalysts for liquid phase glycerol hydrogenolysis in an additive-free system. Catal Commun. 2014;46:98–102. https://doi.org/10.1016/j.catcom.2013.11.031.
138. Xia S, Yuan Z, Wang L, Chen P, Hou Z. Hydrogenolysis of glycerol on bimetallic Pd-Cu/solid-base catalysts prepared via layered double hydroxides precursors. Appl Catal A Gen. 2011;403:173–82. https://doi.org/10.1016/j.apcata.2011.06.026.
139. Liu S, Tamura M, Shen Z, Zhang Y, Nakagawa Y, Tomishige K. Hydrogenolysis of glycerol with in-situ produced H_2 by aqueous-phase reforming of glycerol using Pt-modified $Ir-ReO_x/SiO_2$ catalyst. Catal Today. 2018;303:106–16. https://doi.org/10.1016/j.cattod.2017.07.025.
140. Nakagawa Y, Shinmi Y, Koso S, Tomishige K. Direct hydrogenolysis of glycerol into 1,3-propanediol over rhenium-modified iridium catalyst. J Catal. 2010;272:191–4. https://doi.org/10.1016/j.jcat.2010.04.009.
141. Kurosaka T, Maruyama H, Naribayashi I, Sasaki Y. Production of 1,3-propanediol by hydrogenolysis of glycerol catalyzed by $Pt/WO_3/ZrO_2$. Catal Commun. 2008;9:1360–3. https://doi.org/10.1016/j.catcom.2007.11.034.
142. Gong L, Lu Y, Ding Y, Lin R, Li J, Dong W, Wang T, Chen W. Selective hydrogenolysis of glycerol to 1,3-propanediol over a $Pt/WO_3/TiO_2/SiO_2$ catalyst in aqueous media. Appl Catal A Gen. 2010;390:119–26. https://doi.org/10.1016/j.apcata.2010.10.002.
143. Zhang Y, Zhao X-C, Wang Y, Zhou L, Zhang J, Wang J, Wang A, Zhang T. Mesoporous Ti–W oxide: synthesis, characterization, and performance in selective hydrogenolysis of glycerol. J Mater Chem A. 2013;1:3724–32. https://doi.org/10.1039/C3TA10217C.

144. Tripathi N, Hills CD, Singh RS, Atkinson CJ. Biomass waste utilisation in low-carbon products: harnessing a major potential resource. npj Clim Atmos Sci. 2019;2:35. https://doi.org/10.1038/s41612-019-0093-5.
145. Gandam PK, Chinta ML, Pabbathi NPP, Baadhe RR, Sharma M, Thakur VK, Sharma GD, Ranjitha J, Gupta VK. Second-generation bioethanol production from corncob—a comprehensive review on pretreatment and bioconversion strategies, including techno-economic and lifecycle perspective. Ind Crops Prod. 2022;186:115245. https://doi.org/10.1016/j.indcrop.2022.115245.

Chapter 11
Fuel Ethanol Production from Lignocellulosic Wastes via Bioconversion

Xinchuan Yuan and Mingjie Jin

Abstract The second-generation fuel ethanol production uses lignocellulosic wastes as a raw material, which is abundant and typically treated as waste, addressing the issue of "competing for food with people" caused by the first generation fuel ethanol and showing great significance to national energy security and social sustainability. Lignocellulosic wastes used in the second generation fuel ethanol production requires pretreatment by biological, physical or chemical methods to disrupt its recalcitrant structure and enhance its enzymatic digestibility. The polysaccharide components of the pretreated lignocellulose are then degraded by enzymes into fermentable monosaccharides that can be converted into ethanol by fermentation microorganisms.

Currently, many pretreatment methods as well as strategies for enzymatic hydrolysis and fermentation have been developed for the purpose of biomass conversion. Nevertheless, the key to foster the cellulosic ethanol industry is still figuring out how to effectively disrupt lignocellulosic structure at a low cost with a low energy consumption. This chapter reviews and evaluates a variety of lignocellulose pretreatment methods developed in recent years, including advantages and drawbacks, pretreatment mechanisms, and offers a summary of the current status of the cellulosic ethanol technologies. Furthermore, this chapter also discusses processes of enzymatic hydrolysis and fermentation for fuel ethanol production, as well as a detailed analysis of many factors affecting the process performance. Moreover, the bottleneck of the lignocellulosic ethanol production and future perspective are discussed.

Keywords Fuel ethanol · Lignocellulosic biorefinery · Pretreatment · Enzymatic hydrolysis · Fermentation

---

X. Yuan (✉) · M. Jin
School of Environmental and Biological Engineering, Nanjing University of Science and Technology, Nanjing, China
e-mail: yuanxinchuan@njust.edu.cn; jinmingjie@njust.edu.cn

Z. Fang et al. (eds.), *Production of Organic Acids and Alcohols from Agricultural Residues and Food Wastes*, Biofuels and Biorefineries 13,
https://doi.org/10.1007/978-981-96-1222-2_11

11.1 Introduction

Considering the acceleration of urbanization and the growth of global population, world energy demand has remarkably increased. Fossil fuels are currently still the primary source of energy, while their use has caused serious environmental pollution including the greenhouse effect. Therefore, it is urgent to reduce using traditional fossil fuels and develop clean and renewable fuels. Fuel ethanol is a favorable bulk renewable energy applicable to transportation, with several merits including high-octane number, cleanliness and safety [1]. Fuel ethanol includes the first and second generations according to the types of raw feedstock. The first generation fuel ethanol uses starch based or sugar juice based materials, such as corn and sugarcane as the main raw feedstock, leading to the issue of "competing for food with people", which is not conducive to consolidating national food security and limits the production scale of fuel ethanol [2]. The second generation fuel ethanol (cellulosic ethanol) production uses widely available and inexpensive lignocellulosic biomass as feedstock, primarily from agricultural wastes (e.g. agricultural straw, grass, forest residues, fruit shells and pomace, etc.) [3]. Producing fuel ethanol from agricultural wastes is a practical way to promote the sustainable development of society and realize the valorization of agricultural wastes.

Obtaining high-yield fermentable sugar from agricultural wastes is the key point of cellulosic ethanol production via bioconversion [4]. The intrinsic recalcitrant structure of lignocellulose hinders its inner components from being digested by enzymes or microorganisms, so pretreatment is required to disrupt the lignocellulosic structure. Lignocellulosic pretreatment methods usually consist of three categories: biological, physical and chemical methods [5]. As reported, many of the existing pretreatment methods perform well, but they often have one or more drawbacks, such as high energy consumption, severe pretreatment conditions, high reagent costs and generation of many degradation products [6]. The loose biomass mostly used in traditional biorefinery, which is fluffy and of low density, leads to high costs of logistics and pretreatment [7]. Besides, loose biomass is easily contaminated by fungi, resulting in difficulty in handling and inconsistency in quality.

In view of the unfavorable characteristics of loose biomass, compressing loose biomass into a uniform high-density biomass pellet is considered an effective approach for reducing the cost of logistics and improving the working efficiency of pretreatment [7]. The inner components (such as starch, polysaccharide, protein and lignin) play an important role in biomass molding [7]. Hence, an efficient pretreatment "Densifying Lignocellulosic biomass with Chemicals (DLC)" has been developed, which effectively reduces the logistics' cost and increases the efficiency of bioprocessing, offering a biomass with good digestibility and high fermentability [8, 9]. Enzymatic hydrolysis and fermentation are other essential procedures during cellulosic ethanol production, which are performed independently or simultaneously. They are called Separate Hydrolysis and Co-Fermentation (SHCF) and Simultaneous Saccharification and Co-Fermentation (SSCF), respectively [10]. Compared to the SHCF process, the SSCF process has the advantages of shorter operation time, high sugar consumption, better ethanol yield and a lower risk of contamination [11]. The ethanol production during SSCF is affected by many factors, such as temperature, pH, enzyme dosage, biomass solid loading and initial strain inoculation [12]. The optimal process conditions can be obtained by optimizing process parameters so as to achieve the best fermentation performance. Some strategies, such as pre-hydrolysis, fed-batch of biomass and enzyme [11], biological detoxification [13], strain domestication [14], and surfactant addition [15], are developed to improve the SSCF performance and thereby obtain higher ethanol yield, taking into consideration the issues of high viscosity and strong inhibition during high solid-loading SSCF.

This chapter reviews a variety of pretreatment methods developed in recent years and provides a thorough analysis of their advantages and defects. Furthermore, a wide range of variables influencing fermentation and enzymatic hydrolysis are analyzed, along with pertinent enhancement strategies. The current tendencies and challenges of cellulosic ethanol production are summarized at the end of this chapter.

11.2 Lignocellulosic Feedstocks

According to the report [16], with over 800 million tons of agricultural wastes (such as corn stover, rice straw and wheat straw, etc.) generated in China annually, how to properly dispose of these wastes is a huge challenge. Agricultural waste, as a lignocellulose-based biomass, is regarded as a potential abundant feedstock for biofuels and bio-based chemical production [17]. Lignocellulose is mainly composed of cellulose, hemicellulose and lignin. The content of cellulose is about 30–50%, the content of hemicellulose is about 20–30%, and the content of lignin is about 15–30%. The composition of the three components varies between different types of plants. Plant cell walls are composed of three kinds of structures: intercellular layer, primary wall and secondary wall. Pectin, cellulose, hemicellulose and lignin are the main constituents of plant cell walls, and they are crucial for keeping the cells in their proper form and for increasing their mechanical strength. The intercellular layer is a shared membrane that separates two neighboring cells, in which pectin is the principal ingredient. Cells inside the intercellular layer secrete the primary wall after the intercellular layer has formed, which is mostly composed of cellulose, hemicellulose and pectin. The secondary wall is only found in thick-walled cells, which is a cell wall formed inside the primary wall with a high content of lignin. Cellulose, hemicellulose and lignin are closely intertwined to form a three-dimensional polymer chemical structure, which effectively shields the interior cells and tissues of plants [18].

11.2.1 Cellulose

Cellulose is a long-chain macromolecular polysaccharide that has the greatest distribution and highest content in nature. It is a linear polymer made up of many D-glucopyranose units joined by β-1,4-glucoside linkages [19]. Some parts of these macromolecules are arranged neatly, which are called crystalline regions; the other parts are arranged disorderly, which are called amorphous regions [20]. For the crystalline region, cellulose molecular chains are oriented consistently, and many hydrogen bonds link the molecules to form a highly compact structure that makes it difficult for water molecules to enter and cellulose to hydrolyze. For the amorphous region, molecular chain orientation is poor; there are few hydrogen bonds between the molecules with a low binding strength, so the cellulase can easily hydrolyze the cellulose. Therefore, reducing the crystalline index of cellulose is helpful to promote the enzymatic hydrolysis of cellulose.

11.2.2 Hemicellulose

Hemicellulose is the second most abundant component in lignocellulose, and it is a branched compound polysaccharide composed of several monosaccharides (xylose, arabinose, glucose, mannose, galactose, etc.) residues and 4-*O*-methyl glucuronic acid through β-D-furanose bond, glucuronide bond and ester bond [21]. Unlike the long-chain structure of cellulose, most hemicellulose chains are short. In hardwood, *O*-acetyl-4-*O*-methyl glucuronic acid xylan is the main component of hemicellulose (about 15–30%), and its skeleton is made up of β-D-pyranose residues linked by 1,4-glycosidic bonds; 4-*O*-methyl-α-D-linked by 1,4-glycosidic linkages is another kind of xylose unit [22]. In cork, the main component of hemicellulose is *O*-acetyl galactomannan (about 10–25%), and its main chain is a straight chain or a few branches containing 1,4-β-D-mannose and β-D-glucopyranose units [23]. The primary chain of cork hemicellulose has a xylose unit that bears an α-L-arabinose substituent and a 4-*O*-methyl-α-D-glucuronic acid group [24]. Arabinoxylan, the main component of hemicellulose in grass, consists of the linear backbone of 1,4-β-D-xylopyranose units, in which a single α-L-arabinose residue is composed of 1,4-linked D-xylopyranose residues linked by α-1,2 and/or α-1,3 bonds as side groups [25]. Hemicellulose is usually regarded as an amorphous polysaccharide because of its high branching, and it is easy to degrade into monosaccharide or furfural during strong acid conditions [26].

11.2.3 Lignin

Lignin is one of the most abundant renewable biopolymers on earth and the largest natural source of aromatic compounds. More than 100 million tons of lignin are extracted annually by the wood pulping and other biorefinery industries; the majority of lignin is utilized directly for combustion power production or discarded, with just about 10% being recovered for upgrading [27]. Lignin is a three-dimensional reticular amorphous polymer composed of three primary phenylpropane units/hydroxycinnanol (Guaiacyl (G), Syringyl (S) and *p*-Hydroxyphenyl (H)) joined by a variety of chemical linkages [28, 29]. The primary chemical linkages that connect lignin subunits are C=C and C–C bonds, such as β-1, β-5 and β-β, as well as C–O–C bonds, like β-O-4 and α-O-4. Among them, the content of β-O-4 is the most abundant, which could reach 50%; the content of β-β varies from 3.5% to 25%; and the contents of β-5 and α-O-4 accounts for 4–10% and 3–5%, respectively [30]. Numerous functional groups, including phenolic and aliphatic hydroxyl groups, are also present in lignin subunits, which have an impact on the hydrophobic and hydrogen bonding properties of lignin. These phenolic hydroxyl groups will notably affect the hydrogen bonding ability of lignin and amino acid residues in cellulase [31]. Besides, the lignin-carbohydrate complex, which is bonded by many alkyl/aryl linkages between lignin and hemicellulose, prevents enzymes from coming into touch with the cellulose during enzymatic hydrolysis [32].

11.3 Lignocellulosic Pretreatment Methods

Pretreatment is a crucial step to efficiently decompose lignocellulose and convert it into biofuels and bio-based chemicals [5] (Fig. 11.1). The modification of the structures of the three lignocellulosic components and the effect on the enzymatic accessibility of polysaccharide components are the essential criteria for evaluating the advantages and drawbacks of pretreatment technologies. Besides, efficacy, energy consumption and economic viability are also important indicators to

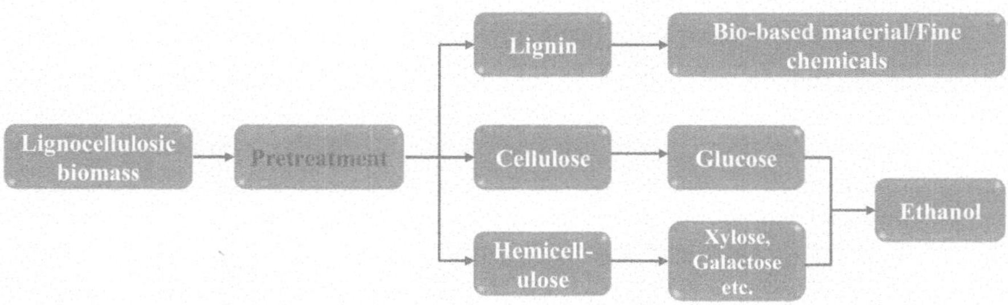

Fig. 11.1 Schematic diagram of the process of converting lignocellulosic biomass into ethanol and bio-based chemicals

determine the feasibility of pretreatment. The pretreatment methods that have been developed can be summarized as follows: biological, physical and chemical methods.

11.3.1 Biological Pretreatment

Biological pretreatment has the advantage of selectively degrading lignin in lignocellulose by using extracellular lignin degradation oxidases released from microorganisms (such as lignin peroxidase and laccase). The effective degradation of lignin is usually realized through the synergistic effect of fungi and bacteria, in which fungi play a major role [33]. It was found that fungi, including brown rot fungi, white rot fungi and soft rot fungi, all biodegrade lignin well in wheat straw. In particular, white rot fungi are the most efficient microorganisms for biomass pretreatment with a strong ability to decompose lignin [34]. The three extracellular ligninolytic enzymes of white rot fungi are laccase, Lignin Peroxidase (LiPs) and Manganese Peroxidase (MnPs) [35].

Although biological pretreatment is an environmentally friendly method that has the advantages of mild conditions, low energy consumption and less dosage of chemical reagents [36], it is time-consuming with lignin wasted, and cellulose and hemicellulose consumed to some extent.

11.3.2 Physical Pretreatment

11.3.2.1 Mechanical Grinding and Extrusion Pretreatment

Mechanical grinding reduces the particle size of biomass and increases the specific surface area of biomass by shearing and grinding. The advantages of this method are below: slightly reduction of the crystallinity and polymerization degree of cellulose; improvement of the enzymatic accessibility of biomass; enhancement of the mass transfer efficiency. However, the drawbacks of mechanical grinding are high energy consumption and low treatment efficiency.

The physical and chemical structure of biomass can be altered by the high shear force, high temperature and pressure created during the screw extrusion pretreatment, which is beneficial to increase the accessible surface area of biomass and enhance the enzymatic hydrolysis performance [37]. The pretreatment efficacy of screw extrusion is influenced by several critical factors, including extrusion speed, temperature, and biomass humidity [38]. Previous studies have demonstrated that the combination of chemical and mechanical pretreatment facilitates the further improvement of treatment efficacy [39]. For instance, Han et al. [40] combined extrusion with ionic liquid pretreatment to treat wood and achieved a glucose yield of 99% and a xylose yield of 99.5%.

11.3.2.2 Microwave and Ultrasonic Pretreatment

Microwave pretreatment uses the constantly changing magnetic field generated from microwave radiation to trigger the vibration of polar chemical bonds in biomass, thus producing a thermal effect on biomass. Microwave radiation improves the digestibility of biomass by changing the structure of cellulose and removing lignin and hemicellulose [41]. Microwave pretreatment has many advantages: short reaction time, rapid and uniform heating, few side effects, easy operation and environmental friendliness [42].

Ultrasonic pretreatment exhibited mechanical effect and sonochemical effect at low frequencies (20–60 kHz) and high frequencies (100–1000 kHz), respectively. These effects facilitate disrupting the crystal structure of cellulose and enhancing the mass and heat transfer effects on the surface of biomass, thus improving the digestibility of biomass [43]. Ultrasonic and microwave pretreatment have several drawbacks, including high equipment costs, low working mass and unsuitability for large-scale production [43, 44].

11.3.3 Chemical Pretreatment

11.3.3.1 Acid Pretreatment

As a technical mature and widely used lignocellulosic pretreatment method, acid pretreatment has the capacity to improve the accessibility of cellulose to enzymes by removing hemicellulose [45]. Sulfuric acid, hydrochloric acid, phosphoric acid and organic acids are frequently applied for acid pretreatment. Acid pretreatment is usually conducted in two ways: dilute acid at high temperatures (>160 °C) and concentrated acid (30–70%) at lower temperatures (<50 °C) [46]. Abud et al. studied the morphological changes of cellulose during dilute sulfuric acid pretreatment [47]. The findings demonstrated that although the cellulose microfibrils' orientation remained unchanged after dilute sulfuric acid pretreatment, spherical particles were produced on the cell surface and the original fiber layered structure vanished. According to the literature [48], lignin undergoes a shift from solid state to molten state when the temperature of dilute acid treatment exceeds the glass transition temperature of lignin, and then condenses into spherical particles rich in lignin that are attached to the biomass surface. Related studies have confirmed that these spherical particles are pseudo-lignin composed of carbohydrates and lignin degradation products [48, 49]. The pseudo-lignin may undergo irreversible unproductive adsorption with cellulase, which will hinder the hydrolysis of cellulose (Fig. 11.2) [50].

Concentrated acid can hydrolyze cellulose and hemicellulose to simple sugars [51]. However, the drawbacks of concentrated acid pretreatment, such as low working mass, severe reactor corrosion and usage of much acid hinder its large-scale application. Dilute acid pretreatment has the advantages of directly

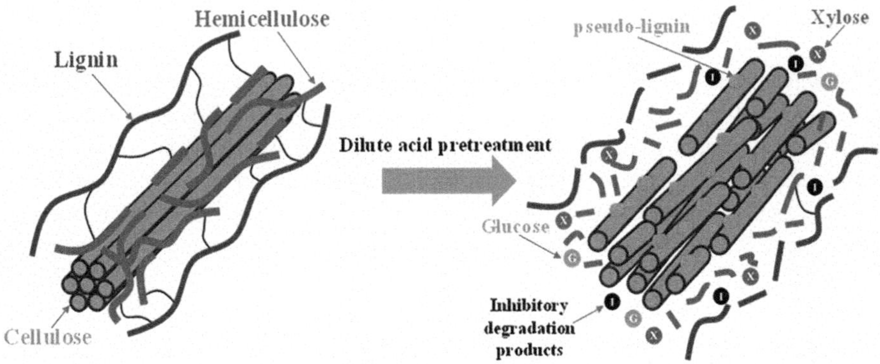

Fig. 11.2 Schematic diagram of dilute acid pretreatment

hydrolyzing hemicellulose into fermentable monosaccharides. For example, Hans et al. [52] found that when the acid dosage of pretreatment was 2.18% (v/v), the solid loading was 14.35% (w/v), and the retention time was 29.49 min, the total reducing sugar yields obtained after enzymatic hydrolysis were 215.28 ± 2.4 and 306.14 ± 5.3 g/kg sugarcane bagasse, respectively. Nevertheless, several shortcomings including high energy consumption, low working mass and severe pretreatment conditions reduce its economic viability in industrial use. In addition, toxic degradation products generated from dilute acid pretreatment, such as organic acid, furfural and 5-Hydroxymethylfurfural (HMF), drastically lower the fermentability of pretreated biomass [53]. Therefore, dilute acid pretreated biomass requires water washing or detoxification to achieve satisfying fermentation performance.

11.3.3.2 Alkaline Pretreatment

Alkaline pretreatment is a method of pretreating biomass with alkaline reagents (such as KOH, NaOH, $Ca(OH)_2$, etc.). Saponification and solvation reactions during alkaline pretreatment break the linkages of lignin carbohydrate complex, accompanied by the removal of acetyl and uronic acid groups in hemicellulose [54]. After most lignin in biomass is removed during alkaline pretreatment, the exposed cellulose expands and causes the fiber bundles to flex or even peel off, thus reducing the crystallinity of cellulose [55] (Fig. 11.3). Alkaline pretreatment is more effective for biomass with low lignin content (e.g., hardwood and grasses) than for biomass with high lignin content (e.g., softwood) [56]. Among many alkaline pretreatment methods, NaOH pretreatment is the most popular and efficient biomass pretreatment, which has strong delignification ability [57]. Mohanasundaram et al. [58] pretreated corn cob with optimum conditions of 1.81% (w/v) NaOH and 5% (w/v) solid loading for 90 min, resulting in 82.03% lignin removal. Compared with NaOH pretreatment, $Ca(OH)_2$ is regarded as another encouraging alkaline pretreatment because of its low-cost and low corrosiveness [59]. Nevertheless, the phenolic compounds

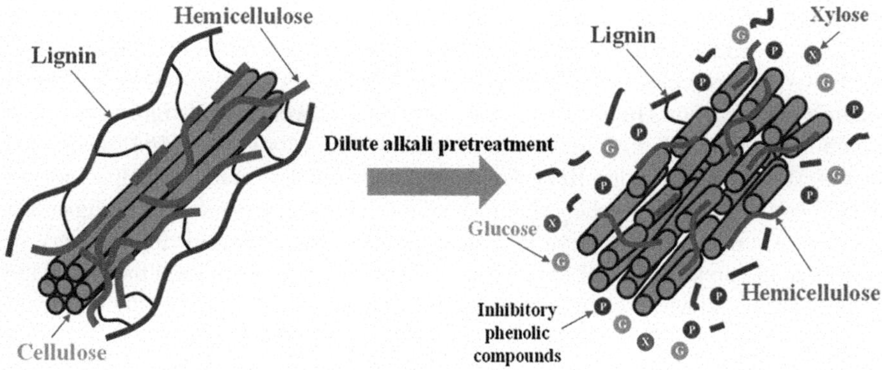

Fig. 11.3 Schematic diagram of alkaline pretreatment

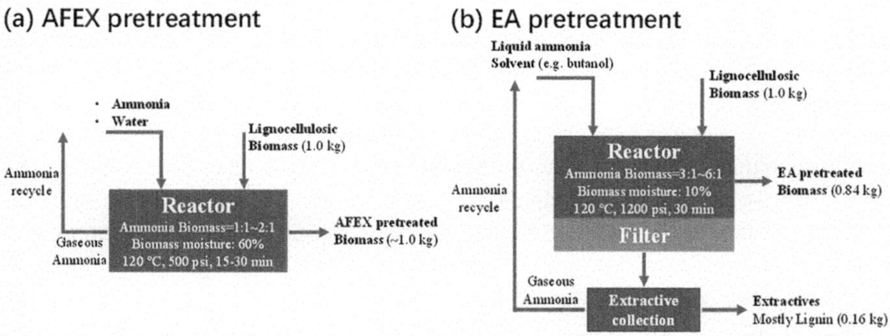

Fig. 11.4 Process diagram of AFEX pretreatment (**a**) and EA pretreatment (**b**). (Reproduced with permission from [64]. Copyright © 2019, Springer Nature)

generated from lignin degradation during alkaline pretreatment severely impair the fermentability of biomass [54].

11.3.3.3 Ammonia Pretreatment

Ammonia-based pretreatment, including Ammonia Fiber Explosion (AFEX) (Fig. 11.4a) and Extractive Ammonia (EA) (Fig. 11.4b), is regarded as one of the mainstream lignocellulosic pretreatment methods and has been extensively studied [60, 61]. During AFEX pretreatment, the biomass is submerged in liquid anhydrous ammonia for a predetermined amount of time at high pressure and temperature, and then the pressure is abruptly reduced to destroy the biomass's structure and increase the enzymatic hydrolysis performance of biomass. AFEX pretreatment is a "dry-to-dry" process with the generation of a few toxic degradation products. AFEX pretreated biomass owns good digestibility and fermentability without water washing or detoxification [62].

Liquid anhydrous ammonia is used in EA pretreatment, which can not only selectively extract lignin from biomass, but also convert natural cellulose I into easily hydrolyzed cellulose III. EA pretreatment consists of three stages: ammonia loading and reaction, lignin extraction, ammonia recycling and lignin recovery, which can extract 40–50% lignin from biomass [61]. Morais et al. [63] developed a COmpacted Biomass with Recycled Ammonia (COBRA) pretreatment, achieving >95% conversion of soluble monomeric and oligomeric sugar from sugarcane bagasse. Ammonia pretreatment is a green method that allows for the recovery of ammonia, the preservation of the biomass's internal components, and the use of the leftover ammonia as a source of nitrogen for fermentation.

11.3.3.4 Steam Explosion Pretreatment and Liquid Hot Water Pretreatment

Steam explosion pretreatment [65]: raw biomass is kept at high temperature (180–240 °C) and high pressure for a short time to make high-pressure steam penetrate into lignocellulose; the steam is then rapidly released to demolish the lignocellulose structure (Fig. 11.5a) [66]. Acetic acid produced by the cleavage of acetyl group during steam explosion will foster autocatalytic reaction (similar to acid hydrolysis), which further facilitates the degradation of hemicellulose and the cleavage of chemical linkages. Borand et al. [67] showed that pinewood pretreated with steam explosion at 190 °C for 10 min could achieve 97.7% glucose conversion and 85.6 xylose conversion after 72 h enzymatic hydrolysis. The drawbacks of this method include severe pretreatment conditions and high operation cost; organic acids, furfural, and phenols produced during the pretreatment, which are inhibitory to the subsequent enzymatic hydrolysis and fermentation.

Liquid Hot Water (LHW) pretreatment [68] pretreats biomass at 150–240 °C in water; with part hemicellulose hydrolyzed and lignocellulose structure disrupted (Fig. 11.5b) [68]. Xu et al. [69] optimized liquid hot water pretreatment of sugarcane bagasse (SCB), followed by high-solids enzymatic hydrolysis, achieving a conversion of 280 mg fermentable sugars/g SCB at pretreatment conditions of 186 °C, 25 min and 12.5% (w/w). Similar to steam explosion, LHW pretreatment also undergoes self-hydrolysis reaction, producing many degradation products (such as organic acids and furfural). Short treatment time and no use of chemical reagents are two advantages of LHW pretreatment. Nevertheless, much sugar loss, high energy and water consumption, and low working mass impede its industrial use.

11.3.3.5 Organic Solvent Pretreatment

Pretreating biomass using organic solvents (e.g. methanol, ethanol, ethylene glycol, and acetone) and sometimes adding organic acids (oxalic acid and salicylic acid) or alkali (NaOH and $Ca(OH)_2$) as catalysts is known as organic solvent pretreatment [70]. Lignin-hemicellulose linkages are dissolved and disrupted during organic

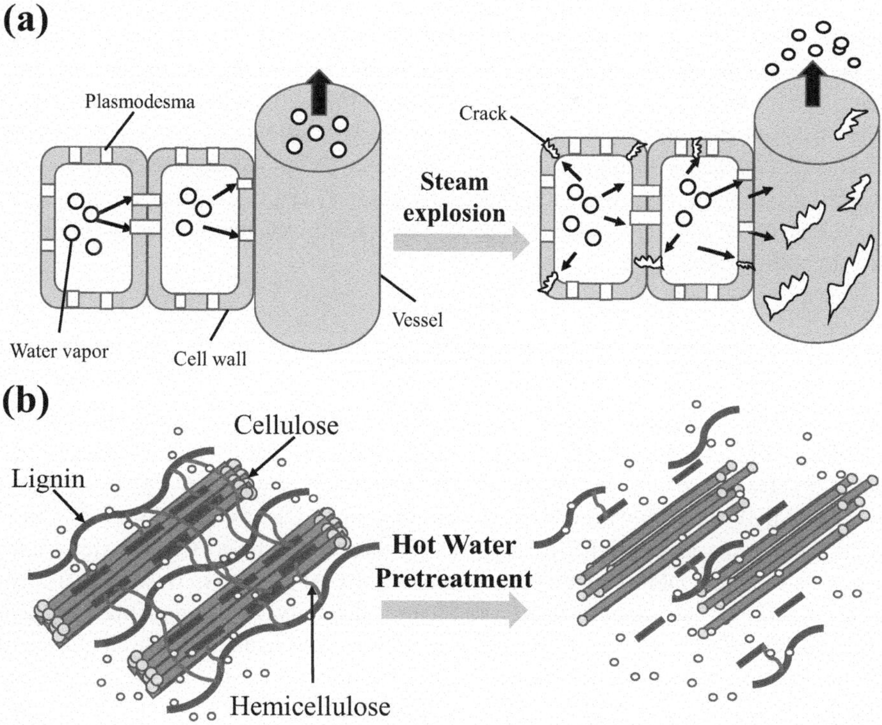

Fig. 11.5 Schematic diagram of steam explosion pretreatment (**a**) and hydrothermal pretreatment (**b**)

solvent pretreatment, exposing cellulose and increasing the digestibility of biomass [54]. Adding acid as catalyst in organic solvent pretreatment is helpful to improve the removal efficiency of lignin and hemicellulose [71]. As reported [72], $AlCl_3$-catalyzed organic solvent pretreatment with different temperatures (150–200 °C) was put forward to remove lignin and hemicellulose, reserve the majority of cellulose, and obtain an ethanol concentration of 59.76 g/L with an ethanol yield of 69.57% in subsequent simultaneous saccharification and fermentation. The lignin dissolved by organic solvents has more active groups with a consistent molecular weight, making it more suitable for value-added usage than lignin obtained by alkali pretreatment [73].

11.3.3.6 Ionic Liquid Pretreatment

Ionic Liquids (ILs) is called "green solvents" and consist of small inorganic anions and large organic cations [74], has the ability to dissolve cellulose and the dissolved cellulose can be regenerated under certain conditions. In addition, ILs can be recovered and recycled after pretreatment [75]. Compared to other organic solvents,

ILs have the following merits: low vapor pressure and non-volatilization; good thermal and chemical stability; anion and anion composition adjustability; friendly to the environment. Nakasu et al. [76] studied the protic IL and treated sugarcane bagasse with monoethanolammonium acetate and [MEA]OAc] at 150 °C for 2 h. It was observed that 60% of lignin is solubilized without loss of carbohydrate and 78.6% and 49% of cellulose and hemicellulose yields during saccharification. Nevertheless, poor mass transfer caused by high viscosity of ILs, inhibition on enzyme activity and high cost of reagent recovery, limit its application in lignocellulosic pretreatment [77].

11.3.3.7 Deep Eutectic Solvents Pretreatment

Deep Eutectic Solvents (DES) is a green and efficient reagent for lignocellulosic component fraction, composed of hydrogen bond donors and hydrogen bond acceptors [78]. Due to the strong hydrogen bond interaction between hydrogen bond donor and hydrogen bond acceptor, the melting point of DES is much lower than that of its single component. DES is easier to prepare, lower in cost and toxicity and better in biodegradability than other organic solvents and ILs. Besides, DES is similar to ILs in adjustability and combination diversity of components [79]. As reported [80], DES can break the hydrogen bonds between cellulose chains and effectively dissolve lignin, but hardly dissolve cellulose. Vigier et al. [81] suggested that the strong cohesion of cellulose prevents the dissociation and recombination of its hydrogen bond network, and the hydroxyl group in choline chloride was able to form a strong hydrogen bond network with cellulose, thus generating a stable DES-cellulose structure. Wang et al. [82] proposed a hydrothermal-DES pretreatment to separate hemicellulose and high-purity lignin simultaneously from bamboo with most of cellulose retained in the residues, achieving 80.1% delignification with 99.49% lignin purity and 98.2% hemicellulose degradation. Increasing the treatment temperature is conducive to reducing the viscosity of DES, thus enhancing the mass transfer efficiency of DES and improving the delignification performance, while it will aggravate the sugar loss [83, 84]. Several drawbacks of DES, such as its high viscosity, low working mass of pretreatment and high reagent recovery cost, hinder its use in large-scale biorefineries.

11.3.3.8 Inorganic Salt Pretreatment

Inorganic salt pretreatment has the advantages of low cost, low corrosiveness and toxicity and mild treatment conditions, has attracted increasing attention [85]. Among inorganic salt compounds, metal halides are more prominent for dissolving cellulose [86]. Halogen ions disrupt the hydrogen bond network and decrease the crystallinity of cellulose by forming complexes with cellulose, the complexes formed from metal cations and water are also helpful to break glycosidic linkages [87]. Besides, H_3O^+ ion generated from metal cation hydrolysis depolymerizes hemicellulose during pretreatment [88]. However, inorganic salt pretreatment needs to be carried out in liquid phase with low working mass. Furthermore, the residual inorganic salt ions after pretreatment have a potential inhibitory impact on enzymes and microorganisms. Therefore, the pretreated biomass requires water washing or detoxification prior to subsequent bioconversion [89]. Lara-Serrano et al. [90] studied the use of inorganic salt hydrates as solvents in the dissolution/precipitation pretreatment of cellulose. The best performance was obtained with $H_4SiW_{12}O_{40}$ (0.05 mol/L) at 140 °C for 300 min, where the cellulose conversion was close to 99% and the glucose yield was 90%.

11.3.4 Ideal Lignocellulosic Pretreatment

Given the merits and drawbacks of traditional lignocellulosic pretreatment methods (Table 11.1), the complexity and diversity of lignocellulosic biomass should be fully considered for developing ideal pretreatment methods. Ideal pretreatment should have the following characteristics: (1) remarkably enhance the digestibility of biomass; (2) minimize the generation of toxic degradation products; (3) minimize the loss of sugar; (4) well preserve the original structure of lignin; (5) low energy consumption and low cost; (6) universality for a variety of biomass; (7) facilitates biomass logistics.

11.3.5 DLC Pretreatment

Several adverse properties of loose biomass, such as low energy density, low bulk density and high logistic cost, are also essential factors restricting the large-scale use of lignocellulosic biomass. As an emerging and integrated system for biomass pretreatment and biomass logistics, "Densifying Lignocellulosic biomass with Chemicals (DLC)" has increasingly attracted attention due to its merits of low severity, generation of fewer toxic degradation products and ease of scale-up, which commendably overcomes the drawbacks of traditional pretreatment methods (Fig. 11.6) [8, 9, 91]. DLC pretreated biomass exhibited high digestibility and fermentability with high bulk density and great durability (highly resistant to microbial contamination). During DLC pretreatment, densification rendered penetration of chemical reagents into the biomass, the reagents react with internal lignocellulosic components during densification and biomass logistics, thus breaking the biomass' structure and facilitating the conversion of biomass to biofuels or chemicals. As reported [8, 9], DLC pretreated biomass after regular autoclave treatment (at 121 °C) offers high-concentration sugar after enzymatic hydrolysis with high fermentability, which was not achievable by traditional pretreated biomass without water washing or detoxification.

Table 11.1 Comparison of typical lignocellulosic biomass pretreatment methods

Pretreatment types	Advantages	Disadvantages
Ammonia	Fewer degradation products; high sugar recovery; reagent can be recycled	High equipment requirements; high ammonia recovery cost
Deep eutectic solvent	High fraction efficiency; mild pretreatment conditions	High reagent viscosity; high recovery cost, low processing efficiency
Dilute acid	High hemicellulose hydrolysis; high digestibility	Harsh pretreatment conditions, high equipment requirements; high contents of degradation products
Dilute alkali	High lignin removal; high cellulose accessibility; reduced cellulose crystallinity	High contents of phenolic degradation products; low sugar recovery
Inorganic salt	High cellulose solubility; low toxicity; mild pretreatment conditions	Low pretreatment efficiency; high reagent recovery cost; strong fermentation inhibition
Ionic liquid	Low reagent vapor and non-volatile; good thermal stability; good chemical stability	High reagent viscosity; high recovery cost; low processing efficiency
Liquid hot water	Short treatment time; no chemicals required	High contents of degradation products; high energy and water consumption; low processing efficiency
Mechanical	Improve enzymatic accessibility; enhance mass transfer efficiency; no reagents required	High equipment and energy cost; low sugar yield
Microwave nuclear radiation	Short reaction time; few side reactions, easy operation and environmental friendliness	High equipment cost; low processing efficiency; low sugar yield
Organic solvent	High removal of lignin and hemicellulose; reagent can be recycled; increase cellulose accessibility	High reagent cost; low reagent safety; high recovery cost
Steam explosion	Low chemical consumption, high biomass digestibility	High contents of degradation products; high energy consumption; low processing efficiency

Low ethanol titer caused by poor fermentability of pretreated biomass is considered as one of the bottlenecks hindering the advancement of cellulosic ethanol industry. To our knowledge, lignocellulosic ethanol concentrations were achieved rarely higher than 60 g/L for most traditional biorefineries without water washing or detoxification (Table 11.2). Most pretreated biomass require solid-liquor separation, washing or detoxification after pretreatment, thus removing a large amount of toxic degradation products but increased treatment cost. Loose biomass used in traditional biorefinery systems owns negative properties, such as low bulk density, easy to be contaminated, fluffy and difficult to handle, which are not conducive to biomass logistics. Besides, Low working mass during pretreatment, high energy consumption, generation of many toxic products severely impeded the development of traditional biorefinery. DLC technology well addressed abovementioned issues, conducted at mild conditions with high biomass loadings, and is easy to scale-up.

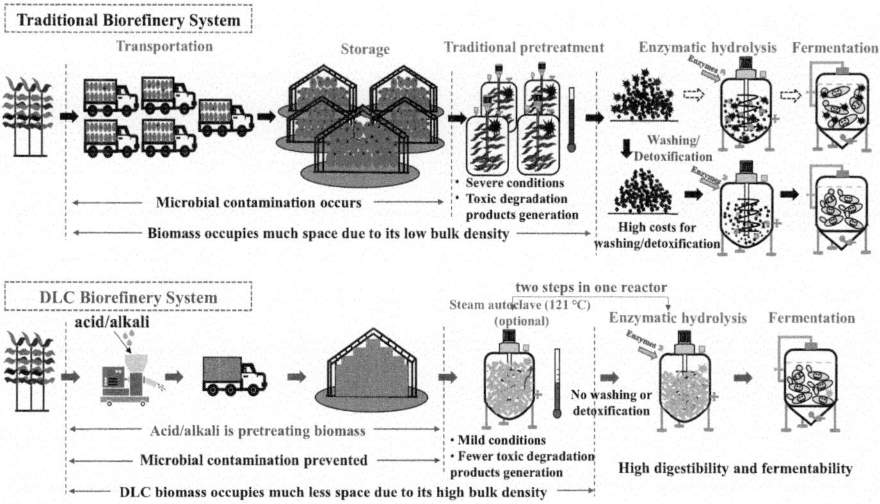

Fig. 11.6 Comparison of traditional biorefinery system and DLC biorefinery system. (Reproduced with permission from [9]. Copyright © 2022, Elsevier)

Notably, an ethanol titer as high as 82.9 g/L was obtained on DLC pretreated corn stover without washing or detoxification [92]. Overall, DLC pretreatment method indicates huge superiority in industrial application compared with traditional pretreatment methods.

Analysis of pretreatment mechanisms concluded that the synergism of chemical and mechanical actions during DLC pretreatment plays a critical role in ensuring superior performance [92]. The internal polysaccharide of biomass is exposed by disrupting the cell wall structure via mechanical actions, and reacts with concentrated reagents and is partially deconstructed. This synergy increased the biomass porosity and accessible breakpoints of crystalline cellulose to cellulase by reducing the crystallinity and polymerization degree of cellulose, digesting amorphous cellulose and removing a large amount of hemicellulose. It is worth noting that most of the inter-unit linkages and non-condensed subunits are well preserved in the lignin of DLC pretreated biomass, without the generation of pseudo-lignin and large amounts of toxic degradation products. This finding indicates the huge potential of valorization for DLC lignin and thus favors boosting the economic viability of lignocellulosic biorefinery.

Table 11.2 Performance of typical biorefining processes for high-titer ethanol production from lignocellulosic biomass

Pretreatment	Feedstock	Biomass loading during pret. (kg dry biomass/m^3 reactor)	Temperature of pret. (°C)	Solid-liquor separation/ washing/detoxification	Ethanol titer (g/L)	Reference
AFEX	Corn stover	80–100	100–140	No	60.3	[93]
Alkaline	Corn stover	33	75	Yes	64.1	[94]
CELF with Tetrahydrofu-ran (THF)	Corn stover	7.5	150	Yes	58.8	[95]
DLC	Corn stover	300–400	121	Yes	110	[91]
DLC	Rice straw	200–300	121	No	82.9	[92]
DMR	Corn stover	76.9	80	Yes	86	[96]
DryAP	Corn stover	60	175	Yes	90.3	[97]
Liquefacted hydrother-mally with acetic acid	Sweet sor-ghum bagasse	100	229	Yes	47.9	[98]
Organic solvent	Poplar	30	200	Yes	59.76	[72]
SPORL	Lodgepole pine	86.9	165	Yes	47.1	[99]
Steam explosion with SO_2	Spruce	70	205	Yes	48	[100]
Steam explosion	Corn stover	50	180	Yes	60.8	[101]
Wet oxidation	Corn stover	60	195	Yes	52	[102]

11.4 Enzymatic Saccharification and Fermentation of Lignocellulose

11.4.1 Overview of Saccharification and Fermentation Process

Cellulosic ethanol production is usually divided into three steps: (1) Biomass pretreatment to improve the enzymatic accessibility of cellulose; (2) Cellulose and hemicellulose hydrolysis to obtain fermentable sugar; (3) Fermentation to convert sugars to ethanol. The steps 2 and 3 carried out separately are called Separate Hydrolysis and Fermentation (SHF), which requires high dosages of enzyme and long operation time to produce high concentrations of fermentable sugar [103]. The steps 2 and 3 performed together are designated as Simultaneous Saccharification and Fermentation (SSF), which greatly shortens the operation time, alleviates sugar inhibition and increases ethanol productivity [104]. The SHF or SSF, which ferments both glucose and xylose, is designated as Separate Saccharification and Co-Fermentation (SHCF) or Simultaneous Saccharification and Co-Fermentation (SSCF).

11.4.2 Factors Affecting SSCF Process

Compared to SHCF process, SSCF process has shorter operation time, better sugar utilization, higher ethanol yield and lower risk of bacterial contamination [105]. The performance of SSCF process is affected by many factors, such as temperature, pH, enzyme dosage, biomass loading and strain inoculation amount.

11.4.2.1 Temperature and pH

Owing to the fact that enzymatic hydrolysis and fermentation during SSCF are performed simultaneously, the operating temperature of SSCF must be compromised between the higher temperature (50 °C) suitable for hydrolysis and the lower temperature (~30 °C) suitable for fermentation [106]. Low temperature reduces enzyme activity and impairs the enzymatic sugar yield; high temperature for SSCF adversely affects microbial activities [107]. In addition, the optimum pH value for cellulases is about 4.8, while the pH value suitable for ethanol fermentation is about 5.5, making it necessary to find a balanced pH value for SSCF to achieve the optimal process performance.

11.4.2.2 Enzyme Dosage and Biomass Loading

Enzyme dosage is another important factor for SSCF. Insufficient enzyme loading leads to a reduction in sugar release during SSCF, and carbon starvation occurs after sugar exhaustion by fermentation, resulting in the death of the fermenting strain [108]. Conversely, adding excessive enzymes in SSCF to maintain high enzyme catalytic performance increases the cost of SSCF [109]. An improved SSCF process is beneficial to achieve better fermentation results under the premise of lower enzyme loading, and thus lower cost of SSCF.

Ethanol titer is one of the key parameters used to evaluate cellulosic ethanol production. The distillation cost of ethanol decreases with the increase in ethanol titer; the cellulosic ethanol production will potentially be economically feasible only when the ethanol titer exceeds 40 g/L [110]. Increasing the biomass loading of SSCF is helpful to increase the ethanol titer, but this will lead to many problems: The high viscosity of the broth at high biomass loading reduces the efficiency of mass transfer and heat transfer [111]; the high contents of inhibitors in the broth at high biomass loading have a strong inhibition on enzymes and fermenting strains [112].

11.4.2.3 Toxic Degradation Products and Lignin-Enzyme Unproductive Adsorption

Soluble degradation products produced from lignocellulose during pretreatment show varying degrees of inhibition on hydrolysis and fermentation [14] (Fig. 11.7). Soluble phenolic compounds induce β-glucosidase inactivation, which leads to the accumulation of cellobiose and eventually weaken the catalytic efficiency of cellobiose [113]. Sugar degradation products (such as furfural and 5-Hydroxymethylfurfural (5-HMF)) and organic acids adversely affect the microbial growth and fermentation [114]. For instance, acetic acid decreases the cell viability of strains by reducing intracellular potential and pH value, substantially disrupting intracellular pH homeostasis, affecting the function of the cell wall and plasma membrane, and inducing oxidative stress and programmed cell death [115].

Different phenolic compounds have varying impacts on cell growth and ethanol production during SSCF. Some compounds (such as cinnamic acid and ferulic acid) present inhibition on cell growth and fermentation; other compounds (such as vanillin, catechol and coniferyl alcohol) at low contents can improve cell viability and stimulate ethanol production, whereas they inhibit ethanol fermentation at high contents [116, 117].

Nonproductive adsorption of lignin and enzymes is another important factor affecting the hydrolysis of lignocellulose, including hydrophobic interactions, hydrogen bonding interactions, and electrostatic interactions. Lignin content, structure, and functional groups are related to the adsorption of lignin to enzymes [118]. Pretreatment is an essential method to overcome the recalcitrance of lignocellulosic biomass and improve enzymatic hydrolysis. Different pretreatment

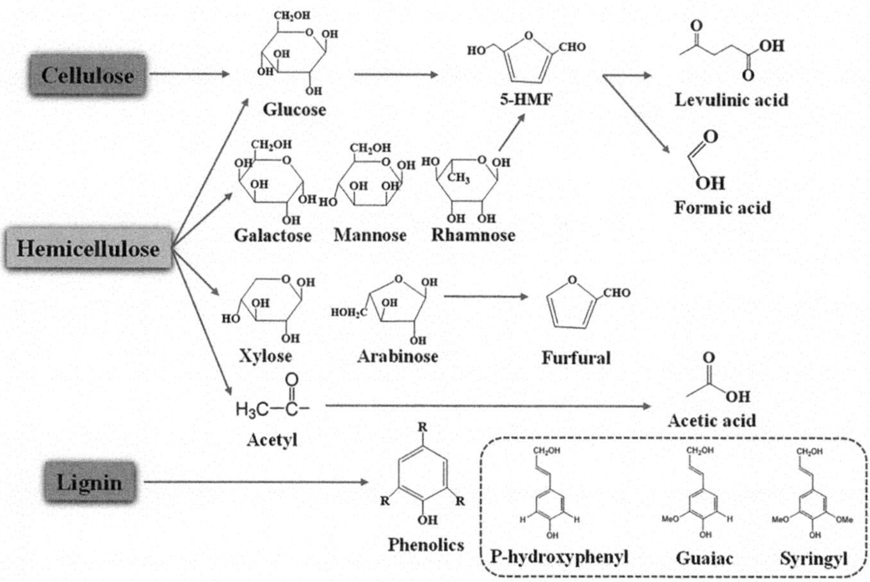

Fig. 11.7 The formation pathway of degradation inhibitors from lignocellulose during pretreatment

methods showed different impacts on content and structure of lignin [119]. For example, alkaline pretreatment can effectively remove lignin from lignocellulose to improve enzymatic hydrolysis [120]. Hydrothermal and acid pretreatment keep most lignin in lignocellulose, resulting in increased nonproductive binding between lignin and enzymes [121]. Pretreatment can also change the structure and functional groups of lignin, thereby affecting the interaction between lignin and enzymes [119]. The degree of protein adsorption by lignin depends on the phenolic hydroxyl content of lignin [121]. The higher the phenolic hydroxyl content of lignin, the stronger its adsorption to enzymes. The negative effect of lignin in enzymatic saccharification can be alleviated by group modification (such as sequential Fenton oxidation and sulfomethylation) [122].

11.4.2.4 Inoculation Amount and Nutrient Addition

The effect of the initial inoculation amounts of microbial strains on the fermentation performance of SSCF cannot be ignored. Low inoculation leads to low stress resistance and cause poor fermentation performance of SSCF at high biomass loadings. High inoculation results in high tolerance to the toxic inhibitors, and thus improves the ethanol yields but increases the cost of SSCF [123].

Besides the inoculation amounts, nutrient addition in the fermentation broth also influences the ethanol yield of SSCF. As reported, amides generated from the ammonolysis reaction during AFEX pretreatment could be used as a nitrogen source in strain growth to maintain cell viability [117]. Adding nitrogen sources (such as yeast extract and tryptone) to the hydrolysate at high biomass loading promotes cell growth and xylose consumption and boosts ethanol yield, indicating that the supplementation of nutrients strengthens the cell tolerance to toxic products [124].

11.4.3 Improved Strategies for SSCF Process

Considering the issues of high broth viscosity and strong inhibition existing in SSCF at high biomass loadings, some targeted strategies such as pre-hydrolysis, fed-batch, biological detoxification, strain domestication and surfactant addition are used to improve SSCF and increase the ethanol yield.

11.4.3.1 Pre-hydrolysis and Fed-Batch Strategy

Pre-hydrolysis is beneficial to reduce the viscosity of broth, and to provide sufficient sugar for initial microbial growth and fermentation [125]. Similarly, the fed-batch strategy can reduce the initial lignocellulosic biomass loading during SSCF, and thus avoid high viscosity [126, 127]. Given that xylose and glucose compete for the same transportation system in yeast and the affinity of transporters for glucose is much higher than that of xylose [128], the transportation of xylose is limited with the presence of glucose, and the effective xylose utilization by *Saccharomyces cerevisiae* can be promoted when the glucose concentration is at a low level. Besides, cell vitality decreases due to the inhibition of toxic products after glucose is exhausted in the late stage of fermentation, resulting in poor xylose utilization. It is reported that xylose utilization can be improved when glucose concentration is at a low level but not exhausted, which is attributed to the induction of transporters and the generation of inducible cofactors of glycolytic enzymes [128, 129]. Fed-batch of enzyme during SSCF contributes to maintain high enzymatic activity for a long time and adjust the glucose and xylose concentrations in the broth, thereby promoting xylose consumption [125].

Overall, the strategy of fed-batch of biomass and enzyme has the following merits: keeping the viscosity of fermentation broth at a low level; making the strain gradually adapt to the toxicity of inhibitors in the broth; improving the xylose utilization by adjusting the concentration of glucose; maintaining good enzyme catalytic activity for a long time, and thus facilitates enzyme dosage reduction.

11.4.3.2 Short-Term Domestication

Short-term stress-resistant domestication of microbial strains is another method to reduce the inhibition of toxic degradation products during SSCF. Short-term domestication is a method to cultivate fermentation strains with pre-adaptive ability under conditions akin to the late stage of SSCF, thus improving the fermentation performance of SSCF. Heterogeneous phenotypes induced by short-term domestication enable strains to have stress-resistant function under specific environmental conditions [130]. During short-term domestication, enzymes and proteins with high metabolic activity produced by yeast under selective pressure can provide enough energy for degrading inhibitors and regulating intracellular pH [131]. Short-term domestication during strain propagation is conducive to enhance tolerance and thus improve sugar utilization for ethanol production. However, strain domestication under high-content inhibitor conditions require much sugar consumption to breed enough yeast cells with good fermentation ability. Furthermore, the uncertainty of domestication efficiency increases with the inhibitor content in the domestication environment [131].

11.4.3.3 Surfactant Addition

Ionic surfactants (e.g., Tween and polyethylene glycol), which have the advantages of high stability, good compatibility and strong solubility, could enhance enzymatic hydrolysis performance and increase the ethanol yield of SSCF [132]. For instance, Tween 80 effectively prevents the unproductive adsorption of cellulase to lignin, thereby facilitates the reduction of enzyme dosage [133]; Polyethylene Glycol (PEG) can reduce the inactivation of cellulase caused by the gas-liquid interface and shear force to promote cellulose hydrolysis, thereby boosting the enzymatic sugar yield [134].

11.5 Conclusions and Future Perspectives

Cellulosic ethanol produced from lignocellulosic wastes is regarded as a vital development direction for fuel ethanol industry. With technologies development in recent years, the techno-economics for lignocellulosic ethanol become intriguing. The issues related to biomass logistics and pretreatment have largely been addressed by DLC pretreatment. DLC biomass pellets could be shipped and stored like corn, and could be a commodity feedstock for fermentation industry. DLC biomass after shipped to the factory could be directly enzymatically hydrolyzed or be hydrolyzed after steam sterilization to highly fermentable sugars, which can be fermented to various products, such as ethanol (Table 11.2). DLC reduces cellulosic ethanol production cost by $>30\%$ [92].

For future R&D, it is anticipated that developing cellulases with higher activities and superior tolerances to thermal environment and toxic degradation products through artificial intelligence, rational design, directional evolution, and etc., will further enhance the hydrolysis efficiency of cellulose and further reduce the cost of enzymes for cellulosic ethanol production. Fermentation strains could be improved for faster xylose consumption, higher product yield, more robust, better thermal tolerance, and metabolism/conversion of toxic degradation products, by means of synthetic biology, metabolic and evolutionary engineering strategies. It would be also critical to produce high-value products from lignin residue to leverage the overall economics for a biorefinery.

Acknowledgement This work was supported by the "National Key R&D Program of China" (No. 2021YFC2101301).

References

1. Barua S, Sahu D, Sultana F, Baruah S, Mahapatra S. Bioethanol, internal combustion engines and the development of zero-waste biorefineries: an approach towards sustainable motor spirit. RSC Sustain. 2023;1(5):1065–84. https://doi.org/10.1039/d3su00080j.
2. Li J, Zhao R, Xu Y, Wu X, Bean SR, Wang D. Fuel ethanol production from starchy grain and other crops: an overview on feedstocks, affecting factors, and technical advances. Renew Energy. 2022;188:223–39. https://doi.org/10.1016/j.renene.2022.02.038.
3. Zhang J, Rentizelas A, Zhang X, Li J. Sustainable production of lignocellulosic bioethanol towards zero waste biorefinery. Sustain Energy Technol Assess. 2022;53:102627. https://doi.org/10.1016/j.seta.2022.102627.
4. Gao W, Li Z, Liu T, Wang Y. Production of high-concentration fermentable sugars from lignocellulosic biomass by using high solids fed-batch enzymatic hydrolysis. Biochem Eng J. 2021;176:108186. https://doi.org/10.1016/j.bej.2021.108186.
5. Zhao L, Sun Z-F, Zhang C-C, Nan J, Ren N-Q, Lee D-J, et al. Advances in pretreatment of lignocellulosic biomass for bioenergy production: challenges and perspectives. Bioresour Technol. 2022;343:126123. https://doi.org/10.1016/j.biortech.2021.126123.
6. Abolore RS, Jaiswal S, Jaiswal AK. Green and sustainable pretreatment methods for cellulose extraction from lignocellulosic biomass and its applications: a review. Carbohydr Polym Technol Appl. 2024;7:100396. https://doi.org/10.1016/j.carpta.2023.100396.
7. Gong C, Meng X, Thygesen LG, Sheng K, Pu Y, Wang L, et al. The significance of biomass densification in biological-based biorefineries: a critical review. Renew Sust Energ Rev. 2023;183:113520. https://doi.org/10.1016/j.rser.2023.113520.
8. Chen X, Yuan X, Chen S, Yu J, Zhai R, Xu Z, et al. Densifying lignocellulosic biomass with alkaline chemicals (DLC) pretreatment unlocks highly fermentable sugars for bioethanol production from corn stover. Green Chem. 2021;23(13):4828–39. https://doi.org/10.1039/D1GC01362A.
9. Yuan X, Chen X, Shen G, Chen S, Yu J, Zhai R, et al. Densifying lignocellulosic biomass with sulfuric acid provides a durable feedstock with high digestibility and high fermentability for cellulosic ethanol production. Renew Energy. 2022;182:377–89. https://doi.org/10.1016/j.renene.2021.10.015.
10. Sharma S, Nair A, Sarma SJ. Biorefinery concept of simultaneous saccharification and co-fermentation: challenges and improvements. Chem Eng Process Process Intens. 2021;169:108634. https://doi.org/10.1016/j.cep.2021.108634.
11. Yuan X, Shen G, Chen S, Chen X, Zhang C, Liu S, et al. Modified simultaneous saccharification and co-fermentation of DLC pretreated corn stover for high-titer cellulosic ethanol production without water washing or detoxifying pretreated biomass. Energy. 2022;247:123488. https://doi.org/10.1016/j.energy.2022.123488.
12. Jia X, Liu D, Lin H, Zhang H, Liang X, Ding K, et al. Enhancing efficiency in ethanol production from high solid corn stover: insights into enzymatic hydrolysis parameters and

cellulase recovery. J Taiwan Inst Chem Eng. 2024;2024:105644. https://doi.org/10.1016/j.jtice.2024.105644.
13. Saini JK, Himanshu, Hemansi, Kaur A, Mathur A. Strategies to enhance enzymatic hydrolysis of lignocellulosic biomass for biorefinery applications: a review. Bioresour Technol. 2022;360:127517. https://doi.org/10.1016/j.biortech.2022.127517.
14. Kim D. Physico-chemical conversion of lignocellulose: inhibitor effects and detoxification strategies: a mini review. Molecules. 2018;23(2):309. https://doi.org/10.3390/molecules23020309.
15. Vignesh N, Chandraraj K. Improved high solids loading enzymatic hydrolysis and fermentation of cotton microdust by surfactant addition and optimization of pretreatment. Process Biochem. 2021;106:60–9. https://doi.org/10.1016/j.procbio.2021.04.002.
16. Liu J, Liu L. Recycling utilization of crop straw resources: technology, mode and main problems. Sustain Dev. 2023;13(2):847–60. https://doi.org/10.12677/SD.2023.132088.
17. Zhang T. Taking on all of the biomass for conversion. Science. 2020;367(6484):1305–6. https://doi.org/10.1126/science.abb1463.
18. Hemati A, Nazari M, Asgari Lajayer B, Smith DL, Astatkie T. Lignocellulosics in plant cell wall and their potential biological degradation. Folia Microbiol. 2022;67(5):671–81. https://doi.org/10.1007/s12223-022-00974-5.
19. Nosrati H, Khodaei M, Alizadeh Z, Banitalebi-Dehkordi M. Cationic, anionic and neutral polysaccharides for skin tissue engineering and wound healing applications. Int J Biol Macromol. 2021;192:298–322. https://doi.org/10.1016/j.ijbiomac.2021.10.013.
20. Wohlert M, Benselfelt T, Wågberg L, Furó I, Berglund LA, Wohlert J. Cellulose and the role of hydrogen bonds: not in charge of everything. Cellulose. 2022;29(1):1–23. https://doi.org/10.1007/s10570-021-04325-4.
21. Rao J, Lv Z, Chen G, Peng F. Hemicellulose: structure, chemical modification, and application. Prog Polym Sci. 2023;140:101675. https://doi.org/10.1016/j.progpolymsci.2023.101675.
22. Hsieh YS, Harris PJ. Xyloglucans of monocotyledons have diverse structures. Mol Plant. 2009;2(5):943–65. https://doi.org/10.1093/mp/ssp061.
23. Gírio FM, Fonseca C, Carvalheiro F, Duarte LC, Marques S, Bogel-Łukasik R. Hemicelluloses for fuel ethanol: a review. Bioresour Technol. 2010;101(13):4775–800. https://doi.org/10.1016/j.biortech.2010.01.088.
24. Martínez-Abad A, Berglund J, Toríz G, Gatenholm P, Henriksson G, Lindström ME, et al. Regular motifs in xylan modulate molecular flexibility and interactions with cellulose surfaces. Plant Physiol. 2017;175:1579–92. https://doi.org/10.1104/pp.17.01184.
25. He H-J, Qiao J, Liu Y, Guo Q, Ou X, Wang X. Isolation, structural, functional, and bioactive properties of cereal Arabinoxylan—a critical review. J Agric Food Chem. 2021;69(51):15437–57. https://doi.org/10.1021/acs.jafc.1c04506.
26. Qaseem MF, Shaheen H, Wu A-M. Cell wall hemicellulose for sustainable industrial utilization. Renew Sust Energ Rev. 2021;144:110996. https://doi.org/10.1016/j.rser.2021.110996.
27. Wang H, Fu F, Li Q, Xi Y, Yang D. Research progress on the preparation of lignin-derived carbon materials and their application in catalysis. CIESC J. 2021;72(9):4445–57. https://doi.org/10.11949/0438-1157.20210090.
28. Linger JG, Vardon DR, Guarnieri MT, Karp EM, Hunsinger GB, Franden MA, et al. Lignin valorization through integrated biological funneling and chemical catalysis. Proc Natl Acad Sci. 2014;111:12013–8. https://doi.org/10.1073/pnas.1410657111.
29. Erfani Jazi M, Narayanan G, Aghabozorgi F, Farajidizaji B, Aghaei A, Kamyabi MA, et al. Structure, chemistry and physicochemistry of lignin for material functionalization. SN Appl Sci. 2019;1(9):1094. https://doi.org/10.1007/s42452-019-1126-8.
30. Parthasarathi R, Romero RA, Redondo A, Gnanakaran S. Theoretical study of the remarkably diverse linkages in lignin. J Phys Chem Lett. 2011;2(20):2660–6. https://doi.org/10.1021/jz201201q.

31. Yuan Y, Jiang B, Chen H, Wu W, Wu S, Jin Y, et al. Recent advances in understanding the effects of lignin structural characteristics on enzymatic hydrolysis. Biotechnol Biofuels. 2021;14(1):205. https://doi.org/10.1186/s13068-021-02054-1.
32. Sethupathy S, Murillo Morales G, Gao L, Wang H, Yang B, Jiang J, et al. Lignin valorization: status, challenges and opportunities. Bioresour Technol. 2022;347:126696. https://doi.org/10.1016/j.biortech.2022.126696.
33. Sharma A, Aggarwal NK. Biological pretreatment: need of the future. In: Sharma A, Aggarwal NK, editors. Water hyacinth: a potential lignocellulosic biomass for bioethanol. Cham: Springer International Publishing; 2020. p. 51–72.
34. Sindhu R, Binod P, Pandey A. Biological pretreatment of lignocellulosic biomass—an overview. Bioresour Technol. 2016;199:76–82. https://doi.org/10.1016/j.biortech.2015.08.030.
35. Hatakka A. Lignin-modifying enzymes from selected white-rot fungi: production and role from in lignin degradation. FEMS Microbiol Rev. 1994;13(2):125–35. https://doi.org/10.1111/j.1574-6976.1994.tb00039.x.
36. Rouches E, Herpoël-Gimbert I, Steyer JP, Carrere H. Improvement of anaerobic degradation by white-rot fungi pretreatment of lignocellulosic biomass: a review. Renew Sust Energ Rev. 2016;59:179–98. https://doi.org/10.1016/j.rser.2015.12.317.
37. Duque A, Manzanares P, Ballesteros M. Extrusion as a pretreatment for lignocellulosic biomass: fundamentals and applications. Renew Energy. 2017;114:1427–41. https://doi.org/10.1016/j.renene.2017.06.050.
38. Yoo J, Alavi S, Vadlani P, Amanor-Boadu V. Thermo-mechanical extrusion pretreatment for conversion of soybean hulls to fermentable sugars. Bioresour Technol. 2011;102(16):7583–90. https://doi.org/10.1016/j.biortech.2011.04.092.
39. Zheng J, Choo K, Bradt C, Lehoux R, Rehmann L. Enzymatic hydrolysis of steam exploded corncob residues after pretreatment in a twin-screw extruder. Biotechnol Rep. 2014;3:99–107. https://doi.org/10.1016/j.btre.2014.06.008.
40. Han S-Y, Park C-W, Endo T, Febrianto F, Kim N-H, Lee S-H. Extrusion process to enhance the pretreatment effect of ionic liquid for improving enzymatic hydrolysis of lignocellulosic biomass. Wood Sci Technol. 2020;54(3):599–613. https://doi.org/10.1111/j.1574-6976.1994.tb00039.x10.1007/s00226-020-01170-9.
41. Bundhoo ZMA. Microwave-assisted conversion of biomass and waste materials to biofuels. Renew Sust Energ Rev. 2018;82:1149–77. https://doi.org/10.1016/j.rser.2017.09.066.
42. Li H, Qu Y, Yang Y, Chang S, Xu J. Microwave irradiation—a green and efficient way to pretreat biomass. Bioresour Technol. 2016;199:34–41. https://doi.org/10.1016/j.biortech.2015.08.099.
43. Bundhoo ZMA, Mohee R. Ultrasound-assisted biological conversion of biomass and waste materials to biofuels: a review. Ultrason Sonochem. 2018;40:298–313. https://doi.org/10.1016/j.ultsonch.2017.07.025.
44. Vani S, Binod P, Kuttiraja M, Sindhu R, Sandhya SV, Preeti VE, et al. Energy requirement for alkali assisted microwave and high pressure reactor pretreatments of cotton plant residue and its hydrolysis for fermentable sugar production for biofuel application. Bioresour Technol. 2012;112:300–7. https://doi.org/10.1016/j.biortech.2012.02.076.
45. Veluchamy C, Kalamdhad AS, Gilroyed BH. Advanced pretreatment strategies for bioenergy production from biomass and biowaste. In: Hussain CM, editor. Handbook of environmental materials management. Cham: Springer International Publishing; 2018. p. 1–19.
46. Heinonen J, Tamminen A, Uusitalo J, Sainio T. Ethanol production from wood via concentrated acid hydrolysis, chromatographic separation, and fermentation. J Chem Technol Biotechnol. 2011;87(5):689–96. https://doi.org/10.1002/jctb.2766.
47. Abud Y, Costa LT, de Souza W, Sant'Anna C. Revealing the microfibrillar arrangement of the cell wall surface and the macromolecular effects of thermochemical pretreatment in sugarcane by atomic force microscopy. Ind Crop Prod. 2013;51:62–9. https://doi.org/10.1016/j.indcrop.2013.08.076.

48. Sannigrahi P, Kim DH, Jung S, Ragauskas A. Pseudo-lignin and pretreatment chemistry. Energy Environ Sci. 2011;4(4):1306–10. https://doi.org/10.1039/C0EE00378F.
49. Shinde SD, Meng X, Kumar R, Ragauskas AJ. Recent advances in understanding the pseudo-lignin formation in a lignocellulosic biorefinery. Green Chem. 2018;20(10):2192–205. https://doi.org/10.1039/C8GC00353J.
50. Zhao X, Bian B, Huang C, Lai C, Song J, Jin Y, et al. Elucidating the nonproductive adsorption mechanism of cellulase with lignin fractions from hydrothermally pretreated poplar using multi-dimensional spectroscopic technologies. Green Chem. 2023;25(16):6383–97. https://doi.org/10.1039/D3GC00907F.
51. Kong-Win Chang J, Duret X, Berberi V, Zahedi-Niaki H, Lavoie J-M. Two-step thermo-chemical cellulose hydrolysis with partial neutralization for glucose production. Front Chem. 2018;6:1–11. https://doi.org/10.3389/fchem.2018.00117.
52. Hans M, Pellegrini VOA, Filgueiras JG, de Azevedo ER, Guimaraes FEC, Chandel AK, et al. Optimization of dilute acid pretreatment for enhanced release of fermentable sugars from sugarcane bagasse and validation by biophysical characterization. Bioenergy Res. 2023;16(1):416–34. https://doi.org/10.1007/s12155-022-10474-6.
53. Solarte-Toro JC, Romero-García JM, Martínez-Patiño JC, Ruiz-Ramos E, Castro-Galiano E, Cardona-Alzate CA. Acid pretreatment of lignocellulosic biomass for energy vectors production: a review focused on operational conditions and techno-economic assessment for bioethanol production. Renew Sust Energ Rev. 2019;107:587–601. https://doi.org/10.1016/j.rser.2019.02.024.
54. Kumar B, Bhardwaj N, Agrawal K, Chaturvedi V, Verma P. Current perspective on pretreatment technologies using lignocellulosic biomass: an emerging biorefinery concept. Fuel Process Technol. 2020;199:106244. https://doi.org/10.1016/j.fuproc.2019.106244.
55. Zlenko DV, Vtyurina DN, Usachev SV, Skoblin AA, Mikhaleva MG, Politenkova GG, et al. On the orientation of the chains in the mercerized cellulose. Sci Rep. 2021;11(1):8765. https://doi.org/10.1038/s41598-021-88040-x.
56. Galbe M, Zacchi G. Pretreatment: the key to efficient utilization of lignocellulosic materials. Biomass Bioenergy. 2012;46:70–8. https://doi.org/10.1016/j.biombioe.2012.03.026.
57. Zhu J, Wan C, Li Y. Enhanced solid-state anaerobic digestion of corn stover by alkaline pretreatment. Bioresour Technol. 2010;101(19):7523–8. https://doi.org/10.1016/j.biortech.2010.04.060.
58. Mohanasundaram Y, Nambissan VD, Gummadi SN. Optimization of sequential alkali/acid pretreatment of corn cob for xylitol production by Debaryomyces nepalensis. Biomass Convers Biorefinery. 2024;14(11):12483–500. https://doi.org/10.1007/s13399-022-03660-1.
59. Chen B-Y, Chen S-W, Wang H-T. Use of different alkaline pretreatments and enzyme models to improve low-cost cellulosic biomass conversion. Biomass Bioenergy. 2012;39:182–91. https://doi.org/10.1016/j.biombioe.2012.01.012.
60. Chundawat SPS, Pal RK, Zhao C, Campbell T, Teymouri F, Videto J, et al. Ammonia fiber expansion (AFEX) pretreatment of lignocellulosic biomass. J Vis Exp. 2020;158:e57488. https://doi.org/10.3791/57488.
61. da Costa Sousa L, Jin M, Chundawat SPS, Bokade V, Tang X, Azarpira A, et al. Next-generation ammonia pretreatment enhances cellulosic biofuel production. Energy Environ Sci. 2016;9(4):1215–23. https://doi.org/10.1039/C5EE03051J.
62. Teymouri F, Laureano-Perez L, Alizadeh H, Dale BE. Optimization of the ammonia fiber explosion (AFEX) treatment parameters for enzymatic hydrolysis of corn stover. Bioresour Technol. 2005;96(18):2014–8. https://doi.org/10.1016/j.biortech.2005.01.016.
63. Morais ARC, Zhang J, Dong H, Otto WG, Mokomele T, Hodge D, et al. Development of an ammonia pretreatment that creates synergies between biorefineries and advanced biomass logistics models. Green Chem. 2022;24(11):4443–62. https://doi.org/10.1039/D2GC00488G.
64. Jin M, Dale BE. AFEX™ pretreatment-based biorefinery technologies. In: Park JM, editor. Handbook of biorefinery research and technology. Dordrecht: Springer Netherlands; 2018. p. 1–16.

65. Yu Y, Wu J, Ren X, Lau A, Rezaei H, Takada M, et al. Steam explosion of lignocellulosic biomass for multiple advanced bioenergy processes: a review. Renew Sust Energ Rev. 2022;154:111871. https://doi.org/10.1016/j.rser.2021.111871.
66. Sui W, Chen H. Extraction enhancing mechanism of steam exploded *Radix Astragali*. Process Biochem. 2014;49(12):2181–90. https://doi.org/10.1016/j.procbio.2014.08.010.
67. Borand MN, Isler Kaya A, Karaosmanoglu F. Saccharification yield through enzymatic hydrolysis of the steam-exploded pinewood. Energies. 2020; https://doi.org/10.3390/en13174552.
68. Suriyachai N, Weerasai K, Upajak S, Khongchamnan P, Wanmolee W, Laosiripojana N, et al. Efficiency of catalytic liquid hot water pretreatment for conversion of corn stover to bioethanol. ACS Omega. 2020;5(46):29872–81. https://doi.org/10.1021/acsomega.0c04054.
69. Xu C, Xiong Y, Zhang J, Li K, Zhong S, Huang S, et al. Liquid hot water pretreatment combined with high-solids enzymatic hydrolysis and fed-batch fermentation for succinic acid sustainable processed from sugarcane bagasse. Bioresour Technol. 2023;369:128389. https://doi.org/10.1016/j.biortech.2022.128389.
70. Cheah WY, Sankaran R, Show PL, Tg. Ibrahim TNB, Chew KW, Culaba A, et al. Pretreatment methods for lignocellulosic biofuels production: current advances, challenges and future prospects. Biofuel Res J. 2020;7(1):1115–27. https://doi.org/10.18331/brj2020.7.1.4.
71. Geng A, Xin F, Ip J-y. Ethanol production from horticultural waste treated by a modified organosolv method. Bioresour Technol. 2012;104:715–21. https://doi.org/10.1016/j.biortech.2011.10.076.
72. Huang X, Fan M, Xie J, Zhong C, Zhang H. High titer ethanol production from poplar by aluminum chloride catalyzed organosolv pretreatment and simultaneous saccharification and fermentation. Ind Crop Prod. 2022;181:114803. https://doi.org/10.1016/j.indcrop.2022.114803.
73. Zhu J, Gong D, Mao J, Wei Q. Research progress of lignocellulosic biomass pretreatment technology. Adv New Renew Energy. 2022;10(4):383–92. https://doi.org/10.3969/j.issn.2095-560X.2022.04.012.
74. Lim JR, Chua LS, Mustaffa AA. Ionic liquids as green solvent and their applications in bioactive compounds extraction from plants. Process Biochem. 2022;122:292–306. https://doi.org/10.1016/j.procbio.2022.10.024.
75. Zhou Y, Zhang X, Yin D, Zhang J, Mi Q, Lu H, et al. The solution state and dissolution process of cellulose in ionic-liquid-based solvents with different hydrogen-bonding basicity and microstructures. Green Chem. 2022;24(9):3824–33. https://doi.org/10.1039/D2GC00374K.
76. Nakasu PYS, Pin TC, Hallett JP, Rabelo SC, Costa AC. In-depth process parameter investigation into a protic ionic liquid pretreatment for 2G ethanol production. Renew Energy. 2021;172:816–28. https://doi.org/10.1016/j.renene.2021.03.004.
77. Pedersen JN, Pérez B, Guo Z. Stability of cellulase in ionic liquids: correlations between enzyme activity and COSMO-RS descriptors. Sci Rep. 2019;9(1):17479. https://doi.org/10.1038/s41598-019-53523-5.
78. Chowdhury T, Chatterjee S, Deshmukh SH, Bagchi S. A systematic study on the role of hydrogen bond donors in dictating the dynamics of choline-based deep eutectic solvents. J Phys Chem B. 2023;127(33):7299–308. https://doi.org/10.1021/acs.jpcb.3c02191.
79. Chen Y, Mu T. Application of deep eutectic solvents in biomass pretreatment and conversion. Green Energy Environ. 2019;4(2):95–115. https://doi.org/10.1016/j.gee.2019.01.012.
80. Xia Q, Chen C, Yao Y, Li J, He S, Zhou Y, et al. A strong, biodegradable and recyclable lignocellulosic bioplastic. Nat Sustain. 2021;4(7):627–35. https://doi.org/10.1038/s41893-021-00702-w.
81. Vigier KDO, Châtel G, Jérôme F. Contribution of deep eutectic solvents for biomass processing: opportunities, challenges, and limitations. ChemCatChem. 2015;7(8):1250–60. https://doi.org/10.1002/cctc.201500134.

82. Wang R, Wang K, Zhou M, Xu J, Jiang J. Efficient fractionation of moso bamboo by synergistic hydrothermal-deep eutectic solvents pretreatment. Bioresour Technol. 2021;328: 124873. https://doi.org/10.1016/j.biortech.2021.124873.
83. Pandey A, Mankar AR, Ahmad E, Pant KK. 10—Deep eutectic solvents: a greener approach towards biorefineries. In: Bhaskar T, Pandey A, editors. Biomass, biofuels, biochemicals. Elsevier; 2021. p. 193–219.
84. Procentese A, Johnson E, Orr V, Garruto Campanile A, Wood JA, Marzocchella A, et al. Deep eutectic solvent pretreatment and subsequent saccharification of corncob. Bioresour Technol. 2015;192:31–6. https://doi.org/10.1016/j.biortech.2015.05.053.
85. Moodley P, Sewsynker-Sukai Y, Gueguim Kana EB. Progress in the development of alkali and metal salt catalysed lignocellulosic pretreatment regimes: potential for bioethanol production. Bioresour Technol. 2020;310:123372. https://doi.org/10.1016/j.biortech.2020.123372.
86. Loow Y-L, Wu TY, Tan KA, Lim YS, Siow LF, Md. Jahim J, et al. Recent advances in the application of inorganic salt pretreatment for transforming lignocellulosic biomass into reducing sugars. J Agric Food Chem. 2015;63(38):8349–63. https://doi.org/10.1021/acs.jafc.5b01813.
87. Deng W, Kennedy JR, Tsilomelekis G, Zheng W, Nikolakis V. Cellulose hydrolysis in acidified LiBr molten salt hydrate media. Ind Eng Chem Res. 2015;54(19):5226–36. https://doi.org/10.1021/acs.iecr.5b00757.
88. Rodriguez Quiroz N, Norton AM, Nguyen HL, Vasileiadou E, Vlachos DG. Homogeneous metal salt solutions for biomass upgrading and other select organic reactions. ACS Catal. 2019;9(11):9923–52. https://doi.org/10.1021/acscatal.9b01853.
89. Marcotullio G, Krisanti E, Giuntoli J, de Jong W. Selective production of hemicellulose-derived carbohydrates from wheat straw using dilute HCl or $FeCl_3$ solutions under mild conditions. X-ray and thermo-gravimetric analysis of the solid residues. Bioresour Technol. 2011;102(10):5917–23. https://doi.org/10.1016/j.biortech.2011.02.092.
90. Lara-Serrano M, Morales-delaRosa S, Campos-Martín JM, Fierro JLG. High enhancement of the hydrolysis rate of cellulose after pretreatment with inorganic salt hydrates. Green Chem. 2020;22(12):3860–6. https://doi.org/10.1039/D0GC01066A.
91. Shen G, Yuan X, Cheng Y, Chen S, Xu Z, Jin M. Densification pretreatment with a limited deep eutectic solvent triggers high-efficiency fractionation and valorization of lignocellulose. Green Chem. 2023;25(20):8026–39. https://doi.org/10.1039/D3GC02776G.
92. Yuan X, Shen G, Chen S, Shen W, Chen X, Liu S, et al. Elucidating the mechanism of densifying lignocellulosic biomass with acidic chemicals (DLC) for lignocellulosic biorefinery. Green Chem. 2023;25(17):6759–73. https://doi.org/10.1039/D3GC01703F.
93. Jin M, Sarks C, Bals BD, Posawatz N, Gunawan C, Dale BE, et al. Toward high solids loading process for lignocellulosic biofuel production at a low cost. Biotechnol Bioeng. 2017;114(5): 980–9. https://doi.org/10.1002/bit.26229.
94. Ren X, Wang J, Yu H, Peng C, Hu J, Ruan Z, et al. Anaerobic and sequential aerobic production of high-titer ethanol and single cell protein from NaOH-pretreated corn stover by a genome shuffling-modified Saccharomyces cerevisiae strain. Bioresour Technol. 2016;218: 623–30. https://doi.org/10.1016/j.biortech.2016.06.118.
95. Nguyen TY, Cai CM, Osman O, Kumar R, Wyman CE. CELF pretreatment of corn stover boosts ethanol titers and yields from high solids SSF with low enzyme loadings. Green Chem. 2016;18(6):1581–9. https://doi.org/10.1039/c5gc01977j.
96. Chen X, Kuhn E, Jennings EW, Nelson R, Tao L, Zhang M, et al. DMR (deacetylation and mechanical refining) processing of corn stover achieves high monomeric sugar concentrations (230 g L^{-1}) during enzymatic hydrolysis and high ethanol concentrations (>10% v/v) during fermentation without hydrolysate purification or concentration. Energy Environ Sci. 2016;9 (4):1237–45. https://doi.org/10.1039/C5EE03718B.
97. Liu G, Zhang Q, Li H, Qureshi AS, Zhang J, Bao X, et al. Dry biorefining maximizes the potentials of simultaneous saccharification and co-fermentation for cellulosic ethanol production. Biotechnol Bioeng. 2018;115(1):60–9. https://doi.org/10.1002/bit.26444.

98. Matsakas L, Christakopoulos P. Fermentation of liquefacted hydrothermally pretreated sweet sorghum bagasse to ethanol at high-solids content. Bioresour Technol. 2013;127:202–8. https://doi.org/10.1016/j.biortech.2012.09.107.
99. Zhou H, Leu S-Y, Wu X, Zhu JY, Gleisner R, Yang D, et al. Comparisons of high titer ethanol production and lignosulfonate properties by SPORL pretreatment of lodgepole pine at two temperatures. RSC Adv. 2014;4(51):27030–8. https://doi.org/10.1039/C4RA03608E.
100. Hoyer K, Galbe M, Zacchi G. The effect of prehydrolysis and improved mixing on high-solids batch simultaneous saccharification and fermentation of spruce to ethanol. Process Biochem. 2013;48(2):289–93. https://doi.org/10.1016/j.procbio.2012.12.020.
101. Liu Z-H, Chen H-Z. Simultaneous saccharification and co-fermentation for improving the xylose utilization of steam exploded corn stover at high solid loading. Bioresour Technol. 2016;201:15–26. https://doi.org/10.1016/j.biortech.2015.11.023.
102. Varga E, Klinke HB, Réczey K, Thomsen AB. High solid simultaneous saccharification and fermentation of wet oxidized corn stover to ethanol. Biotechnol Bioeng. 2004;88(5):567–74. https://doi.org/10.1002/bit.20222.
103. Cannella D, Jørgensen H. Do new cellulolytic enzyme preparations affect the industrial strategies for high solids lignocellulosic ethanol production? Biotechnol Bioeng. 2014;111 (1):59–68. https://doi.org/10.1002/bit.25098.
104. Sudiyani Y, Dahnum D, Burhani D, Putri AMH. Chapter 10—Evaluation and comparison between simultaneous saccharification and fermentation and separated hydrolysis and fermentation process. In: Second and third generation of feedstocks. Elsevier; 2019. p. 273–90.
105. Taherzadeh MJ, Karimi K. Enzyme-based hydrolysis processes for ethanol from lignocellulosic materials: a review. Bioresources. 2007;2(4):707–38.
106. Zhu J-Q, Zong Q-J, Li W-C, Chai M-Z, Xu T, Liu H, et al. Temperature profiled simultaneous saccharification and co-fermentation of corn stover increases ethanol production at high solid loading. Energy Convers Manag. 2020;205:112344. https://doi.org/10.1016/j.enconman.2019.112344.
107. Naghshbandi MP, Tabatabaei M, Aghbashlo M, Gupta VK, Sulaiman A, Karimi K, et al. Progress toward improving ethanol production through decreased glycerol generation in Saccharomyces cerevisiae by metabolic and genetic engineering approaches. Renew Sust Energ Rev. 2019;115:109353. https://doi.org/10.1016/j.rser.2019.109353.
108. Nguyen TY, Cai CM, Kumar R, Wyman CE. Overcoming factors limiting high-solids fermentation of lignocellulosic biomass to ethanol. Proc Natl Acad Sci. 2017;114(44): 11673–8. https://doi.org/10.1073/pnas.1704652114.
109. Klein-Marcuschamer D, Oleskowicz-Popiel P, Simmons BA, Blanch HW. The challenge of enzyme cost in the production of lignocellulosic biofuels. Biotechnol Bioeng. 2012;109(4): 1083–7. https://doi.org/10.1002/bit.24370.
110. Zacchi G, Axelsson A. Economic evaluation of preconcentration in production of ethanol from dilute sugar solutions. Biotechnol Bioeng. 1989;34(2):223–33. https://doi.org/10.1002/bit.260340211.
111. Battista F, Gomez Almendros M, Rousset R, Boivineau S, Bouillon P-A. Enzymatic hydrolysis at high dry matter content: the influence of the substrates' physical properties and of loading strategies on mixing and energetic consumption. Bioresour Technol. 2018;250:191–6. https://doi.org/10.1016/j.biortech.2017.11.049.
112. Zhu J-Q, Qin L, Li W-C, Zhang J, Bao J, Huang Y-D, et al. Simultaneous saccharification and co-fermentation of dry diluted acid pretreated corn stover at high dry matter loading: overcoming the inhibitors by non-tolerant yeast. Bioresour Technol. 2015;198:39–46. https://doi.org/10.1016/j.biortech.2015.08.140.
113. Kim Y, Ximenes E, Mosier NS, Ladisch MR. Soluble inhibitors/deactivators of cellulase enzymes from lignocellulosic biomass. Enzym Microb Technol. 2011;48(4):408–15. https://doi.org/10.1016/j.enzmictec.2011.01.007.
114. Wang X, Bai X, Chen D-F, Chen F-Z, Li B-Z, Yuan Y-J. Increasing proline and myo-inositol improves tolerance of *Saccharomyces cerevisiae* to the mixture of multiple

lignocellulose-derived inhibitors. Biotechnol Biofuels. 2015;8(1):142. https://doi.org/10.1186/s13068-015-0329-5.
115. Pan S, Jia B, Liu H, Wang Z, Chai M-Z, Ding M-Z, et al. Endogenous lycopene improves ethanol production under acetic acid stress in *Saccharomyces cerevisiae*. Biotechnol Biofuels. 2018;11(1):107. https://doi.org/10.1186/s13068-018-1107-y.
116. Chundawat SPS, Vismeh R, Sharma LN, Humpula JF, da Costa Sousa L, Chambliss CK, et al. Multifaceted characterization of cell wall decomposition products formed during ammonia fiber expansion (AFEX) and dilute acid based pretreatments. Bioresour Technol. 2010;101(21):8429–38. https://doi.org/10.1016/j.biortech.2010.06.027.
117. Qin L, Li X, Liu L, Zhu J-Q, Guan Q-M, Zhang M-T, et al. Dual effect of soluble materials in pretreated lignocellulose on simultaneous saccharification and co-fermentation process for the bioethanol production. Bioresour Technol. 2017;224:342–8. https://doi.org/10.1016/j.biortech.2016.11.106.
118. Yoo CG, Meng X, Pu Y, Ragauskas AJ. The critical role of lignin in lignocellulosic biomass conversion and recent pretreatment strategies: a comprehensive review. Bioresour Technol. 2020;301:122784. https://doi.org/10.1016/j.biortech.2020.122784.
119. Li M, Guo C, Luo B, Chen C, Wang S, Min D. Comparing impacts of physicochemical properties and hydrolytic inhibitors on enzymatic hydrolysis of sugarcane bagasse. Bioprocess Biosyst Eng. 2020;43(1):111–22. https://doi.org/10.1007/s00449-019-02209-3.
120. Kim JS, Lee YY, Kim TH. A review on alkaline pretreatment technology for bioconversion of lignocellulosic biomass. Bioresour Technol. 2016;199:42–8. https://doi.org/10.1016/j.biortech.2015.08.085.
121. Ko JK, Kim Y, Ximenes E, Ladisch MR. Effect of liquid hot water pretreatment severity on properties of hardwood lignin and enzymatic hydrolysis of cellulose. Biotechnol Bioeng. 2014;112(2):252–62. https://doi.org/10.1002/bit.25349.
122. Ying W, Xu G, Yang H, Shi Z, Yang J. The sequential Fenton oxidation and sulfomethylation pretreatment for alleviating the negative effects of lignin in enzymatic saccharification of sugarcane bagasse. Bioresour Technol. 2019;286:121392. https://doi.org/10.1016/j.biortech.2019.121392.
123. Wang X, Li B-Z, Ding M-Z, Zhang W-W, Yuan Y-J. Metabolomic analysis reveals key metabolites related to the rapid adaptation of *Saccharomyce cerevisiae* to multiple inhibitors of furfural, acetic acid, and phenol. OMICS. 2013;17(3):150–9. https://doi.org/10.1089/omi.2012.0093.
124. van Dijk M, Mierke F, Nygård Y, Olsson L. Nutrient-supplemented propagation of Saccharomyces cerevisiae improves its lignocellulose fermentation ability. AMB Express. 2020;10(1):157. https://doi.org/10.1186/s13568-020-01070-y.
125. Jin M, Lau MW, Balan V, Dale BE. Two-step SSCF to convert AFEX-treated switchgrass to ethanol using commercial enzymes and *Saccharomyces cerevisiae* 424A(LNH-ST). Bioresour Technol. 2010;101(21):8171–8. https://doi.org/10.1016/j.biortech.2010.06.026.
126. Koppram R, Tomás-Pejó E, Xiros C, Olsson L. Lignocellulosic ethanol production at high-gravity: challenges and perspectives. Trends Biotechnol. 2014;32(1):46–53. https://doi.org/10.1016/j.tibtech.2013.10.003.
127. Hodge DB, Karim MN, Schell DJ, McMillan JD. Model-based fed-batch for high-solids enzymatic cellulose hydrolysis. Appl Biochem Biotechnol. 2009;152(1):88–107. https://doi.org/10.1007/s12010-008-8217-0.
128. Qiu Y, Wu M, Bao H, Liu W, Shen Y. Engineering of *Saccharomyces cerevisiae* for co-fermentation of glucose and xylose: current state and perspectives. Eng Microbiol. 2023;3(3):100084. https://doi.org/10.1016/j.engmic.2023.100084.
129. Zan X, Sun J, Chu L, Cui F, Huo S, Song Y, et al. Improved glucose and xylose co-utilization by overexpression of xylose isomerase and/or xylulokinase genes in oleaginous fungus *Mucor circinelloides*. Appl Microbiol Biotechnol. 2021;105(13):5565–75. https://doi.org/10.1007/s00253-021-11392-x.

130. López-Maury L, Marguerat S, Bähler J. Tuning gene expression to changing environments: from rapid responses to evolutionary adaptation. Nat Rev Genet. 2008;9(8):583–93. https://doi.org/10.1038/nrg2398.
131. Nielsen F, Tomás-Pejó E, Olsson L, Wallberg O. Short-term adaptation during propagation improves the performance of xylose-fermenting *Saccharomyces cerevisiae* in simultaneous saccharification and co-fermentation. Biotechnol Biofuels. 2015;8(1):219. https://doi.org/10.1186/s13068-015-0399-4.
132. Zheng T, Jiang J, Yao J. Surfactant-promoted hydrolysis of lignocellulose for ethanol production. Fuel Process Technol. 2021;213:106660. https://doi.org/10.1016/j.fuproc.2020.106660.
133. Qin L, Zhao X, Li W-C, Zhu J-Q, Liu L, Li B-Z, et al. Process analysis and optimization of simultaneous saccharification and co-fermentation of ethylenediamine-pretreated corn stover for ethanol production. Biotechnol Biofuels. 2018;11(1):118. https://doi.org/10.1186/s13068-018-1118-8.
134. Li H, Wang C, Xiao W, Yang Y, Hu P, Dai Y, et al. Dissecting the effect of polyethylene glycol on the enzymatic hydrolysis of diverse lignocellulose. Int J Biol Macromol. 2019;131:676–81. https://doi.org/10.1016/j.ijbiomac.2019.03.131.

Chapter 12
Methane Production from Biowastes via Solid-State Anaerobic Digestion

Yiqing Yao, Ziyan Zhou, and Xiao Yan

Abstract Treatment and resource utilization of biowastes have emerged as environmental and economic challenges. Among the waste management strategies, solid-state anaerobic digestion (SS-AD) stands out for its ability to offer a multitude of positive environmental benefits. SS-AD is particularly suited for the anaerobic digestion of high-solid-containing substrates (Total solids (TS) concentration greater than 25%), such as agricultural waste. Continuous research and technological innovation are essential to further improve the efficiency and feasibility of methane produced from biowastes via SS-AD, promoting the sustainable utilization of biowaste resources. This chapter provides a comprehensive overview of methane production from biowastes via SS-AD, focusing on its technological principles, influencing factors, optimization strategies, applications, and prospects. By summarizing the effects of temperature, C/N ratio, total solid content, inoculation, and other factors on the produced from biowastes via SS-AD, it offers a reliable solution to enhance methane production rate of SS-AD.

Keywords Anaerobic digestion · Methane · Production efficiency · Technological influence · Factor

12.1 Introduction

The steady decline in fossil fuels reserves, accompanied by unpredictable price fluctuations, underscores the fragility of global energy infrastructure. Simultaneously, the uncontrolled accumulation of organic waste unbalance the Earth's resource, which consequently exacerbates environmental problems and climate

---

Y. Yao (✉) · Z. Zhou (✉) · X. Yan
College of Mechanical and Electronic Engineering, Northwest A&F University, Yangling, China

Northwest Research Center of Rural Renewable Energy, Exploitation and Utilization of Ministry of Agriculture, Northwest A&F University, Yangling, China
e-mail: dzhtyao@126.com; cholmolhari@icloud.com

Z. Fang et al. (eds.), *Production of Organic Acids and Alcohols from Agricultural Residues and Food Wastes*, Biofuels and Biorefineries 13,
https://doi.org/10.1007/978-981-96-1222-2_12

change [1, 2]. Accordingly, the need for alternative energy sources requires urgent attention and sustainable solutions. Biomass is indeed one of the main sources of renewable energy and has gained significant global interest due to its potential to provide sustainable and environmentally friendly energy solutions. Biowaste means any biodegradable organic waste of vegetable and/or animal origin, susceptible to biological environment, especially food waste, poultry waste, and crop stalks. The biowaste is made up of complex carbohydrates-polymers (hemicellulose and cellulose), lipids, protein, and aromatic polymers that can decompose under both anaerobic and aerobic conditions [3–5]. Biomass energy refers to the renewable energy obtained from biological wastes such as plant residues, animal feces, food residues, and sewage sludge via thermal conversion, composting, and anaerobic digestion (AD) [6, 7]. The effective management of biowastes presents a significant challenge in sustainable waste management and renewable energy production [8]. Biogas production through AD can indeed play a crucial role in reducing reliance on fossil fuels and curbing greenhouse gas (GHG) emissions. When integrated with other bioenergy production technologies, it creates a synergistic effect that enhances the overall sustainability and efficiency of renewable energy systems [9, 10].

AD is a biological process in which microorganisms break down organic matter (such as food waste, animal manure, sewage sludge, and other biomass) in the absence of oxygen [11]. Conventional AD processes typically carried out in liquid phase reactors, do encounter significant challenges when processing biowastes with high solid content. These challenges include difficulties in mixing, mass transfer limitations, and issues related to process stability. In recent years, SS-AD has emerged as a promising alternative for treating various types of biological waste, particularly those with high solid content. This approach has shown significant potential in enhancing methane production and addressing multiple environmental, energy, and agricultural challenges. The total solid content (TS%) in SS-AD can be as high as 40%, which is suitable for the digestion of agricultural waste with high solid content such as straw. However, the SS-AD process faces specific limitation of methane generation, especially the need for more inoculants, longer retention time, and the inefficiency in digestion caused by the accumulation of volatile organic acids and ammonia. Thus, the strategies and approaches to mitigate issue and improve methane production efficiency of SS-AD are urgently needed.

In this chapter, the authors describe the SS-AD process, the main influencing factors, and strategies to improve the efficiency of SS-AD methane production, contributing to advancing sustainable waste management strategies and renewable energy production.

12.2 Reviews of SS-AD

SS-AD is a process that decomposes organic waste materials with low moisture content (typically less than 85%) in the absence of oxygen [12]. Based on the TS% of the substrate, AD can be classified into different types. When the TS% of the

Table 12.1 Comparison of solid-state (SS) and liquid-state (LS) AD

	SS-AD	LS-AD
Material characteristics	(Total solids) TS < 50%	(Total solids) TS < 15%
Key technology	Mass transfer	Homogeneity
Engineering characters	Simple	Complex
Energy consumption	Low	High
Energy density	High	Low
Technological factors	Solid material handling	Water treatment
Secondary pollution	No problem	More problems

substrate is below 15%, the process is classified as LS-AD [13]. Compared to LS-AD, SS-AD offers several advantages, including a higher organic load ratio, smaller reactor volume, lower heating energy requirements, higher volumetric methane productivity, and reduced wastewater generation [14]. SS-AD has significant advantages in biowaste treatment and energy conversion, making it a technology for future development. It offers a sustainable solution for waste management while simultaneously producing renewable energy in the form of biogas. So far, a series of studies have been carried out on SS-AD, particularly focusing on lignocellulosic biomass with high solid content properties [15, 16]. However, SS-AD, which utilizes lignocellulosic biomass as feedstock, still faces several challenges, including low methane production, process instability, and low-end product value [17, 18]. Therefore, a better understanding of the fundamental mechanisms affecting SS-AD methane productivity is critical for optimizing the process. Ongoing research and development efforts aim to address these challenges and enhance SS-AD technology for broader implementation and increased sustainability in waste management practices. Overall, the goal is to optimize SS-AD to make it a more efficient and reliable technology for converting organic waste into renewable energy and valuable byproducts (Table 12.1).

12.2.1 Principles and Processes of SSAD

The process of SS-AD involves a series of biochemical reactions carried out by diverse consortia of bacteria and archaea. These microorganisms collaboratively convert organic matter into carbon dioxide (CO_2) and methane (CH_4) through Redox reactions. Similar to the AD process, SS-AD proceeds through multiple steps, each characterized by intricate interactions among various microbial communities. The SS-AD can be categorized into four stages based on the cooperative behavior of microbial communities, which are shown in Fig. 12.1.

The hydrolysis and acid production processes in SS-AD exhibit faster reaction rates and are primarily regulated by acidogenic bacteria [19]. Conversely, methane (CH_4) generation, which is typically the rate-limiting step, requires specific environmental conditions for optimal performance [20, 21].

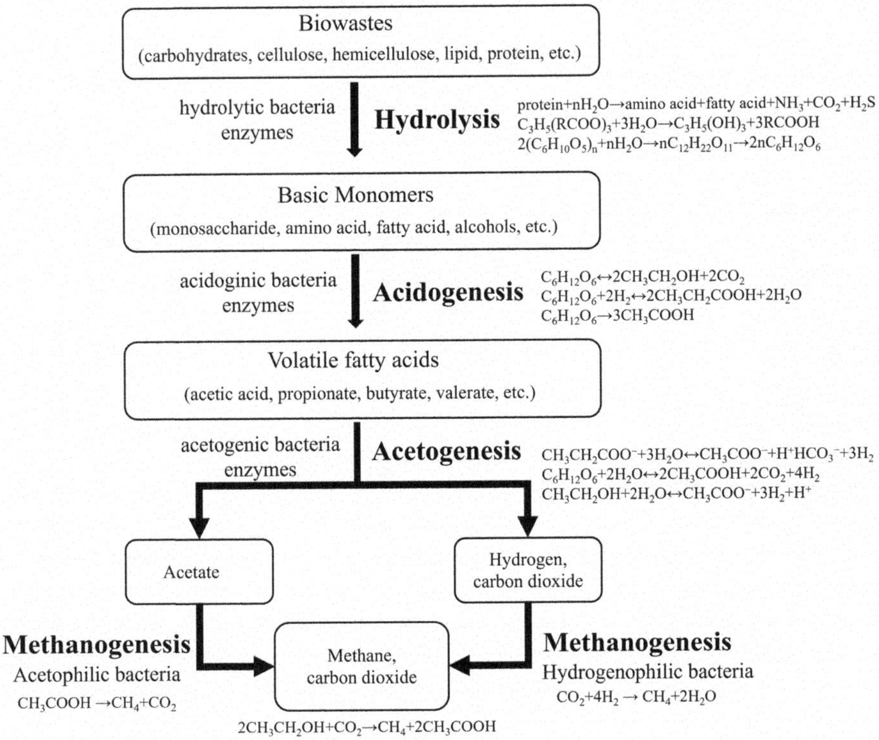

Fig. 12.1 SS-AD biomass decomposition stages (hydrolysis, acidogenesis, acetogenesis, and methanogenesis) and key compounds

12.2.1.1 Hydrolysis

Hydrolysis is the initial step in the SS-AD process and is typically the rate limiting step. During this stage, complex organic molecules such as carbohydrates, cellulose, hemicellulose, fats, and proteins are broken down into simpler compounds such as glucose, mannose, fatty acid and glycerol by hydrolytic bacteria. These bacteria secrete specific enzymes that catalyze the hydrolysis of corresponding complex compounds. For example, glycolytic bacteria are responsible for converting carbohydrates into sugars, while proteolytic bacteria cleave proteins to release amino acids.

The efficiency and extent of hydrolysis depend significantly on the properties of the substrate. Additionally, the hydrolysis rate is influenced by several key operating parameters of the SS-AD process, including pH, temperature, solid concentration. The optimal conditions for these parameters are crucial to enhance the activity of hydrolytic enzymes and improve the overall efficiency of the hydrolysis stage.

12.2.1.2 Acidogenesis

The products of hydrolysis are further metabolized by acidogenic bacteria in the acidogenesis stage. During this stage, acidogenic bacteria ferment the soluble organic compounds generated from hydrolysis into volatile fatty acids (VFAs), alcohols, and other intermediate products. This stage is characterized by the production of various organic acids such as acetic acid, propionic acid, and butyric acid. These intermediate products are crucial as they serve as substrates for the subsequent stages of the SS-AD process. The efficient conversion of hydrolysis products into VFAs and other compounds by acidogenic bacteria is essential for maintaining the overall progression and stability of the anaerobic digestion process.

12.2.1.3 Acetogenesis

In the acetogenesis stage, acetogenic bacteria metabolize the intermediate products generated during the acidogenesis stage. These bacteria convert VFAs and alcohols into acetic acid, hydrogen, and carbon dioxide through various biochemical pathways. Acetogenesis is an essential step in the SS-AD process because acetic acid is a primary precursor for the subsequent methanogenesis stage. Efficient acetogenesis ensures the availability of acetic acid, which is necessary for methanogenic archaea to produce methane. Therefore, the effective functioning of acetogenic bacteria is vital for the overall success of methane production in the anaerobic digestion process.

12.2.1.4 Methanogenesis

Methane (CH_4) generation is the final and important stage of the SS-AD process. Methanogenic archaea play a key role in this stage, known as methanogenesis. Methanogens utilize hydrogen, carbon dioxide, and acetate produced in the previous stages to generate methane (CH_4) and carbon dioxide (CO_2) through anaerobic respiration. Acetoclastic and hydrogenotrophic methanogenesis are the two primary pathways involved in methanogenesis. In acetoclastic pathway, methanogens such as *Methanosarcina* and *Methanosaeta* directly convert acetate into methane and carbon dioxide. This process is highly depended on the efficient breakdown of acetate. In hydrogenotrophic pathway, methanogens such as *Methanobacterium* and *Methanospirillum* utilize hydrogen and carbon dioxide to produce methane. This pathway is important for the consumption of hydrogen, which helps maintain the balance of electron donors and acceptors in the AD process.

12.2.2 Current Advancements in SS-AD

Biomass energy accounted for the largest share of renewable energy sources contributing approximately 70% in 2017. Among the various sources of biomass energy, agricultural waste is significant due to its potential for high energy production. Agricultural straw and residues alone are capable of producing up to 82 exajoules (EJ) of energy, which can meet about 14% of the global energy demand [22]. Given that the total solids in agricultural waste ranges between 50% and 80%, these materials are considered to be the most abundant and suitable raw material for SS-AD. Therefore, SS-AD holds significant potential for expansion and development in the treatment of raw materials with high organic load and low water demand, such as food waste, crop straw, food processing waste, and other biowastes.

Countries such as Denmark, Germany, France, and Sweden have researched biogas production from SS-AD as early as the 1980s, primarily focusing on the treatment of municipal organic waste [23]. Many states in the United States, for example, California, require waste diversion or targets, which have led waste managers and municipalities to concentrate on reducing the organic fraction of municipal solid waste (OFMSW) disposal to meet these targets [24]. However, AD treatment for OFMSW requires significant investment and technical expertise, making its capital and operating costs higher than composting and landfill options. Compared to European countries, the U.S. has struggled to be competitive in AD due to low landfill dumping fees and low energy prices [25–28].

Internationally, AD has been used for decades, especially in rural areas, to produce biogas for cooking and lighting fuel. In rural areas of China and India, numerous home-scale biogas digesters are employed for waste treatment and gas production. In the United States, many livestock ranches are adopting AD to mitigate the environmental impact of manure lagoons while capturing methane for energy production. In recent years, EU energy and climate policies have promoted the biogas production by incentivizing the use of renewable resources [29]. The European market has shown a growing preference for dry digestion systems over wet systems, with the percentage of dry systems increasing since 2005 [30]. For SS and high-solids AD systems, the digester contents are maintained at 20–40% TS (equivalent to 60–80% MC). It is found that the biogas yield and rate of the single-stage dry system are as high or even larger than that of the LS-AD system. There are two commercial-scale designs most used, one inoculates the feedstock by mixing it with partially digested material, and the other is mixed by high-pressure biogas injection. The Dranco process was developed in the late 1980s. It is a high-solids, single-stage anaerobic digestion system that operates at thermophilic temperature (Fig. 12.2a). The Valorga process was developed in 1981 to treat organic solid waste and accept household waste after proper separation of the stubborn part. A high solid digester is an OFMSW with 25–30% TS adjusted, diluted using steam heating and process water, and fed as needed. According to raw materials and economic conditions, choose to use a medium temperature system or a thermal system (Fig. 12.2b). Unlike the other two digesters, the Kompogas system uses a horizontal plug flow

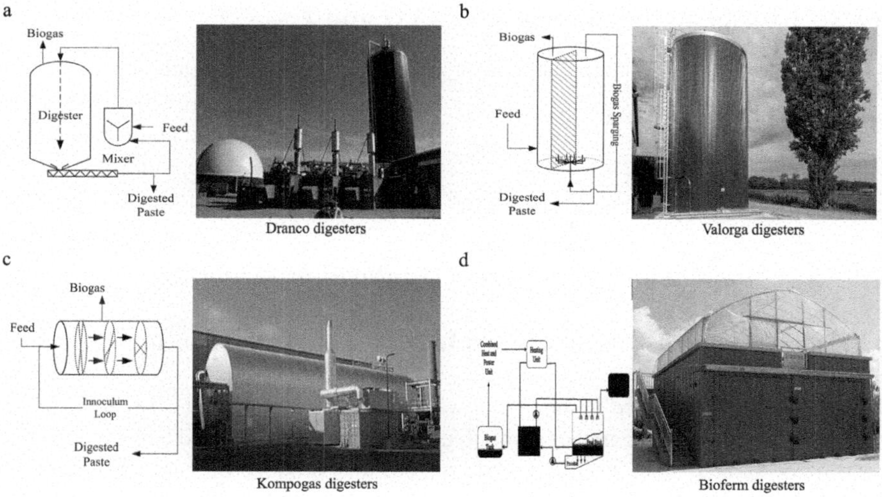

Fig. 12.2 Continuous system designs (**a**–**c**) and intermittent system design (**d**) of SS-AD. (**a**–**c** were adapted with permission from [31], Copyright © IWA Publishing 2005. **d** was adapted with permission from [32], Copyright © MDPI 2018)

digester with an internal rotor to assist in degassing and homogenization (Fig. 12.2c). Compared with the continuous SS-AD process, the intermittent SS-AD process developed relatively late and began to be commercially applied in the early 1990s. There are mainly German Bioferm (Fig. 12.2d), BEKON and Wehrlewerk company's Bioferm, BEKON and Biopercolat SS-AD process. Bioferm is mainly used in the treatment of organic solid waste with water content below 75%. The main feature of the process is that the raw material is added to the reactor without stirring or flipping, and without supplementary water, and the raw material does not need to do any pretreatment after entering the reactor. Chinese SS-AD industry is also gradually expanding, with more and more enterprises engaging in this field of research development and application of related technologies. At the same time, the Chinese government has introduced a series of supportive policies, including financial support and preferential taxation, to foster the growth of the SS-AD industry. These policies have provided a strong support for the rapid development of the SS-AD industry in China. However, despite significant progress in SS-AD technology, challenges remain, such as geographical and economic limitations affecting the promotion and application. Nonetheless, SS-AD offers better economic and environmental benefits for solid waste treatment compared to other treatments (e.g., combustion, landfills). To maximize these benefits and expedite the development of SS-AD, it would be wise to leverage the wealth of global experience available to address potential issues and accelerate technological advancements.

12.3 Merits and Demerits of SS-AD and LS-AD

12.3.1 Summary of the Merits and Demerits

See Table 12.2.

12.3.2 Applicability in Different Application Scenarios

12.3.2.1 Municipal Wastewater and Industrial Waste Treatment

LS-AD is commonly used in municipal wastewater treatment plants. The high water content of sewage sludge makes it suitable for LS-AD. This method helps reduce the volume of sludge, stabilize the organic matter, and produce biogas, which can be used for energy generation.

Industries that generate high amounts of organic solid waste, such as food processing, breweries, and paper mills, can benefit from SS-AD. The method can handle the high solid content of industrial organic waste streams, converting them into biogas for energy use.

12.3.2.2 Agricultural Applications

In agricultural settings, LS-AD is often used for the treatment of manure and slurry. These substrates have a high moisture content, making them ideal for wet digestion. The biogas produced can be used on-site for heating, electricity generation, or as a vehicle fuel, while the digestate can be applied as a nutrient-rich fertilizer.

For agricultural residues such as crop straw, husks, and other lignocellulosic materials, SS-AD is often more applicable. These materials typically have low moisture content and high lignin content, which makes them less suitable for LS-AD. SS-AD can effectively break down these tough organic materials, producing biogas and nutrient-rich digestate.

Table 12.2 Merits and demerits of solid-state (total solids $<50\%$) anaerobic digestion over liquid-state (total solids $<15\%$) anaerobic digestion

Merits	Demerits
High solid content	Longer retention times
Reduce reactor volume	Limited mixing
Less energy and water consumption	Potential clogging
Lower leachate production	Process monitoring
Enhanced methane production	Microbial adaptation

12.3.2.3 Solid Waste Management

LS-AD is also effective for processing food waste, which typically has high moisture content. This method is suitable for large-scale food waste processing facilities, such as those operated by municipalities or food processing industries. The biogas produced can be utilized for energy, and the digestate can be used as a soil conditioner.

SS-AD is well-suited for solid waste management, especially for the treatment of OFMSW and yard waste. The lower moisture content of these feedstocks makes SS-AD a more appropriate choice. This method can be used in urban areas where solid waste collection is significant, providing a means to reduce landfill use and generate renewable energy.

12.4 Critical Factors Influencing the SS-AD Process

Several critical factors influence the performance and efficiency of the SS-AD process. These factors affect substrate degradation, biogas production, process stability, and overall system performance. Understanding and optimizing these factors are essential for maximizing biogas yield and ensuring the success of SS-AD systems.

12.4.1 Temperature

Optimal temperature ranges for SS-AD typically vary between mesophilic (25–45 °C) and thermophilic (45–65 °C) conditions, depending on the microbial consortia and feedstock characteristics [33]. Compared to mesophilic (25–45 °C) AD, thermophilic (45–65 °C) AD accelerates the hydrolysis of raw materials, resulting in a shorter start-up time and hydraulic retention time (HRT). Additionally, since the optimal growth temperature for methanogens is 55 °C [34], the methane production rate of high-temperature SS-AD is also relatively higher. In contrast, poor stability is a major obstacle for thermophilic (45–65 °C) SS-AD. Microorganisms in the digestive system showed higher environmental sensitivity at elevated temperatures, leading to reduced community diversity and richness. Excessive temperatures can adversely affect the stability of the microbial ecosystem, making it more susceptible to fluctuations and disturbances [35].

12.4.2 pH

pH levels within the range of 6.5–8.0 are generally favorable for microbial growth and methanogenic activity in SS-AD. Methanogens are very sensitive to pH and perform optimally around a pH of 7.0. However, hydrolytic and acidogenic bacteria perform better at slightly more acidic pH between 5.5 and 6.5 [36]. Acidogenic bacteria are more tolerant to a broader pH range [37]. Fluctuations in pH can disrupt microbial communities and affect process stability.

12.4.3 C/N Ratio

An appropriate carbon-to-nitrogen (C/N) ratio promotes microbial activities and helps maintain a stable digestive system in SS-AD. In SS-AD, the degradation rate of organic matter is closely related to the C/N ratio. The change in the carbon-nitrogen ratio may affect the stability of the SS-AD reactor. A high C/N ratio (higher than 30:1) may cause microbial community imbalance in the reactor, producing volatile organic acids and other products that are not conducive to the digestive process. However, a low C/N ratio (10–20:1) may lead to too high nitrogen concentration in the reactor, produce harmful gases such as ammonia, and affect the stability and safety of the reactor. Changes in the ratio of carbon to nitrogen also affect the quality of the digestive fluid produced by SS-AD. A proper C/N ratio (20–30:1) helps produce a nutrient-rich digestive solution that can be used as a quality organic fertilizer or soil amendment.

12.4.4 Substrate Characteristics

The composition and characteristics of the feedstock significantly influence SS-AD performance, such as moisture content, C/N ratio, organic matter content, lignocellulosic content, and particle size impact substrate biodegradability, microbial activity, and biogas production. High lignocellulosic content (higher than 40%) and low moisture content (lower than 40%) can impede microbial access to nutrients and substrate, resulting in reduced biogas yield and process inefficiency. Easily degradable organic substances, such as starch and protein, are quickly metabolized by microorganisms to produce methane gas. Conversely, the substances consisted of intricate structures, like cellulose and lignin, necessitate prolonged durations and complex microbial metabolic pathways, impacting the rate and efficiency of the digestion process. Furthermore, substrates may contain resistant substances such as lignin, polymers, and chemical residues, which could inhibit microbial activity and hinder the digestion process, consequently reducing substrate degradation rates and gas production yields. Understanding and optimizing these factors are crucial for enhancing the efficiency and effectiveness of SS-AD systems. Adjusting feedstock characteristics to balance moisture content, C/N ratio, and particle size, and pre-treating lignocellulosic materials can help improve substrate biodegradability, microbial activity, and biogas yield.

12.4.5 *Total Solids*

Water plays a vital role in the AD system by dissolving and transporting nutrients, promoting mass transfer, and diluting materials to prevent the inhibitory concentration of reactants. Maintaining appropriate TS levels is essential for creating a balanced environment that supports microbial growth and metabolic activity, which in turn enhances substrate degradation and digestive efficiency. When TS levels are optimal (20–40%), the system can maintain good support and structure, ensuring adequate stomatal space and aeration. This environment is conducive to microbial proliferation and activity, facilitating efficient substrate breakdown and improving the overall performance of the anaerobic digestion process. However, excessively high TS levels (higher than 40%) can lead to poor contact and poor ventilation between substrate particles, which hinder microbial access to substrates and slow down the degradation process. Also, high TS levels (higher than 40%) cause the accumulation of volatile fatty which lower the pH of the system, inhibiting microbial activity and leading to process instability. This can reduce the methane content of the biogas, affecting its quality and energy value.

12.4.6 *Anaerobic Microorganisms*

The performance of SS-AD is heavily influenced by microbial consortia present in the inoculum and their capacity to degrade complex organic compounds under solid-state conditions. Consequently, the selection and acclimation of robust microbial communities that can efficiently hydrolyze, ferment, and perform methanogenesis on solid substrates are essential for optimizing biogas production.

Anaerobic microorganisms decompose organic matter, including carbohydrates, proteins, and fats in a substrate to produce volatile fatty acids (VFAs) and other intermediate products. These intermediates are subsequently converted by other microorganisms into end products such as methane gas and carbon dioxide. An appropriate microbial community can improve the methane yield, increase the substrate degradation rate, and reduce the accumulation of volatile fatty acids in the reactor.

12.5 Strategies to Enhance the Methane Production Efficiency of SS-AD Process

To improve the efficiency of methane production in SS-AD processes, several strategies can be implemented to optimize substrate degradation, microbial activity, and methane production to maximize resource utilization.

12.5.1 Pretreatment

In order to change the stubborn structure of the substrate and improve the digestibility of SS-AD, various pretreatment strategies have been introduced. Depending on the type of substrate, different types of pretreatment methods can be used.

12.5.1.1 Physical Pretreatment

For lignocellulosic biomass, size reduction through grinding and shredding of feedstock increases the surface area and improves microbial access to nutrients, thereby enhancing substrate degradation and methane production efficiency [14]. Importantly, physical pretreatment does not produce any toxic compounds that could inhibit the AD process. Mshandete et al. found that biogas production from sisal fiber waste is proportional to the particle size [38]. Similarly, Dahunsi et al. performed mechanical pretreatment on six different types of lignocellulosic biomass and observed a 22% increase in methane production [39].

Thermal pretreatment involves treating organic waste at high temperatures before AD. This process helps to break down complex organic molecules, damage cellular structures, and improve the biodegradability of waste. Thereby thermal pretreatment promotes substrate hydrolysis and improves the overall efficiency of the AD process.

The freeze-thaw pretreatment method is a novel, non-polluting approach for enhancing AD. This method leverages the mechanism of water expansion during freezing, which damages the biomass structure, thereby improving its biodegradability. By simulating natural or outdoor low-temperature conditions, straw can be subjected to freeze-thaw in a cold climate. Results have shown that the maximum methane yield is obtained at a temperature of 20 °C with a 48 h [40].

12.5.1.2 Chemical Pretreatment

Chemical pretreatment primarily functions by removing the lignin or hemicellulose from biomass, enhancing its biodegradability. This method is categorized based on the type of chemicals including acidic, alkaline, oxidizing and organic solvent treatment. The main acids involved in chemical pretreatment include sulfuric acid, hydrochloric acid, formic acid and nitric acid. Common chemicals used in alkaline pretreatment processes are sodium hydroxide and ammonia. Acid pretreatment can disrupt the van der Woll bonds, hydrogen bonds, and covalent bonds of the substrate, resulting in biomass decomposition and cell lysis. This process effectively eliminates the hemicellulose portion of the lignocellulosic biomass. The study found that the effect of dilute sulfuric acid pretreatment on water hyacinth and observed a 131% increase in biogas production compared to the control group [41].

Alkaline pretreatment dignifies the substrate, thereby increasing the porosity and reducing the degree of polymerization of lignocellulosic biomass. In the AD of silage, pretreatment with 5% NaOH was found to increase methane production by 38% [42]. However, although NaOH pretreatment can effectively increase biogas production, care must be taken to avoid the formation of Na^+ ions during the pretreatment process, which can inhibit AD process. Yao et al. found that pretreatment of wheat straw with urea was beneficial for increasing methane yield. After 1% and 3% urea treatments, the time required to reach a stable state (50% methane yield) was reduced by 2 days compared to untreated straw [43].

12.5.1.3 Biological Pretreatment

Due to its lower energy consumption and milder process conditions, biological pretreatment is considered an environmentally friendly alternative to physical or chemical processes. Biological pretreatment involves the use of microorganisms or enzymes to break down complex organic compounds in the substrates before the AD process. This method utilizes the natural biological processes of microorganisms to improve the biodegradability of organic waste.

Microbial pretreatment involves inoculating organic waste with specific microorganisms, such as bacteria, fungi, or algae, capable of degrading complex organic compounds [44]. These microbes produce enzymes, including cellulases, proteases, and lipases, which break down polymers like cellulose, proteins, and lipids into simpler compounds that are more easily digested by anaerobic microbes during subsequent digestion. Fungal pretreatment involves inoculating organic waste with specific fungi, such as white rot fungi (e.g., *Phanerochaete chrysosporium*) or brown rot fungi (e.g., *Trametes versicolor*), capable of degrading lignocellulosic materials. These fungi produce enzymes like lignin peroxidase, manganese peroxidase, and laccase, which break down the lignin and cellulose in the waste, making them more accessible to anaerobic microorganisms during subsequent digestion [45, 46]. For instance, the fungus Trametes versicolor was used to pretreat corn silage, resulting in significantly increased methane production [47]. Constructing microbiomes and using them for substrate pretreatment is an effective method to improve biomass degradation performance [48]. In one study, microbial consortia isolated from decomposing sawdust increased methane production significantly by 72.6% compared to a control treatment [49].

Enzyme pretreatment involves the application of commercial enzymes, such as cellulase, protease, and amylase, to catalyze the breakdown of complex organic compounds in organic waste. These enzymes can be directly added to waste or produced by microorganisms cultured on-site. Enzyme pretreatment is commonly used to improve the digestibility of lignocellulosic materials, including agricultural residues and food waste, by hydrolyzing cellulose and hemicellulose into fermentable sugars. During the reaction, the enzyme increases substrate accessibility and enhance mass transfer rate [50]. In a study on enzyme pretreatment of corn stalk (enzyme loading of 30 FPU/g substrate, solid content of 60 g/L, 24 h of incubation time), cellulase was found to increase the hydrolysis rate of the substrate, resulting in a 36.9% increase in biogas yield [51].

12.5.1.4 Combined Pretreatment

Combined pretreatment involves integrating multiple pretreatment methods to enhance the effectiveness of organic waste treatment processes. By utilizing a combination of different pretreatment techniques, synergistic effects can be achieved, leading to improved biodegradation, increased biogas production, and overall better performance of waste treatment systems. Specifically, combining biological and thermal pretreatment methods can enhance the breakdown of organic matter. Thermal pretreatment disrupts cell structures and releases intracellular components, making these materials more accessible to microbial degradation in subsequent biological treatment processes. This enhanced accessibility promotes microbial activity and accelerates biochemical reactions, thereby optimizing the efficiency of the treatment system. Consequently, the integration of thermal and biological pretreatment methods can result in more effective organic waste treatment, evidenced by improved rates of biodegradation and increased biogas yields.

12.5.2 Inoculum

The inoculation of enriched and acclimated microbial communities in SS-AD reactors can significantly improve process stability, reduce start-up time, and enhance methane production efficiency. Supplementation with specific inoculants can target and enhance specific metabolic pathways, thereby accelerating substrate degradation and boosting methane production.

12.5.2.1 Biological Inoculum

Biological inoculants are microorganisms with specific functions introduced into a specific environment, such as soil, water, or biological systems to enhance certain processes or functions. In the context of SS-AD, biological inoculants are often used to improve microbial populations, promote biological activity, accelerate the breakdown of organic matter, increase methane production, improve process stability, and reduce odors. Biological inoculants can be sourced from naturally occurring microbial communities, specialized microbial cultures, or commercially available products formulated for specific applications. Choosing the right biological inoculant for SS-AD depends on various factors, including waste composition, desired process results, and environmental conditions. Selecting the dominant strains suitable for AD environment can significantly speed up the digestion process. For instance, some microbial strains can degrade cellulose, thus accelerating the decomposition of cellulose in organic waste and improving digestion efficiency. Ammonia-oxidizing bacteria can convert ammonia into nitrate, thereby reducing ammonia accumulation during AD and improving system stability. Acidogens play a crucial role in the initial stages of AD, where they hydrolyze complex organic compounds into simpler molecules such as sugars, amino acids, and fatty acids. Species like *Clostridium* and *Bacteroides* are effective acidogens that can accelerate the breakdown of organic matter. Methanogens are essential microorganisms responsible for the final stage of AD. They convert intermediate products, including acetate, hydrogen, and carbon dioxide into methane gas. Adding methanogenic archaea inoculants can promote methane production and increase overall biogas production. It has been reported that the lag period of the AD process can be shortened by the enrichment of *Methanobacterium* and *Methanosarcina* [52]. Furthermore, certain microorganisms, such as acetogenic bacteria, play a crucial role in regulating pH levels within the AD system. These bacteria consume excess hydrogen and acetate, which helps prevent acidification and maintaining optimal conditions for methanogenesis. Yu et al. found that the VFAs concentration in the system decreased significantly in the early stage of SS-AD after a single inoculation with an acid-resistant bactericide. This study highlighted the impact of the initial microbial community structure on the performance of SS-AD of corn straw [53].

12.5.2.2 Additive

Soil has the potential for interspecific electron transfer (Fig. 12.3), buffering capacity, and nutrient elements required for microbial growth. In addition, the multiple cations present in the soil help to enhance the balance of the microbial community, thereby increasing the efficiency of AD and improving methane production [54, 55].

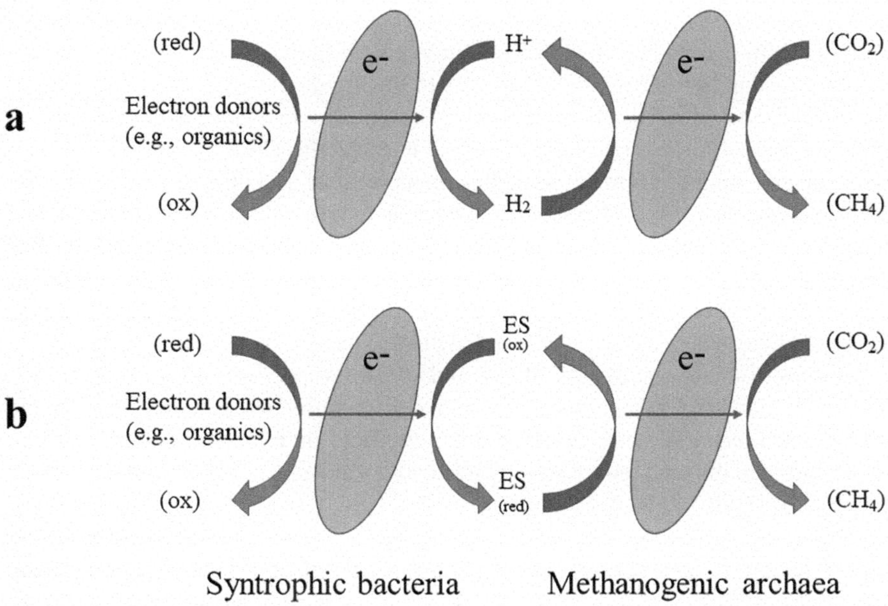

Fig. 12.3 Interspecific electron transfer potential of soils

Yao et al. investigated the effect of soil addition on the AD of cow manure by incorporating powdered Y soil (from Washington State University (WSU) Dairy Center in Pullman, WA, USA) and B soil (from the wheat field) at different feedstock-to-soil (F:S) ratios of 5:1, 2.5:1 and 1:1. The results showed that the maximum total methane production was achieved at the F:S ratio of 2.5:1 with Y soil addition, yielding 149.7 L/kg VS, which was 25.4% higher than that of the control which produced 119.4 L/kg VS. Meanwhile, soil addition primarily stimulated the growth of bacteria *Ochrobactrum* and *Clostridium* and archaea *Methanosaeta* and *Methanosarcina* (Fig. 12.4a) [56]. In another study, Yao et al. explored the benefits of soil addition for accelerating the AD initiation process. They added soil samples to poplar processing residues (PPRs) following NaOH pretreatment. The study revealed that the maximum methane yield of 96.1 L/kg VS was obtained at a PPR-to-soil (P:S) ratio of 2.5:1 after NaOH pretreatment, which was 29.9% and 36.1% higher than that of PPRs (74.0 L/kg VS) and without NaOH pretreatment (70.6 L/kg VS), respectively (Fig. 12.4b) [57]. On the other hand, with the widespread application of SS-AD in the treatment of lignocellulosic biomass, a new method combining SS-AD with simultaneous urea treatment and soil addition has been proposed. By adopting this approach, the total methane yield the total methane yield reached a maximum of 409.6 L/kg VS, which was 22.6% higher than SS-AD with urea pretreatment alone and 76.8% higher than SS-AD with simultaneous urea treatment alone, respectively (Fig. 12.4c) [58]. In the SS-AD study of corn stalk, Yao et al., added soil to the digester at a corn straw-to-soil of 2.5: 1. This intervention led to increased daily methane production and improved methane content. The study found that methane production was highest when NaOH content was 1%, achieving a yield of 344 L/kg VS (Fig. 12.4d) [59].

Biochar can enhance AD through several mechanisms: (1) alkaline pH buffer medium [60]; (2) adsorption of AD inhibitors [61, 62]; (3) conversion of macromolecular substances to dissolved substances [63]; (4) nutrients provision (Ca, Mg, P, K) [64]; (5) microbial biofilms support for bacteria and methanogens because of its high porosity [65, 66]; (6) promote direct interspecific electron transfer between bacteria and methanogens to stimulate methanogens [66, 67]. In addition, the inclusion of biochar in AD systems has been shown to reduce the concentration of total VFA and propionic acid and shorten the lag period of AD at all studied temperatures [64]. For instance, under thermophilic conditions (35 °C), the addition of pine sawdust biochar (pyrolysis at 650 °C) to the dry AD system of food waste increased hydrogen yield by 31.0% and methane yield by 10.0%, while reducing the lag phases of the two stages by 36.0% and 41.0%, respectively [65]. Furthermore, Yao et al. found that in conditions of a high concentration of ammonia nitrogen, the addition of biochar effectively delayed the threshold time for VFAs accumulation, contributing to the total removal of organic carbon and enhancing the stability of AD process [68] (Fig. 12.5).

The addition of metallic materials, such as Fe_3O_4, cerium dioxide, nano-iron, and zero-valent iron, can accelerate the metabolism of AD microorganisms by promoting direct interspecific electron transfer (DIET) between microorganisms [69, 70]. Particularly, zero-valent iron is commonly used as electrode material in anaerobic

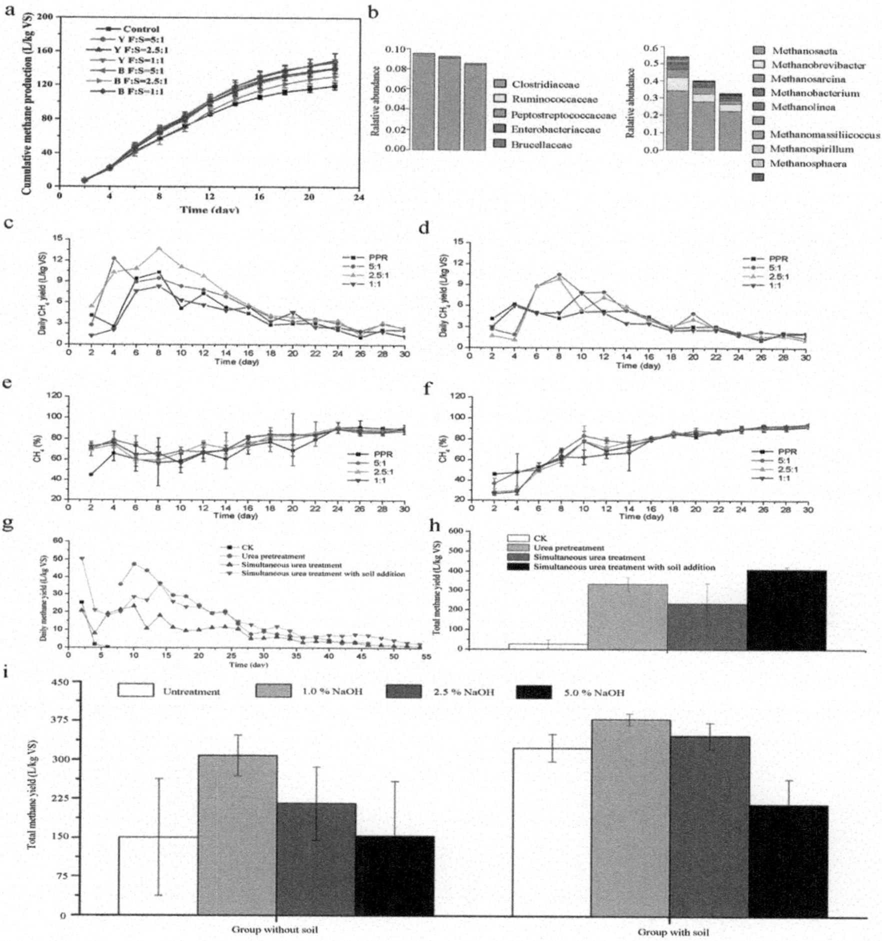

Fig. 12.4 Cumulative methane production of the control and experiments with different F:S ratios (**a**); Taxonomic composition of microbial communities from the control and experiments with Y soil at the feedstock:soil (F:S) ratio of 2.5:1 and B soil at the F:S ratio of 5:1 (**b**); Daily methane yield (L/kg VS) with NaOH pretreatment (**c**); Daily methane yield (L/kg VS) without NaOH pretreatment (**d**); Methane content (%) with NaOH pretreatment (**e**); Methane content (%) without NaOH pretreatment (**f**); Daily methane yield of untreated and different treated rape straws (**g**). CK Sample with no urea treatment and no soil addition; Urea pretreatment sample with urea treatment before AD and no soil addition; Simultaneous urea treatment urea treatment along with AD and no soil addition; Simultaneous urea treatment urea treatment along with AD and soil addition (**h**); Total methane yield of untreated and different treated rape straws (**i**); Total methane yield of T0 and simultaneous NaOH treatment. (**a** and **b** were adapted with permission from [56], Copyright © 2019 Elsevier; **c**–**f** were adapted with permission from [57], Copyright © 2015 BioMed Centra; **g** and **h** were adapted with permission from [58], Copyright © 2016 BioMed Centra; **i** was adapted with permission from [59], Copyright © 2018 Elsevier)

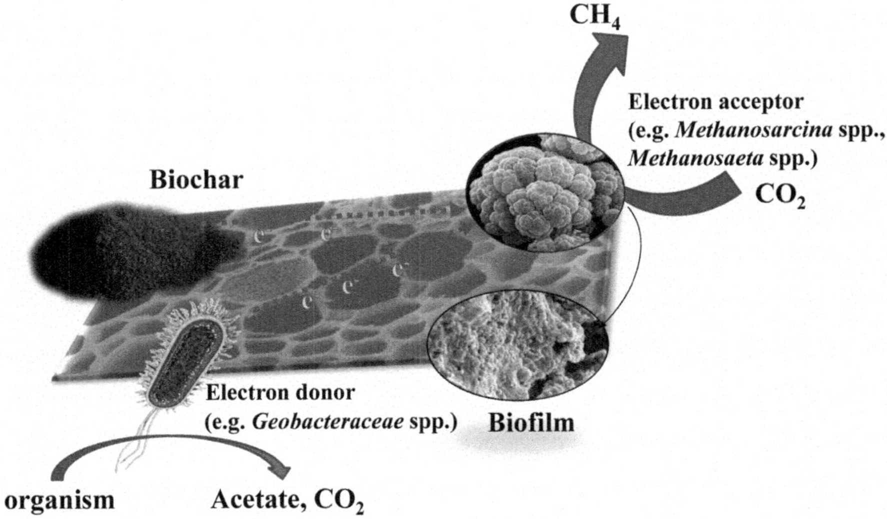

Fig. 12.5 Biochar promotes direct interspecific electron transfer between bacteria and methanogens

wastewater treatment. It has been found to enhance the ability of bacteria to produce VFAs and methanogenic archaea to produce methane [71]. Xu et al. demonstrated that the addition of Fe_3O_4 increased the secretion of electron shuttling components associated with extracellular polymers, thus promoting the efficiency of direct interspecific electron transfer, improving propionate decomposition, and boosting methane generation [72].

12.5.3 Nutrient Optimization

Balancing the carbon-to-nitrogen ratio in the feedstock is crucial for microbial growth, metabolic activity, and methane synthesis during the AD process. Adjusting the feedstock composition or supplementing with nitrogen-rich materials can optimize the C/N ratio, thereby enhancing methane production efficiency. Himmel et al. adjusted the total carbon-to-nitrogen ratio of rice straw substrate through the addition of nitrogen, which increased the biogas yield during the process of SS-AD [73]. Moreover, anaerobic co-digestion of crop straw and animal manure has been shown to effectively increase biogas production. This practice leverages the complementary properties of different feedstocks, optimizing nutrient availability and improving overall digestion performance [74]. Zheng et al. studied the effects of varing C/N ratios on the dry thermophilic anaerobic co-digestion of sewage sludge and kitchen waste. The results showed that the cumulative methane production reached 165.68 ± 11.88 mL/g VS at a C/N ratio of 11, which represented a 1.23-fold increase compared to the methane yield from the single digestion of sewage sludge [75].

12.5.4 *Process Control and Monitoring*

Proper process control and monitoring are necessary for the efficient operation of SS-AD and for maximizing biogas production. By dynamically monitoring key parameters such as temperature, pH, gas composition, substrate water content, particle size distribution, and organic matter content, it is possible to make timely adjustments to maintain system stability and optimize methane yield in the SS-AD process. In general, maintaining the temperature between 35 and 55 °C is crucial for supporting microbial activity and promoting biogas production during the SS-AD process. This temperature range ensures that the microbial communities responsible for digestion are active and efficient. In addition, regular stirring or mixing keeps the substrate well mixed and promotes uniform contact with microorganisms. By keeping the substrate evenly distributed, mixing ensures that microorganisms have consistent access to the organic material, improving the efficiency of digestion. Additionally, regular mixing helps to release trapped gases within the substrate, improving biogas recovery and preventing the formation of gas pockets that could disrupt the digestion process.

12.6 Conclusion and Scope for Future Research

SS-AD offers significant contributions to sustainable waste management and renewable energy production. Its ability to process high-solids feedstocks, reduce water usage, and produce valuable by-products makes it an attractive option for various application scenarios. This chapter provides clear evidence of the potential of SS-AD to treat biological waste and generate clean energy. By outlining strategies to optimize the SS-AD process and improve the efficiency of converting various solid substrates to methane, it aims to help reduce environmental pollution, lower greenhouse gas emissions, and diversify energy sources. Future research should focus on optimizing process parameters, exploring co-digestion strategies, developing advanced reactor designs, and integrating SS-AD with other waste management technologies. Additionally, understanding microbial communities, conducting life cycle assessments, and evaluating economic feasibility are crucial for the widespread adoption and success of SS-AD systems. Establishing supportive policies and regulatory frameworks will further promote the implementation of this technology, contributing to a more sustainable and circular economy.

Acknowledgments This work led to a unique opportunity, and words of appreciation and thanks are entirely due to the contributors, as they responded generously with characteristic rigor and patience for deadlines. Their constructive criticisms and suggestions enrich the content of this book. Hopefully, the result will fully justify the hard work they have put into preparing the retrospective chapter.

References

1. Yan Z, Song Z, Li D, Yuan Y, Liu X, Zheng T. The effects of initial substrate concentration, C/N ratio, and temperature on solid-state anaerobic digestion from composting rice straw. Bioresour Technol. 2015;177:266–73. https://doi.org/10.1016/j.biortech.2014.11.089.
2. Rouches E, Escudie R, Latrille E, Carrere H. Solid-state anaerobic digestion of wheat straw: impact of S/I ratio and pilot-scale fungal pretreatment. Waste Manag. 2019;85:464–76. https://doi.org/10.1016/j.wasman.2019.01.006.
3. Hasan MR, Anzar N, Sharma P, Malode SJ, Shetti NP, Narang J, et al. Converting biowaste into sustainable bioenergy through various processes. Bioresour Technol Rep. 2023;23:101542. https://doi.org/10.1016/j.biteb.2023.101542.
4. Jiménez-Rosado M, Perez-Puyana V, Romero A. The use of biowaste for the production of biodegradable superabsorbent materials. Curr Opin Food Sci. 2023;49:100975. https://doi.org/10.1016/j.cofs.2022.100975.
5. Sheer A, Fahad Sardar M, Younas F, Zhu P, Noreen S, Mehmood T, et al. Trends and social aspects in the management and conversion of agricultural residues into valuable resources: a comprehensive approach to counter environmental degradation, food security, and climate change. Bioresour Technol. 2024;394:130258. https://doi.org/10.1016/j.biortech.2023.130258.
6. Antar M, Lyu D, Nazari M, Shah A, Zhou X, Smith DL. Biomass for a sustainable bioeconomy: an overview of world biomass production and utilization. Renew Sust Energ Rev. 2021;139:110691. https://doi.org/10.1016/j.rser.2020.110691.
7. Sherwood J. The significance of biomass in a circular economy. Bioresour Technol. 2020;300:122755. https://doi.org/10.1016/j.biortech.2020.122755.
8. Czekała W. Selective collection and management of biowaste from the municipal sector in Poland: a review. Appl Sci. 2023;13(19):11015. https://doi.org/10.3390/app131911015.
9. Cusenza MA, Longo S, Guarino F, Cellura M. Energy and environmental assessment of residual bio-wastes management strategies. J Clean Prod. 2021;285:124815. https://doi.org/10.1016/j.jclepro.2020.124815.
10. Maroušek J, Strunecký O, Kolář L, Vochozka M, Kopecký M, Maroušková A, et al. Advances in nutrient management make it possible to accelerate biogas production and thus improve the economy of food waste processing. Energy Sour A Recov Utiliz Environ Effects. 2024;46:9379–88. https://doi.org/10.1080/15567036.2020.1776796.
11. Uddin MM, Wright MM. Anaerobic digestion fundamentals, challenges, and technological advances. Phys Sci Rev. 2023;8(9):2819–37. https://doi.org/10.1515/psr-2021-0068.
12. Ajayi-Banji A, Rahman S. A review of process parameters influence in solid-state anaerobic digestion: focus on performance stability thresholds. Renew Sust Energ Rev. 2022;167:112756. https://doi.org/10.1016/j.rser.2022.112756.
13. Brown D, Shi J, Li Y. Comparison of solid-state to liquid anaerobic digestion of lignocellulosic feedstocks for biogas production. Bioresour Technol. 2012;124:379–86. https://doi.org/10.1016/j.biortech.2012.08.051.
14. Yang L, Xu F, Ge X, Li Y. Challenges and strategies for solid-state anaerobic digestion of lignocellulosic biomass. Renew Sust Energ Rev. 2015;44:824–34. https://doi.org/10.1016/j.rser.2015.01.002.
15. Xu F, Shi J, Lv W, Yu Z, Li Y. Comparison of different liquid anaerobic digestion effluents as inocula and nitrogen sources for solid-state batch anaerobic digestion of corn stover. Waste Manag. 2013;33(1):26–32. https://doi.org/10.1016/j.wasman.2012.08.006.
16. Xu F, Wang ZW, Tang L, Li Y. A mass diffusion-based interpretation of the effect of total solids content on solid-state anaerobic digestion of cellulosic biomass. Bioresour Technol. 2014;167:178–85. https://doi.org/10.1016/j.biortech.2014.05.114.
17. Yang L, Li Y. Anaerobic digestion of giant reed for methane production. Bioresour Technol. 2014;171:233–9. https://doi.org/10.1016/j.biortech.2014.08.051.
18. Shi J, Wang Z, Stiverson JA, Yu Z, Li Y. Reactor performance and microbial community dynamics during solid-state anaerobic digestion of corn stover at mesophilic and thermophilic

conditions. Bioresour Technol. 2013;136:574–81. https://doi.org/10.1016/j.biortech.2013.02.073.
19. Xu L, Peng S, Dong D, Wang C, Fan W, Cao Y, et al. Performance and microbial community analysis of dry anaerobic co-digestion of rice straw and cow manure with added limonite. Biomass Bioenergy. 2019;126:41–6. https://doi.org/10.1016/j.biombioe.2019.04.026.
20. Paritosh K, Yadav M, Kesharwani N, Pareek N, Parthiba Karthyikeyan O, Balan V, et al. Strategies to improve solid state anaerobic bioconversion of lignocellulosic biomass: an overview. Bioresour Technol. 2021;331:125036. https://doi.org/10.1016/j.biortech.2021.125036.
21. Jimenez J, Latrille E, Harmand J, Robles A, Ferrer J, Gaida D, et al. Instrumentation and control of anaerobic digestion processes: a review and some research challenges. Rev Environ Sci Biotechnol. 2015;14(4):615–48. https://doi.org/10.1007/s11157-015-9382-6.
22. World Bioenergy Association. Global bioenergy statistics. World Bioenergy Association; 2023. https://www.worldbioenergy.org/uploads/231219%20GBS%20Report.pdf.
23. Li Y, Park SY, Zhu J. Solid-state anaerobic digestion for methane production from organic waste. Renew Sust Energ Rev. 2011;15(1):821–6. https://doi.org/10.1016/j.rser.2010.07.042.
24. Rapport J, Zhang R, Jenkins BM, Williams RL, Governor AS, Adams LS, et al. Current anaerobic digestion technologies used for treatment of municipal organic solid waste. Integrated Waste Management Board Public Affairs Office Publications; 2014.
25. G.S. Consultants. Task 6 report: Review of MSW landfill regulations from selected states and countries, in Landfill Facility Compliance Study. California Environmental Protection Agency, Integrated Waste Management Board; 2004. https://www2.calrecycle.ca.gov/Publications/Details/1079.
26. G.S. Consultants. Task 7 report: Study of emerging technologies in waste management for MSW landfills, in Landfill Compliance Study. Sacramento, CA: California Integrated Waste Management Board; 2003. https://www2.calrecycle.ca.gov/Publications/Details/1058.
27. Zhang Z. In: P.I.E.R.P. Program, editor. Lessons learned on California dairy digesters. California Energy Commission; 2007.
28. Williams RB. Biofuels from municipal wastes; Background Discussion Paper. California Biomass Collaborative 4th Annual Forum 2007. http://biomass.ucdavis.edu/materials/reports%20and%20publications/2007/2007_Annual_Forum_Background_Paper.pdf.
29. Scarlat N, Dallemand J-F, Fahl F. Biogas: developments and perspectives in Europe. Renew Energy. 2018;129:457–72. https://doi.org/10.1016/j.renene.2018.03.006.
30. De Baere L. Partial stream digestion of residual municipal solid waste. Water Sci Technol. 2008;57(7):1073–7. https://doi.org/10.2166/wst.2008.078.
31. Vandevivere P, Baere L, Verstraete W. In: Mata-Alvarez J, editor. Types of anaerobic digester for solid wastes. Barcelona, Spain: IWA Publishing; 2002. p. 111–40. https://doi.org/10.2166/9781780402994.
32. Fu Y, Luo T, Mei Z, Li J, Qiu K, Ge Y. Dry anaerobic digestion technologies for agricultural straw and acceptability in China. Sustainability. 2018;10(12):4588. https://doi.org/10.3390/su10124588.
33. Croce S, Wei Q, D'Imporzano G, Dong R, Adani F. Anaerobic digestion of straw and corn stover: the effect of biological process optimization and pre-treatment on total bio-methane yield and energy performance. Biotechnol Adv. 2016;34(8):1289–304. https://doi.org/10.1016/j.biotechadv.2016.09.004.
34. Chen H, Chang S. Dissecting methanogenesis for temperature-phased anaerobic digestion: impact of temperature on community structure, correlation, and fate of methanogens. Bioresour Technol. 2020;306:123104. https://doi.org/10.1016/j.biortech.2020.123104.
35. Mao C, Feng Y, Wang X, Ren G. Review on research achievements of biogas from anaerobic digestion. Renew Sust Energ Rev. 2015;45:540–55. https://doi.org/10.1016/j.rser.2015.02.032.
36. Ward AJ, Hobbs PJ, Holliman PJ, Jones DL. Optimisation of the anaerobic digestion of agricultural resources. Bioresour Technol. 2008;99(17):7928–40. https://doi.org/10.1016/j.biortech.2008.02.044.

37. Hagos K, Zong J, Li D, Liu C, Lu X. Anaerobic co-digestion process for biogas production: progress, challenges and perspectives. Renew Sust Energ Rev. 2017;76:1485–96. https://doi.org/10.1016/j.rser.2016.11.184.
38. Mshandete A, Björnsson L, Kivaisi AK, Rubindamayugi MST, Mattiasson B. Effect of particle size on biogas yield from sisal fibre waste. Renew Energy. 2006;31(14):2385–92. https://doi.org/10.1016/j.renene.2005.10.015.
39. Dahunsi SO. Mechanical pretreatment of lignocelluloses for enhanced biogas production: methane yield prediction from biomass structural components. Bioresour Technol. 2019;280: 18–26. https://doi.org/10.1016/j.biortech.2019.02.006.
40. Deng Y, Qiu L, Yao Y, Qin M. A technology for strongly improving methane production from rice straw: freeze-thaw pretreatment. RSC Adv. 2018;8(40):22643–51. https://doi.org/10.1039/c8ra03692f.
41. Sarto S, Hildayati R, Syaichurrozi I. Effect of chemical pretreatment using sulfuric acid on biogas production from water hyacinth and kinetics. Renew Energy. 2019;132:335–50. https://doi.org/10.1016/j.renene.2018.07.121.
42. Xie S. Evaluation of biogas production from anaerobic digestion of pig manure and grass silage. Civil Engineering, NUI Galway; 2012.
43. Yao Y, Bergeron AD, Davaritouchaee M. Methane recovery from anaerobic digestion of urea-pretreated wheat straw. Renew Energy. 2018;115:139–48. https://doi.org/10.1016/j.renene.2017.08.038.
44. Rouches E, Herpoël-Gimbert I, Steyer JP, Carrere H. Improvement of anaerobic degradation by white-rot fungi pretreatment of lignocellulosic biomass: a review. Renew Sust Energ Rev. 2016;59:179–98. https://doi.org/10.1016/j.rser.2015.12.317.
45. Wagner AO, Lackner N, Mutschlechner M, Prem EM, Markt R, Illmer P. Biological pretreatment strategies for second-generation lignocellulosic resources to enhance biogas production. Energies (Basel). 2018;11(7):1797. https://doi.org/10.3390/en11071797.
46. Patel H, Gupte A, Gupte S. Effect of different culture conditions and inducers on production of laccase by a basidiomycete fungal isolate Pleurotus ostreatus HP-1 under solid state fermentation. BioRes. 2009;4(1):268–84. https://doi.org/10.15376/biores.4.1.268-284.
47. Tišma M, Planinić M, Bucić-Kojić A, Panjičko M, Zupančič GD, Zelić B. Corn silage fungal-based solid-state pretreatment for enhanced biogas production in anaerobic co-digestion with cow manure. Bioresour Technol. 2018;253:220–6. https://doi.org/10.1016/j.biortech.2018.01.037.
48. Zhang Q, He J, Tian M, Mao Z, Tang L, Zhang J, et al. Enhancement of methane production from cassava residues by biological pretreatment using a constructed microbial consortium. Bioresour Technol. 2011;102(19):8899–906. https://doi.org/10.1016/j.biortech.2011.06.061.
49. Ali SS, Abomohra AE, Sun J. Effective bio-pretreatment of sawdust waste with a novel microbial consortium for enhanced biomethanation. Bioresour Technol. 2017;238:425–32. https://doi.org/10.1016/j.biortech.2017.03.187.
50. Romero-Güiza MS, Vila J, Mata-Alvarez J, Chimenos JM, Astals S. The role of additives on anaerobic digestion: a review. Renew Sust Energ Rev. 2016;58:1486–99. https://doi.org/10.1016/j.rser.2015.12.094.
51. Wang S, Li F, Wu D, Zhang P, Wang H, Tao X, et al. Enzyme pretreatment enhancing biogas yield from corn stover: feasibility, optimization, and mechanism analysis. J Agric Food Chem. 2018;66(38):10026–32. https://doi.org/10.1021/acs.jafc.8b03086.
52. Li Y, Wang Z, Li T, Jiang S, Sun Z, Jiang H, et al. Changes in microbial community and methanogenesis during high-solid anaerobic digestion of ensiled corn stover. J Clean Prod. 2020;242:116479. https://doi.org/10.1016/j.jclepro.2019.118479.
53. Lu Y, Zhang Q, Wang X, Zhong H, Zhu J. Effects of initial microbial community structure on the performance of solid-state anaerobic digestion of corn stover. J Clean Prod. 2020;260: 121007. https://doi.org/10.1016/j.jclepro.2020.121007.

54. Kato S, Hashimoto K, Watanabe K. Methanogenesis facilitated by electric syntrophy via (semi)conductive iron-oxide minerals. Environ Microbiol. 2012;14(7):1646–54. https://doi.org/10.1111/j.1462-2920.2011.02611.x.
55. Federer CA, Hornbeck JW. The buffer capacity of forest soils in new England. Water Air Soil Pollut. 1985;26:163–73. https://doi.org/10.1007/BF00292066.
56. Yao Y, Li H, Qiu L, Shi J, Chen S, Qiu Y, et al. Facilitated methanogenesis involved in anaerobic digestion of dairy manure by soil. J Clean Prod. 2019;236:117640. https://doi.org/10.1016/j.jclepro.2019.117640.
57. Yao Y, Luo Y, Li T, Yang Y, Sheng H, Virgo N, et al. Using the properties of soil to speed up the start-up process, enhance process stability, and improve the methane content and yield of solid-state anaerobic digestion of alkaline-pretreated poplar processing residues. Biotechnol Biofuels. 2014;7(1):160. https://doi.org/10.1186/s13068-014-0160-4.
58. Yao Y, Chen S. A novel and simple approach to the good process performance of methane recovery from lignocellulosic biomass alone. Biotechnol Biofuels. 2016;9:115. https://doi.org/10.1186/s13068-016-0530-1.
59. Yao Y, Zhou J, An L, Kafle GK, Chen S, Qiu L. Role of soil in improving process performance and methane yield of anaerobic digestion with corn straw as substrate. Energy. 2018;151:998–1006. https://doi.org/10.1016/j.energy.2018.03.069.
60. Linville JL, Shen Y, Ignacio-de Leon PA, Schoene RP, Urgun-Demirtas M. In-situ biogas upgrading during anaerobic digestion of food waste amended with walnut shell biochar at bench scale. Waste Manage Res. 2017;35(6):669–79. https://doi.org/10.1177/0734242x17704716.
61. Masebinu SO, Akinlabi ET, Muzenda E, Aboyade AO. A review of biochar properties and their roles in mitigating challenges with anaerobic digestion. Renew Sust Energ Rev. 2019;103:291–307. https://doi.org/10.1016/j.rser.2018.12.048.
62. Fagbohungbe MO, Herbert BMJ, Hurst L, Ibeto CN, Li H, Usmani SQ, et al. The challenges of anaerobic digestion and the role of biochar in optimizing anaerobic digestion. Waste Manag. 2017;61:236–49. https://doi.org/10.1016/j.wasman.2016.11.028.
63. Pan J, Ma J, Liu X, Zhai L, Ouyang X, Liu H. Effects of different types of biochar on the anaerobic digestion of chicken manure. Bioresour Technol. 2019;275:258–65. https://doi.org/10.1016/j.biortech.2018.12.068.
64. Jang HM, Choi YK, Kan E. Effects of dairy manure-derived biochar on psychrophilic, mesophilic and thermophilic anaerobic digestions of dairy manure. Bioresour Technol. 2018;250:927–31. https://doi.org/10.1016/j.biortech.2017.11.074.
65. Sunyoto NMS, Zhu M, Zhang Z, Zhang D. Effect of biochar addition on hydrogen and methane production in two-phase anaerobic digestion of aqueous carbohydrates food waste. Bioresour Technol. 2016;219:29–36. https://doi.org/10.1016/j.biortech.2016.07.089.
66. Sun C, Liu F, Song Z, Wang J, Li Y, Pan Y, et al. Feasibility of dry anaerobic digestion of beer lees for methane production and biochar enhanced performance at mesophilic and thermophilic temperature. Bioresour Technol. 2019;276:65–73. https://doi.org/10.1016/j.biortech.2018.12.105.
67. Capson-Tojo G, Girard C, Rouez M, Crest M, Steyer JP, Bernet N, et al. Addition of biochar and trace elements in the form of industrial $FeCl_3$ to stabilize anaerobic digestion of food waste: dosage optimization and long-term study. J Chem Technol Biotechnol. 2018;94:505–15. https://doi.org/10.1002/jctb.5797.
68. Qiu L, Deng YF, Wang F, Davaritouchaee M, Yao YQ. A review on biochar-mediated anaerobic digestion with enhanced methane recovery. Renew Sust Energ Rev. 2019;115:109373. https://doi.org/10.1016/j.rser.2019.109373.
69. Yan W, Qian T, Soh YNA, Zhou Y. Micro-level evaluation of organic compounds transformation in anaerobic digestion under feast and famine conditions assisted by iron-based materials—revealing the true mechanism of AD enhancement. Environ Int. 2020;135:105362. https://doi.org/10.1016/j.envint.2019.105362.
70. Boscaro ME, Canaver Marin DF, Camila da Silva D, Maintinguer SI. Effect of Fe_3O_4 nanoparticles on microbial diversity and biogas production in anaerobic digestion of crude

glycerol. Biomass Bioenerg. 2022;160:106439. https://doi.org/10.1016/j.biombioe.2022.106439.

71. Li Y, Ma Y, Zhan J, Zhang Y, Zhao Z, Zhao Z. Combining metal-microbe and microbe-microbe dual direct electron transfer on Fe(0)-cathode of bio-electrochemical system to enhance anaerobic digestion of cellulose wastewater. Chinese Chem Lett. 2022;33(6):3106–12. https://doi.org/10.1016/j.cclet.2021.09.097.
72. Xu Y, Wang M, Yu Q, Zhang Y. Enhancing methanogenesis from anaerobic digestion of propionate with addition of Fe oxides supported on conductive carbon cloth. Bioresour Technol. 2020;302:122796. https://doi.org/10.1016/j.biortech.2020.122796.
73. Himmel ME, Ding SY, Johnson DK, Adney WS, Nimlos MR, Brady JW, et al. Biomass recalcitrance: engineering plants and enzymes for biofuels production. Science. 2007;315(5813):804–7. https://doi.org/10.1126/science.1137016.
74. Shen F, Li H, Wu X, Wang Y, Zhang Q. Effect of organic loading rate on anaerobic co-digestion of rice straw and pig manure with or without biological pretreatment. Bioresour Technol. 2018;250:155–62. https://doi.org/10.1016/j.biortech.2017.11.037.
75. Zheng Y, Wang P, Yang X, Lin P, Wang Y, Cheng M, et al. Process performance and microbial communities in anaerobic co-digestion of sewage sludge and food waste with a lower range of carbon/nitrogen ratio. BioEnerg Res. 2021;15(3):1664–74. https://doi.org/10.1007/s12155-021-10357-2.

Index

Z. Fang et al. (eds.), *Production of Organic Acids and Alcohols from Agricultural Residues and Food Wastes*, Biofuels and Biorefineries 13,
https://doi.org/10.1007/978-981-96-1222-2

M

N

Q

R

S

The manufacturer's authorised representative in the EU is Springer Nature Customer Service Centre GmbH, Europaplatz 3, 69115 Heidelberg, Germany. If you have any concerns regarding our products, please contact ProductSafety@springernature.com

Printed and bound by CPI Group (UK) Ltd, Croydon, CR0 4YY
30/09/2026
02225302-0016